12th International Conference of Archaeological Prospection

12th - 16th September 2017

The University of Bradford

Edited by Benjamin Jennings, Christopher Gaffney, Thomas Sparrow and Sue Gaffney

12th International Conference of Archaeological Prospection

12th - 16th September 2017

The University of Bradford

Edited by Benjamin Jennings,
Christopher Gaffney,
Thomas Sparrow and Sue Gaffney

Archaeopress Archaeology

ARCHAEOPRESS PUBLISHING LTD
Gordon House
276 Banbury Road
Oxford OX2 7ED

www.archaeopress.com

ISBN 978 1 78491 677 0
ISBN 978 1 78491 678 7 (e-Pdf)

Cover: Image created by © Thomas Sparrow, Bradford Visualisation,
University of Bradford, 2017 (with acknowledgements below).

Cover image: Fountains Abbey in North Yorkshire, with a GPR survey at the eastern end, showing the graves of the monks. Data for the image is a combination of multiple projects being co-ordinated by the School of Archaeological and Forensic Sciences, University of Bradford. The point cloud is part of the Curious Travellers project (www.visualisingheritage.org); data is derived from crowd sourced and data-mined imagery, processed using Structure from Motion (SfM) to produce a 3D point cloud. The GPR data was collected by Mike Langton, Mala Geoscience. The aerial image (after Google, DigitalGlobe 2017) is draped over LiDAR (after Environment Agency copyright and/or database right 2015. All rights reserved).

Printed in England by Oxuniprint, Oxford
This book is available direct from Archaeopress or from our website www.archaeopress.com

Contents

Introduction

The 12th International Conference of Archaeological Prospection saw a return to the University of Bradford, the host for the 1st ICAP conference in 1995. Much has changed in the world of archaeological prospection since that inaugural event, but many things have also remained constant. Perusing the abstract list from the 1st conference reveals that there were no less than six presenters who returned to present, and many more of the initial group returned to attend, at this, the 12th conference.

For the 12th International Conference of Archaeological Prospection a number of key themes were targeted, divided into six conference sessions:

- Techniques and new technological developments
- Applications and reconstructing landscapes and urban environments
- Integration of techniques and inter-disciplinarity, with focus on visualisation and interpretation
- Marine, inter-tidal and wetland prospection techniques and applications
- Low altitude prospection techniques and applications
- Commercial archaeological prospection in the contemporary world

Many of the presentations in the Techniques and technological developments session highlighted the use of automated process in the filtering and recognition of data, and also the use of vehicles for the rapid capture of high resolution data.

The Applications and reconstruction of landscapes and urban environments session highlighted contemporary research in a wide range of locations and temporal settings from around the world, from Stonehenge to Mexico, and from Northern Plains earthlodges to 19th century landscape gardens. The variety of presentations amply demonstrates the applicability of prospection techniques to a wide range of situations and purposes, and highlights the expansion seen in archaeological prospection since the 1st ICAP conference in 1995.

Within the integration of techniques and visualisation session a number of presentations detailed the use of simultaneous data capture, and how such techniques have been integrated with specific research programmes to enhance the understanding of archaeological sites.

The special sessions on Marine, inter-tidal and wetland prospection and Low altitude prospection focussed on new and emerging technologies broadening the horizon for archaeological prospection. The use of underwater vehicles and aerial vehicles for data capture in the form of marine seismic data and LiDAR is detailed through a number of case studies. These demonstrate the novel use of emerging technologies for archaeological prospection, and the success of these applications will certainly lead to the growth of this field within coming years.

A special session on commercial archaeological prospection combined a number of presentations from commercial practitioners in the field with a workshop session covering a range of key issues and standards relating to practices within both commercial and research archaeological prospection.

The success of the 12th International Conference of Archaeological Prospection is due to both the presenting contributors and attending delegates. We also recognise the effort made by both the Organising Committe and Scientific committee, and thank the members for their dedication in organising the event, and for the prompt review and comments on all of the scientific papers.

On behlaf of both the ICAP 2017 Organising Committee and the Scientific Committee we extend sincerest thanks to all of the presenters and attendees at the conference, and very much look forward to the 13th conference in 2019.

Ben Jennings, Chris Gaffney,
Thomas Sparrow and Sue Gaffney

Bradford, July 2017

12th International Conference of Archaeological Prospection Committee Members

Organising Committee

Kayt Armstrong	*Durham University*
Cathy Batt	*University of Bradford*
Hannah Brown	*Magnitude Surveys*
Adrian Evans	*University of Bradford*
Chris Gaffney	*University of Bradford*
John Gater	*GSB / Sumo*
Chrys Harris	*University of Bradford*
Ben Jennings	*University of Bradford*
Mike Langton	*Mala*
Mark Newman	*National Trust*
Armin Schmidt	*University of Bradford*
Tom Sparrow	*University of Bradford*
Roger Walker	*Geoscan Research*

Scientific Committee

Kayt Armstrong	*Durham University*
Catherine Batt	*University of Bradford*
Christoph Benech	*University of Lyon*
Hannah Brown	*Magnitude Surveys*
Dave Cowley	*Historic Scotland*
Rinita Dalan	*Minnesota State University Moorhead*
Mahmut Drahor	*Dokuz Eylül University*
Adrian Evans	*University of Bradford*
Jörg Faßbinder	*Bavarian State Department of Monuments and Sites*
Tomasz Herbich	*Institute of Archaeology and Ethnology Polish Academy of Sciences*
ChrisGaffney	*University of Bradford*
Ben Jennings	*University of Bradford*
Neil Linford	*Historic England*
Paul Linford	*Historic England*
Cornelius Meyer	*Eastern Atlas*
Philip Murgatroyd	*University of Bradford*
Wolfgang Neubauer	*University of Vienna & Ludwig Boltzmann Institute for Archaeological Prospection and Virtual Archaeology*
Apostolos Sarris	*Foundation for Research and Technology*
Armin Schmidt	*University of Bradford & GeodataWIZ*
Immo Trinks	*University of Vienna & Ludwig Boltzmann Institute for Archaeological Prospection and Virtual Archaeology*
Gregory Tsokas	*Aristotle University of Thessaloniki*

The use of digital mobile technologies for geoarchaeological survey: the examples of the Pinilla del Valle raw materials project

Ana Abrunhosa[1,2], João Cascalheira[1], Alfredo Pérez-González[3], Juan Luís Arsuaga[4, 5] and Enrique Baquedano[2, 6]

[1]ICArEHB - Interdisciplinary Center for Archaeology and Evolution of Human Behaviour, University of Algarve, Portugal; [2]MAR - Museo Arqueológico Regional, Alcalá de Henares, Spain; [3]CENIEH - Centro Nacional de Investigación sobre la Evolución Humana, Burgos, Spain; [4]Departamento de Paleontología, Facultad de Ciencias Geológicas, Universidad Complutense de Madrid (Spain); [5] ISCIII - Centro Universidad Complutense de Madrid-Instituto de Salud Carlos III de Investigación sobre la Evolución y Comportamiento Humanos, Madrid, Spain; [6]I.D.E.A. - Instituto de Evolución en África, Madrid, Spain

ana.abrunhosa@gmail.com

Archaeology and Apps

There is a general tendency to reduce the use of paper and to simplify and unify different tasks in the smallest number of electronic devices possible. That is especially important when it comes to archaeological field work, such as survey, that imply walking long distances while recording data and recovering samples that testify the same data. Application Software – also known as apps – have been developed with educational and scientific purposes, many of them with a full version only unlocked by payment of a fee. In the field of Geology there are smartphone and tablet Apps with, among others dictionaries, mineral guides, petrographic databases. For example, the British Geological Survey has an interactive geological mapping of the UK available on the App iGeology and iGeology 3D with non-commercial use. The Geological Survey of Denmark and Greenland developed aFieldWork, an App dedicated to the description of geological localities. Although there is a variety of applications available, none of them directly targets the needs of geoarchaeological surveys, and particularly the recording of variables specific for raw material sources characterization.

Archeosurvey

Archaeology in the digital era is increasingly relying on DIY open access, open source or for-profit digital technology adapted to archaeology projects (Morgan and Eve 2012, Motz and Carrier 2012). In 2014, Cascalheira et al. developed the ArcheoSurvey App, a custom freeware app designed for field survey of archaeological sites. The App can be downloaded to any Android-powered device and it combines several features already present on a smartphone such as GPS, camera and internet connection to record and create a locally stored and/or an online database.

Here, we present the Archeosurvey – Raw Material Edition application, how it works in the field, the analyzed traits and the Pinilla del Valle survey project as a case study using the app. The new edition of the app came from the necessity to perform geoarchaeology surveys in the Lozoya river valley (Madrid, Spain) in order to study the lithic raw material sources of the Pinilla del Valle archaeological sites. With each record the App allows to register: surveyor name, site id, sample id, toponym, micro toponym, position of the raw material, visibility, access, rock type, Munsell rock color, geomorphology, size of available raw material, observations, GPS coordinates, its position on Google Maps and a photo. The database is organized in a .txt file that can instantly be stored on a cloud system if there is internet connection or easily be imported to an Excel spreadsheet. This file containing the GPS coordinates is also ready for a GIS integration.

Pinilla del Valle Raw Material Survey

Pinilla del Valle is an archaeological Upper Pleistocene site in a karst complex of Upper Cretaceous dolomites intensely used by Neanderthals. It is located in the Lozoya river valley at about 1100 m a.s.l., within the National Park of the Guadarrama Mountain Range in the Iberian Central System (Madrid, Spain).

For the study of lithic raw material provenance of Pinilla del Valle sites the objectives in the field during survey are very different, and so is the analysed and recovered data. Instead of looking for possible sites we are looking for possible exploited sources of rocks/minerals in Prehistoric periods. For that reason, a new version of the app was created to answer the needs of the geological survey since most of the descriptive fields had to be changed to meet the criteria of geoarchaeological analysis in the field.

The Lozoya river valley is marked by a wide variety of lithic resources that are present in

Figure 1: QR code for download of the ArcheoSurvey mobile application.

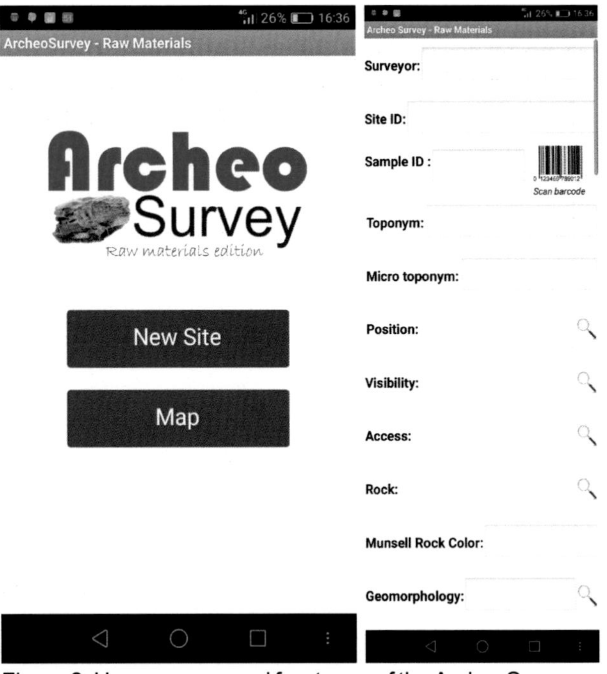

Figure 2: Home screen and frontpage of the ArcheoSurvey - Raw Material Edition mobile application.

the archaeological context. The region is also geologically very complex with a variety of lithic resources in primary and secondary position. To perform geoarchaeological surveys first we did a geological bibliographic study. The information collected allowed to organize and perform punctual field surveys to understand the distribution and characterization of the available knapable lithic resources in their different positions – primary (i.e. rock outcrops) and secondary (i.e. fluvial or slope deposits). Geoarchaeological surveys are especially important for the study of lithic raw material procurement because secondary sources are seldom mapped on geological cartography and these are usually the most used sources exploited by prehistoric groups (Clarkson & Bellas 2014).

Discussion

The main advantages of a paperless recording method by means of a mobile application are ecological and cost/efficiency related. The reduction of paper and number of devices by combining functions in one leads to reduced costs that allow for an optimization of the reduced financial support archaeological research projects face. Overall, for the Pinilla del Valle raw material survey the app allowed an increase of efficiency with less human error and cost, optimizing time that can be used to analyze and interpret the collected data.

Bibliography

Allen-Willems, R. (2012) *Designing the Digital Archaeological Record: collecting, preserving, and sharing archaeological information*. Thesis for the Degree of Master of Arts in Anthropology. Northern Arizona University.

Cascalheira, J, Gonçalves, C. and Bicho, N. (2014) Smartphones and the use of customized apps in archaeological projects. *The SAA Archaeological Record* **13** (5): 20-25.

Cascalheira, J., Gonçalves, C. and Bicho, N. (in press) A Google-based freeware solution for archaeological field survey and on-site artifact analysis. *Advances in Archaeological Practice*.

Clarckson, C. and Bellas, A. (2014) Mapping stone: using GIS special modelling to predict lithic source zones. *Journal of Archaeological Science*. **46**: 324–333.

Morgan, C. and Eve, S. (2012) DIY and digital archaeology: what are you doing to participate?. *World Archaeology*. **44**(4):521-537.

Motz, C. F., Carrier, S. C. (2013) Paperless Recording at te Sangro Valley Project. In G. Earl, T. Sly, A. Chrysanthi, P. Murrieta-Flores, C. Papadopoulos, I. Romanowska, and C. D. Wheatley (eds.) *Archaeology in the Digital Era. CAA2012 Proceedings of the 40th Conference in Computer Applications and Quantitative Methods in Archaeology. Southampton. United Kingdom. 26-29 March 2012*. Amsterdam: Amsterdam University Press.

A multi-methodological approach on a historic wall structure of Heptapyrgion fortress thessaloniki greece: a case study

Dimitrios Angelis[(1)], Panagiotis Tsourlos[(1)], Gregory Tsokas[(1)], George Vargemezis[(1)] and Georgia Zacharopoulou[(2)]

[(1)]Department of Geophysics, School of Geology, Aristotle University of Thessaloniki, Greece; [(2)]Hellenic Ministry of Culture, Ephorate of Antiquities Thessaloniki, Greece

angelisd@geo.auth.gr

Introduction

Heptapyrgion (Fortress of Seven Towers), which is known as Yedikule, is a fortress that stands at the highest point of the Acropolis of Thessaloniki and was built in various construction phases from Early Christian – Early Byzantine period up to the years of Ottoman rule. During late 19th century the fortress was converted into a prison until 1989. Today, the fortress comprises one of the most important monuments of the city (UNESCO monument) and is under the responsibility of the Ephorate of Antiquities, of the City of Thessaloniki (Greek Ministry of Culture and Sports). During the last decades a systematic archaeological study and restoration of the fortress began and continues to this day.

In this work, a combination of geophysical methods was used to study a part of the external wall of the P3 fortress tower of Heptapyrgion (Fig. 1). This part of the wall has been identified as facing moisture and structural problems, thus it is crucial to conduct an investigation of the internal structure that will provide the necessary information for developing future restoration plans. Geophysical survey, and more specifically, the methods of electrical resistivity tomography (ERT) and ground penetrating radar (GPR) have been chosen as non-destructive techniques due to the archaeological importance of these historic walls. This approach proved to be very efficient since the combined interpretation of different types of data provided useful information about the internal structure of the wall.

Figure 1: Location map of the Heptapyrgion fortress (top) and of the study area Tower P3 (bottom).

Data Collection and Processing

The collection of data carried out in 8 parallel profiles with the first located 0.5m above the ground and the spacing between each profile set at 0.2m. For the acquisition of ERT data an IRIS SYSCAL-PRO resistivity meter with 24 electrodes spaced 0.4m apart was used. The galvanic contact established using special bentonite mud electrodes. The GPR data were collected using a RAMAC GPR (MALA Geoscience) with a 500MHz centre frequency, as considered a good appropriate compromise between penetration and resolution

Figure 2: a) 1st and last profile line of survey grid; b) GPR data gathering with 500MHz antenna; c) ERT data collection (bentonite mud electrodes).

(Fig. 2). Measurement interval was set to 0.04m and maximum recording time at 70ns.

For the processing procedure DC2DPRO (Kim 2009) and RADPROfwin (Kim 2004) were used. The ERT data were inverted and the produced results are inner wall sections of the "true" resistivity. For the processing of GPR data, start time correction, dewow and background removal were performed. Further processing steps involved custom gain suitably varying vs the depth along with band pass filtering, filtering in wavenumber domain, predictive deconvolution and as a final step trace interpolation. To transform the time scale to depth scale a test scan in a nearby (and of same material) thick wall was performed. For the test we used a metal plate at the back side of this wall and velocity was calculated to be approximately 0.11m/ns.

Furthermore, to achieve a better visualization of the areas suffering from moisture a Hilbert transform was performed to the processed GPR data as this type of transformation is often more suitable for better visualization of areas with strong reflection or no reflection (Goodman and Piro 2013).

Results and Conclusion

In Figure 3 we present the processed radargrams of Line 1, 3 and 5 (first row), the Hilbert transformation results (second row) and finally the respective ERT inverted lines using a resistivity rainbow scale (third row).

As can been seen, there is a very good agreement between the ERT and GPR sections. The moisture area appears as a low resistivity area in ERT sections and as a strong attenuation area in the respective GPR sections. The same attenuation is depicted more clearly in the Hilbert transform processed data and clearly dominates in the left and eastern part of the survey wall.

Furthermore, in the GPR processed radargrams we need to mention the strong reflection that appears in all profiles and is marked with blue colour. This reflection is probably related to a different construction phase of the tower. Also, using the colour, orange colour the lateral air reflections are presented which are apparent in most of profile lines.

In conclusion, a fully non-destructive pilot survey was conducted which proved very useful in accessing the internal structure of the wall. Different material phases were identified and mapped (structural, dry-wet phases). Also, these first results are very promising in evaluating the effectiveness of previous restoration work in support of future decision making processes.

Finally, the combination of geophysical methods that implemented proved to be a good choice as each method complements the other.

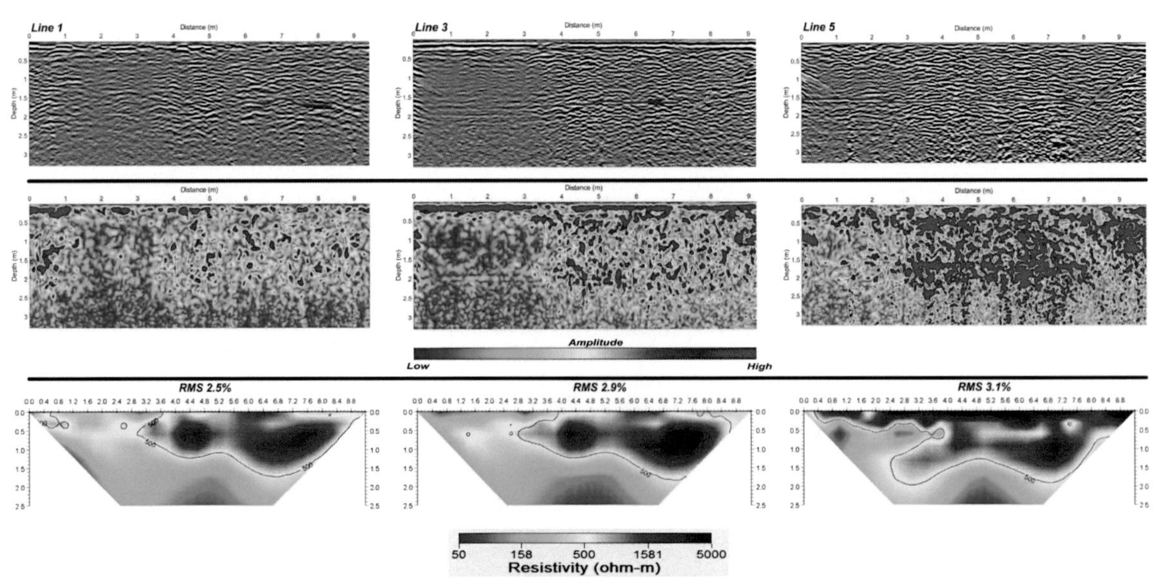

Figure 3: 1st row: Interpreted GPR processed radargrams of Lines 1,3 and 5. Showing the possible second construction phase and the lateral air reflections. 2nd row: The Hilbert transformed GPR data showing the moisture area (blue – green colour). 3rd row: The respective ERT sections showing the moisture area (blue – green colour).

Bibliography

Goodman, D. and Piro, S. (2013). *GPR Remote Sensing in Archaeology.* Heidelberg: Springer.

Kim, J. H. (2004) *RADPRO/GPR. User's guide*, South Korea: KIGAM.

Kim, J. H. (2009) *DC2DPro-2D Interpretation System of DC Resistivity Tomography. User's Manual and Theory.* South Korea: KIGAM.

Acknowledgments

The present study was carried out under the framework of the Decision F42/51198/1651/29-05-2012 of the Greek Ministry of Culture. The authors would like to thank the personnel of the Ephorate of Antiquities, of the City of Thessaloniki (Greek Ministry of Culture and Sports) and in particular the Architect Mr. Ioannis Giannakis for facilitating the survey. Also, would like to thank the AUTH Applied Geophysics MSc students for their help during the fieldwork.

Settling selection patterns and settlement layout development in the Chalcolithic Cucuteni culture of north-eastern Romania. Interpretation and presentation of prospection results

Andrei Asăndulesei[(1)], Felix-Adrian Tencariu[(1)], Mihaela Asăndulesei[(1)] and Radu-Ștefan Balaur[(1)]

[(1)]Interdisciplinary Research Department – Field Science, "Alexandru Ioan Cuza" University from Iași, Romania

andrei.asandulesei@yahoo.com

Introduction and Argument

The Cucuteni culture, regarded as *the last great Eneolithic civilisation of Old Europe*, has been investigated since the 19th century. The culture is part of the notable Cucuteni-Trypillia Cultural Complex, which stretches from south-eastern Transylvania to north-eastern Romania, the Republic of Moldova, and to the forest-steppe of the Ukraine, covering a surface of approximately 350,000 km^2. From a chronological point of view, the culture's evolution spans from 4600 to 3600/3500 cal BC in the Romanian area. This interval is mostly known for fine, good-quality pottery, predominantly with polychrome painting, and also for anthropomorphic wares, revealing the sophisticated artistic and aesthetic ideas held by these communities. There are more than 1800 sites on Romanian territory, whether hilltop or lowland settlements, compact or scattered, seasonal or permanent, main or secondary, small, medium or large in size, clearly showing the extremely dynamic character of these communities.

Although there is a long history of research concerning this culture and numerous trial excavations have been made in many settlements that chronologically span three great phases (A, A–B, B), in only a handful of sites has archaeological research generated a completed planimetric image (viz. Hăbășești, Târpești, Trușești). In addition, investigations were very rarely extended outside the natural limits of the settlements, or to the outer complex systems considered to be fortification or delineation works. It is difficult to propose hypotheses concerning the cultural landscape of the Eneolithic period based solely on the minimal information gathered almost exclusively from archaeological excavations. Accordingly, even though there have been discussions regarding these topics, in certain environmental conditions, a great number of aspects essential for understanding the behaviour of the Cucutenian communities are far from known.

Integrated Approaches

Considering the above, the generalisation for the Cucuteni culture of non-invasive investigations based on integrating the main prospecting methods (Airborne Laser Scanning, aerial photography and geophysical surveying) represents a practical approach for interpreting many of the aspects mentioned above. It also provides new opportunities for understanding the complex evolution of the Cucutenian and other prehistoric settlements. Such a methodology, based on the integration of the main non-invasive prospecting methods, is still weakly represented in Romanian archaeology, even though it has justifiably become a cornerstone of archaeological research worldwide.

Study Area and Case Studies

For this paper, we selected three case studies from Moldavian Plateau for which the measurements were recently completed, located on hilltops or low terraces, distributed chronologically along all of the three great phases of the Cucuteni culture: 1. Războieni, *Dealul Mare* (phase A), 2. Ripiceni, *Holm* (phase A-B), 3. Brătești, *Chicera* (phase B) (Fig. 1).

Războieni, *Dealul Mare* it is located in the territory of the village of Războieni, Ion Neculce commune, on the border with Filiași village. The settlement is located on a promontory running in a NW–SE direction, strongly affected by active landslides. The promontory is situated approximately 500 m south of the village of Filiași, on the right bank of the Valea Oilor/Recea brook (a left-bank tributary of Bahluieț River), part of the Prut catchment basin. The Valea Oilor brook forms a loop in this area, surrounding the promontory along three of its sides.

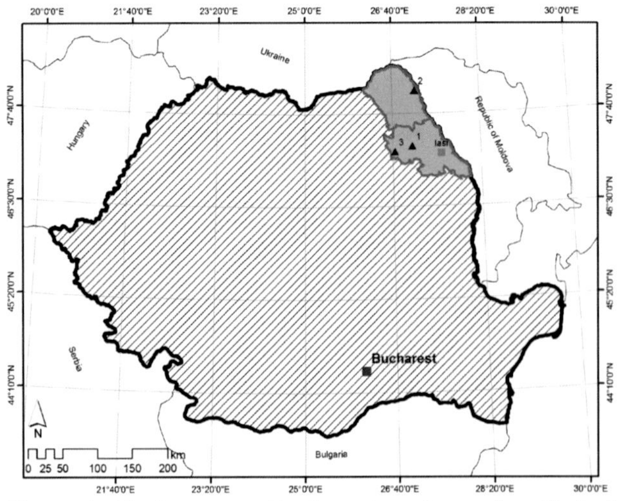

Figure 1: Study area – North-eastern Romania with Iași and Botoșani counties marked; Case studies: 1. Războieni, *Dealul Mare*; 2. Ripiceni, *Holm*; 3. Brătești, *Chicera*.

Ripiceni, *Holm* archaeological site is located in the north-eastern part of the Ripicenii Noi village, on the right bank of the Prut River. The settlement sits on a backslope with an elevation of approximately 82 m. The current geomorphological situation is strongly modified by some anthropic interventions in this area. The site is already more than half destroyed because of the floods from NNE side.

The site of Brătești, *Chicera* is situated at the south-eastern edge of Brătești village, on a plateau of the steep northward-facing slope of Chicera Hill. It is a hilltop settlement, with a relative altitude 65 m and an absolute one of 340 m.

Results and Discussion

The interpretation of all the features identified from one or more methods in a GIS environment offers a detailed image of their shapes, sizes and distributions for each case study (Fig. 2).

Based on research results, a completely new planimetric organisation of the site of *Dealul Mare* has been revealed, different from what was known, until now, from the professional archaeological literature of Romania. The existence of a fortified area in the NNW, with a high density of archaeological structures, certainly represents the initial core of the settlement from which the habitation extended towards the SSE as the number of inhabitants increased (Asăndulesei 2017). A similar novel evolution was recently observed in magnetometry surveys conducted by our team also in the settlement from Ripiceni, *Holm*, Botoșani County, from the second phase of the Cucuteni culture. A fortification system following the same pattern of two large parallel ditches and a third

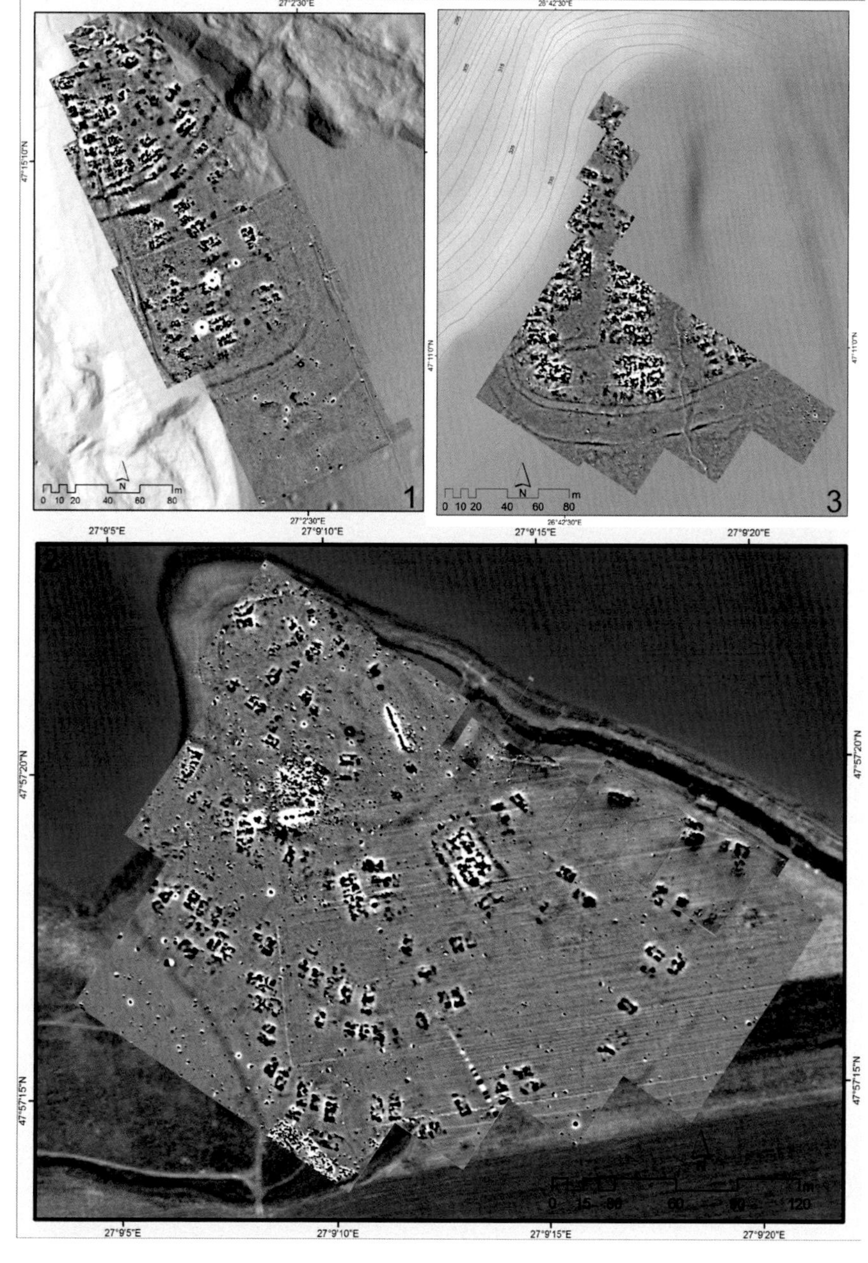

Figure 2: 1) Războieni, *Dealul Mare* magnetic map (-15/+15 nT, white/black) superimposed on ALS derived DEM; 2) Ripiceni, *Holm* magnetic map (-15/+15 nT, white/black) superimposed on orthorectified image; 3) Brătești, Chicera magnetic map (-15/+15 nT, white/black) superimposed on DEM.

narrower to the inner part was also documented at Brătești, *Chicera* site.

The buildings in the fortified group towards the NNW in *Dealul Mare* site are placed in a semicircle, with the highest density in the SW part along the path of the ditches. Although the settlement is naturally protected by steep slopes on the northern, eastern and western sides, these areas seem to have been enclosed, probably with palisades, perhaps due in part to the risk of people and animals falling down the slopes or to block access of the herds to those areas.

Another core element of originality in the Cucuteni culture revealed by our results is the presence of consistent habitation outside the main fortified areas, with houses placed in rows (for *Dealul Mare*) or semicircles (as at Ripiceni site). Apparently, heavily burned features detected in *Chicera* site, which can be attributed to burnt houses, are also arranged in rows. It is worth mentioning the large size of the structures, unusual for the Romanian Chalcolithic.

The habitations outside the fortified area are also surrounded by ditches. While their functionality is difficult to establish, what is certain is that their dimensions (narrower and shallower) prove they did not have a defensive character. Perhaps their role was symbolic and apotropaic, to isolate the settlement.

Final remarks

This study utilises a number of integrated techniques of remote sensing in a novel way (in terms of complexity and diversity) to document for the first time the spatial organisation of Eneolithic settlements belonging to the Cucuteni-Trypillia cultural complex. New understanding of the behaviour of Eneolithic communities, which are distributed through eastern Romania, the Republic of Moldova, and Ukraine, are introduced to the professional literature regarding the typology of the fortification systems, the existence of ritual or boundary ditches, and the presence of habitations outside fortified areas. Given our results from these sites and others from the Moldavian plateau, we can argue that internal spatial organization of Cucutenian settlements was determined by the landform of the environment around the settlement.

Bibliography

Asăndulesei, A. (2017) Inside a Cucuteni Settlement: Remote Sensing Techniques for Documenting an Unexplored Eneolithic Site from Northeastern Romania. *Remote sensing* **9**(1): 41.

Acknowledgements

This study was supported by the Partnership in Priority Domains project PN-II-PT-PCCA-2013-4-2234 No. 314/2014 of the Romanian National Research Council, *Non-destructive approaches to complex archaeological sites. An integrated applied research model for cultural heritage management.*

Monitoring marine construction zones through the iterative use of geophysics and diving

P A Baggaley[(1)], L H Tizzard[(1)] and S H L Arnott[(1)]

[(1)]Wessex Archaeology, Salisbury, United Kingdom

p.baggaley@wessexarch.co.uk

Over the last 15 years offshore renewable developments have given archaeologists access to large areas of seafloor which would not otherwise have been subjected to archaeological investigation. The heritage assets within these areas are comprised of remains of vessels, aircraft and associated debris through to historic landscapes as well as remains deriving from the subsequent history of the British Isles and its inhabitants' exploitation of the sea (English Heritage 2002). While the larger assets are often known about and can be considered from the early stages of a development, the smaller or more ephemeral assets may not be discovered until the construction process is underway.

The primary tools for investigating these sites are repeated, high resolution, geophysical surveys, which are complemented by diver and ROV (Remotely Operated Vehicles) survey (Fig. 1), although due to the cost and H&S requirements of diving/ROV surveys, the geophysical survey techniques play a larger role in the identification and investigation of these heritage assets than in terrestrial environments. However, the subsequent ground truthing of archaeological interpretations as part of the site investigation process is crucial to increasing our archaeological knowledge of the area under investigation and improving the accuracy of future archaeological interpretations of geophysical data because it is often impossible to determine the true nature of an asset through the geophysical data alone.

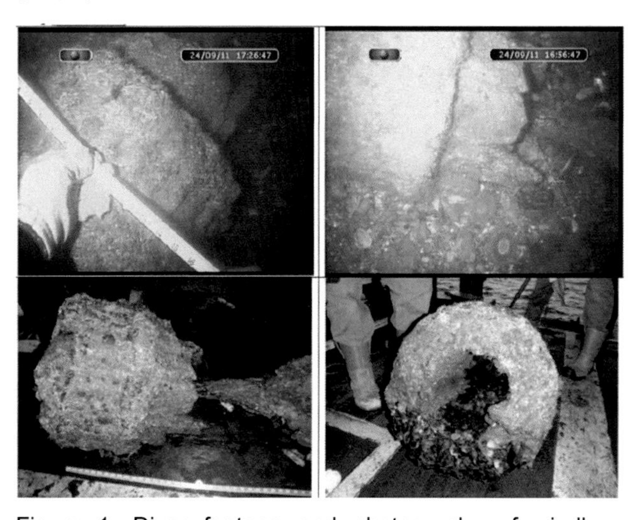

Figure 1: Diver footage and photographs of windlass recovered from seafloor after being located during marine UXO/archaeology survey.

The framework for the protection of the marine historic environment has developed considerably over the past decade, which has led to archaeological contractors developing methodologies to rapidly assess the likely presence of archaeological assets which may be impacted. This archaeological advice begins with the Environmental Impact Assessment, to help avoid important assets where possible, and then continues through the construction phase of the development to help mitigate chance discoveries. Finally, archaeological contractors are also involved in the post-construction monitoring of developments in order to assess how any changes in the physical environment may be affecting known archaeological sites. This process often leads to repeated geophysical surveys over many years as the development progresses.

This has led to the mitigation of the impacts of offshore renewable energy developments and their associated infrastructure on the marine historic environment being guided through repeated geophysical survey which allow archaeologists to continuously refine their previous interpretations and to monitor sites in dynamic environments. The detailed requirements for this iterative approach has been developed over many years through collaboration with stakeholders such as the Crown Estate, developers and curators.

While the initial assessment of the archaeological resource within an offshore development area at EIA stage will identify coherent wrecks and large sites, the data are of a relatively low resolution in order to provide an affordable level of data coverage of a large area within a reasonable timeframe. This is sufficient for the initial assessment of the area since the design of the development is not fixed and any constraints identified at this stage can normally be avoided through engineering options. However, as the project progresses and the engineering design is frozen, the archaeological mitigation works need to be covered by a Written Scheme of Investigation (WSI) which sets out the need for further, targeted, archaeological assessments using higher resolution geophysical data, acquired during the site investigation process, to identify and characterise smaller sites (Wessex Archaeology 2010). This allows archaeologists to refine their understanding of what material is present within the development area and to adjust their advice accordingly and use a judgement led process to select the best option for mitigation.

This paper will show examples of how archaeological sites are identified and studied using marine site investigation techniques, along with mitigation strategies which evolve for each site as more information is gathered. In particular, this paper will

Figure 2: High resolution sidescan sonar data acquired at 500kHz and 50m range for a marine UXO/ archaeology survey. The data shows numerous targets standing proud of the seafloor with the object on the right hand side of the image measuring approximately 1.5m across.

provide details of the level of archaeological input that can be provided at the site investigation stage, often conducted using geophysical data which are primarily acquired for UXO assessments such as sidescan sonar data (Fig. 2).

By achieving such synergies between high resolution geophysical surveys for UXO and archaeological investigations archaeologists are proving to be an integral and cost effective part of the site investigation process (Fig. 3).

The consideration of archaeological assets is now routinely undertaken in UK waters during the development of offshore renewable schemes, initially as part of the planning process and continuing through to the construction phase and post-construction monitoring. As such archaeological input is required both prior to, and post - consent, being granted, with the archaeological assessment of geophysical data key to providing cost effective archaeological advice in order to reduce the risk of chance discoveries which may adversely impact the scheme in terms of increased costs and delays to the construction programmes.

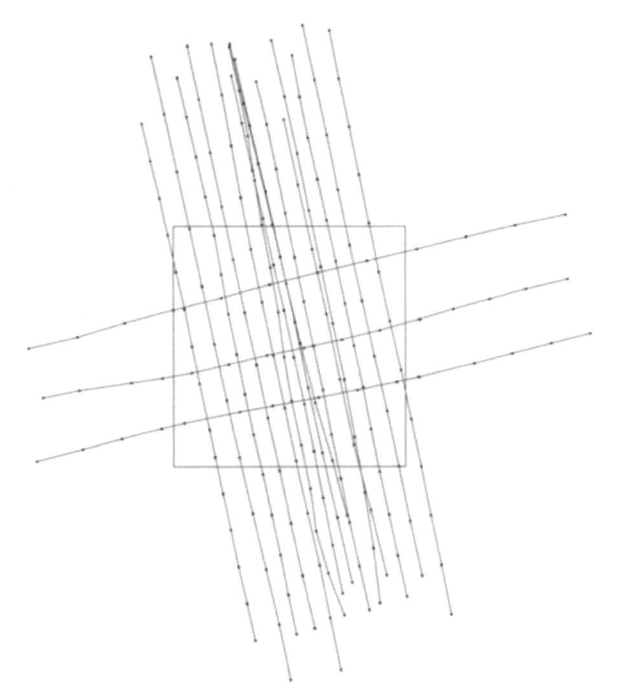

Figure 3: Geophysical survey line plan from marine UXO/ archaeology survey within 150m x 150m box around a proposed Wind Turbine Generator location. Survey lines at approx. 10m spacing north-south and 30m spacing east-west.

Bibliography

Roberts, P. and Trow, S. (2002) *Taking to the Water: English Heritage's Initial Policy for the Management of Maritime Archaeology in England.* Swindon: English Heritage.

Wessex Archaeology (2010) *Model Clauses for Archaeological Written Schemes of Investigation: Offshore Renewable Projects.* London: The Crown Estate.

Geophysical studies in Maya sites of the Caribbean coast, Quintana Roo, Mexico

Luis Barba[1], Jorge Blancas[1], Agustín Ortiz[1], Patricia Meehan[2], Roberto Magdaleno[2] and Claudia Trejo[2]

[1]Laboratorio de Prospección Arqueológica, Instituto de Investigaciones Antropológicas. UNAM; [2]Subdirección de Conservación e Investigación. Coordinación Nacional de Conservación del Patrimonio Cultural. INAH

barba@unam.mx

The archaeological sites on the Mexican Caribbean coast include Tulum and Tancah, some of the most important Postclassic period Maya sites. The Tulum's monumental centre is located in an elevated karstic bedrock formation facing the Caribbean Sea and surrounded by a defensive wall, while Tancah is placed 5km to the North of Tulum. After some years of monitoring the behaviour of some of the main structures of these sites, such as that of Structure 16 of Tulum, which needs restoration because of cracks. Similarly, in the archaeological site of Tancah, Structure 12 presented similar cracks.

The project team hypothesized that their instability problems could be produced by underground characteristics of the limestone bedrock or perhaps, to buried earlier construction phase structures.

Anomalies detected could be represented in either radargrams or time/depth slice maps depending on the answers being sought. Ground verification of the GPR data was accomplished by excavating a number of test pits in order to verify the presence of some archaeological resources. The architectural remains identified were subsequently consolidated according to international standards and their building system.

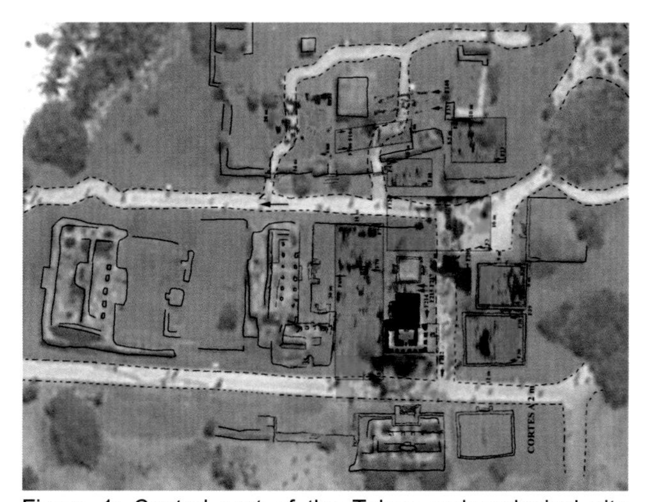

Figure 1: Central part of the Tulum archaeological site showing the 2 m deep slices. Red zones represent underground cavities.

Methodology

The methodology used in this project was the use of archaeological prospection techniques, which provided continuous feedback during the field data acquisition work (Barba and Ortiz 2001). Georadar studies were proposed to determine if the anomalous behaviour of the structures was related to the underground conditions. Studies began with examination of aerial images and the topographic survey of the monumental centre using an Ashtech differential GPS.

The common-midpoint (CMP) is advisable for data collection during the GPR reflection mode. The information obtained from CMP record is invaluable in the interpretation of the data (Blancas 2000).

CMP setup used in these study involved recording a GPR reflection trace with the transmitting and receiving antennae separated by twenty centimetres, then moving the antennae away from each other at twenty centimetre intervals to record additional traces. This procedure provides a set of traces with increasing offset distance. The produced 3D model allowed us to understand the drainage patterns for the site. Then, several georadar lines with GSSI SIR3000 and 200 MHz antenna along the walkways and open spaces were carried out as a first approach to know the conditions of the subsoil, to depth of 5m. As a result, it was evident that in most of the surveyed lines wide reflections displayed the heterogeneity of the underground conditions.

More detailed studies in grids with a 400 MHz antenna and electrical resistivity survey with a Geoscan RM15, were then conducted in specific areas to verify the presence and characteristics of underground anomalies detected by the previous lines, and in the first superficial layers to have more detail.

Geophysical studies indicate that at the depth of 1.5 to 2.5 m below the surface there is a softer stratum with anomalies. First results showed important reflections in areas close to the damaged structures.

Detailed studies confirmed the presence of underground cavities below 1.5 m depth with wide reflections and higher values of electric resistivity. These anomalies have been interpreted as dissolution cavities in one of the limestone layers (Fig. 1).

Geophysical studies performed around Tancah structure 12 show patterns similar to that of Tulum structure 16, with reflections showing limestone cavities, and near-structure anomalies.

Tancah and Tulum, are on the same karstic formation as T'Ho, the ancient great capital of the northern

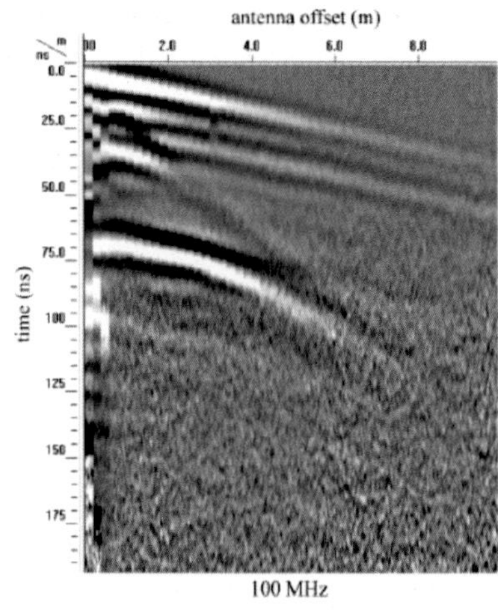

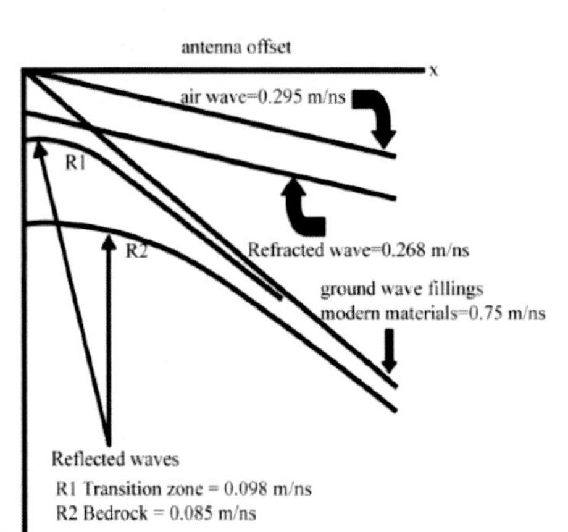

Figure 2: CMP data of velocity information and parameters as separation and antenna frequency (Figure 3, 2009a:29).

Maya area, (De Landa, 1982) that currently houses the city of Mérida, Yucatan. In Merida, studies were carried out previously by the Archaeological Prospecting Laboratory, IIA-UNAM, on the paved surface of the present city, where we detected with GPR and the CMP data collection (Blancas, 2000), "three main layers within subsoil: -Paved modern materials and shallow fillings. -Transition zone. Corresponds to mixture of weathered limestone and clays. -Limestone bedrock" (Barba *et al.* 2009a, 29)

With the data acquired with GPR and CMP it was possible to detect the velocity of reflection in the different layers of the subsoil at a resolution of 100 MHz, for example, the transition zone is 0.98 m/ns, bedrock 0.085m/ns (Barba et al 2009a) (Fig. 2).

The exploration study carried out in Mérida, shows us subsoil conditions, suggesting that it is not only the Caribbean coast, which remains prone to architectural structures presenting the same deteriorations, as those referred to at Tulum and Tancah. The studies in the historical centre of Mérida show reflections at depth of 1m, areas with cavities and fractures, even the presence of "cenotes" (sinkhole) (Barba *et al.* 2009a, 29).

The instability of these underground cavities has produced the observed cracks and differential sinking of the structures in Tulum and Tancah

Figure 3: Survey near to structure 16 in Tulum, Mexico indicating cavities in the subsoil.

archaeological sites and the hypothesis of damage because the presence of earlier construction phases has been rejected. As a result, the information recovered in these two sites and Mérida city allows us to predict that most archaeological sites and even modern structures, built in Yucatán, will in all probability suffer the consequences of the presence of the underground dissolution cavities that produce terrain instability (Fig. 3).

The coastal zone of Quintana Roo is exposed to greater erosion due to the waves of the sea in contact with the limestone rock, which it is possible to see in Tulum; However, is not exclusive to the coast, the study carried out in Mérida shows that in more continental places, the subsoil is eroding, a factor detected with GPR that limestone has wear in the most superficial layers, in addition to cavities and cenotes, In the area of the historic centre, indicating similar patterns to which we must pay attention, as these factors threaten the pre-Hispanic, colonial and present-day architectural structures throughout the Yucatan peninsula.

Bibliography

Blancas, V. J. (2000) *Principio y aplicaciones del método de Radar de Penetración Terrestre (GPR).* México: Tesis Licenciatura, Facultad de Ingeniería, UNAM.

De Landa, F. D. (1982) *Relación de las cosas de Yucatán.* México: Editorial Porrúa, 12th ed.

Barba, L. and Ortiz, A. (2001) A methodologycal approach to the study of archaeological remains in urbano contexts. *Proceedings of the 32th International Symposium on Archaeometry. CD ROM, IIA, UNAM, México.*

Barba, L., Blancas, J., Ortiz, A. and Ligorred, J (2009a) GPR detection of karst and archaeological targets below the historical centre of Merida, Yucatan, Mexico. *Studia Universitatis Babes-Bolyai, Geologia*, **54**(2):27-31.

Barba, L., Blancas, J. and Ortiz, A. (2009b) Preservation of Buried Archaeological Patrimony Through Georadar Studies. The case of the Historical Center of the Merida City, Yucatan, Mexico. *ISAP News*, **21**:2-5.

Investigations of Esie steatite structures using geophysical, petrological and geotechnical techniques

A M Bello[(1)], V Makinde[(2)], O Mustapha[(3)] and M Gbadebo[(4)]

[(1)]Department of Science Laboratory Technology SLT, Physics Unit, Kwara State Polytechnic, Ilorin; [(2)]Associate Professor, Department of Physics, Faculty of Physical Science, Federal University of Agriculture FUNAAB Abeokuta; [(3)]Professor/Dean, Faculty of Physical Sciences, Federal University of Agriculture, FUNAAB Abeokuta; [(4)]Associate Professor, Department of Geology, Federal University of Agriculture, FUNAAB Abeokuta

abdulmajeedfabello@gmail.com

In Esie, which is located about 495km from Abuja, the national capital city of Nigeria at latitude 8.21330 and longitude 4.89770 (Fig. 1), we have more than eight hundred (800) steatite or soapstone sculptures. These statues have been reputed to be the largest assemblage of sculptures in Sub-Saharan Africa and they vary in height from about 14cm to 104cm (see Fig. 2). These steatite sculptures represent human beings in different positions of activities with facial marks and striations depicting Blackman from such backgrounds as the former Kebbi, Nupe, Benin and Yoruba Kingdoms. The different forms of these structures with for example beads, royal costumes as well as various cultural and agricultural tools reveal the hitherto existence of a highly developed and complex civilization in this part of Africa. It is important to point out that the uniqueness of the artworks, the awesomeness and mysteries that surrounded them caused the then British Colonial Government of Nigeria to establish the first National Museum in Nigeria at Esie in 1945.

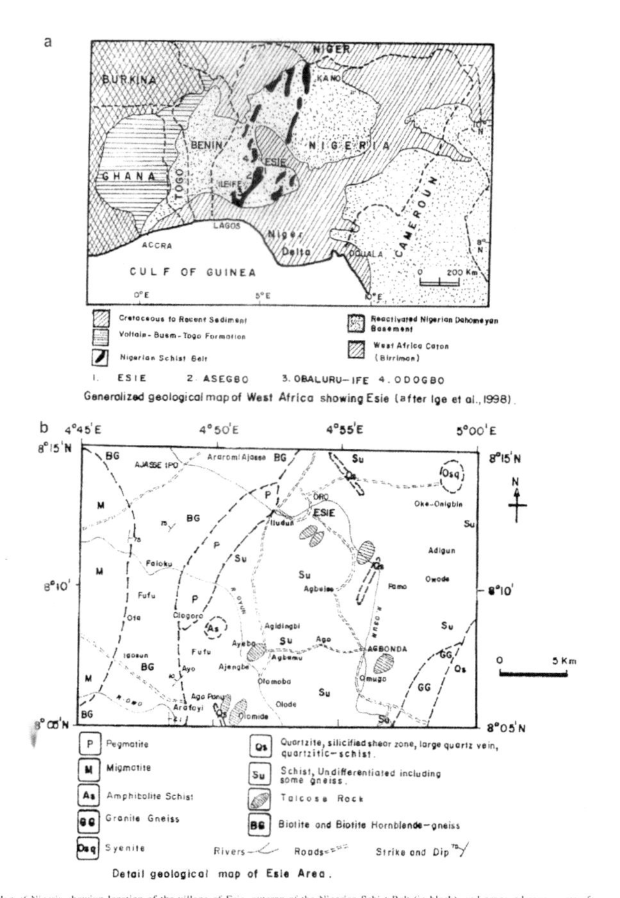

Figure 1: Geological Map of Esie Area (Courtesy : O.A Ige and Samuel E. Swanson)

The origin of these sculptures has a mythical concept of human petrifaction and it is enigmatic. In this work, integrated geophysical techniques involving Electrical Resistivity Tomography, (ERT), magnetic method as well as petrological and geotechnical techniques to assess other sources of burial of soaptone structures around Esie. The results of interpretations of the geophysical data collected revealed that areas of high resistivity coincide with the areas of high magnetic readings while areas of low resistivity go along with magnetic low areas in the study area. The inverse models of the 2D ERT imaging are similar to the total magnetic intensity maps. These indicate that more steatite structures in the form of those excavated within the present location of Esie Museum are likely to be found around Igbo-Ilowe, Igbo-eji and Oko-odo around Esie.

The study area lies within the Schist Belt of the Precambrian Basement Complex of South western Nigeria. This area is situated to the west of Effon psammite that runs NNE-SSW from Effon Alaaye, Ekiti State through Ilesa in Osun State to Oreke in Kwara State where it is calcitic. According to the field work results, the geology of Esie is essentially made up of two rock types namely the talc schist and Steatite or soapstone with a minor occurrence of quartz-mica schist and the pegmatitic intrusion. In this work, field work and petrological studies

Figure 2: Image of the statues under consideration.

Table 1: Summary of the geotechnical laboratory test results.

Sample Label / Test		ESIE I LOC 2	ESIE 2 LOC 4	IJAN I	MUSEUM I LOC 5	MUSEUM 2
Natural Water Content (%)		10.5	6.3	7.1	8.1	6.2
Density (g/cm^3)	Bulk	1.50	1.41	1.61	1.44	1.43
	Dry	1.36	1.32	1.51	1.33	1.35
Specific Gravity		2.64	2.60	2.55	2.62	2.65
Grading Test %Composition	GRAVEL	2	9	12	9	1
	SAND	65	58	58	57	70
	SILT	13	15	3	17	11
	CLAY	20	18	27	17	18
% Passing 0.075mm		32.5	33.5	30.0	34.0	29.0
Atterberg Limits (%)	LL	33.0	17.5	36.0	14.0	15.0
	PL	15.0	10.9	19.0	7.1	8.1
	PI	18.0	6.6	17.0	6.9	6.9
Compaction Test	MDD (g/cm^3)	1.85	1.88	1.80	1.87	1.88
	OMC (%)	14.0	10.0	14.0	10.0	11.0
Shear Box Test Proctor (^{0})		34	34	33	34	32
Shear Strength C (kN/m^2)		20	10	20	10	10

were carried out around Esie, Igbo-Ilowe, Igbo Eji, Oko Odo and Ijan. The results of these exercises revealed the talc-schist rock type occurs as a very low line exposure at approximately 15 metres on the bearing of 3300 to the palace of Onijan in Ijan community. The exposure has an area of approximately 20m2 around a spot that is defined by longitude 4055/ 32.9// E and Latitude 80 11/7.4// N. The talc-schist is also seen as a pocket of rocks within the residual rocks of an extensively lateritised rock in a location that is defined by Longitude 4054/ 00.4//E and Latitude 8021/ 16.4// N. The talcose schist is poorly laminated. These poorly defined lamination planes of the talc schist is inferred to be a result of its relatively soft nature. The strike directions of the rock vary from 1500 to 1550 while the dip directions are to the east within the range of 450 to 750. Mineralogically, the rock is dominated by clay minerals (Talc) (70%). Closely associated with this are quartz (15%) and crystals of green tourmaline (10%).

The quartz-mica schist was found in-site as a road cut exposure around Oko-Odo, and as a road cut exposure midway between Ijan and Esie communities. It is closely associated with pegmatite rock by body and covers an area of less one square meters, 1m2. Mineralogical, this extensively weathered rock is made of quartz (35%), muscovite (30%), biotite (10%) and others (25%). Structurally, the strike and dip directions of the rock are 1180 / 300S, 1800 / 340S, 1000/420S (essentially).

Also, the pegmatites observed on the field in the study area are closely associated with the talc schist and the quartz-mica schist. The pegmatites are relatively zoned within Ijan especially in front of the palace with coordinate of Longitude 4055/ 32.9// E and Latitude 8011/ 7.4 // N and unzoned close to Esie town. The pegmatite rock is also seen approximately 260m south of National Museum, Esie. The lengths of the pegmatites vary from over 20m in Ijan to over 50m in Esie and have a width of 10m. Mineralogically, the pegmatite is composed of quartz es which vary in colour from smoky to milky (40%), muscovite (30%), biotile (10%), black tourmaline (10%), plagioclase (5%) other minerals (5%).

Structurally, the pegmatites like the talc schist have been fractured. The fracture directions are 0640(5), 0580, 0980 (6), 1620 (11), 1040 (20), 1740, 1800 (10), 0380 (9), 0680 (10), 0580 (11), 0700, 0160 (54), 0620 (4) mainly. From the measured joint directions, the joints in this area are essentially in the NNE-SSW direction while subordinate directions is SEE-NWW (Fig. 3). Some geotechnical properties of some soils obtained from the subsurface specifically at Esie and environs were natural water content, bulk and dry density, percentage composition test in form of maximum dry density (MDD), and optimum moisture content (MCC) and shear strength test. The results of the determination of these geotechnical parameters shown in attached table 1 suggest that the geologic materials derived from the Esie Schist Belt have variable geotechnical properties. Nevertheless a composition of the result obtained for the soil derived from Ijan with similar works of Ijan talc-schist can be (very much) used to produce steatite or soapstone ceramic structures.

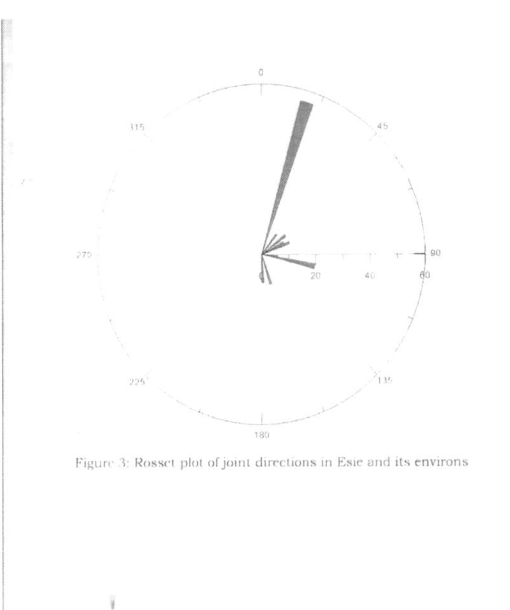

Figure 3: Rosset plot of joint directions in Esie and its environs

Revealing the topography of the Ancient Kition (Larnaka, Cyprus): an integrated approach

Christophe Benech[(1)], Marine Audebert[(2)], Antoine Chevalier[(2)], Lionel Darras[(1)], Sébastien Flageul[(2)], Sabine Fourrier[(3)], Alexandre Rabot[(3)], Fayçal Réjiba[(2)], Cyril Schamper[(2)] and Alain Tabbagh[(2)]

[(1)]UMR5133 Archéorient, Univ Lyon, Lyon, France; [(2)]UMR7619 Métis, Sorbonne Universités, Paris, France; [(3)]UMR5189 HiSoMA, Univ, Lyon, Lyon, France

Christophe.benech@mom.fr

The ancient Kition, now covered by the modern city of Larnaka, is located on the south-eastern coast of Cyprus. Founded during the late Bronze Age (13th century BC), it is one of the oldest cities in the eastern Mediterranean with a continuous occupation since their creation to the present day. During its history, Kition has experienced periods of expansion and regression. Its importance, from the political and economic point of view, makes the city a privileged witness to the diplomatic and commercial relations of the Mediterranean basin at all times, from the age of the empires of the second millennium BC to the height of the Levant in the nineteenth century.

The study of the ancient topography of the city and its evolution is of major scientific interest (Yon 2009), but also of an emergency to better manage and protect the archaeological heritage of the city. The excavation, too often limited and subjected to the vagaries of the destruction and reconstruction of buildings in the modern city, does not make it possible to deal with the problems of general organization of the ancient urban space. It is necessary to go beyond a site-based approach (these windows are often limited to the dimensions of a plot explored by the excavation) to understand the global organization of the city. The use of geophysical methods can play an essential role in the recontextualization of the knowledge provided by ancient and current excavations, by bringing new data about the ancient topography.

For this purpose, the French archaeological mission of Kition (co-financed by the French Foreign Ministry and the HiSoMA laboratory) in collaboration with the Cyprus Department of Antiquities have, since 2008, produced a GIS (Geographic Information System) listing and mapping all the discoveries made in Larnaka. The Cyprus Department of Antiquities also included in this database all the unusual remains unearthed during the excavations and recorded in their archives. This program provided a tool for the rational management of the heritage, as well as maps for each chronological period.

Over the past ten years, geophysical methods have gradually adapted to urban recognition. They have shown their capacity to bring new and original information about the history of the subsoil of cities (Lepore and Silani 2011). The objective of this project will be to show that a reasoned and systematic exploration by geophysical prospecting can constitute an essential contribution to the knowledge of the ancient topography of Kition and hence of its history. The geophysical survey was focused on two objectives: the exploration of the western Necropolis of Kition located in the modern neighbourhood of Pervolia and the identification of the layout of the Bronze Age fortification whose only very few segments (Fig. 1) have been identified by archaeological soundings (mainly in the Northern part).

Methodology

Different methods were implemented following the urban context: magnetic method (G858 Geometrics), EMI (electromagnetic induction, CMD Mini Explorer and GEM-2) and GPR (bistatic ground penetrating radar system from Radarteam) in the open field of the eastern Necropolis; GPR and a new electrostatic system for the survey in the streets in the case of the fortification.

The electrostatic device developed makes it possible to measure the apparent resistivity of the subsoil without galvanic contact. This new version has 4 channels of potential measurement. The system developed aims at exploring the urban sub-soil at medium depth (a few meters) adapted to the city of Larnaca for mapping the layout of the ancient fortifications: the spacing between the poles of injection and measurement makes it possible to record simultaneously 4 measurements of apparent resistivity up to approximately 1, 2, 3 and 5m depth (Fig. 2).

The design of such a device met very particular constraints: it was therefore necessary to design a

Figure 1: The Bronze Age fortification in the archaeological area of Katari (Northern neighbourhood of Larnaka).

Figure 2: The electrostatic system in use in the streets of Larnaka (2016).

specific measurement system, since the resistivity meters available on the market could not be used.

The technical characteristics of the resistivity meter are as follows:

- Injection voltage: 16.5 to 196.5 V (variable)
- Frequency selection: 0.98 KHz to 31.25 KHz
- Number of measuring channels 1 to 5 (differential channels)

Around 20 streets have been surveyed by electrostatic and GPR methods according to the former hypothesis of the layout of the fortification (Nicolaou 1976).

Results

The high conductivity of the subsoil of Larnaka dramatically limited the depth investigation of the GPR to the most recent layers of the city. On the other hand, the electrostatic survey delivered interesting results both on the archaeological and environmental context of the foundation of the ancient city. Some electrostatic profiles provided direct information on the location of the fortification in agreement with the archaeological hypothesis and helped to identify a "geophysical signature" of the fortification wall (Fig. 3). Others delivered largely environmental information by measuring the contrast between the more resistant plate where the city was laid out and the conductive surrounding due to the ancient lagoon area (Morhange *et al.* 1999).

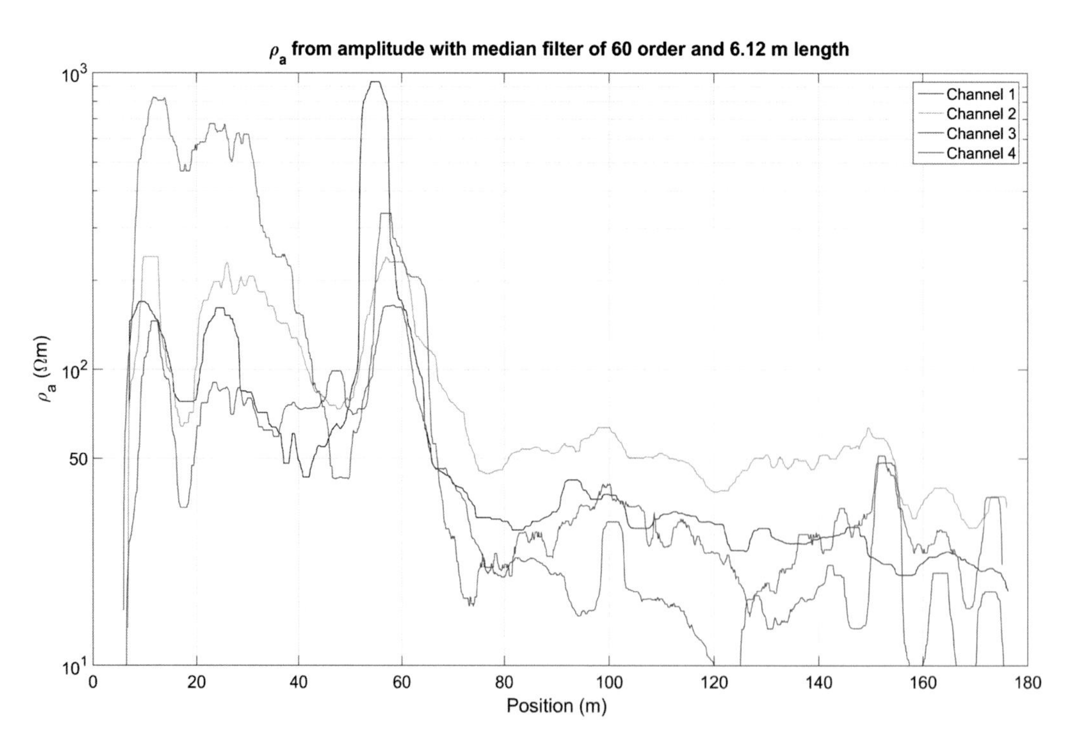

Figure 3: Example of an electrostatic profile from the streets of Larnaka with the measurements of the 4 channels.

Bibliography

Lepore, G. and Silani, M. (2011) Senigallia Urban Archaeology Project: new strategies of research and urban planning, in: *17 International Conference on Cultural Heritage and New Technologies*, Vienna, K. Fischer Ausserer-Museen der Stadt Wien, 2012, XVII, 1 – 15.

Morhange, C., Goiran, J.-P., Bourcier, M., Carbonel, P., Kabouche, B., Le Campion, J., Prone, A., Pyatt, F.B., Rouchy, J.-M., Sourisseau, J.-C.and Yon, M. (1999). 3000 ans de modifications des environnements littoraux à Kition Bamboula (Larnaca, Chypre, Méditerranée Orientale). *Quaternaire* **10**: 133–149.

Nicolaou, K. (1976) *The Historical topography of Kition*, Studies in Mediterranean archaeology. Göteborg: P. Aström.

Yon, M. (2009) Identifier les espaces urbains à Kition (Chypre). In S. Helas and D. Maezoli (eds.), *Phönizisches und punisches Städtwesen. Akten der internationalen Tagung in Rom vom 21. Bis 23.* Februar 2007, *Iberia Archaeologica* **13** (Mayence, 2009): 101-114.

Acknowledgements

This project was funded thanks to the support of the University Lyon 2 and the French Archaeological Mission of Kition and Salamine.

3D induced polarization and electrical resistivity tomography surveys from an archaeological site

Meriç Aziz Berge[1] and Mahmut Göktuğ Drahor[1]

[1]Dokuz Eylül University, Department of Geophysical Engineering, İzmir, Turkey

meric.berge@deu.edu.tr

Introduction

Electrical resistivity tomography (ERT) applications are recently one of the most used geophysical methods just after magnetic and GPR in archaeological prospection. This geophysical tool provides reliable images of subsurface resistivity distribution that shows the buried archaeological relics. Many applications that reveal the success of the ERT method can be found in the literature of archaeological prospection. However, the induced polarization (IP) method has very limited case histories on archaeological site investigations. The pioneer studies were started by Aspinall and Lynam (1968, 1970), but unfortunately the technique had not been applied in investigation of subsurface archaeological relics for a long time. However, some useful results are recently seen (e.g. Florsch *et al.* 2012, Leucci *et al.* 2014). The integrated usage of IP tomography (IPT) and ERT will provide satisfactory results in defining of some buried conductive bodies such as concentrated bronze materials, slag heaps, ditches, trenches including anthropogenic materials and other oxidised metallic objects that can be found on specific archaeological sites. In addition, some wall structures that consist of mud brick material and rammed earth (pisé) walls, basement and other structures may produce significant IP anomalies according to their mineral content and metallic concentration rates. Due to the time consuming data acquisition of the IPT method, the researchers would not usually prefer this method to map buried archaeological structures. At this point, new resistivity systems which are capable to measure time-domain IP data with resistivity data simultaneously from the survey field could be used in combined ERT and IPT surveys. Besides, tomographic evaluation of both data has become routine in many processing software packages. Therefore, the capability of the IPT method was investigated at Šapinuwa archaeological site, central Anatolia, Turkey.

Site Description and Data Acquisition

ERT and IPT methods were applied in Šapinuwa (a Hittite Empire city) where integrated geophysical applications have been systematically used since 2012. In recent survey periods, a small scale IPT investigation together with ERT as means of 3D measurement was performed (Fig. 1.a). The clayey rich soil of the investigated site and walls made by volcanic rocks and limestones describe the site characteristics of the investigation area. Also, mud brick and rammed earth can be frequently used as construction materials. Claystone, sandstone and conglomerate form the bedrock at the site and in its surroundings (Drahor *et al.* 2015).

AGI-Supersting resistivity and time-domain IP measurement system was used in the field, and standard two-dimensional tomographic data of these methods was collected. We preferred the

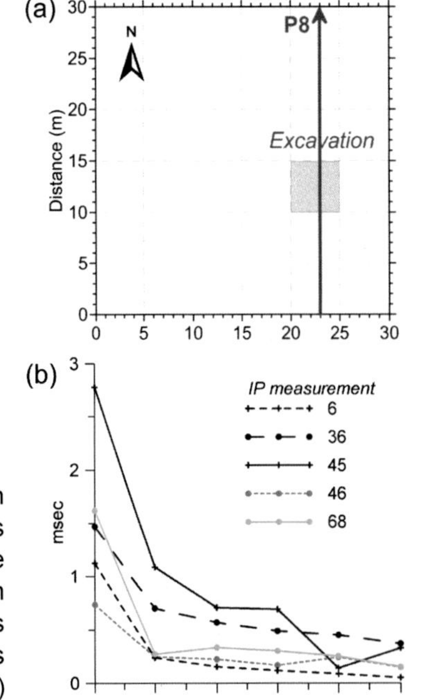

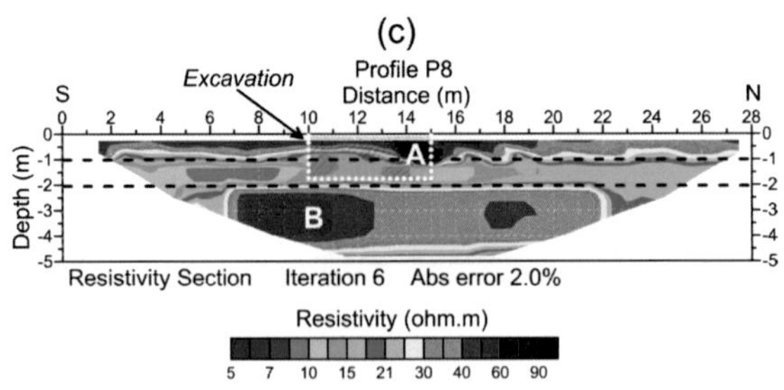

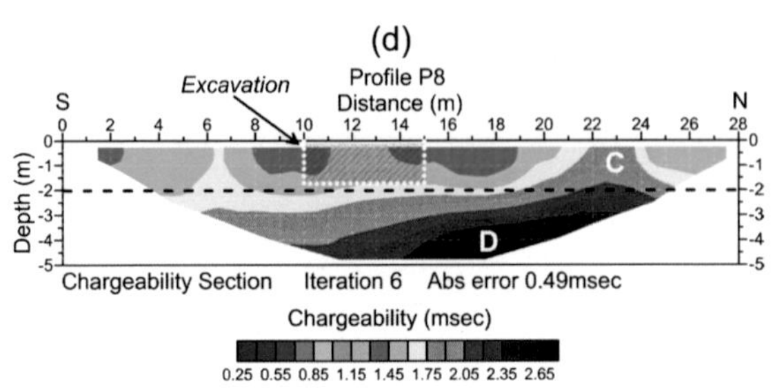

Figure 1: Acquisition plan (a). Examples of IP decay curve characteristics from various measurements (b). 2D inversion results of ERT (c) and IPT (d) obtained on profile P8.

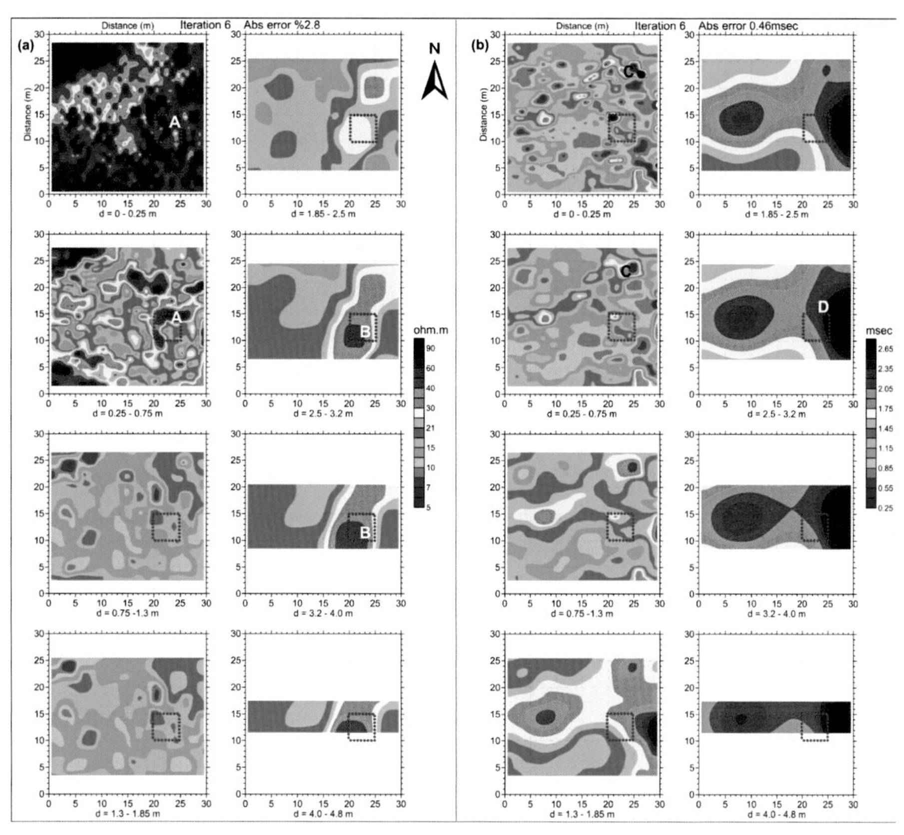

Figure 2: Depth slices of ERT (a) and IPT (b) obtained from 3D inversion results. Blue dashed rectangles show the excavation area.

Wenner-Schlumberger array during ERT and IPT data acquisition stage using 1 m electrode and traverse intervals. IP data was measured for 1 second cycles when the current is off (Fig1b). Before the inversion process, the recorded decay of IP data was checked to remove unwanted values from the data sets. Dahlin (2002) noted that bad IP data could arise from the effects of the poor contact resistance between stainless steel electrodes and soil. However, if one could provide good contact resistance, the use of stainless steel electrodes reveals satisfactory IP data. In order to obtain the tomographic images of subsurface resistivity and chargeability distribution we processed both data sets by using Res2Dinv and Res3Dinv software.

Results

In order to compare the results of ERT and IPT surveys, two dimensional resistivity and chargeability sections of profile P8 are given in Figure 1.c and .d, respectively. The resistivity section presents a three-layered subsurface model as shown by dashed lines. The first meter of the section represents a resistive character followed by a comparatively conductive layer down to 2 meters' depth. In the deeper part of the section resistivity values increase once again (Fig. 1.c). We could define a possible archaeological relic (shown by A) with high resistivity values (>90 ohm/m) in the shallow part of the ERT result. The conductive layer found between depths of 1 and 2 meters shows poor resistivity variations (5-15 ohm/m) which prevents us from interpreting them as possible mud brick and rammed earth walls and materials. Finally, resistivity values of the third layer interestingly increase, which could be defined as a deeper structure (marked by B). The IPT section of same measured profile shows relatively moderate and high chargeability characteristics of the subsurface. The discontinuity of these two different chargeability layers was defined approximately around 2 meters' depth (shown by dashed line). In the IPT model, highly chargeable structures (C and D) are seen on the shallow and deeper parts of the section (Fig. 1.d).

In the ERT depth slice, a three-layered subsurface model could be seen (Fig. 2.a). In order to compare with the previous 2D section, we marked the aforementioned structures on the depth slices. The first two depth slices (0.25 and 0.75 meters) present high resistivity values and possible structural plans of shallow subsurface relics. The slices shows low resistivities down to 1.85 meters. Once again, a resistive part of the section (marked B) appears in the deeper slices. According to IPT results, shallow chargeable structures (e.g. C) could be explained with clayey rich soil material that can contain some oxidised metal objects (Fig. 2.b). In addition, chargeability values increased towards to deeper slices. These high values could be associated with the clayey rich anthropogenic soils, possibly distributed bronze slags and/or rammed earth materials in the basement of archaeological structures.

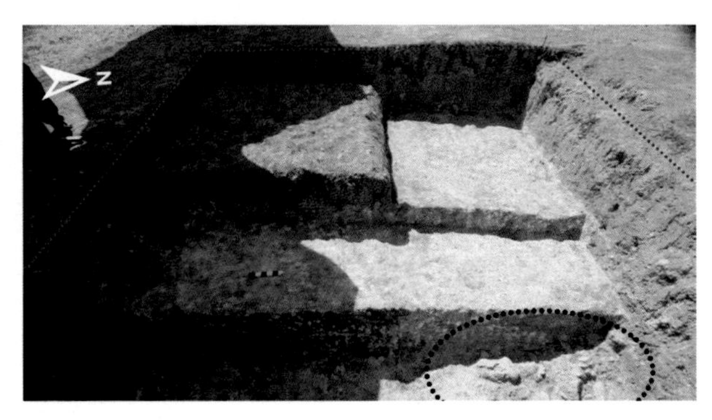

Figure 3: Photographs from the excavation conducted on the specified area.

A test excavation located on structure A, which shows regular extension in the second depth slice of ERT was performed by archaeologists after the geophysical surveys (Fig. 1.a and Fig. 2.a). Excavation revealed a very disturbed topsoil down to approximately 0.7 m from the surface. Interestingly, a clay and hydrated lime mixture layer was revealed after 1 m depth. Only a number of stone pieces (shown by ellipses) found in approximately 1.5 m were unearthed under this layer (Fig 3). This layer corresponds to conductive zone in ERT results until 1.85 meters' depth. In order to verify the deeper structures, the future excavation will be continued in the 2017 field campaign.

Bibliography

Aspinall, A. and Lynam, J. T. (1968) Induced polarization as a technique for archaeological surveying. *Prospezioni Archeologiche* **3**: 91-93.

Aspinall, A. and Lynam, J. T. (1970) An induced polarization instrument for the detection of near-surface features. *Prospezioni Archeologiche* **5**: 67-75.

Dahlin, T. (2002) Measuring techniques in induced polarisation imaging. *Journal of Applied Geophysics* **50**: 279-298.

Drahor, M. G., Berge, M. A., Öztürk, C. and Ortan, B. (2015) Integrated geophysical investigations at a sacred Hittite Area in Central Anatolia, Turkey. *Near Surface Geophysics* **13**: 523-543.

Florsch, N., Llubes, M. and Téreygeol, F. (2012) Induced polarization 3D tomography of an archaeological direct reduction slag heap. *Near Surface Geophysics* **10**: 567-574.

Leucci, G., De Giorgi, L. and Scardozzi, G. (2014) Geophysical prospecting and remote sensing for the study of the San Rossore area in Pisa (Tuscany, Italy). *Journal of Archaeological Science* **52**: 256-276.

Looking for the ancient Nile banks and their relationship with a Neolithic site: the example of Kadruka (Sudan)

Yves Bière[1], Pierrick Matignon[2], Ludovic Bodet[3] and Julien Thiesson[3]

[1]UMR5133 - Archéorient. Environnements et sociétés de l'Orient Ancien. Lyon France; [2] Freelance archaeologist; [3]UMR 7619 – Milieux environnementaux, transferts et interactions dans les hydrosystèmes et les sols, UPMC, Paris, France

julien.thiesson@upmc.fr

Introduction

The village of Kadruka is located in the Wadi-el-Khowi, made of former courses of the Nile. It lies in the east bank of the actual river, south of the third Cataract. The banks of the Wadi-el-Khowi were occupied since the Mesolithic with a climax in the settlement around the Neolithic period. In this context, the site of Kadruka consists of a large density of funerary sites and settlements in which those dated to the Neolithic constitute the largest group. The erosion considerably abraded the settlement sites which can only be identified by the presence of artifacts spread over the ground surface. In contrast, cemeteries, which were established on mounds, appear to be much more preserved and their excavation is a unique opportunity to understand societies living in the region between the VIth and IVth millennia BC.

The SFDAS (Section Française des Antiquités du Soudan) have managed several rescue excavations in the area (Reinold 1998, Alarashi *et al.* 2016). The actual campaigns are mainly driven by the quick expansion of agricultural activities around the site.

During these field surveys, about fifty Neolithic sites were identified, whether cemeteries or settlements. Concerning the mounds, about 700 graves were excavated and approximately dated from 4800 to 4000 BC.

The material culture and funerary practices of these graves seem to reflect both the homogeneity of these populations as well as the quick evolution of their social organization. The layout of the cemeteries exhibit the growing role of the hierarchy through time. During the Neolithic, chiefdoms appeared with the emergence of some dominating figures. These societies were the basis of the first Protohistoric states, preparing the appearance of the first Kingdoms. The aims of this study are:

- The evaluation and the characterization of the settlements by magnetic surveys (not shown here)
- The study of the relationship between settlements and burial mounds by reconstructing the paleo-embankment using geophysical and geomorphological survey.

Material and Methods

The study of the direct surroundings of the funeral mound (Fig. 1) was assessed with a detailed magnetic survey, completed with a G858 from Geometrics in dual sensor mode (vertical gradient geometry). The surveyed area covered with that technique is about 33,100 m^2. Data were acquired in continuous mode along parallel profiles separated by 1m.

This study was carried out in order to complete a wide mesh EMI prospection using the CMD explorer from GF instruments. This device has one transmitter and three receivers. The measurements were achieved using the Vertical Coplanar (VCP) or Horizontal Magnetic Dipole (DMH) mode. The three Tx-Rx spacings are 1.5, 2.5 and 4.5 m and the working frequency is 10 kHz. The total area covered with this technique is about 135000 m2 (Fig. 1).

Figure 1 shows the prospected areas and how they are distributed around the excavations.

Results

The magnetic maps are shown in Figure 2. It appears that very few features could be identified. The eastern area presents very faint patterns which could not be identified as archaeological remains. However, the western area exhibits linear patterns which could be linked to recent human activities, such as the cultivations of the fields near the prospected areas or desert tracks.

The CMD map is shown on Figure 3 for the second Tx-Rx spacing. The electrical conductivity results show lateral changes from west to east. Three areas can be discerned. The first one, to the west is

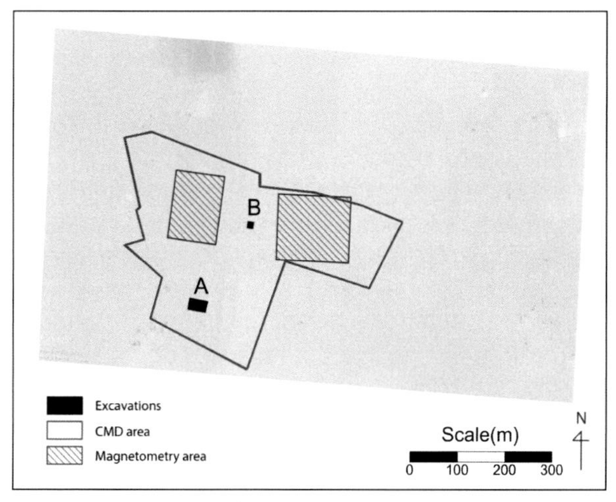

Figure 1: Location of the areas prospected around the excavated sites: a) settlement [green], b) funeral mound [red].

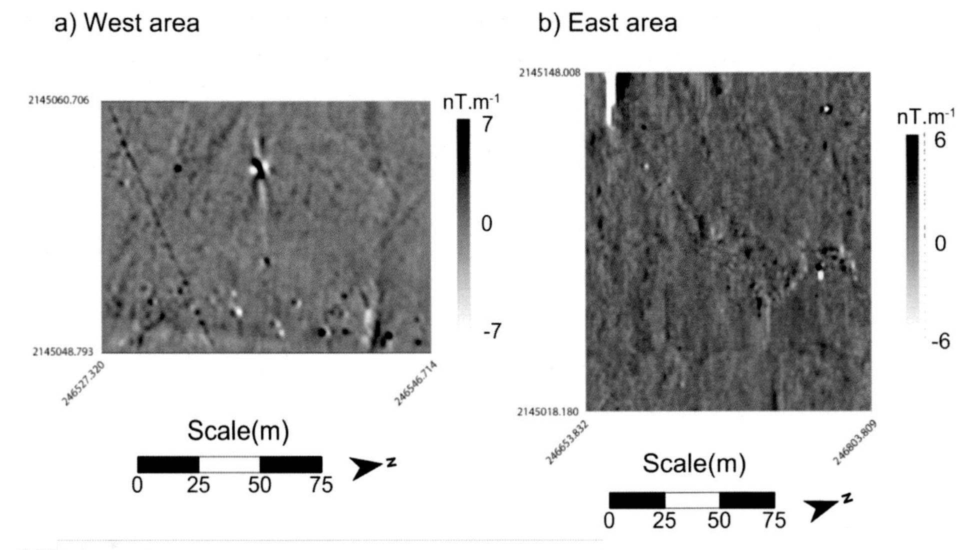

Figure 2: Results of magnetic survey: a) western part of the funeral mound, b) eastern part of the funeral mound.

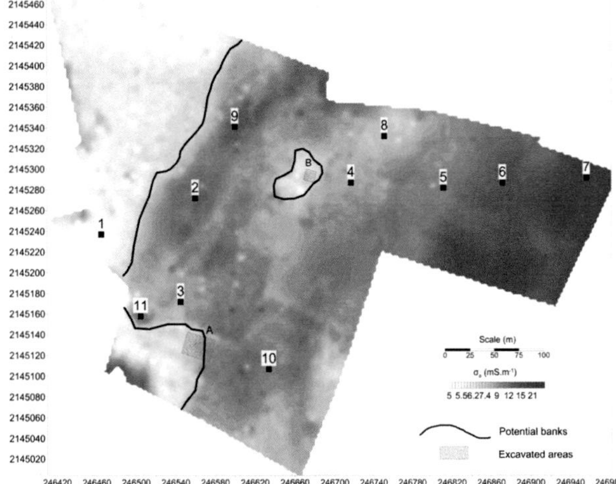

Figure 3: Results of the CMD survey with the location of the cores of the geomorphological study.

an area characterized by low electrical conductivity values. The second area, on the eastern part of the survey has the highest electrical conductivity values. These two areas are quite homogeneous. Between them, the central area shows intermediate values with short wave length spatial variations.

Discussion

The magnetic maps covered areas where the apparent electrical conductivity changes quickly. If there were remains, they appear to be very abraded, mainly by the erosion, and the magnetic signals do not present indications of significant changes, which could be caused by the lateral changes in the geometry of the subsurface. The electrical conductivity exhibits interesting patterns especially showing that the funeral mounds appear to be located over a less conductive area, (Fig. 3, area B). It is also the case of the settlement which is excavated at the south (Fig. 3, area A). Nevertheless, the funerary mound is separated from the settlement by a more conductive band. The black line corresponds to the 6.7 mS.m^{-1}. This line could be hypothesized as an ancient bank. According to these results, we could assume that the funeral mound was separated from the settlement by a channel of the ancient Nile river.

In addition, the first on-field comparisons of the geophysical results with the cores made on the site (numbered on Figure 3) highlight that equal conductivity (for example for cores 6 and 9) occurs with various geomorphologies (more silty in Core 6 and with salinization traces in core 9). This last point indicates that the changes linked to the actual agricultural activities also affect the properties of site soil. Then, the geophysical results should be handled carefully in order to reconstruct the Neolithic landscape.

Conclusions

In a very difficult context we use geophysical survey to extend the knowledge around excavations. Despite some good geophysical results (especially in electrical conductivity), the geophysical signal alone could not delineate trends linked to soil texture changes from the remains of contemporary irrigated areas. The next step is to reconstruct the ancient landscape based on the full results of both geomorphology and geophysics.

Bibliography

Alarashi, A., Khalidi, L., Gourichon, L., Baumann, L., Langlois, O. and Chambon P. (2016) *From raw materials to finished objects: Stages of artifact production as grave offerings in the Middle Neolithic Kadruka 23 cemetery (Northern State, Sudan)* 23rd biennal meeting of the society of Africanist archaeologist Toulouse France

Reinold, J. (1998) Le Néolithique de Haute Nubie. Traditions funéraires et structures sociales, *Bulletin de la Société française d'égyptologie* **143**: 19-40.

Acknowledgements

The authors wish to thank the SFDAS and especially Dr Marc Maillot for receiving us in Khartoum.

The forgotten castle of the Ciołek family in Żelechów, Mazowieckie province, Poland

Wojciech Bis[(1)], Tomasz Herbich[(1)] and Robert Ryndziewicz[(1)]

[(1)]Institute of Archaeology and Ethnology, Polish Academy of Sciences, Warsaw, Poland

herbich@iaepan.edu.pl

The feature presented in this article lies in the Garwolin district (Mazowieckie province), north-west of the city square and the parish church in Żelechów, on the southern side of the fish pond and in the curve of the Żelechówka stream. It was first noted when analyzing LiDAR data. A digital model of the ground reveals a structure resembling a quadrangular fortified structure (Fig. 1). A historical source query indicates the presence in this area of the family seat of the Ciołek family, lords of Żelechów. Field walking confirmed the existence of a *fortalicium*, also referred to as a castle. The presence of this complex had been suggested in literature, but was located tentatively in the northern part of the town, on the site of the standing stone-masonry manor of the late 18th century.

The building was first mentioned by Jan Długosz in his *Liber Beneficjorum* (AD 1470–1480) as a *praedium militare*, which was presumably a knightly grange, possibly with a palisade around it. Its fortification may have been dictated by the political situation beyond the eastern borders of the Kingdom of Poland at the end of the 15th and in the beginning of the 16th century AD, animosity with neighbors or a growing trend for grand fortified family seats. The term *fortalicia* occurs for the first time in 1515 when Andrzej Ciołek *de Zelechow et Borzecz* gifted his stepbrother Feliks *de Zelanka*, starost of Łuków, half the town and the *fortalicia* along with the suburbs and a few villages (Wierzbowski 1910, 2380). A few years later, in 1523, King Sigismund the Old issued an order mentioning the Chełmno castellan and Łuków starost Jan *Czolek de Zelechow* and the town of *Zelechow cum fortalitio* and villages *Vola Zelechowska, Ostrozenye* and *Lomnycza* (Wierzbowski 1910, 4208) in the context of what may have been a family feud to be resolved by the king. No mention of the place seems to exist after 1523. It was destroyed presumably during the Swedish Deluge of 1655, probably at the end of the year, just like the castle in nearby Stężyce, which was ruined by the Swedish troops on the night of December 24 (Nagielski *et al.* 2015, 440).

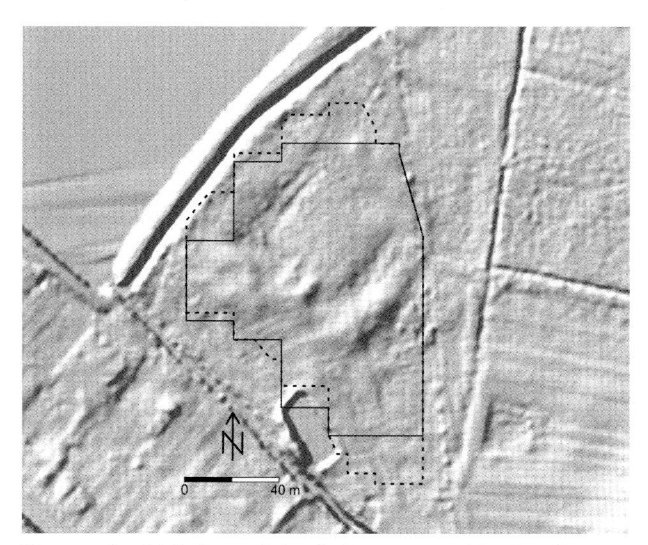

Figure 1: Digital model of the area with the feature interpreted as a quadrangular fortified complex. Dashed line indicates the extent of the magnetic survey, solid line that of the electrical resistivity prospection.

Noninvasive research using the magnetic and electrical resistivity methods confirmed the presence of the feature and clarified its plan (Fig. 2). The magnetic survey was carried out with a Geoscan Research FM256 fluxgate gradiometer. Measurements were taken in parallel mode, in a 0.25 x 0.50m grid. The same grid was used for the electrical resistivity profiling, using in this case a Geoscan Research RM15 instrument and a twin-probe array with AM=0.5m spacing of the probes. The measurement step was 0.5m along profiles spaced 1m apart.

The magnetic map reveals a quadrangular structure that is almost square in plan with the corners aligned with the cardinal directions. The parallel arrangement of linear anomalies characterized by high amplitudes or prevalence of positive values of magnetic field intensity can be interpreted as the foundation of an inner and outer enclosure wall constructed of red brick. If true, the interpretation would suggest a castle complex with the outer walls approximately 80–85m long. The distance between the lines would have been about 11m (north-western and south-western wings) and about 9m (north-eastern and south-eastern wings). Anomalies reflecting inner divisions (corresponding to partition walls) are also discernible on the map. The irregularity of these anomalies suggests a differentiated size of the chambers and could also reflect layers of red-brick rubble of different thickness, which could be responsible for the irregular image. The outlines of two rectangular features, projecting from the line corresponding to the outer walls of the complex, were observed in the south-western and north-western wings. In the area of what would have been the castle courtyard, the survey revealed highly magnetic anomalies in the southern corner, by the inner edge of the north-eastern wing and in the central area. These could be architectural remains as well.

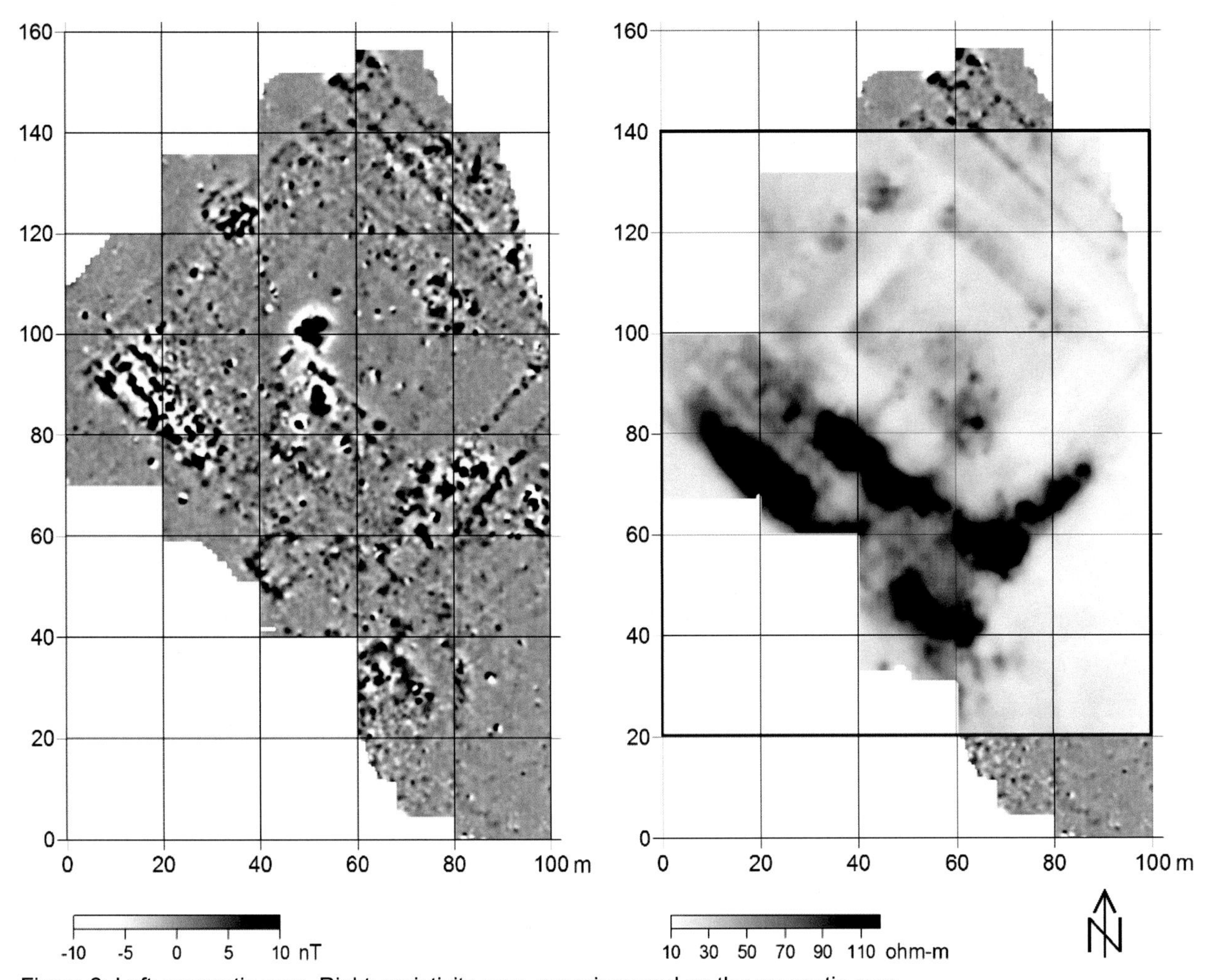

Figure 2: Left: magnetic map. Right: resistivity map, superimposed on the magnetic map.

The area proved to be greatly differentiated in terms of the electrical resistivity results (range from 10 to 300ohm/m). The arrangement of anomalies corresponded to the magnetic mapping results, giving rise to a plan outlining a square structure with sides 80m long. Linear anomalies with raised resistivity indicated the position of the defensive walls in places corresponding to magnetic anomalies. The outer edge of the structure on the south-eastern, north-eastern and north-western sides corresponded to the position indicated by magnetic surveying. The only discrepancy that can be seen concerns the south-western side where linear anomalies corresponding to the enclosure wall and inner partitions are accompanied on both the inside and outside by long anomalies of maximum resistivity values (up to 300ohm/m). These anomalies could reflect stone cobbles over a large area or else a thick layer of dumped rubble. This is the highest-lying part of the castle and it may be assumed that the ground here conceals the best preserved masonry remains.

The resistivity survey gave better results than the magnetic mapping in the area of the castle courtyard. An anomaly about 7–8m wide was recorded along the inner enclosure wall. It could be linked to the presence of a moat around an evident square feature measuring 40 x 40m. A perpendicular structure connecting the two lines of walls, seen in the central section of the southwestern side, may be interpreted tentatively as a gate bridge.

The survey results suggest the presence of two castles in Żelechów (Fig. 3). The first and smaller one was of late medieval origin. It had a moat around the walls and a towered gate entrance on the southwestern side, leading to the courtyard. The north-eastern wing functioned as a living and household space. In the second phase (in the 16th? century), a much bigger castle was constructed on the spot. It was of regular shape with building architecture in all of the wings. No moat could be traced around this larger precinct. The orientation was slightly different compared to the earlier complex, about 4–5 degrees to the west.

The noninvasive survey carried out in Żelechów identified the location of the Ciołki family castle, later enlarged, in Żelechów. The castle was known from written sources, which failed to mention its precise location. Excavation is planned to confirm

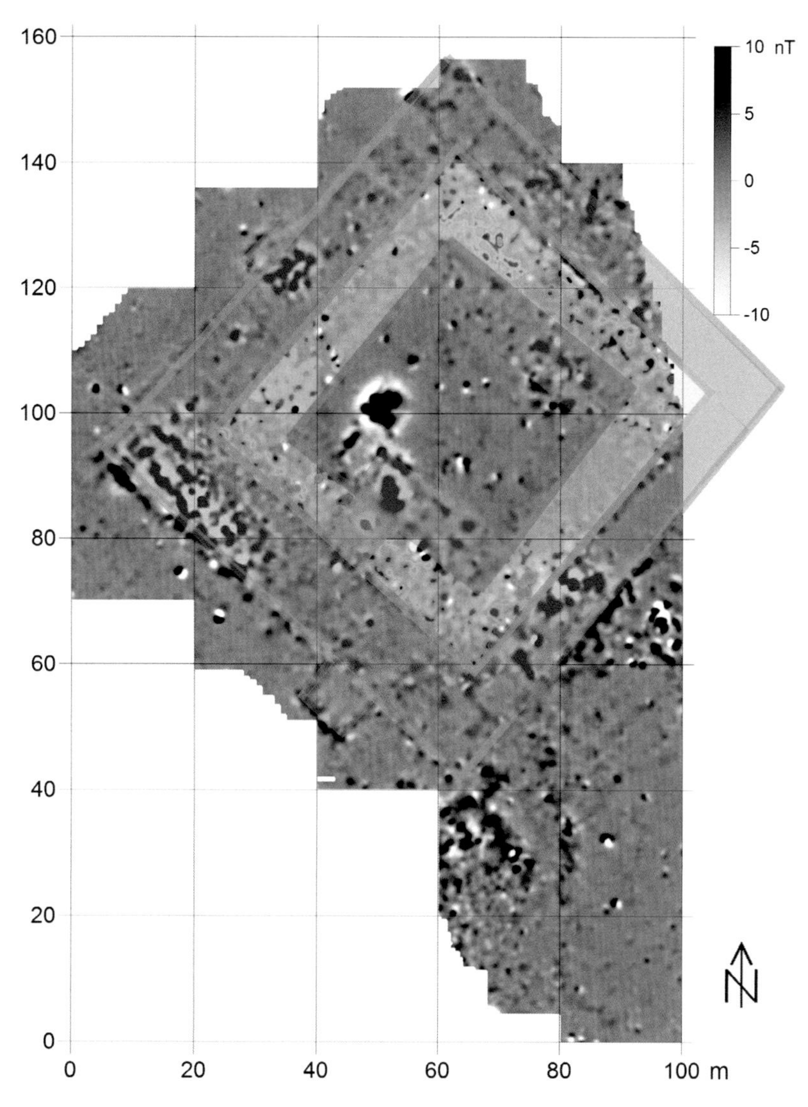

Figure 3: Reconstruction of building phases. Earlier phase in blue, later phase in red.

the presence of architecture on the site, verifying the tentative archaeological interpretation put forward on the grounds of this survey.

Bibliography

Wierzbowski, T. (ed.) (1910). *Metricularum Regni Poloniae Summaria*, vol. 4.

Nagielski, M., Kossarzecki, K., Przybyłek, Ł. and Haratym, A. (2015) *Zniszczenia szwedzkie na terenie Korony w okresie potopu 1655-1660*, Warszawa: Wydanie I.

Acknowledgments

The authors owe the source query on the site to Michał Zbieranowski from the Institute of History, PAS, in Warsaw.

Living in a post-workhorse world: observations learnt from rapidly acquired electromagnetic induction surveys in Ireland (...when magnetometry just won't do...)

James Bonsall[1, 2]

[1]Earthsound Geophysics Ltd., Claremorris, Ireland; [2]Sligo Institute of Technology, Sligo, Ireland

james@earthsound.net

Flogging a (Dead?) Workhorse in 21st Century Ireland

Clark's description of magnetometry as the 'workhorse' of archaeological geophysics reflects the technique's unique ability to rapidly identify a wide range of archaeological features (Clark 1996). Despite recent technological developments that allow GPR, earth resistance and EMI instruments to economically map increasingly larger areas; as a passive technique, magnetometry will rightly continue to offer the most rapid form of geophysical assessment (Bonsall and Gaffney 2016). Geophysicists in Ireland have historically preferred magnetometry over other techniques, while its use in Northern Ireland has been mixed or limited due to that region's predominantly volcanic geology. A reappraisal of geophysical data from development-led Irish projects demonstrated that a reliance on magnetometry led to the non-detection of several types of small-medium sized archaeological features and large enclosure sites (Bonsall *et al.* 2014a). These were due in part to poorly specified and inappropriate methods or sampling strategies used to fulfil the (often economically determined) aim of rapidly-acquiring data over large areas, which strongly impacted on the success or otherwise of those surveys. In these instances, magnetometry was commissioned regardless of the archaeological feature type under investigation and little consideration was given – especially when prospecting for unknown sites – to the response of magnetometry on (in)appropriate soils and geologies. In many cases, this resulted in the loss of crucial information that was important in the assessment of geophysical data acquired by development-led projects (Bonsall 2014).

Adaptive Procurement & Prospection Strategies

Transport Infrastructure Ireland (TII, formerly known as the National Roads Authority), responsible for frequently commissioning large linear corridor assessments in Ireland, have recognised that an over-reliance on magnetometry is not beneficial in the long-term and that alternative – but appropriate – methods and techniques should be used when required. TII archaeologists - and others, including Northern Ireland's Historic Environment Division - use a procurement and guidance document (Bonsall *et al.* 2014b) to help determine the most appropriate technique(s) for a given site, area, soil, geology, etc. under investigation. Common geological challenges to magnetometry encountered in both Ireland and Northern Ireland (e.g. weak contrasts for archaeological features on Carboniferous Limestone and widespread peatlands; a variable and often noisy background on Limestone Tills; strong background contrasts on Basalt and other igneous geologies) have been acknowledged by a number of both commercially-driven and research-led projects, which resulted in a decrease in the frequency of 'magnetometry-only' assessments and a comparative increase in the use of other techniques.

The TII procurement strategy requires the collection of 0.5m x 1m resolution 'earth resistance or resistivity data' for sites/geologies where it is (frequently) deemed appropriate, but adopts a *laissez-faire* attitude to probe geometry, array selection and the acquisition of resistivity data. Individual contractors may choose between different earth resistance array options or apparent electrical resistivity calculated from EMI quadrature measurements, selecting the most appropriate technique(s) or instrument(s) for the given depth of investigation, surface soil dryness, ground cover or speed of acquisition. This 'hands-off' approach has increased the collection of electrical resistivity data on Irish infrastructure projects and had a wider impact on survey choices beyond both TII infrastructure assessments and the Irish border.

Earthsound Geophysics Ltd. have favoured the use of EMI to rapidly acquire multi-depth apparent electrical conductivity/resistivity (and/or apparent magnetic susceptibility) data to assess large areas for infrastructure and research projects, collecting >500 ha of high resolution EMI data in the last 5 years. A key outcome is a large reference collection of anomalies and responses to different archaeological, environmental and geological features. In many cases, the speed of development-led projects has also ensured a large comparative dataset of excavation records that provide a measure of success for the survey results. These grey literature outputs have generated insights regarding the 'best' and 'worst' outcomes of EMI use across Ireland and Northern Ireland for a range of different site types, soils and geologies.

Contribution of EMI-driven Grey Literature

In many cases, such as the discovery of an early medieval enclosure at Castlesize (Co. Kildare) by magnetometry and EMI, the assessment of narrow infrastructure corridors clearly benefited from the application of two complimentary techniques across areas of limestones, tills and alluvial deposits on the edge of the River Liffey (Bonsall *et al.* 2016). However, surveys over limestone and peatland continue to be mixed and include notable failures of false positive and false negative responses: neither magnetometry nor EMI (both at 0.5m x 0.25m resolution) were capable of identifying the c.500 inhumations found at Barnhall (Co. Roscommon) although the surrounding medieval enclosure was clearly mapped (Bonsall and Gimson 2015a) and neither technique identified the majority of 42 burnt spreads / burnt mounds of stone (*fulachta fiadh*) along a 73 ha linear corridor in Co. Mayo (Bonsall and Gimson 2015b).

Larger features - typically enclosure sites – have been identified easily by EC_a and/or MS_a EMI surveys. Post-medieval garden archaeology in the form of a sunken parterre (Gimson 2014) and the remains of a (probable Neolithic) passage tomb at the base of Dublin's Hellfire Club, were both clearly defined by EMI survey (Bonsall and Gimson 2014). On the A4 Enniskillen Bypass (County Fermanagh), comprehensive magnetometry and EMI surveys identified archaeologically prospective near-surface gravels and islets within a wider wetland landscape (Bonsall and Gimson 2016).

Aitken (1959) described the magnetic properties of near-surface Basalts across Northern Ireland's Antrim Plateau as "too violent to permit archaeological work". Recent infrastructure and development-led projects covering >100 ha over these Basalts required non-intrusive investigation; EMI EC_a and/or MS_a proved to be useful for the prospection of previously unknown archaeological and palaeo-environmental features such as Ballyscullty hilltop enclosure (Bonsall & Gimson 2015c), as well as post-medieval field systems and palaeochannels, although notable failures include false negative response for a Late Mesolithic circular structure and an Iron Age ring ditch at Drumakeely (Bonsall and Gimson 2013). The EMI response to Basalt geology was also found to mimic the morphology of circular enclosing features in several areas (Gimson and Bonsall 2014), echoing the experience of 'non-unique' archaeological-derived anomalies observed in magnetometry surveys.

Conclusion

Northern Ireland's statutory Historic Environment Division now specifies the use of EMI - where appropriate – instead of magnetometry/earth resistance assessments, based on the generally positive (and rapid) contribution made by EMI surveys across the region's challenging soils. The growing use of EMI in Ireland and Northern Ireland demonstrate that the technique has made a significant contribution to the assessment of sites and landscapes under investigation. Its use, where appropriate, is increasingly relevant although the poor response on limestone and peats to small-scale features (inhumations, post-pit/-hole features) and some burnt mounds of stone is – as with magnetometry - a hindrance. EMI research must now focus on resolving its (in)ability to detect small-scale feature types (even at a very high resolution of data capture) or accept it as a limitation of the technique.

Bibliography

Aitken, M. J. (1959) 'News and Notes - Magnetic Prospecting: An Interim Assessment'. *Antiquity* **33**(131): 205-207.

Bonsall, J. (2014) *A reappraisal of archaeological geophysical surveys on Irish road corridors 2001-2010.* Thesis submitted for the degree of Doctor of Philosophy, Archaeological and Environmental Science, School of Life Sciences, University of Bradford.

Bonsall, J. and Gaffney, C. (2016) 'Change is Good: Adapting Strategies for Archaeological Prospection in a Rapidly Changing Technological World' In F. Boschi (ed.) *Looking to the Future, Caring for the Past: Preventive Archaeology in Theory and Practice.* Bologna: Bononia University Press, 41-58.

Bonsall, J., Gaffney, C. and Armit, I. (2014a) 'A Decade of Ground Truthing: Reappraising Magnetometer Prospection Surveys on Linear Corridors in light of Excavation evidence 2001-2010', In H. Kammermans, M. Gojda and A. Posluschny (eds.) *A Sense of the Past: Studies in current archaeological applications of remote sensing and non-invasive prospection methods.* Oxford: Archaeopress, 3-16.

Bonsall, J., Gaffney, C. and Armit, I. (2014b) *Preparing for the future: A reappraisal of archaeo-geophysical surveying on National Road Schemes 2001-2010.* University of Bradford report for the National Roads Authority of Ireland.

Bonsall, J. and Gimson, H. (2013) *A26 Dualling, Glarryford to A44 (Drones Road) Junction, Ballymena, Co. Antrim.* Earthsound Archaeological Geophysics. Unpublished report No. EAG 238, November 2013.

Bonsall, J. and Gimson, H. (2014) *The Hellfire Club, Hellfire Archaeological Project, Montpelier Hill, South County Dublin.* Earthsound Archaeological Geophysics. Unpublished report No. EAG 249, June 2014.

Bonsall, J. and Gimson, H. (2015a) *N61 Coolteige (Phase 1) Road Project, County Roscommon, Stage (i) i Archaeological Geophysical Survey.* Earthsound Archaeological Geophysics. Unpublished report No. EAG 253, January 2015.

Bonsall, J. and Gimson, H. (2015b) *N5 Westport to Turlough Road Project; Archaeology Consultancy Services Contract (Phase i & ii).* Earthsound Archaeological Geophysics. Unpublished report No. EAG 270, June 2015.

Bonsall, J. and Gimson, H. (2015c) *Proposed Development, Ballysculty, Muckamore, County Antrim*. Earthsound Archaeological Geophysics. Unpublished report No. EAG 271, July 2015.

Bonsall, J. and Gimson, H. (2016) *A4 Enniskillen Bypass, County Fermanagh: Archaeo-geophysical Survey*. Earthsound Archaeological Geophysics. Unpublished report No. EAG 299, August 2016.

Bonsall, J., Hogan, C. and Gimson, H. (2016) *M7 Osberstown Interchange & R407 Sallins Bypass Scheme, County Kildare: Archaeological Consultancy Services Contract Stage (i) I Geophysical Survey*. Earthsound Archaeological Geophysics. Unpublished report No. EAG 291, September 2016.

Clark, A. J. (1996) *Seeing beneath the soil: prospecting methods in archaeology*. rev. edn. London: Batsford.

Gimson, H. (2014) *Formal Gardens, Castle Coole, Enniskillen, Co. Fermanagh*. Earthsound Archaeological Geophysics. Unpublished report No. EAG 242, April 2014.

Gimson, H. & Bonsall, J. (2014) *A31 Magherafelt Bypass, Co. Londonderry*. Earthsound Archaeological Geophysics, Unpublished report No. EAG 254, September 2014.

The magnetic signature of Ohio earthworks

Jarrod Burks[(1)]

[(1)]Ohio Valley Archaeology, Inc., Columbus, Ohio

jarrodburks@ovacltd.com

The Middle Ohio River Valley in the Midwest, United States is rich in ancient earthwork sites (e.g. Squier and Davis 1848; Lilly 1937; McLean 1879). Ditch-and-embankment enclosures smaller than 100 meters across and in the forms of circles and squares with rounded corners (i.e. squircles) abound; unique geometric forms also are present, including a quatrefoil, ovals, an ellipse, crescents, and many more. The largest enclosures range up to 450 meters across and commonly occur in earthwork complexes with other large forms (e.g. Lepper 2005). These larger complexes often consist of embankments lacking ditches, some running on literally for miles. The state of Ohio alone has nearly six hundred documented sites containing earthen enclosures (Fig. 1) (Burks 2010; Fowke 1902; Mills 1914), all likely dating to a period from about 300 BC to AD 400. Two hundred years of ploughing has flattened the vast majority of Ohio's earthwork sites and most have been lost over the last century or more since they were first documented in the 1800s. Geophysical survey and aerial photo analysis are beginning to reverse this problem; in fact, at least seven previously undocumented earthwork sites have been found in the last decade.

Over the last 15 years I have had the opportunity to collect magnetic data on portions, or all, of at least three dozen different earthwork sites in Ohio (e.g., Burks and Cook 2011; Burks 2012, 2014a, 2014b). The surveys have been conducted with fluxgate gradiometers, including a Geoscan Research FM 256 and a Foerster Instruments 4-probe Ferex DLG 4.032 cart-based system. Data densities have ranged from eight to twenty readings per meter (i.e. 50 cm or 100 cm transect spacing with 8-10 readings per meter). Data processing was all done with Geoscan Research's Geoplot software.

Not unexpectedly, ditch-and-embankment enclosures are the easiest to detect (Fig. 2). The fill in ditches produces elevated magnetic gradients while the ploughed out remains of the accompanying embankments are usually less magnetic. Ditch fill magnetic gradients often vary based on soil type, with more strongly magnetic ditches being found in floodplain contexts where the soil-filled ditches penetrate into sand or gravel deposits. Other, minor details of earthwork morphology also emerge in the magnetic data. Embankment width varies little from enclosure to enclosure, but there are at least three classes of ditch width: narrow, moderate, and wide varieties. Gateway width is fairly consistent regardless of size, though larger enclosures do tend to have slightly larger gateways.

The embankment walls of large enclosures, those lacking accompanying ditches, tend to be harder to detect and vary considerably in how they appear in the magnetic gradient data (Fig. 3). Some are outlined by slightly stronger magnetic gradient values, while others have fill that is consistently more magnetic throughout. In nearly all cases the instruments are detecting variability in magnetic susceptibility. Augmented/reconstructed embankments have a very consistent magnetic signature. Most unusual are the embankments with complex dipolar magnetic signatures (e.g. the High Bank Works in Ross County). Some of this variability is related to plough damage—those that are more eroded or flattened tend to be represented by simple outlines. Complex embankment fills, especially those containing magnetically contrasting layers or lenses of sandy gravels mixed with silty loams, not surprisingly produce rather complicated magnetic anomalies. Most difficult to detect are the sets of long, parallel embankment walls that often run from the large earthwork complexes towards the nearest nearby stream. Most have been so thoroughly ploughed away that they simply are no longer detectable in magnetic gradient data, and

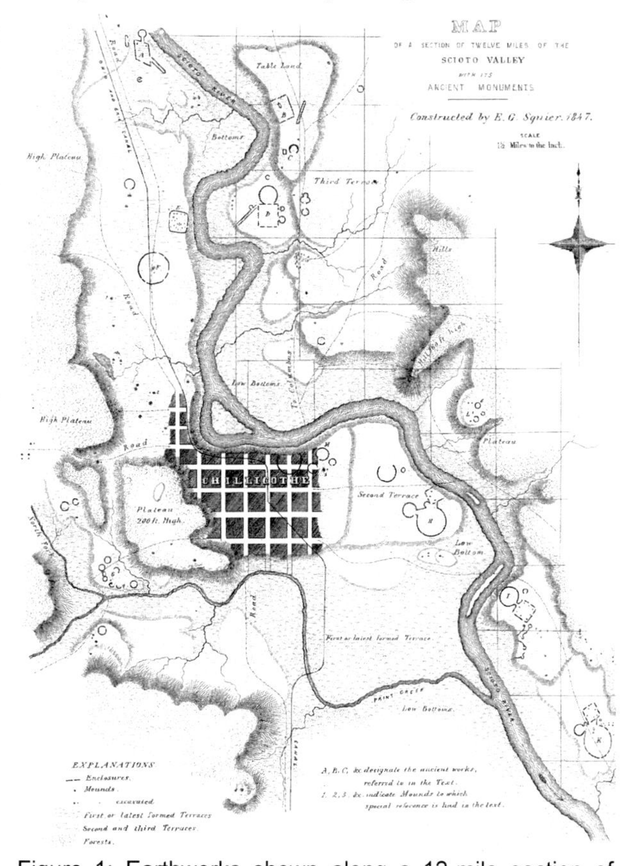

Figure 1: Earthworks shown along a 12-mile section of the Scioto River around Chillicothe, Ohio (from Squier and Davis 1848: Plate 2).

they often defy detection through other geophysical means, as well, though some are visible in early twentieth century aerial photographs.

In addition to detecting embankments and ditches, magnetic surveys at Ohio earthworks also reveal many different types of accompanying features. Some of these are clearly associated with enclosures based on their arrangements or positioning. Large thermal pits (for cooking?) and borrow pits are common, including some just outside gateways or positioned at key geometric locations within and outside of enclosures. Post circles and other building footprints are rare in Ohio magnetic data, but have been found at several sites, suggesting that perhaps they rarely are detected rather than being rarely present. Pavements of gravel and stone rarely appear in magnetic gradient data, but they have appeared in surveys with ground-penetrating radar.

The mapping and excavation of mounds and enclosures in Ohio played a key role in the nineteenth century development of American archaeology as a scientific discipline. As plowing quickly erased most of Ohio's earthworks and the numbers of archaeologists working in the state diminished into the twentieth century, the locations of many earthwork sites were lost. The growth of geophysics in Ohio archaeology, however, is bringing many old earthwork sites back to light. It is also rekindling both public and scholarly interest in what has to be one of the larger concentrations of ancient earthworks on the planet. With continuing magnetic surveys, it is becoming increasingly apparent that the homogenized drawings of earthworks from the nineteenth century in fact mask considerable variability in size, shape, and construction technique. Some of this magnetic variability is also related to factors of geomorphic setting and preservation. Undoubtedly, Ohio has many more undocumented enclosure sites yet to be revealed.

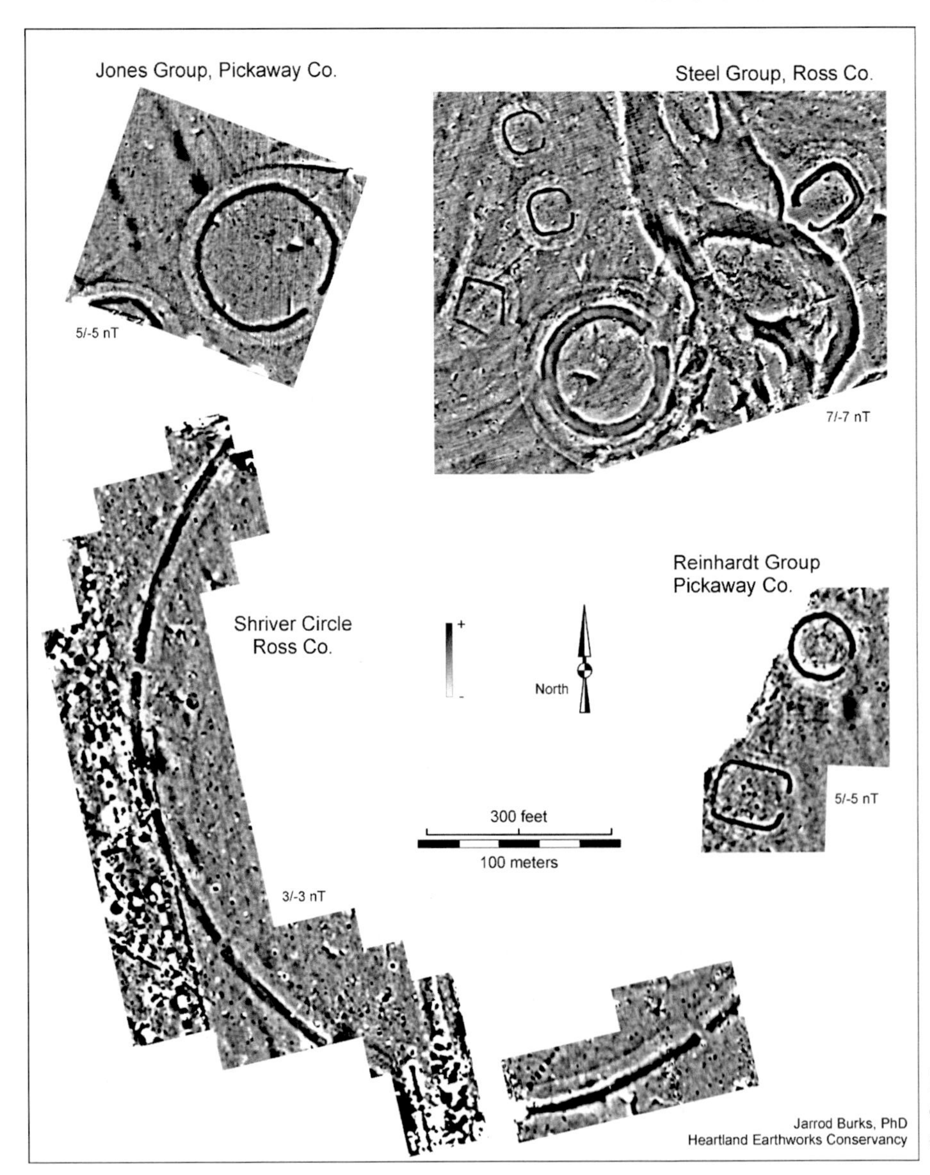

Figure 2: Examples of Ohio ditch-and-embankment enclosures in magnetic data.

Anderson Works,
Ross County

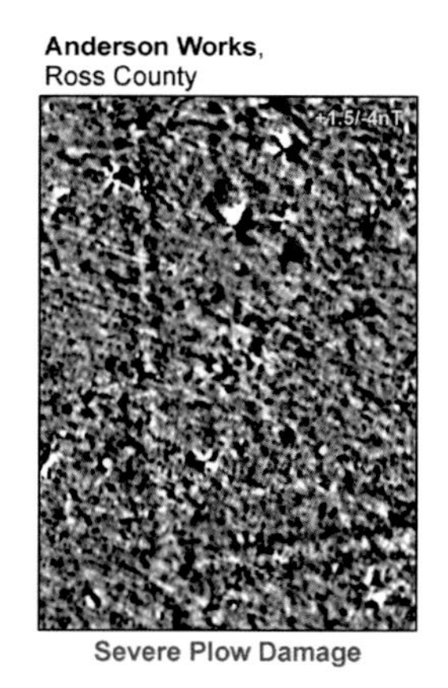

Severe Plow Damage

Hopeton Works,
Ross County

+4.5/-4.5nT

Moderate Plow Damage

High Bank Works, Octagon
Ross County

+4.5/-4.5nT

Moderate Plow Damage

Figure 3: Examples of embankments in magnetic data from Ohio earthworks lacking ditches.

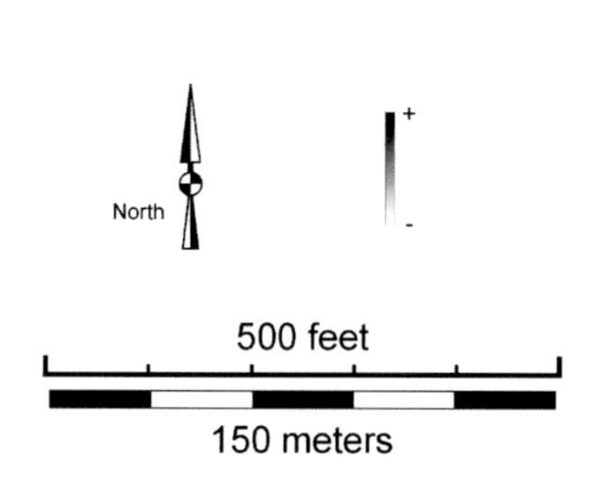

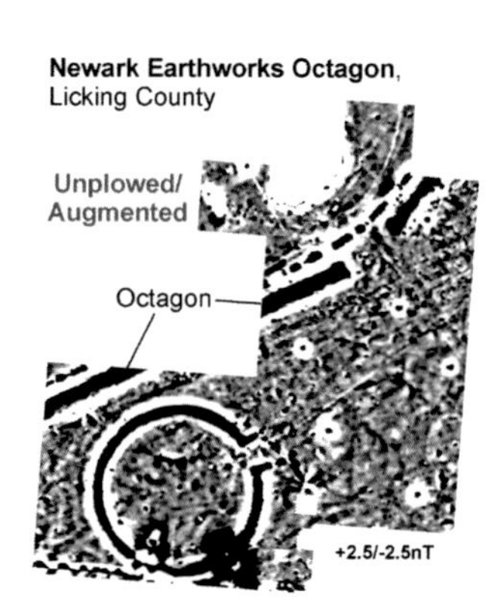

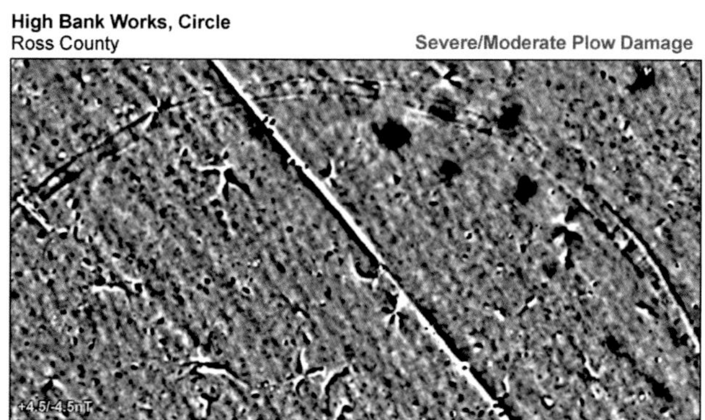

Bibliography

Burks, J. (2010) Recording Earthworks in Ohio-historic Aerial Photography, Old Maps and Magnetic Survey. In D. C. Cowley, R. A. Standring, and M. J. Abicht (eds.) *Landscapes through the Lens: Aerial Photographs and the Historic Environment.* Oxford: Oxbow Books, 77-87.

Burks, J. (2012) Ohio's Great Serpent Mound Surveyed. *ISAP News* **32**:6-7.

Burks, J. (2014a) Geophysical Survey at Ohio Earthworks: Updating Nineteenth Century Maps and Filling the "Empty" Spaces. *Archaeological Prospection* **21**:5-13.

Burks, J. (2014b) Recent Large-Area Magnetic Gradient Surveys at Ohio Hopewell Earthwork Sites *ISAP News* **39**:11-13.

Burks, J. and Cook, R. A. (2011) Beyond Squier and Davis: Rediscovering Ohio's Earthworks Using Geophysical Remote Sensing. *American Antiquity* **76**(4):667-689.

Fowke, G. (1902) *Archaeological History of Ohio.* Columbus: Ohio State Archaeological and Historical Society.

Lepper, B. T. (2005) *Ohio Archaeology: An Illustrated Chronicle of Ohio's Ancient American Indian Culture.* Wilmington, Ohio:Orange Frazer Press.

Lilly, E. (1937) *Prehistoric Antiquities of Indiana.* Indianapolis: Indiana Historical Society.

MacLean, J. P. (1879) *The Mound Builders.* Cincinnati: Robert Clarke & Company.

Mills, W. C. (1914) *Archaeological Atlas of Ohio.* Columbus: The Ohio State Archaeological and Historical Society.

Squier, E. G. and Edwin H. Davis, E. H. (1848) *Ancient Monuments of the Mississippi Valley.* Contributions to Knowledge, vol. 1. Washington, D.C.: Smithsonian Institution.

Three hundred miles in the footsteps of Vespasian ... and the Ancient Monuments Laboratory

Paul Cheetham[(1)], Dave Stewart[(1)] and Harry Manley[(1)]

[(1)]Department of Archaeology, Anthropology & Forensic Science, Faculty of Science and Technology, Bournemouth University, Talbot Campus, Fern Barrow, Poole, Dorset, UK - BH12 5BB

pcheetham@bournemouth.ac.uk

According to Suetonius in his *Lives of the Twelve Caesars*, Vespasian, as the general of the *Legio II Augusta*, thrust his legion westward through the county of Dorset as part of the conquest of the lowland areas of Britain in the decade after the Claudian invasion of AD 43. He is credited by Suetonius for subduing two tribes and twenty towns, as well as capturing the Isle of Wight. Vespasian's military success in the Dorset countryside set him on trajectory that would see him becoming Emperor in AD 69.

For many years it was postulated that there should be a Roman invasion base somewhere in Dorset from which this phase of Vespasian's campaign was conducted. Norman Field, a schoolteacher and amateur archaeologist, suggested the possible site of such a base at Lake Farm near Wimborne Minster, with his trial excavations in the 1960s indicating he might be correct (Field 1992). As a result of the finding, in the late 1970s and early 80s, one of the most significant large-scale geophysical surveys undertaken at the time by the Ancient Monuments Laboratory (AML) proved the presence of not just a small fort, but unexpectedly, a large 12Ha Roman vexillation fortress. The survey and interpretation were untaken by none other than Andrew David, Alastair Bartlett and Tony Clark, and the survey figures prominently in Tony Clark's *Seeing Beneath the Soil* (Clark 1990: 139-141).

Further limited excavation was undertaken on the site, but the AML survey remained the main source of evidence for the size, layout and orientation of the fortress. That this survey provided a level of interpretive detail hard to exceed today, despite the constraints of the surveying technology available at the time, is a testament to the expertise of the AML team (Fig. 1). In 2016 a team from Bournemouth University took on the challenge of building on that seminal AML survey, by resurveying the areas and, more importantly, extending the survey into areas of the fortress and its immediate hinterland not covered by the original surveys.

The surveys have covered an area in excess of 40 hectares, and required walking more than 300 statute miles (326 Roman miles or 483km) in the process. While not adding major anomalies in the areas already surveyed by the AML team, the increased coverage does clarify the overall size and layout of the fortress and reveals extensive evidence of extramural activity (Fig. 2). It is this extramural activity that has led to a reinterpretation of the fortress in relation to the military road system. The results show that there appears not to have been a road approaching from the south and directly entering the fortress by its southern gateway. However, there is a clear road from the east lined with anomalies of an 'industrial' character. This suggests that the original supply line into the fortress was from the east, approaching up the valley of the Stour, utilising the river to bring

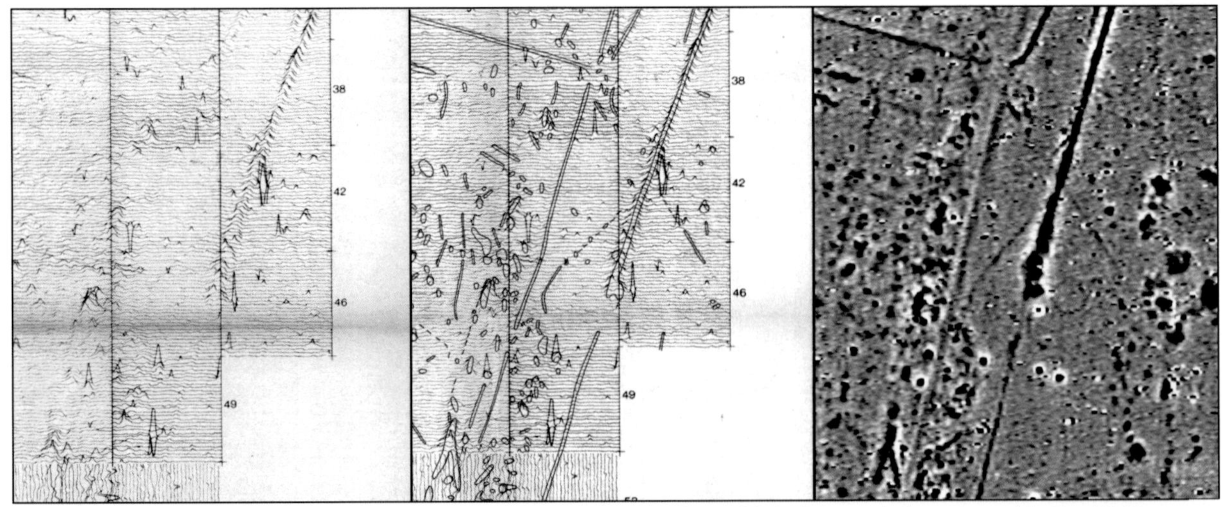

Figure 1: An example of the original 1980 AML x-y plotter survey data with its interpretation alongside the 2016 survey results presented as a greyscale image. Note that all the major anomalies have been picked up in the AML survey, but not the subtle annular anomaly (an Iron Age roundhouse?) in the centre of the area, and the industrial area outside the east of fortress beyond the AML's survey area.

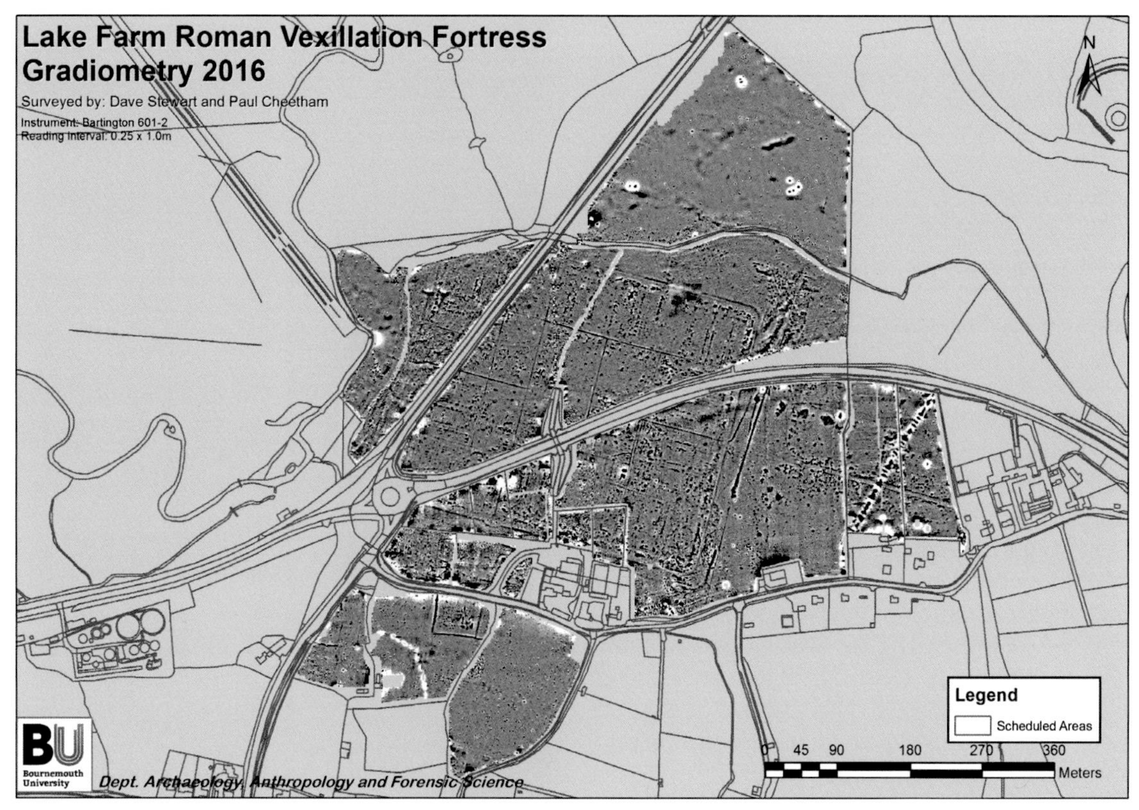

Figure 2: Fluxgate gradiometer survey of the fortress and surrounding areas. The 2016 surveys revealed the site of the military bathhouse situated directly outside the south gateway of the fortress blocking any direct road into the fortress from the south, and the 'industrial' areas lining the road extending east from the fortress. Bartington 601-2 at 0.25x1m reading intervals. Black positive. Plotting Range -5 to +5nT.

supplies inland to the fortress and not overland, as has been previously assumed, from the Roman port that developed at Hamworthy, on the banks of Poole Harbour.

This research delivers both a celebration of, and an accolade to, the pioneers of archaeological geophysics in the UK in their discovery of this fortress, while showing how the more recent geophysical surveys have changed our understanding of the role of the fortress and its relationship to the Roman port at Hamworthy. Both surveys confirm that it is geophysical techniques that have provided the step changes in our understanding of this monument.

Bibliography

Clark, A. (1990) *Seeing Beneath the Soil*. London: Batsford.

David, A. and Thomas, B. (1980) *Lake Farm 1980*, AML Report on Magnetometer Survey, Report No. 21/80.

Field, N. (1992) *Dorset and the Second Legion*. Tiverton: Dorset Books.

Identification of buried archaeological features through spectroscopic analysis

Yoon Jung Choi[1, 2], Johannes Lampel[4], David Jordan[3], Sabine Fiedler[2] and Thomas Wagner[1]

[1]Max Planck Institute for Chemistry, Satellite Remote Sensing Group, Mainz, Germany; [2] Institute of Geography, Johannes Gutenberg University Mainz, Germany; [3]School of Natural Sciences and Psychology, Liverpool John Moores University, UK; [4]Institute of Environmental Physics, University of Heidelberg, Germany

y.choi@mpic.de

Buried archaeological features are mainly noticeable and identifiable by their colour. This study investigates the spectral features of archaeological remains from the visible to the near infrared region in order to identify remains which might not show characteristic colour. To do this, a modified principal component analysis (PCA) which can identify spectral features of archaeological remains among natural soils is introduced. This method is applied to spectra gathered from an archaeological site in Calabria, Italy.

Spectra were gathered from 6 soil profiles in Calabria, Italy, where one pit contained an archaeological stratum, as shown in Figure 1. These spectral measurements were taken with the Analytical Spectral Devices (ASD) spectrometer with artificial halogen light. Previous studies (Choi *et al.* 2015) have shown that the colour of the soil is not the necessarily the main feature which can be used to separate the spectral characteristics between archaeological and natural soils and, therefore, the wavelength range of 400 to 1000 nm is considered for the PCA.

Figure 2 shows the PCA result at this wavelength range where a clear separation between the natural soil cluster and archaeological soils is observed. Here, the spectra of natural soils are located on the left side with PC1 values less than 0.5 and the archaeological soils are mainly situated on the right side of the figure. The results show that the PCA method can separate archaeological features from natural soil spectra. To further develop the PCA method, a simple assumption was made. We assumed that the main spectral characteristic features (first two principal components) of natural soils should be similar to any kind of natural soils, but fairly different from the spectra of archaeological remains. Therefore, the difference (D) between the original spectrum S and a modified spectrum S', which represents the principal component (PC) values of a group of natural soils (N_{soil}), is calculated. If the difference between the original and the

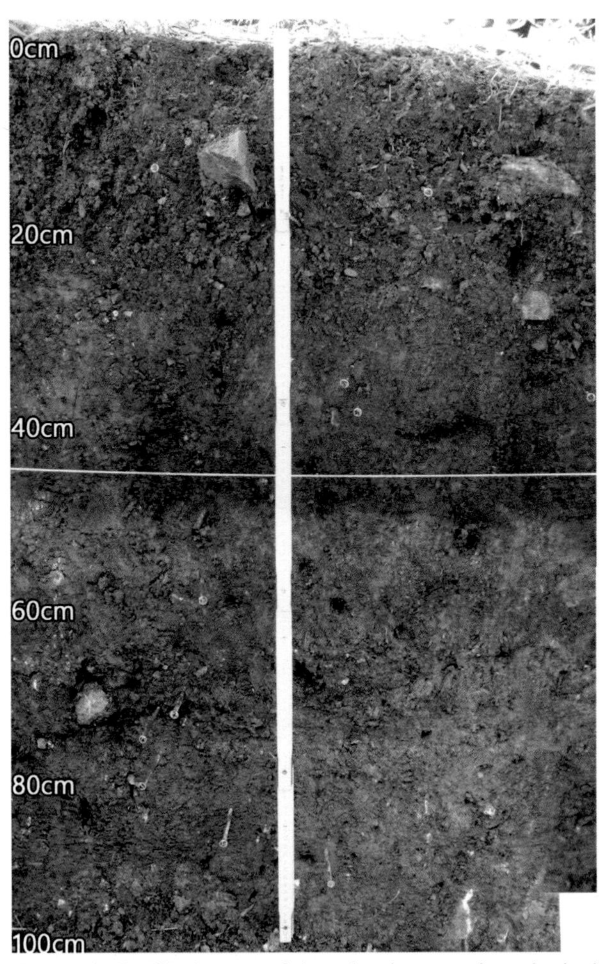

Figure 1: Profile image of the pit where archaeological strata were observed. A red-coloured stratum is observed at 30 to 70 cm depth with a thin burned soil shown in black just below the stratum.

modified spectrum is large, then the spectrum does not contain spectral features of natural soils and thus represents an archaeological material. Thus, the ratio between the D values of natural soils (D-nat) and archaeological soils (D-arch) should be larger than 1 in order to indicate archaeological material. Since the D value depends on the group of natural soils (N_{soil}) used for the S' calculation, two different types of N_{soil} will be compared in this study; a group of natural soil spectra gathered from a nearby region from the archaeological site in Calabria, Italy, and a group of natural soil spectra from the ICRAF-ISRIC Soil VNIR Spectral Library, where around 4000 natural soil spectra were gathered globally, to test if the technique can be applied universally.

Figure 3 summarises the D calculation results for the archaeological site in Calabria, Italy, using two different N_{soil}. The D value for natural soils is slightly smaller when N_{soil} is used from Calabria region (around 0.09) than from the ISRAF-ISRIC (slightly above 0.1). The D value for archaeological soils is much larger for N_{soil} from Calabria (approximately 0.4). This indicates that the method works better

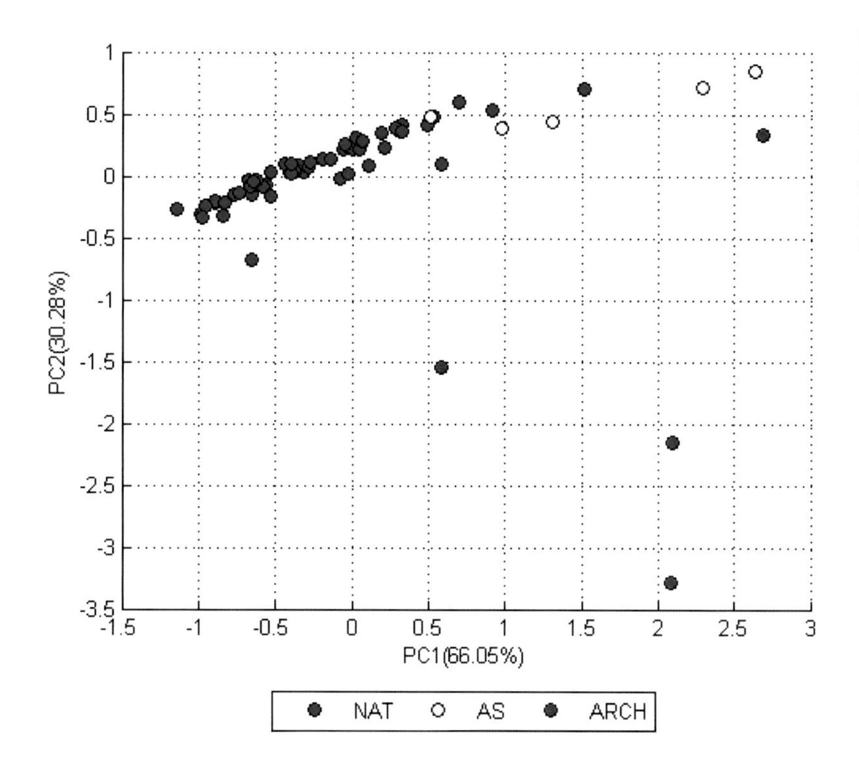

Figure 2: PC1-PC2 score plot for a wavelength range of 400 to 1000 nm showing the separation between natural soils and archaeological soils. NAT represents the natural soil spectra, AS the spectra of the red archaeological stratum and ARCH the spectra of burned soils and ceramics.

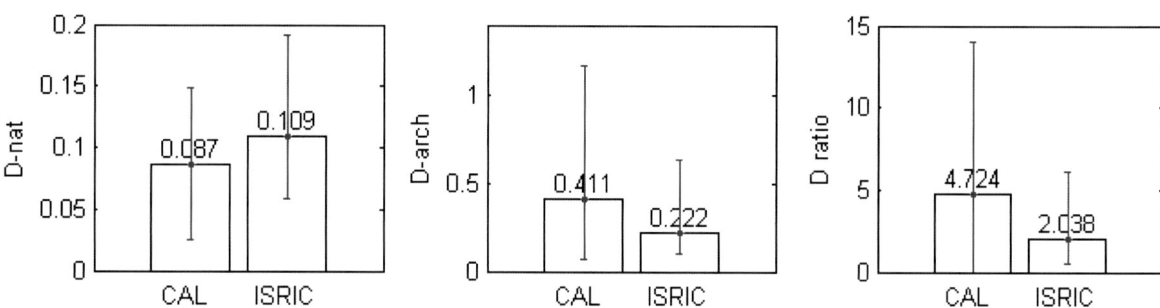

Figure 3: D values (D-nat: natural soils, D-arch: archaeological soils, D-ratio: ratio between D-arch and D-nat) when Nsoil is used from the natural soils of Calabria, Italy (CAL) and when it is used from the ICRAF-ISRIC spectral library (ISRIC).

when N_{soil} contains similar soil types. When N_{soil} is from Calabria, the D-ratio value is around 4, indicating that the archaeological materials clearly stand out from natural soils. If the N_{soil} is used from ICRAF-ISRIC, the ratio is only around 2. However, this value is still large enough to separate the archaeological spectra from natural soils. The results show that when a sufficient amount of natural soil spectra are collected from an unknown site, one can expect clear archaeological features with large D-ratio values above 4. However, even if natural soil spectra were not gathered (or it remains unclear whether the spectra represent natural soils or not), the method can still be applied by using the N_{soil} from ICRAF-ISRIC global spectral library and expect a returned D-ratio value above 2 to be an archaeological feature.

The study will present further results of various D values when the method is applied to other archaeological sites with different S' calculation conditions. By doing so, a D-ratio value which can be applied in a universal way for identifying archaeological spectra will be investigated.

We express our thanks to the University of Groningen and the National Museum of Hungary for their help and support during the fieldwork.

Bibliography

Choi, Y. J., Lampel, J., Jordan, D., Fiedler, S. and Wagner, T. (2015) Principal component analysis (PCA) of buried archaeological remains by VIS-NIR spectroscopy. *Archaeologia Polona* **53**: 412-415.

World Agroforestry Centre (ICRAF) and ISRIC - World Soil Information (2010) ICRAF-ISRIC Soil VNIR Spectral Library. Nairobi, Kenya: World Agroforestry Centre (ICRAF). Available at http://africasoils.net/.

Integration of ground-penetrating radar and magnetic data to better understand complex buried archaeology

Lawrence B Conyers[(1)]

[(1)]Department of Anthropology, University of Denver, Denver, Colorado, USA

lconyers@du.edu

Introduction

Geophysical archaeologists have long relied on an analysis from multiple geophysical sensors to provide an interpretation of buried archaeological features and their surrounding matrix. Sometimes this is done by "data fusion" (Kvamme 2006) but often relies on a direct comparison of maps produced by different datasets (Piro *et al.* 2000), with a visual analysis of features or anomalies that appear to exist (or not) in a study area. While this can often be informative, a comparison of ground-penetrating radar (GPR) reflection features with those obtained by magnetic gradiometry are often not spatially comparable or indeed do they overlap at all, and are actually exhibiting properties of the ground that are very different both locally and regionally. While producing images of very different variables of the ground, the two datasets (GPR and gradiometry) are actually quite comparable and informative, but not when directly "matched up" in space. A few examples are given below where surveys of one method were complemented and enhanced with specific analyses of the other.

Ground-penetrating radar has the ability to view geological and archaeological units in three-dimensions that show the boundaries between layered materials in the ground with differences in relative dielectric permittivity (RDP) (Conyers 2013). The differences in RDP are almost wholly the product of differences in porosity and permeability of layers, and therefore their ability to retain or shed water (Conyers 2012). These differences in units can be viewed in reflection profiles and those profiles then re-sampled and mapped in slices, in order to show the distribution of important units in the ground (Conyers 2016a).

Magnetic gradiometry is measuring the differences in the magnetic susceptibility of units in the ground or their thermo-remnant magnetism (Aspinall *et al.* 2009; Fassbinder 2015). Those factors that affect the readings obtained by the modification of the earth's magnetic field by buried materials over space are very different than those obtained by GPR. Magnetic readings are also constrained by an inability to differentiate depth, while they are also modified by the depth of the materials in the ground, their geometry and the magnetic field orientation at different locations on the surface of the earth.

These very different factors that influence the way GPR and magnetic images are created need not be a constraining factor, as often GPR profiles and maps can be used to visualize buried features and stratigraphy in three-dimensions, and then their magnetic "signatures" can be used to understand their constituents. However, any direct correlation of GPR and magnetic gradiometry maps will often be disappointing, as they are displaying very different spatial locations of buried materials (Conyers 2016b).

Hollister Site, Connecticut, USA

In the Connecticut River Valley of New England a 17th century Colonial farming community was discovered using geophysics and excavated in the summer of 2016. This farmstead was founded soon after the Massachusetts Bay Colony by the earliest Pilgrims from England to the New World, and contains artefacts from both those earliest colonists and the indigenous tribes, with whom they apparently had amicable relations (at least for a time). This site was hypothesized by a surface scatter of pottery and other artefacts that suggested an occupation existed somewhere in the area. The landowners are also direct descendants of these earliest colonialists, and retained a verbal and partial written history of where their ancestors may have lived long ago.

The GPR slice from about 50-75 cm depth over a large area shows a complex series of reflections with very high amplitude, which were generated from glacial till and post-glacial fluvial units (Fig. 1), which are the "bedrock" in this area, below any cultural layers. On those deposits a soil was formed when the nearby Connecticut River degraded to its present level sometime in the Holocene, and produced a fill terrace. It was on this surface that the living surface on which Indian peoples and the Colonialists constructed their dwellings and other structures. The cellars of buildings, Indian people's pit structures, wells and possible storage pits, which were filled by fine-grained sand from post-abandonment flood episodes, are visible in the GPR slice. They can be seen as white areas of no radar reflection produced by a uniform fill sediment, with the surrounding stratigraphic units being very high amplitude (Fig. 2). In contrast, the magnetic map (Fig. 1) shows a large number of recent metal objects near the surface, none of which help in understanding the buried buildings. If only the magnetic map had been produced in this area, there would have been no discovery of the important buried features below.

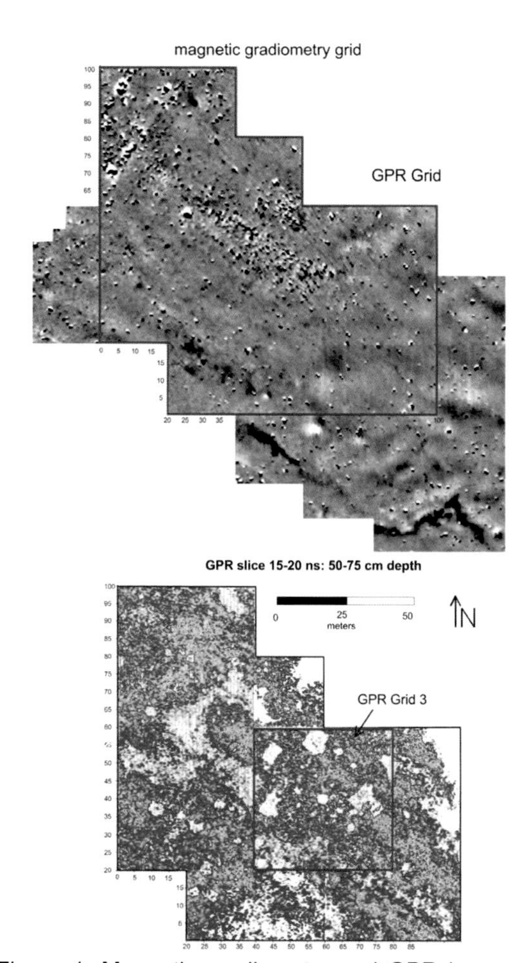

Figure 1: Magnetic gradiometry and GPR images from a large area in Connecticut, with the magnetic map showing mostly surface scatters of metal objects. The GPR slice shows many buried house cellars and other features built on this historic landscape now buried below recent alluvium. Grid 3 outline on the GPR map is shown in detail in Figure 2.

While the cellars are visible in the GPR slice maps, the magnetic map of this area appears to be mostly worthless. But it provides important information that can be used to interpret the GPR images. Many GPR reflection profiles were visualized, and the corresponding magnetic values were taken directly from the gridded magnetic readings used to produce the map (Fig. 3). They were then compared directly along multiple profiles in order to understand the magnetic properties of features and layers visible with GPR. When this is done the cellars are quite visible in profile, and the magnetic readings show that the material that filled in the cellars (and likely artefacts and burned material directly on their floors) display high magnetic readings. This provides evidence that there was likely a burning event, either intentionally or accidentally, at the time this community was abandoned and the cellars will filled with burned objects large enough to be visible in the GPR profiles (Fig. 3) and many smaller objects not visible with GPR, which have high thermo-remnant magnetism.

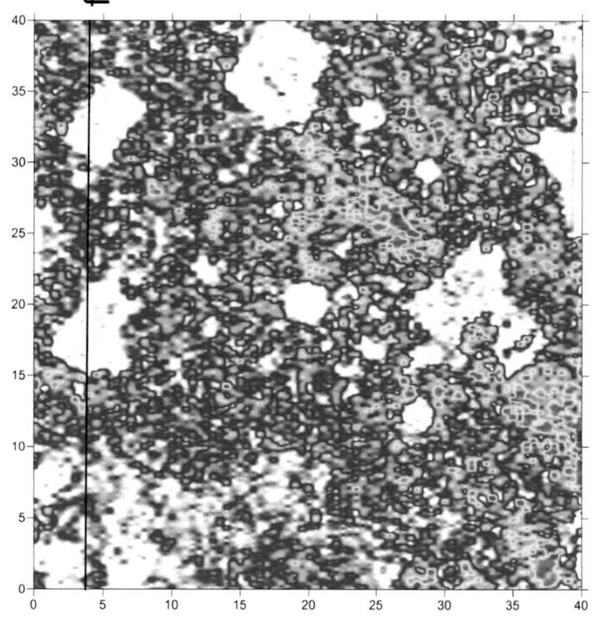

Figure 2: Grid 3 GPR amplitude map showing the cellars of historic buildings in white, with some storage pits and wells in the circular forms.

Conclusions

In this case, while the GPR amplitude maps and reflection profiles were more than sufficient to map this buried Colonial farmstead, the magnetic information provides important additional information on the nature of the materials visible with GPR. The magnetic map, when viewed alone, is not helpful at mapping this buried site, but within those data are important values that when extracted and compared to the GPR profiles, tell a good deal. This is one example of GPR mapping being very important in the interpretation of this site with magnetic data playing a secondary but still important role. There are many other examples, which I have analyzed from elsewhere in the world where the magnetic maps are much more illustrative than the GPR, but once those images are produced they can direct the interpreter to look at the GPR in greater detail. The key to this kind of "data fusion" is that the two methods are often not directly compatible but parts of each dataset can be used together with great benefit. This can be a iterative process with both GPR and magnetic data analyzed, re-imaged and then interpreted again as data are extracted and compared. Magnetically-visible features can often be put into correct three-dimensional space using the GPR, or the GPR features visible in profiles or maps can be further interpreted with magnetic analysis. This deliberate merging of information then produces a much more holistic interpretation.

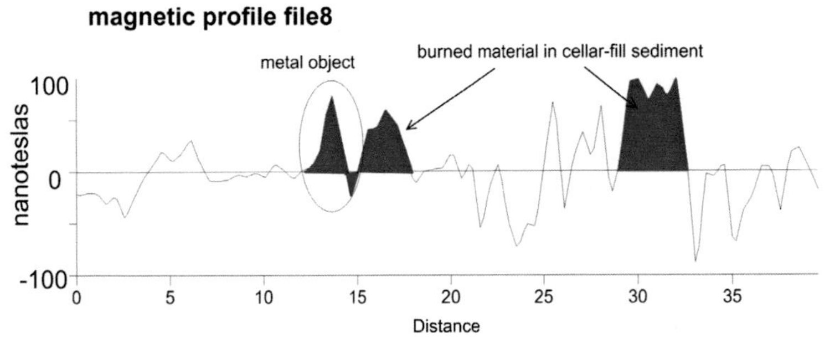

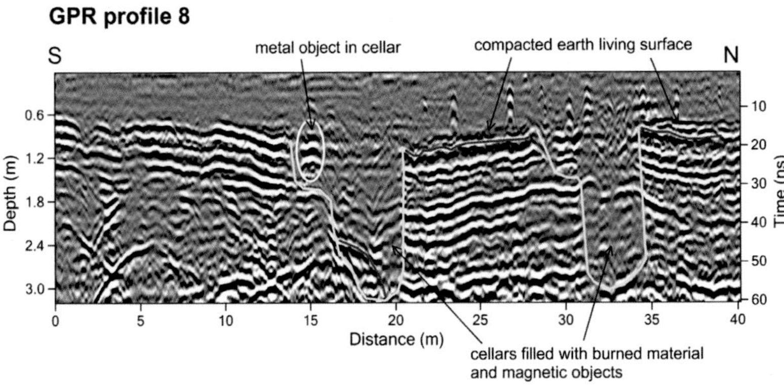

Figure 3: GPR reflection profile 8 (location in Figure 2) showing the incised cellars that were dug into the surrounding strata. The magnetic readings that correspond to this GPR profile show the types of material that filled in the cellars.

Bibliography

Aspinall, A., Gaffney, C., and Schmidt, A. (2009) *Magnetometry for Archaeologists (Vol. 2).* Latham, Maryland: Rowman and Littlefield Publishers, Alta Mira.

Conyers, L. B. (2012) *Interpreting Ground-penetrating Radar for Archaeology.* New York: Altamira Press, Routledge, Taylor and Francis Group.

Conyers, L. B. (2013) *Ground-penetrating Radar for Archaeology, Third Edition.* Latham, Maryland: Rowman and Littlefield Publishers, Alta Mira Press.

Conyers, L. B. (2016a) Ground-penetrating radar mapping using multiple processing and interpretation methods. *Remote Sensing* **8**: 562

Conyers, L. B. (2016b) *Ground-penetrating Radar for Geoarchaeology.* London: Wiley-Blackwell Publishers.

Fassbinder, J. W. (2015) Seeing beneath the farmland, steppe and desert soil: magnetic prospecting and soil magnetism. *Journal of Archaeological Science* **56**: 85-95.

Kvamme, K. L.(2006) Integrating multidimensional geophysical data. *Archaeological Prospection* **13**:57-72.

Piro, S., Mauriello, P., and Cammarano, F. (2000) Quantitative integration of geophysical methods for archaeological prospection. *Archaeological Prospection* **7**: 203–213.

Acknowledgements

Very many thanks to Maeve Herrick and Jasmine Saxon for lending me their fabulous dataset from Connecticut for this analysis. Also to David Wilbourn of Terrasurveyor for magnetic data corrections and processing. Ken Kvamme and Jarrod Burks were also helpful in magnetic data interpretation.

Archaeological prospection of Medieval harbours in the North Atlantic

Joris Coolen[(1)], Natascha Mehler[(1)], Dennis Wilken[(2)], Ronny Weßling[(3)], John Preston[(4)], Tina Wunderlich[(2)] and Peter Feldens[(5)]

[(1)]Centre for Baltic and Scandinavian Archaeology, Schleswig, Germany; [(2)]Institute of Geosciences, Christian-Albrechts-Universität Kiel, Germany;[(3)] Crazy Eye Perspectives, Vienna, Austria; [(4)]School of Geosciences, University of Edinburgh, UK; [(5)] Leibniz Institut für Ostseeforschung Warnemünde, Rostock, Germany

joris.coolen@univie.ac.at

Harbours were of vital importance for the coastal and seafaring communities that settled across the North Atlantic from the late 8th century AD onwards. Harbours and landing sites not only played a key role in the Norse settlement of Shetland, the Faroe Islands, Iceland and Greenland, but they were also essential to sustain the sub-Arctic colonies and their contact with mainland Scandinavia and the rest of Europe. Despite the lack of towns or other large settlements, these islands played an important role in the economic history of northern Europe. For instance, Greenland supplied Europe with walrus ivory and pelts, while Iceland traded wool, fish and gyrfalcons.

However, the medieval harbours of the Norse colonies were generally simple landing places or anchorages that provided natural protection from wind and waves. Besides the frequently used settlement harbours, there were small commercial centres that were only temporarily used during the summer months. As permanent harbour infrastructure was rare, the harbours are hard to trace archaeologically. Quite often, trading activity from the past is only testified by place names or written sources. From an archaeological point of view, the elusive harbours of the North Atlantic are a challenging study (Mehler *et al. 2015*). To understand what the Norse harbours may have looked like and how they were used, we need to follow an interdisciplinary approach.

Following Christer Westerdahl's (2011) concept of the maritime cultural landscape, harbours are only one of many categories of material and immaterial remains of maritime culture. To understand the role of a harbour and its potential shift over time, we therefore need to include the natural topography as well as cultural elements of the maritime landscape, such as churches, coastal settlements, boat sheds and cairns in our study.

The international research project Harbours in the North Atlantic AD 800-1300 (HaNoA) has taken up

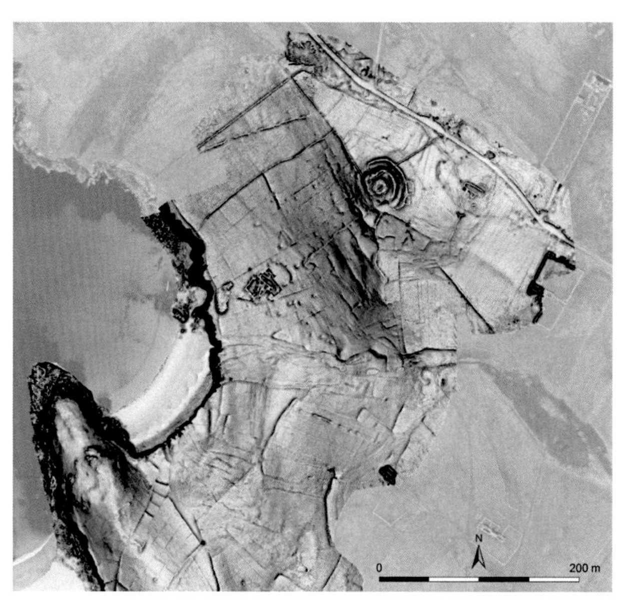

Figure 1: Sky view factor of a surface model created from low altitude aerial imagery showing a rich archaeological landscape near Underhoull on Lunda Wick bay, island of Unst, Shetland. Data: Crazy Eye; Esri / DigitalGlobe.

the challenge to identify and reconstruct medieval harbours and landing sites in the Norse settlement areas across the North Atlantic. It is part of a large research programme on harbours from the Roman period to the Middle Ages across Europe, funded by the German science fund DFG[1].

Using both established archaeological, geophysical, geomorphological and geographical methods as well as more experimental methods and modelling approaches, the project aims to develop a detailed understanding of harbour use, their relationship with the hinterland and the practical aspects of harbour and trading activities.

During multiple field campaigns, geophysical, archaeological and photographic surveys were carried out on selected case-study sites across the study area, both on land and under water. We present two of those case studies in this paper:

Lunda Wick, Shetland

The bay of Lunda Wick is located on the west coast of Shetland's northernmost island, Unst, and presents a naturally sheltered anchorage. A large number of archaeological features can be found along the coast of this embayment, including an Iron Age broch, several Norse longhouses, a medieval church and graveyard and overlapping field systems of various age (Fig. 1). The head of the bay is marked by two sandy beaches divided by a rocky promontory. At one of the beaches, the remains of two small *nausts* (boat sheds) were

1 http://www.spp-haefen.de/en/home/ (accessed 26 January 2017).

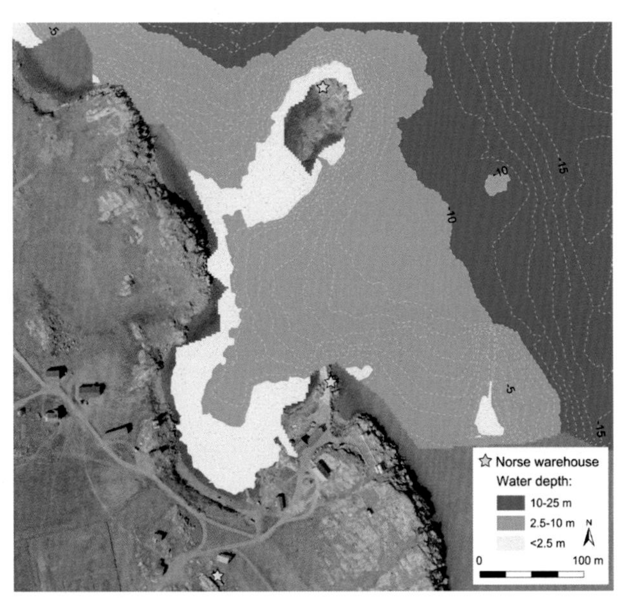

Figure 2: Bathymetry map based on seismic measurements in Igaliku fjord, Greenland. As the sea level is believed to have risen 3 m per 1000 years in the late Holocene (Mikkelsen *et al.* 2008), the bathymetry data can be used to reconstruct the tentative outline of the Norse harbour. Data: CAU Kiel; Nunagis/Asiaq.

discovered in the 1960s. A series of OSL samples that were taken for the HaNoA project seems to confirm that they were constructed in the Late Norse period. The OSL analysis gives evidence of seemingly continuous sand accumulation at least since the 13th century AD, with arguably heightened activity at the onset of the Little Ice Age.

Igaliku, South Greenland – the Harbour of Garðar

As well as coastal erosion or accumulation processes, the North Atlantic harbours were affected by relative sea level change, mainly caused by postglacial rebound or isostatic subsidence. While in Norway remains of medieval harbours are now found several meters above the present shoreline, Greenland has experienced significant sea level rise since the middle Holocene. Remains of the Norse harbours may therefore be located several meters under water in the present day.

A bathymetry survey using marine reflection seismic and shallow-water echo sounding as well as sidescan sonar measurements were carried out in Igaliku in south Greenland to reconstruct the probable coastline in Norse settlement times, map the potentially shippable areas and search for submerged features related to the Norse harbour. A sediment core from the estuary of a small stream north of Igaliku was taken to perform a sea level analysis to constrain sea level rise in the area.

Igaliku has been identified as Garðar, the seat of the bishop of Greenland from the early 12th until the 15th century AD. The extremely large byres and barns, which could host up to 100 cattle, give evidence of the bishopric's remarkable wealth. The harbour of Igaliku was a central node in the connection between Greenland, Iceland and Norway. Several large buildings along the shore and on small islands at the head of the fjord were probably used as warehouses to store tithes and trading goods. The bathymetry derived from the seismic survey shows that one of the small islands, which hosts the remains of a warehouse, was probably still connected to the shore in the Norse period, when estimates of sea level rise are taken into account (Fig. 2). The warehouse thus stood at the head of a promontory, which formed a natural jetty.

Although all of the case study sites were already known to archaeologists, and most of them had been surveyed at least to some extent, the prospection data provide a more consistent and contiguous representation of the near-shore areas and their historic landscape and allow for new archaeological insights.

Bibliography

Mehler, N., Gardiner, M., Dugmore, A. and Coolen, J. (2015) The elusive Norse harbours of the North Atlantic: why they were abandoned, and why they are so hard to find. In: Th. Schmidts & M. Vučetić (eds.) *Häfen im 1. Millennium AD. Bauliche Konzepte, herrschaftliche und religiöse Einflüsse.* RGZM Tagungen 22 (= Interdisziplinäre Forschungen zu den Häfen von der Römischen Kaiserzeit bis zum Mittelalter in Europa 1), Mainz: Verlag des Römisch-Germanischen Zentralmuseums: 313-321.

Mikkelsen, N., Kuijpers A. and Arneborg, J. (2008) The Norse in Greenland and late Holocene sea-level change. *Polar Record* **44/228**: 45-50

Westerdahl, C. (2011) The maritime cultural landscape. In: B. Ford and A. Catsambis (eds) *The Oxford Handbook of Maritime Archaeology*. Oxford: Oxford University Press: 733-762.

Re-visiting Sutton Hoo: revealing new elements of the princely burial ground through ground and aerial remote sensing

Alexander Corkum[1], Cathy Batt[1], Jamie Davis[2], Chris Gaffney[1], Mike Langton[3] and Thomas Sparrow[1]

[1]University of Bradford, Bradford, UK; [2]Ohio Valley Archaeology, Inc., Columbus, OH, USA; [3] MALÅ Geoscience, Sundbyberg, Sweden

Accorkum@student.bradford.ac.uk

Introduction

The iconic burial site at Sutton Hoo consists of 14 extant Anglo-Saxon mounds central to the survey area (Fig. 1) with an additional two groups of 8th-10th century execution burials. In total, 18 mounds have been identified; extensive fieldwork has been undertaken at Sutton Hoo with intermittent survey from 1938-2011, resulting in the excavation of ship burial mounds (Kendrick *et al.* 1939), the production of research bulletins (Carver 1993), and early use of multi-technique geophysical survey. Despite continued interest and archaeological investigation, Sutton Hoo remains an enigmatic cultural resource, with many questions regarding site composition and integrity left unanswered.

The aim of the research design was to provide additional context for existing data sets through a phased geophysical survey utilizing high resolution data collection strategies. Beginning with terrestrial laser scanning, drone based LiDAR, and photogrammetry, the topography of the site was collected to record both documented and undocumented archaeological features. These data could be further used as a benchmark for further remote sensing techniques to record change in the site over time. An area GPR survey followed the micro-topographic survey in order to cover the entire Princely Burial Grounds with a single high resolution geophysical technique which can be analysed in conjunction with the past excavations and smaller geophysical surveys. Promising anomalies representative of excavated and unexcavated mounds identified through the GPR survey will then be targeted in a focused 2.5D ERT survey. This research design will identify new avenues of research through focused geophysical investigation as well as generate public engagement on site.

Figure 1: Plan of the site surveyed

A survey area was selected which incorporates all of the extant mounds, and expands upon previous geophysical surveys. The ambiguity of previous geophysical survey (Carver and Hummler 2014) suggests a need for additional work with modern field techniques and equipment. With a minimum of six potentially unexcavated mounds, a unique opportunity for a high resolution GPR and 2.5 dimensional Earth Resistivity Tomography survey with topographic correction exists which can add complementary information on mound construction and composition.

Micro-topographic Survey

Sutton Hoo poses some unique challenges for topographic survey, including: medium length dense grass, roaming wildlife, constant visitors on the property, and subtle topographic features. A series of competing methods for topographic survey was chosen for this project: the Faro X330 terrestrial laser scanner, a Vulcan Raven X8 drone with Velodyne HDL 32 laser array, A DJI Inspire with a FC550, and as a benchmark, the GNSS data from the GPR survey. While analysis of these data is ongoing (Fig. 2), it is hoped that the use of these methods will identify new unrecorded features, record the exact dimensions of known archaeology, and establish the value of each technique on this type of site when compared with its peers.

Area GPR Survey

A MALÅ miniMIRA (400MHz) ground penetrating radar system was chosen for this project to efficiently collect high resolution data over nearly the entire project area, this was supplemented with a single antenna 450Mhz MALÅ GroundExplorer radar system over space confined areas or features deemed inappropriate to traverse with the miniMIRA

Table 1: Survey techniques for topographic measurement

Technique	Collection Strategy	Time to complete Collection	Estimated resolution of Point Cloud
Faro X330	Base stations with reference spheres	16hrs	0.02 m
Velodyne HDL 32	East-west zig-zag with reference targets	1hr	0.10 m
DJI Inspire w/FC550	North-south and east-west zig-zag with reference targets	1hr	0.03 m
GNSS Data	Zig-zag	16hrs	0.75 m

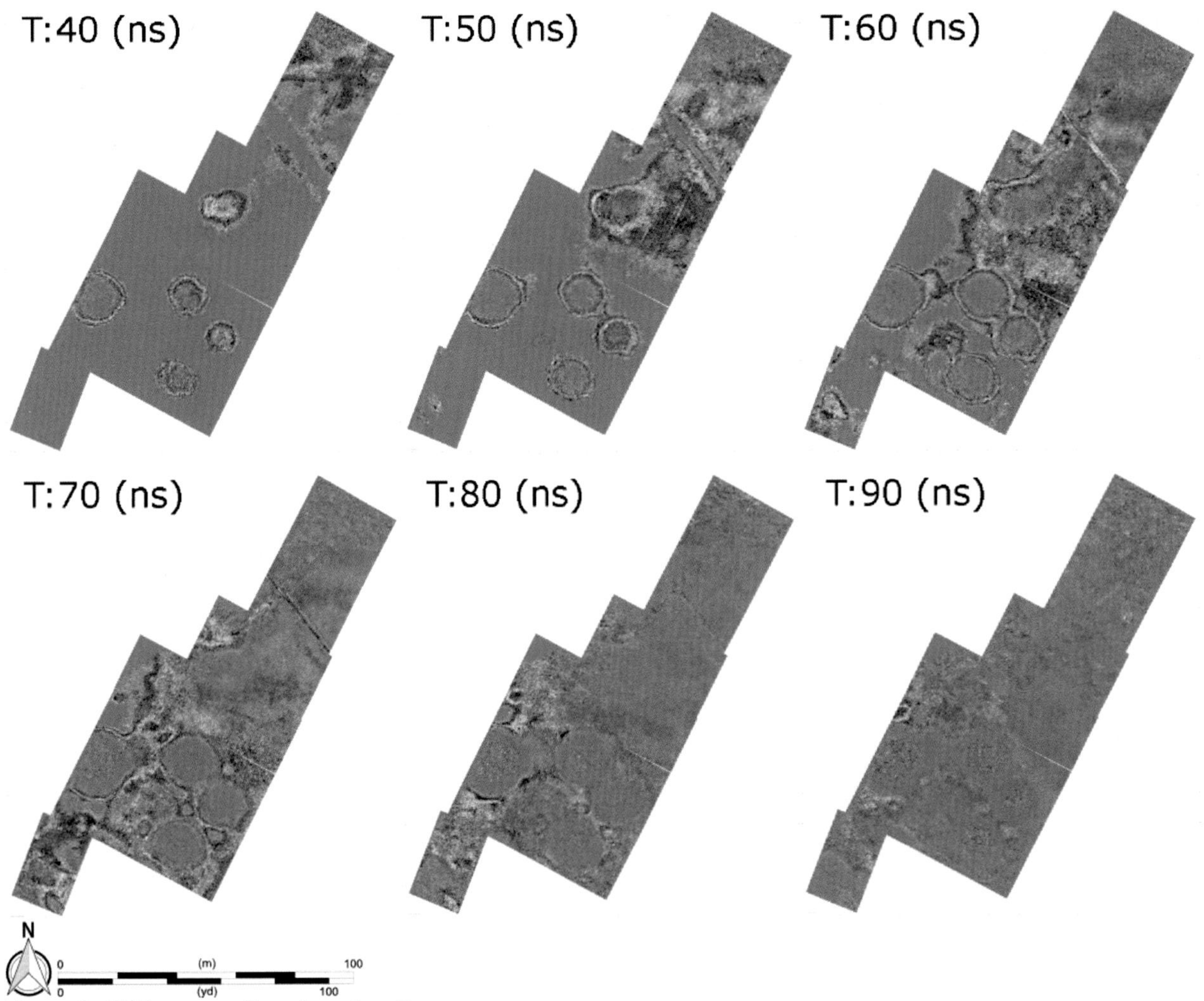

Figure 2: GPR survey slices from the site.

system. Data collection with the miniMIRA finished in May 2017 (Fig. 3) utilized a survey density of 8 x 8 cm using a RTK GNSS system to reference the data in real time. Data collected with the GroundExplorer were collected as zig-zag traverses with a 0.5 m transect separation and topographically corrected using a Trimble R10 RTX GNSS system.

Electrical Resistivity Tomography (ERT) Survey

Scheduled for June 2017, areas that have been revealed as significant during both the micro-topographic and area GPR surveys will be subject to an ERT survey. Due to the desired data resolution and time constraints, a relatively small area will be surveyed. The electrode and transect separation will be governed by the size of the feature being surveyed while confined to a maximum separation of 0.5 m. Survey will be conducted with the ZZ-Geo FlashRes64, which generates 60,000 data points per transect on full survey mode, from which arrays can be extracted for processing and interpretation. Electrode elevation will be recorded with a Trimble R10 GNSS system in handheld mode for use in topographic correction.

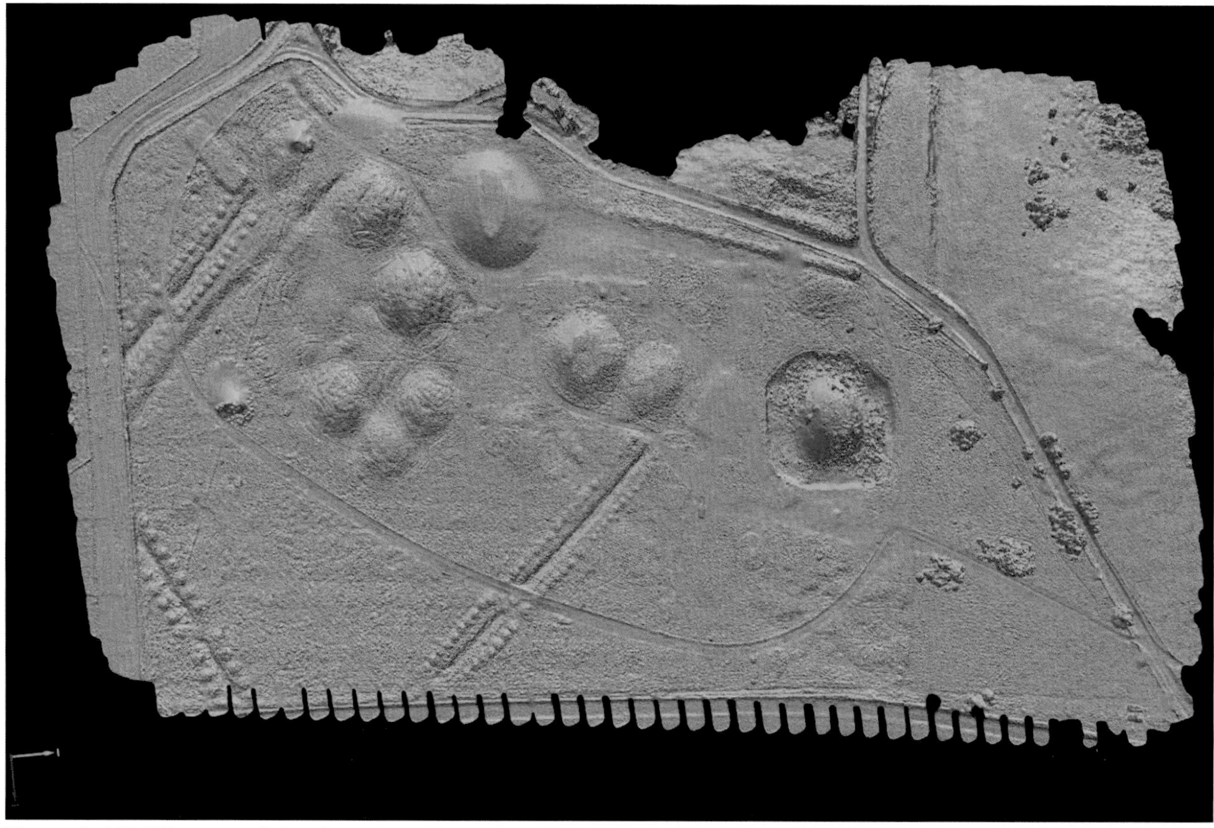

Figure 3: LiDAR scans of the site.

Conclusions

While there is still significant of work yet to be completed, there have been some interesting early results. The sheer density and volume of the various point cloud data is computationally intensive and has proven challenging to work with. Additionally, there are number of anomalies identified within the mounds in the GPR data which could prove interesting targets for the high density ERT survey. Further work is pending with processing and analysis to follow.

Bibliography

Carver, M. (1993) *Sutton Hoo Research Committee bulletins 1983-1993.* Woodbridge: Boydell.

Carver, M. and Hummler, M. (2014) Sutton Hoo Research Plan.

Kendrick, T. D. (ed.) (1939) The Sutton Hoo Finds. *The British Museum Quarterly*, ii-136.

Using geophysical techniques to 'dig deep' at Grave Creek mound for cultural resource management

Alexander Corkum[(1)], Cathy Batt[(1)], Jamie Davis[(2)], Chris Gaffney[(1)], and Thomas Sparrow[(1)]

[(1)]University of Bradford, Bradford, UK; [(2)]Ohio Valley Archaeology, Inc., Columbus, OH, USA

Accorkum@student.bradford.ac.uk

Introduction

Grave Creek Mound (Fig. 1), a designated National Historic Landmark, is located in the town of Moundsville in Marshal County, West Virginia, USA. Built between 300 and 200 BC by the Adena culture during the Early Woodland period. When this conical-shaped earthen mound was first documented in 1838, it measured roughly 21 meters tall and 90 meters in diameter, and was surrounded by a ditch (Schoolcraft 1845, Squier and Davis 1848). It is the largest and only remaining mound in an area that once boasted a large complex of earthen structures. Archaeologists in the 1800s carried out a number of modifications to the mound. They discovered two burial chambers within it, tunnelled two (possibly brick-lined) shafts through the burial chambers (one from the top, one from the side), built a temporary wooden structure at its summit, and excavated a rotunda at its base. The location and present condition of the archaeological features and the historic modifications were unknown prior to survey.

The Delf Norona Museum, caretaker of the Grave Creek Mound Archaeological Complex, asked the University of Bradford to map archaeological features within the mound using non-invasive geophysical techniques. To help the museum more effectively maintain and preserve the site, a research design was developed with the goal of efficiently mapping the extant archaeology within the mound and surrounding area with ground penetrating radar, earth resistivity tomography, and drone based photogrammetry.

Figure 1: Location of Grave Creek Mound.

Fieldwork at Grave Creek Mound was completed in July of 2016. Forty-three lines of GPR data at 0.75 m transect separation and thirty-three lines of ERT data at 1.5 m separation were collected on the mound. An additional twenty-two 20 x 20 m grids and two 10 x 40 m grids of GPR data at 0.5 m transect spacing were collected in the area surrounding the mound. The drone based photogrammetry survey was flown with a DJI Inspire with 80% photograph overlap in both the north-south and east-west orientations resulting in a high density point cloud from which a 3 cm DEM was derived (Fig. 2).

Ground Penetrating Radar- Area Survey

The area survey was conducted with Sensors & Software Noggin 500 (500 MHz) and GSSI SIR3000 (450 MHz) GPR systems. Two systems with different central peak frequencies were used due to time constraints, which presented challenges when integrating and combining the datasets for analysis. All data were combined and processed in ReflexW 8.0.

Earth Resistivity Tomography on Mound

A FlashRes64 ERT system was used on the "full survey" setting to collect thirty-three lines of data, each with approximately 60,000 data points at 1.5 m electrode and transect separation to ensure measurement depths would exceed the maximum height of the mound while still encompassing the surrounding area. The ERT electrode positions were located using the Trimble R10 GNSS system in a hand-held configuration. It was not time appropriate to collect each individual electrode height measurement with the Trimble R10, so electrodes were systematically measured along the line, and the elevation of the intervening electrodes was extracted from the photogrammetry derived DEM. The raw ERT data were then extracted into array types, collated into a 3D file in Res2DINVx64, and inverted with topography in Res3DINVx64.

Ground Penetrating Radar on Mound

A Mala GX160 (160MHz) GPR system was used in conjunction with the Trimble R10 GNSS system. The NMEA string recorded through the MALA GX controller was the sole method of recording location, as the wheel encoder was found unreliable due to slopes often in excess of 40° on the mound. A separation of 0.75 m was chosen in order to

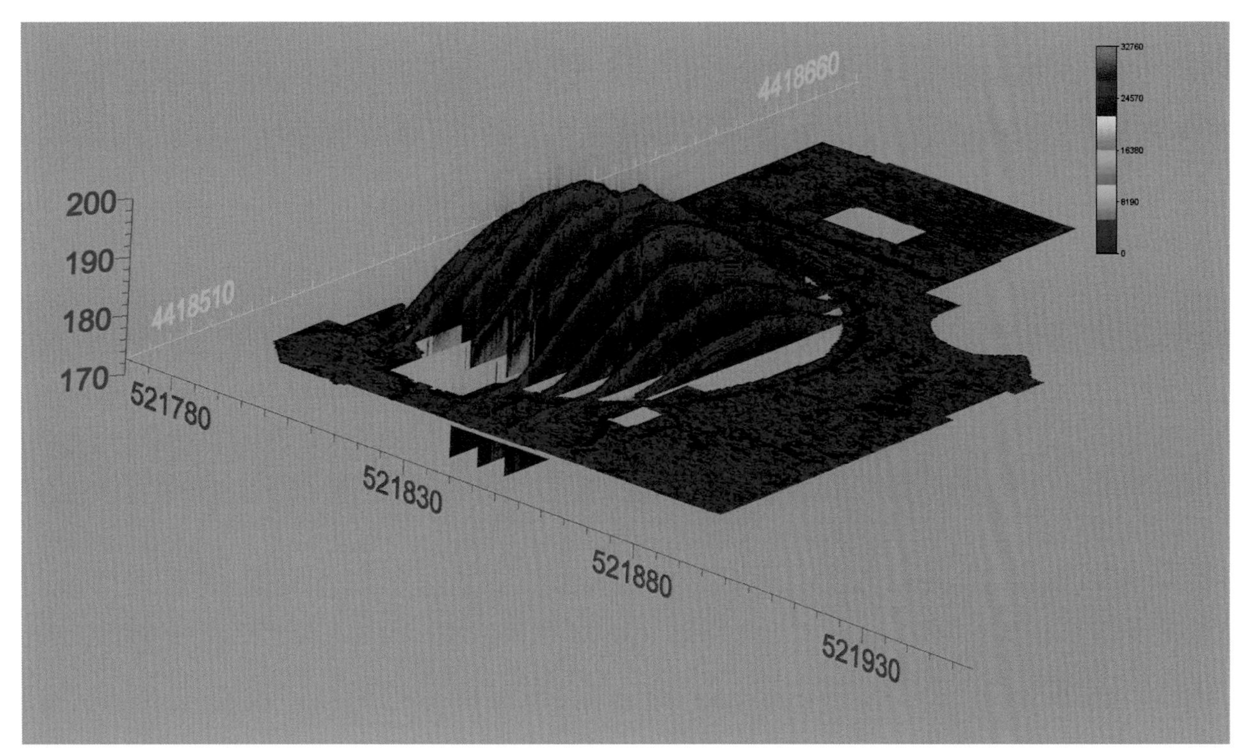

Figure 2: GPR sections of Grave Creek Mound.

maximize the coverage of the mound in the time allotted for survey. The additional modification of survey technique included the removal of the harness from the unit and the mounting of the GX controller directly to the GPR so that two people holding ropes attached to either side could pull the system up and over the mound while maintaining its orientation. Despite initial issues involving the vertical accuracy of the GNSS data, the individual lines were combined, processed, and topographically corrected in ReflexW 8.0. The transect separation was found to be less than ideal when combined into a 3D file, though ultimately suitable for the scale of the features being investigated.

Results

Analysis of the data is still ongoing, but a number of anomalies have been identified which may prove useful in clarifying the history and context of Grave Creek Mound.

The area surrounding the mound has seen multiple historic modifications which resulted in a generally noisy dataset (Fig. 3). Though ephemeral, a broad linear anomaly approximately 8 m wide, along with a number of rectilinear anomalies are likely a historic road and associated historic building foundations, both visible in a 1954 aerial photograph. There also appears to be the remnants of a ditch encircling the mound, which was described in historic documents and identified in a 1976 excavation. It was estimated to be approximately 10 m wide and 1.5 m deep (Grantz 1984). However, the noise created by both historic and modern disturbance makes the extent of this prehistoric ditch difficult to determine.

On the mound the top burial chamber, as indicated in historical documents, appears evident in the ERT data as a large, roughly cuboid high resistance anomaly. In addition, a collapsed shaft extending from the top of the mound down, which was filled in the 1950s appears to be present in the ERT data manifesting as a cone shaped high resistance anomaly. Neither anomaly is present in the GPR data (Fig. 3), likely due to the stone slabs covering the summit.

Conclusions

Geophysical survey at Grave Creek Mound proved to be problematic in the collection strategy, the processing, and analysis of the data. The sheer scale and steep topography resulted in sacrifices in terms of resolution for both the ERT and GPR, and the physical modification of field equipment to better facilitate data collection. Nevertheless, the fieldwork has revealed new information about the mound and surrounding area. Though not within the scope of the original research design, additional survey techniques such as twin-probe earth resistance or EM might help clarify the ditch location and extent that was identified in the area survey.

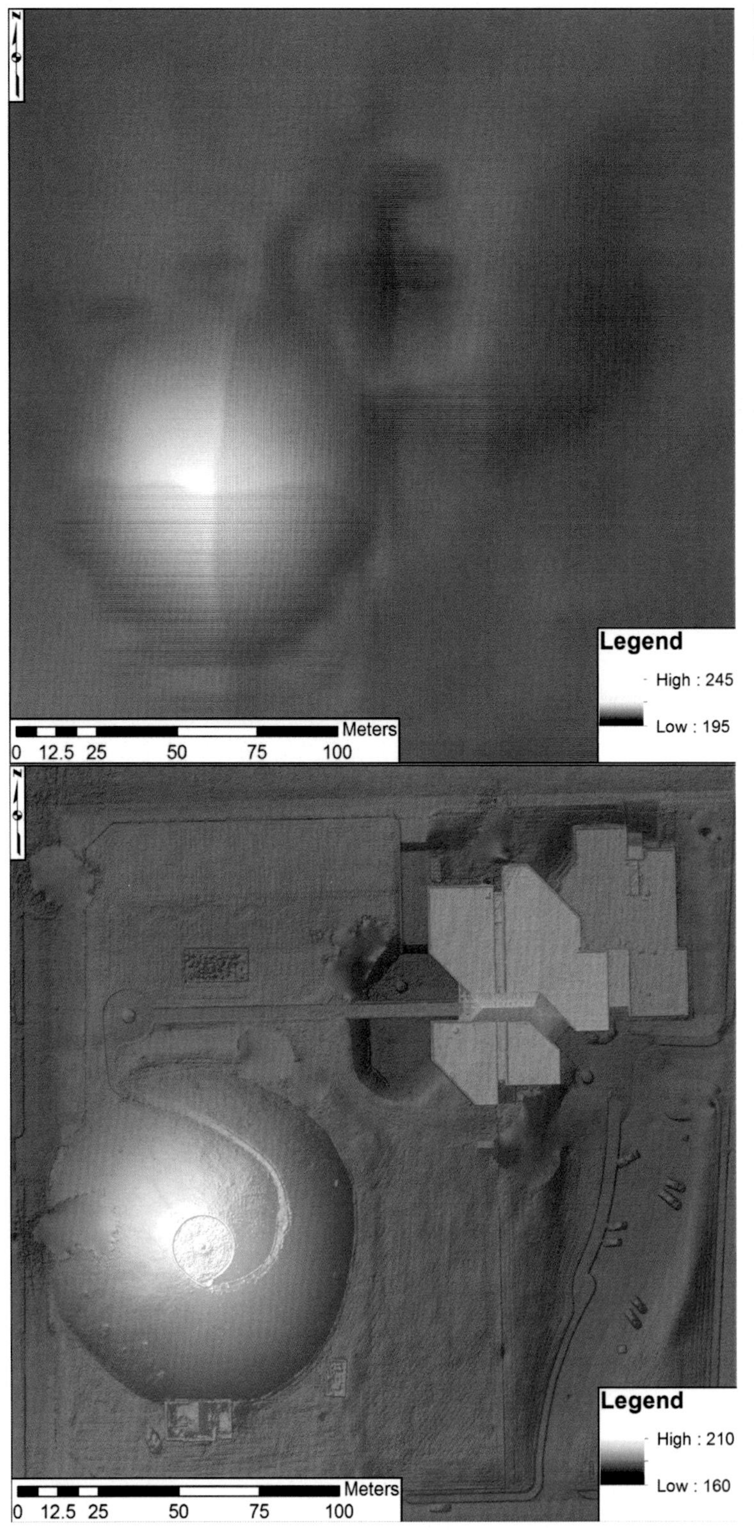

Figure 3: LiDAR image of Grave Creek Mound.

Bibliography

Grantz, D. L. (1984) Grave Creek Mound NRHP Nomination Form. In: N. P. Service (ed.) United States Department of Interior.

Schoolcraft, H. R. (1845) *Observations Respecting the Grave Creek Mound in Western Virginia: The Antique Inscription Discovered in Its Excavation and the Connected Evidences of the Occupancy of the Mississippi Valley During the Mound Period and Prior to the Discovery of America by Columbus*. Bartlett & Welford.

Squier, E. G. and Davis, E. H. (1848) *Ancient monuments of the Mississippi Valley*. New York, Cincinnati: Bartlett & Welford, J.A. & U.P. James.

Geological and pedological artefacts within UK magnetic gradiometer data for archaeological prospection

Edward Cox[1] and Rebecca Davies[1]

[1]SUMO Geophysics for Archaeology and Engineering, Upton-Upon-Severn, UK

edward.cox@stratascansumo.com

Magnetic gradiometry is widely used within the UK and Europe as the most suitable method to rapidly identify a wide range of archaeological features (Gaffney & Gater, 2003). The effectiveness of these surveys can occasionally be prohibited by unfavourable geological and pedological conditions and much excellent work has been completed on the subject (Bonsall *et al.* 2012, Campana & Dabas 2011). A guide produced by Historic England (Historic England, 2008) aims to advise potential clients on magnetic response over a selection of geological environments, however there are always exceptions and in some cases archaeology can still be found within the most unsuitable of terrains.

We will present magnetic data collected over a wide range of geological environments in the UK and discuss how geology typically appears in magnetic data; how in some terrains careful data collection and processing can overcome difficult conditions. Additionally we will present 'surprising' anomalies that appear archaeological but are in fact geological in origin. As a final remark we hope to demonstrate that the most extreme limitations of data collection are anthropogenic in origin rather than geological.

Bibliography

Bonsall, J., Gaffney, C. and Armit, I. (2012) A Decade of Ground Truthing: Reappraising Magnetometer Prospection Surveys on Linear Corridors in Light of Excavation Evidence 2001-2010. *In H. Kamermans, M. Gojda, and A. G. Posluschny (eds.) A Sense of the Past: Studies in current archaeological applications of remote sensing and non-invasive prospection methods.* Oxford: Archaeopress, 3-16.

Campana, S. and Dabas, M. (2011) Archaeological Impact Assessment: The BREBEMI Project (Italy). *Archaeological prospection* **18**: 139-148.

Gaffney, C. and Gater, J. (2003) *Revealing the Hidden Past: Geophysics for Archaeologists.* Stroud: Tempus.

Historic England. (2008). Geophysical Survey in Archaeological Field Evaluation. [Online]. [Accessed 31 January 2017]. Available from:

https://content.historicengland.org.uk/images-books/publications/geophysical-survey-in-archaeological-field-evaluation/geophysics-guidelines.pdf/

Moving beyond an identification of 'ferrous': a re-interpretation of geophysical surveys over WW1 practice trenches on salisbury plain

Nicholas Crabb[(1)], Paul Baggaley[(1)], Lucy Learmonth[(1)], Rok Plesničar[(1)] and Tom Richardson[(1)]

[(1)]Wessex Archaeology, Salisbury, United Kingdom

n.crabb@wessexarch.co.uk

Introduction

Geophysical survey techniques are commonly deployed to investigate the archaeology of former military sites, especially now that former military land is increasingly being used for commercial developments, as the UK government seeks to release this land for housing. In particular, gradiometer surveys have proven to be an extremely useful tool for mapping features identified on aerial photographs and historic mapping (Gaffney *et al.* 2004; Masters and Stichelbaut 2009). These datasets reveal the outline of trenches and associated structures, but also have the potential to reveal a great deal more regarding their former function and use.

This paper discusses three case studies where WW1 remains have been identified in geophysical survey data and investigated by intrusive works. It is argued that the nature and strength of these geophysical anomalies could be used to detail more specific interpretations. In each of these examples subsequent archaeological evaluations have aided the reinterpretation of these anomalies, the correlation of which may have implications for how we interpret such sites in the future.

Salisbury Plain

All three case studies presented here lie within Salisbury Plain, which has had a military presence since 1898. At the beginning of WW1, and during the inter-war period, practice trenches were excavated across this area to train troops in the construction of trenches and learn trench warfare tactics, whilst also increasing fitness and building team spirit prior to operational deployment on the Western Front (Brown 2004). Two of the sites are no longer within MOD practice areas and are under consideration for proposed housing developments.

Bulford

A gradiometer survey was carried out at Bulford ahead of a proposed housing development. Aside from two exceptionally clear ring-ditches, a large concentration of ferrous responses was recorded in the east of the site (Fig. 1.a). This was interpreted as modern debris; the large response in the centre of this was thought to be related to a large buried ferrous object.

Subsequent archaeological evaluation provided further detail regarding the precise nature of these anomalies with numerous military features recorded. This included a large collection of horseshoes, and other iron work, along with possible practice trenches

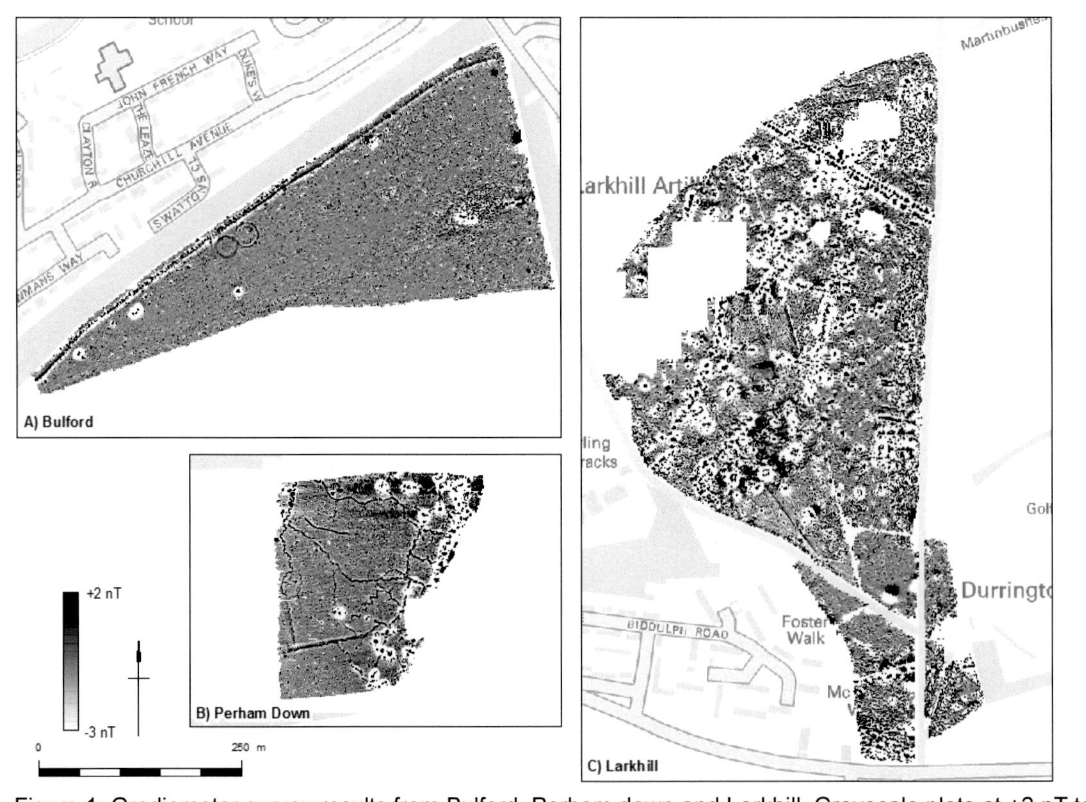

Figure 1: Gradiometer survey results from Bulford, Perham down and Larkhill. Greyscale plots at +2 nT to -3 nT.

and/or firing positions. During the excavation of a particular trench, a large quantity of spent WWII anti-tank rounds were unearthed. This trench lay within an area of increased magnetic response and a reinterpretation of the geophysical results in the light of the discovery of unexploded ordnance (UXO), indicated that these anomalies probably coincided with further undiscovered military *materiel*.

The majority of the UXO was located towards the eastern side of the Site, amongst which were parts of an armoured vehicle and it has been suggested that the eastern side of the Site may have been used as an anti-tank range during the 1940s. This is likely to correspond with the pattern of ferrous anomalies detected in the eastern part of the gradiometer survey data.

Perham Down

WWI practice trenches were created at Perham Down as a replica of part of the Somme Battlefield in early 1917. Contemporary field plans provide exceptionally useful information, detailing the position of the various components, such as shelters, latrines and dug-outs, as well as the function of the various trenches. The trenches do not show as surface features, but are clearly visible as crop marks on aerial photography. A gradiometer survey targeted part of this trench network and aimed to confirm its location.

In their developed form, trench systems were composed of three distinct elements: a front line, support trenches and a reserve line, all of which were connected by a further series of communication trenches (McOmish *et al.* 2002, 140). These elements, as well as a number of additional features, can all be identified within the gradiometer survey results in considerable detail. A series of large strong ferrous anomalies is visible along the line of many of the trenches, most of which are located at junctions of support and communication trenches (Fig. 1.b). Aided by information on a 1915 Field map, these were interpreted as structural elements of the trench network. Shelters, latrines, a possible kitchen, a first aid and command post and a possible machine gun post were hypothesised. Subsequent excavation confirmed the function of some of these features. Buried metal, particularly larger pieces, is clearly reflected in the results as strong dipolar responses; however, their presence obscures some detail of weaker anomalies.

Larkhill

At Larkhill, an extensive complex of practice trenches is visible as subtle earthworks on aerial photographs. Numerous pits are also visible, suggesting the area was used for land mine practice after the trenches went out of use. Gradiometer survey undertaken at the site has shown it to be covered by high concentrations of ferrous responses, with a number of coherent forms indicative of trenching and military activity (Fig. 1.c). As the site had also been used as an artillery range, it was suggested that much of the ferrous response related to metal ammunition casings and shrapnel. However, several surprising additional features were uncovered during subsequent archaeological investigations of the area.

Subsequent archaeological evaluation has shown that the majority of features observed in the western part of the Site were military in origin including practice trenches from late 19th century onwards. Further archaeological mitigation works have revealed an extensive range of WWI features, many of which relate to the strongest ferrous responses (+/- 10 – 100 nT) recorded in the gradiometer surveys. For example, many of the practice trenches were found to be reinforced and lined with corrugated iron sheets and a number of UXO have also been discovered, indicating that live ammunition was used in training. In addition to these trenches several entrances to subterranean tunnels were also discovered along with a motorbike and a car, both buried on the site.

Conclusions

Gradiometer surveys of WW1 practice trenches often give good and distinctive results, however more specific details of the trenches are harder to identify. Whilst a simple acknowledgment of stronger ferrous anomalies is often noted, the three case studies presented show that a more detailed interpretation of strong ferrous or dipolar responses is possible if you have an understanding of how the military trained their troops and planned their trenches. In the cases above this has been achieved largely through a reinterpretation in hindsight. Following subsequent excavation, it has been possible to correlate similar types of response with specific attributes of the WW1 practice trenches.

Bibliography

Brown, M. (2004) A Mirror of the Apocalypse: Great War Training Trenches. *Sanctuary* **33**: 54–58.

Gaffney, C., Gater, J., Saunders, T. and Adcock, J. (2004) D-Day: Geophysical Investigation of a World War II German Site in Normandy, France. *Archaeological Prospection* **11**: 121–128.

McOmish, D., Field, D. and Brown, G. (2002) *The Field Archaeology of the Salisbury Plain Training Area*. Swindon: English Heritage.

Masters, P. and Stichelbaut, B. (2009) From the Air to Beneath the Soil - Revealing and Mapping Great War Trenches at Ploegsteert (Comines-Warneton), Belgium. *Archaeological Prospection* **16**(4): 279–285.

In search of the lost city of Therouanne: a new integrated approach

Michel Dabas[(1)], François Blary[(2)], Laurent Froideval[(3)] and Richard Jonvel[(4)]

[(1)]ENS UMR777, Paris, France; [(2)]Université libre de Bruxelles EA4282, Bruxelles, Belgique; [(3)]CNRS UMR6143, Caen, France; [(4)]Société des Antiquaires de Picardie, Amiens, France

michel.dabas@ens.fr

Introduction

June the 20th, 1553: during a siege of seven weeks, the army of Emperor Charles the Fifth assaulted the fortified town of Thérouanne, a French enclave in the territory of the Burgundian Netherlands (today Nord Pas-de-Calais, region Hauts-de-France). Charles the Fifth immediately ordered the complete destruction of this city. A few months later, of the ancient capital of the Gallic tribes of Morins, also an episcopal city funded by saint Omer himself, the fortress of François the First, nothing more remained, besides some figurative representations. Even salt spread all over the past city so that no crop could grow for generations to come...

Nevertheless, the site, now known locally as the "Old town", returned to agriculture. Only a small town developed in the south of the former city, at the foot of the hillside which occupied the medieval surrounding walled enclosure, underlined by profound ditches which are still perfectly readable in the landscape with their polygonal shape. The quite particular interest ensues from its tragic end. Only today is archaeology able to highlight its architectural wealth as depicted in numerous paintings and engravings.

Previous Work

Despite numerous archaeological excavations (Camille Enlart, Roland Delmaire, Honoré Bernard), partially published, very little is known from this site, except perhaps the area of the cathedral. Most of the artefacts need to be studied or even listed. It was also necessary to create a synthesis of the many archaeological excations since those excavations.

From 2014, on the initiative of the Ministry of Culture, François Blary launched a project ("PCR Thérouanne, antique and medieval city") in order to share all available information. A multidisciplinary team (archaeologists, historians, art historians, topographers, geologists, geophysicists, specialists in LiDAR Surveying) was established (35 participants to this day). The general objectives are to raise the state of archaeological knowledge, and to place it in a wider historic and archaeological context; to establish the inventory of the archaeological furniture (miscellaneous artefacts, statuary and stone corpuses, etc.); to survey in the area of the Old town and its immediate neighbourhood (electric and magnetic geophysical surveys, microtopography, archaeological endoscopy, etc.). These studies would lead to several specific excavations specifically to establish links between various data.

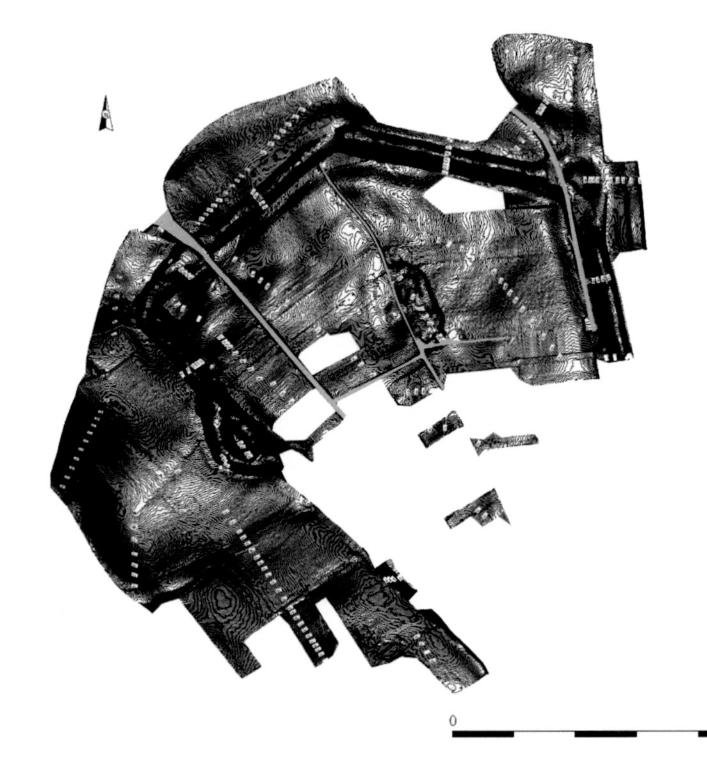

Figure 1: Micro-topography (dGPS and Tacheometer).

Figure 2: Apparent Electrical Resistivity channel2 (0-1m, ARP®).

Surveys

Topographical field surveys were performed (R. Jonvel) all over the area enclosed by the medieval ditches except the southern part occupied by the present city. The area was scanned with a dGPS and height points measured over a semi-regular grid (3m). 56 ha were surveyed (70546 points) and a DTM with 10cm level curves was derived. Moreover, it was possible to derive the shape of the ditches where trees are numerous by using an electronic tacheometer (Fig. 1).

Three geophysical campaigns were undertaken in 2014, 2015 and 2016 by Geocarta, Paris. During the first survey, continuous electrical surveying (ARP® Geocarta), magnetometry and electromagnetic surveying were performed on a single 1ha plot giving exceptional results. All methods (resistivity for 3 depths of investigations, vertical magnetic pseudo-gradient, electrical conductivities and magnetic susceptibilities for 6 depths of investigations) performed equally well over the site and could

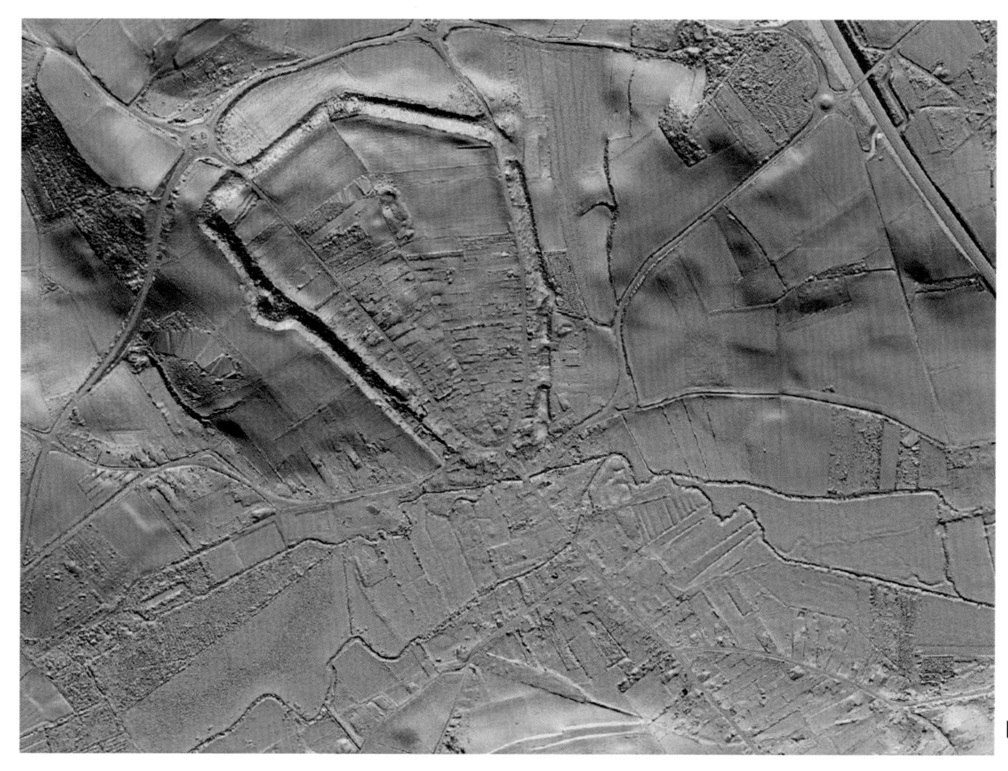

Figure 3: LiDAR data.

delineate the medieval layout of the town around the cathedral, individuate city blocks and even define the inner structures of some houses. As it was stressed to the archaeologists that geophysical maps from different methods do not bring the same information, the survey was extended to the other plots using both resistivity and magnetometry (from 2014 to 2016: 9ha for ARP® and 11ha for magnetometry were performed)(Fig. 2). By the end of 2017, probably all areas inside the area delimited by the medieval ditch will be surveyed (total area available to survey is 17ha). All data sets are now shared and displayed through a web-GIS designed by Geocarta (GCserver).

After LiDAR technology was chosen in 2016, the knowledge of a detailed micro-topography over a wider area than the town by itself appears very quickly. The objective was to rapidly cover a 100 km² territory, in perfect complementarity with the other techniques and while mitigating their limits. The recourse to the airborne LiDAR thus took into account all these parameters adapted to the specific needs for this research (the laser ground coverage density was extended to 24 pts/m2 inside the city and 9 pts/m² outside over an area of 100km2). Data were acquired and processed by researchers from CNRS (UMR 6143 M2C CNRS – Université de Caen Normandie). Although the work is not yet completed, the obtained LiDAR image shows a great quantity of reliefs: they reveal the structuring of the city, the organization of ramparts and position of the former towers (Fig. 3).

The work of integrating of all these new data is still on-going in a collaborative web-GIS. Besides the interpretation with archaeologists and comparison with old drawings plans, this work opens new possibilities concerning the combined use of LiDAR data and geophysical data. The rather limited area of this town and the conditions of its rapid destruction makes Thérouanne a unique case study for other periods, for example the Roman town of Wroxeter (Buteux *et al.* 2000) or the Iberian site of Ullastret (Garcia-Garcia *et al.* 2016).

Bibliography

Garcia-Garcia, E., de Prado, G. and Principal J. (2016) *Monografies d'Ullastret3 Working with buried remains at Ullastret (Catalonia*). Barcelona: editions de Museu d'Arqueologia de Catalunya.

Buteux, S., Gaffney, V., White R. and Van Leusen, M. (2000) Wroxeter hinterland project and geophysical survey at Wroxeter, *Archaeological Prospection*,**7**(2): 69-80.

Subsurface geophysical approaches to understanding Northern Plains earthlodges

Rinita A Dalan[1], George R Holley[2], Kenneth L Kvamme[3], Mark D Mitchell[4] and Jay Sturdevant[5]

[1,2]Minnesota State University Moorhead, Moorhead, MN, USA; [3]University of Arkansas, Fayetteville, AR, USA; [4]Paleocultural Research Group, Arvada, CO, USA; [5]National Park Service, Midwest Archeological Center, Lincoln, NE, USA

dalanri@mnstate.edu

Earthlodges (timbered dwellings with earthen mantles) in the prehistoric Northern Great Plains of North America are found in large, well-defined villages after 1400 AD. Many villages are reported to have more than 100 lodges (Wood 1967: 24-28). These multi-family dwellings change in form from long-rectangular (which may not have been earth covered) to circular, four-post earth-mantled lodges (Fig. 1), with great variability in size and internal features indicating that they encode significant information about status and identity.

Typical earthlodges are shallow semi-subterranean with a formal entrance, central hearth, and a variety of internal features. Such features might include framed facilities that are permanent or temporary, such as beds, stables (after 1750 AD), or shrines and earthen altars. Localized activities include food preparation (processing and cooking), sleeping, tool maintenance, and storage.

In the absence of modern cultivation, earthlodge depressions are visible from the surface, often revealing dimensional data, sometimes features such as entrances and pits, and occasionally earthen altars (Wilson 1934; Wood 1993). Villages that have been disturbed and those with deep-sedentism have lodge outlines obscured, buried, or borrowed (Kvamme 2007) limiting visual examination and at times geophysical investigations (Kvamme 2007:218).

Remote sensing has been a critical player in the study of Northern Plains villages, aiding in mapping lodge locations and understanding village layout and structure. This has included DEMs acquired through Lidar or drones, aerial photography showing crop marks, surveys using thermal or infrared sensors and surface geophysical surveys (Kvamme & Ahler 2007). Surface geophysical surveys have been highly successful in confirming that depressions represent lodges and revealing lodges where there are not surface indications, and in mapping entrances, hearths, storage pits, and even posts, as well as fortification ditches, bastions, and other village elements. Magnetometry has played a key role, but multiple-method surveys also involving electrical resistance, ground-penetrating radar (GPR) and electromagnetic conductivity have provided enhanced interpretations.

Broader questions posed by archaeologists about earthlodges, relating to house-building traditions (Blakeslee 2005), gendered-space (Pauls 2005), ceremonial use (Wood 1967), and political economy (Mitchell 2013), among others, however, require that subsurface variability is documented before such questions can be adequately addressed. The focus of our paper is on subsurface geophysical investigations at the household level to supplement costly excavations for gathering subsurface information.

Subsurface magnetic susceptibility investigations, integrated with other remote sensing data and excavations, derive from three earthlodge villages. Work at the Biesterfeldt site in eastern North Dakota (Dalan *et al.* 2011) provided the foundation for our approach. Subsurface susceptibility investigations of earthlodges were extended in multiple field seasons at each of two sites in the Missouri River region of North Dakota. The Sakakawea site is located within the Knife River Indian Villages National Historic Site (KNRI) and Chief Lookings Village (CLV) is near the confluence of the Heart and Missouri rivers. The most recent Sakakawea fieldwork was conducted in conjunction with an

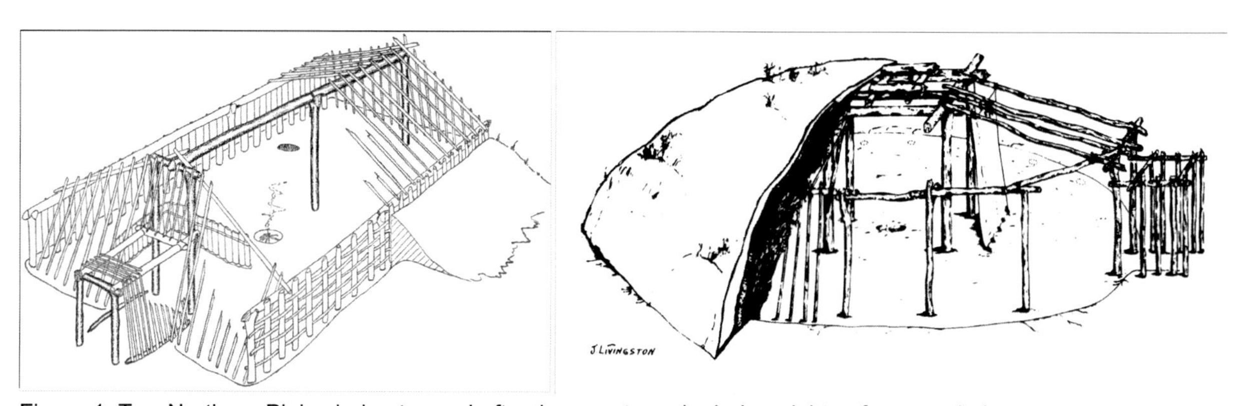

Figure 1: Two Northern Plains lodge types. Left: a long-rectangular lodge; right: a four-post lodge.

Figure 2: Magnetic susceptibility surveys of excavation walls at Chief Lookings Village.

"ArcheoBlitz" event allowing middle-school students the opportunity to conduct supervised archeological field research. The collaborative project at CLV is focused on comparing material practices of families residing in two distinctly different types of lodges (long-rectangular vs. circular four-post).

Subsurface magnetic susceptibility studies at these sites included downhole tests, surveys along excavations walls and floors (Figure 2), and even "cutbank geophysics" (Dalan *et al.* 2017) along natural exposures. The Bartington MS2H downhole sensor was deployed along transects, and also in grids covering entire lodges (Figure 3). The Bartington MS2K sensor was used within excavations and along the cutbank. In some cases, samples were collected for soil magnetic studies

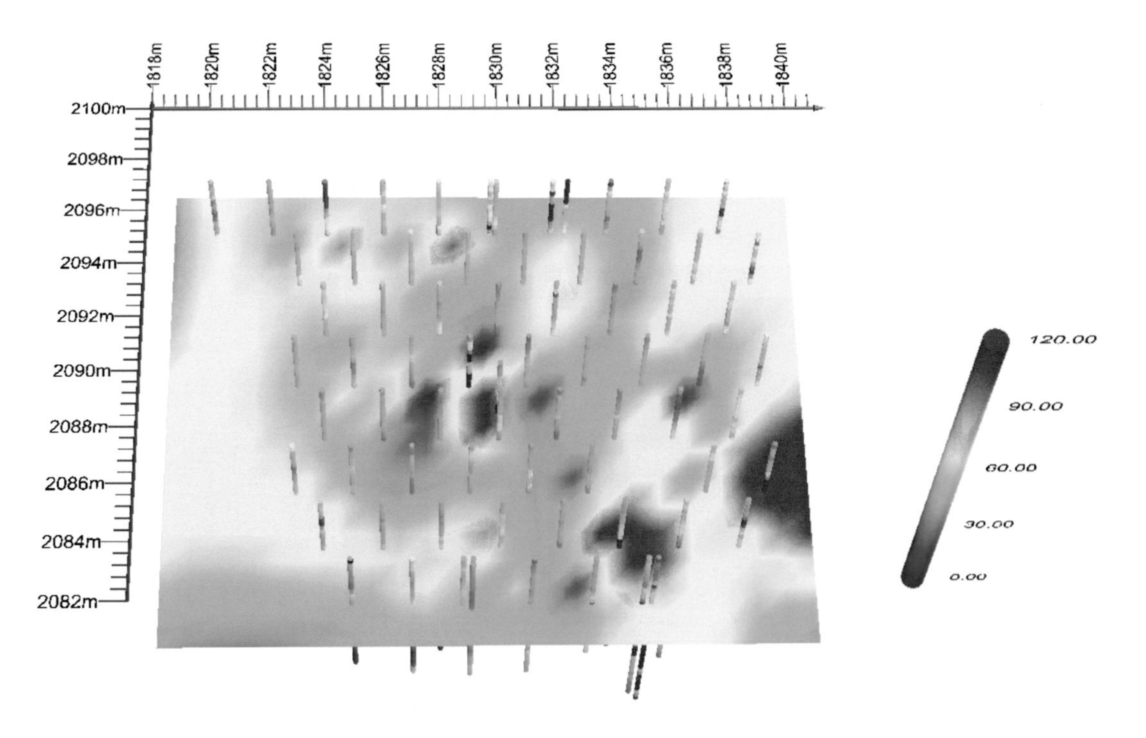

Figure 3: Grid of downhole tests over an earthlodge at the Sakakawea Village. Fence at 32 cm below surface (prepared using TerraSurveyor3D) showing internal lodge structure.

in the lab to provide additional information on the origin of susceptibility contrasts.

Integrating subsurface geophysical data with other remote sensing studies, excavations, observations from cores, and soil magnetic studies has proved useful in unraveling the complexities of Northern Plains earthlodges. Subsurface geophysical data are complimentary to other remote sensing efforts, corroborating locations identified through those methods, but providing additional information on horizontal, and especially vertical, limits of lodges and internal and external features. GPR and downhole susceptibility have been a particularly effective pairing, with down-hole susceptibility surveys yielding information on gradual changes with depth and entire volumes, as opposed to the abrupt interfaces mapped by GPR, and, in some cases documenting more complex layering. In understanding cases of rebuilding, this pairing has been particularly important. Downhole susceptibility also provides an opportunity to map thin surfaces at depth beyond the resolution of surface applications. Thermal-related features are readily detected and susceptibility can be used as a proxy for the quantity of ash and charcoal and for targeting features appropriate for radiocarbon dating. Susceptibility surveys on exposed faces provide information on soil changes that may be invisible to the excavator. Our goal is to build a quantitative catalogue of magnetic signatures for feature types and activities across lodges and sites to create testable models regarding lodge morphology, function, and archaeological contexts.

Bibliography

Blakeslee, D.J. (2005) Middle Ceramic period earthlodges as the products of craft traditions. In D.C. Roper and E.P. Pauls (eds.) *Plains Earthlodges: Ethnographic and Archaeological Perspectives*. Tuscaloosa: The University of Alabama Press, 82-110.

Dalan, R., Sturdevant, J., Wallace, R., Schneider, B. and De Vore, S. (2017) Cutbank Geophysics: A New Method for Expanding Magnetic Investigations to the Subsurface Using Magnetic Susceptibility Testing at an Awatixa Hidatsa Village, North Dakota. *Remote Sensing* **9**(2): 112.

Dalan, R.A., Bevan, B., Goodman, D., Lynch, D., De Vore, S., Adamek, S., Martin, T., Holley, G. and Michlovic, M. (2011) The Measurement and Analysis of Depth in Archaeological Geophysics: Tests at the Biesterfeldt Site U.S.A. *Archaeological Prospection* **18**: 245-265.

Kvamme, K.L. (2007) Geophysical mappings and findings in Northern Plains village sites. In S.A. Ahler and M. Kay (eds.) *Plains Village Archaeology: Bison-hunting Farmers in the Central and Northern Plains*. Salt Lake City: The University of Utah Press, 210-222.

Kvamme, K.L. & Ahler, S.A. (2007) Integrated Remote Sensing and Excavation at Double Ditch State Historic Site, North Dakota. *American Antiquity* **72**: 539–561.

Mitchell, M.D. (2013) *Crafting History in the Northern Plains: A Political Economy of the Heart River Region, 1400-1750*. Tucson: University of Arizona Press.

Pauls, E.P. (2005) Architecture as a source of cultural conservation: Gendered social, economic, and ritual practices associated with Hidatsa earthlodges. In D.C. Roper and E.P. Pauls (eds.) *Plains Earthlodges: Ethnographic and Archaeological Perspectives*. The University of Alabama Press: Tuscaloosa, 51-74.

Wilson, G.L. (1934) The Hidatsa Earthlodge. *Anthropological Papers* **33** (pt 5). New York: American Museum of Natural History.

Wood, W.R. (1967) An Interpretation of Mandan Culture History. *River Basin Survey Papers, Bureau of American Ethnology, Bulletin* **198**. Washington DC: Smithsonian Institution.

Wood, W.R. (1993) Integrating Ethnohistory and Archaeology at Fort Clark State Historic Site, North Dakota. *American Antiquity* **58**(3): 544-559.

Geophysics in Iraqi Kurdistan: discovering the origins of urbanism

Lionel Darras[1], Christophe Benech[1] and Régis Vallet[2]

[1]UMR5133-Archéorient – CNRS/Université Lyon2, Lyon, France; [2]CNRS - IFPO, Erbil, Iraq

lionel.darras@mom.fr

Introduction

Excavations started in 2015 on the sites of Girdi Qala and Logardan, in the governorate of Suleymaniah in Iraqi Kurdistan. The scientific purpose of this new project is to study the formation of complex societies, the appearance of territorial polities and long-term intercultural processes. Indeed, despite recent developments (Kopanias and MacGinnis 2016), southern Kurdistan remains poorly documented, although it seems an ideal laboratory for investigating these research questions. It is no exaggeration to say that the region is at the very heart of the Near East, a crossroad between northern and southern Mesopotamia as well as between Mesopotamia and Iran (Fig. 1). The project is more specifically focused on the Chalcolithic period and on the Bronze Age, two periods for which the redefinition of cultures on a regional basis is a major issue. The data already collected are starting to change the picture of the formation of proto-urban societies. In this central Zagros piedmont, the Uruk culture expansion began as early as the Late Chalcolithic 2 (LC2), around 3900 BC (instead of 3600 BC as expected), contemporary with south Mesopotamian Early Uruk. A major craft district, at the foot of an indigenous site (Girdi Qala Main Mound), a residential enclave (Girdi Qala North Mound) and a political and religious center (Logardan) document the Uruk presence (Vallet 2015, Vallet *et al.* in press, Vallet in preparation).

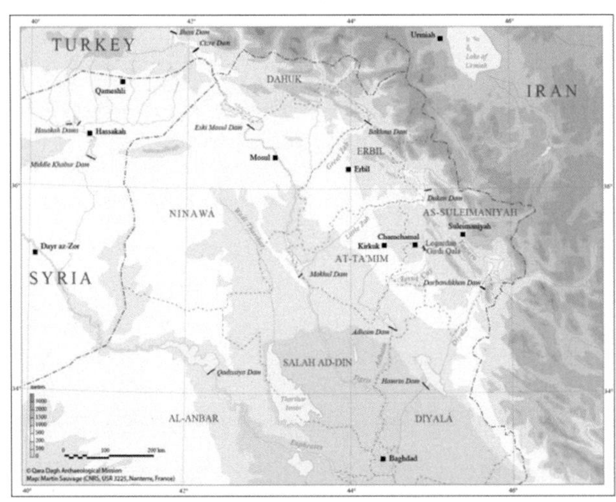

Figure 1: Geographical location of Girdi Qala and Logardan in northern Iraq.

Besides traditional surveys and trenches, we undertook a full geophysics coverage of both sites, which proved essential to guide the excavations as well as to protect those unique sites from a quick destruction by modern agriculture.

Geophysical Survey

The magnetic survey has been carried out with a cesium gradiometer (Geometrics) over a surface of 5ha on the site of Girdi Qala and 3ha on the site of Logardan measuring every 100ms along profiles 1 meter apart. All the data were processed with the "open-source" software WuMapPy (Marty *et al.* 2015). These surveys contribute to the renewal of Iraqi archaeology in this region: the international archaeological missions have intensively used the geophysical approach since 2011 (see recently Kepinski *et al.* 2015, Mühl and Fassbinder 2016) and we can hope for the future to build a unique dataset of Bronze Age sites. The excavations, which immediately followed the magnetic survey, gave us the opportunity to characterize the chalcolithic occupation by an integrated approach and a best interpretation of the magnetic anomalies.

On the site of Girdi Qala, the interest of the archaeologists was at first focused on the main tell which might be occupied by a Bronze Age citadel. The magnetic map (Fig. 2) clearly shows a linear positive anomaly surrounding the top of the tell and which can be interpreted as a fortification even if it is still too early to confirm a Bronze Age dating. The internal organisation is more complex: this area was occupied until the Islamic period: the levels of this latter period are mainly characterised by the presence of pottery kilns and poorly preserved walls with stone foundations (Vallet 2015). The magnetic map is therefore the reflection of this ultimate occupation where the site was no more than a small village reduced to few domestic dwellings.

The excavations carried out at the bottom of the tell showed that the occupation of the tell dates back to the late Chalcolithic period (Fig. 3). The different archaeological soundings confirmed that the anomalies visible on the magnetic map are linked to this same period. The spatial organization of the Chalcolithic occupation is divided between a handcraft area with an impressive set of pottery kilns, a domestic area to the northwest (north mound) and between both an apparently empty area whose function is still discussed. The most striking results were obtained on the north mound, where the geophysical survey was able to define the precise limit of the early-middle Uruk colony (the most ancient known in the near-east), which proved crucial to successfully obtain the protection of this unique site.

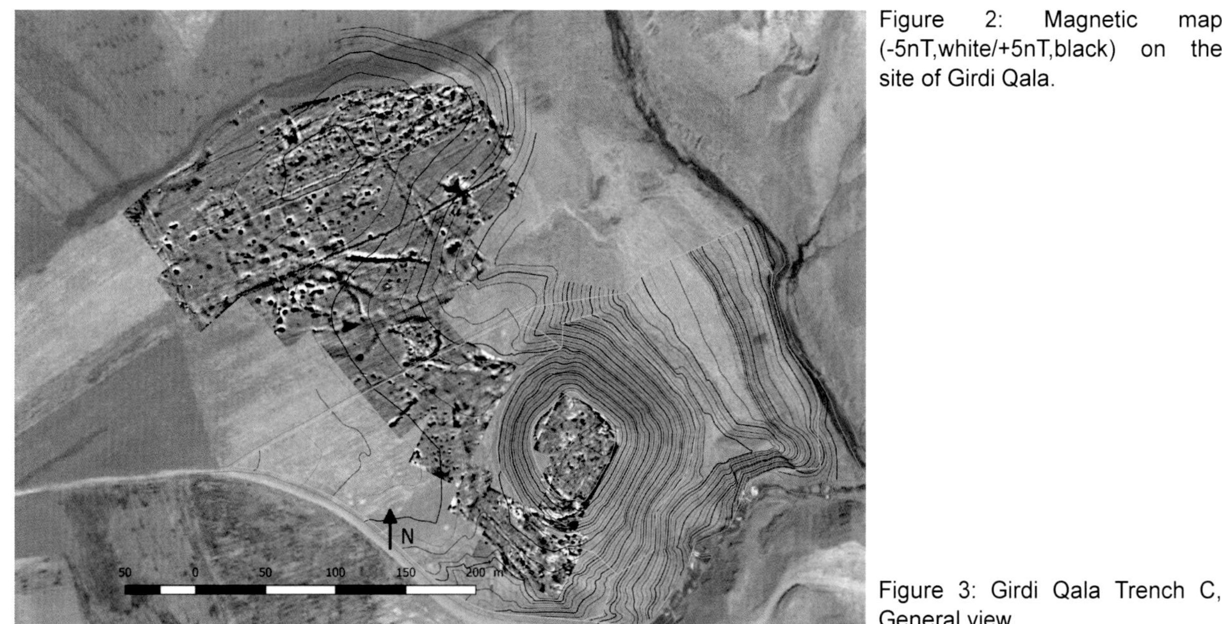

Figure 2: Magnetic map (-5nT,white/+5nT,black) on the site of Girdi Qala.

Figure 3: Girdi Qala Trench C, General view.

In Logardan, 1.5km to the north, the magnetic survey was mainly focused on the upper part of the site, settled on a natural hill of 27m high, with the purpose of identifying the organisation of this eyrie-like site. The results revealed a set of linear anomalies identified by the excavations as a stone ramp of 2m width, dating from the 4th millennium and giving access from the south-western slope to the top. At the top of the tell, the magnetic survey revealed a very contrasting organization, divided between a set of pits in the Northern part, a built area to the south and west and an empty area to the East. In 2016, extensive excavations based on those results have revealed a 4th millennium monumental acropolis at the very top and scattered 3rd millennium public buildings on the upper terrace more to the east.

Conclusion

The two first campaigns carried out on both sites of Girdi Qala and Logardan revealed original and promising results on the Chalcolithic and Bronze Age occupations in this region and can be considered as an important renewal of our knowledge of these periods in Iraqi Kurdistan. The use of the geophysical methods combined with archaeological survey and excavations enabled an efficient strategy of exploration, and a preliminary characterization of Chalcolithic settlements through the magnetic map. The identification of the maximal extension of the sites will also help the Iraqi authorities to delimitate the archaeological area which has to be protected from the development of agriculture in the neighbourhood.

Bibliography

Kepinski, C., Tenu, A., Benech, C., Clancier, P., Hollemaert, B., Ouraghi, N. and Verdellet C. (2015) Kunara, petite ville des piedmonts du Zagros à l'âge du Bronze : rapport préliminaire sur la première campagne de fouille, 2012 (Kurdistan irakien), *Akkadica* **136**(1): 51-88.

Kopanias, K. and MacGinnis, J. (2016) *The Archaeology of the Kurdistan Region of Iraq and Adjacent Regions.* Oxford: Archeopress.

Mühl, S. and Fassbinder, J. W. (2016) Magnetic investigations in the Shahrizor Plain: Revealing the unseen in survey prospections, in K. Kopanias, J. MacGinnis and Panepistēmio Athēnōn (eds.) *The archaeology of the Kurdistan region of Iraq and adjacent regions.* Oxford: Archaeopress, 241-248.

Vallet, R., Baldy, J. S., Naccaro, H., Rasheed, K., Saber, S. A. and Hamarasheed, S. J. (June 2017, In press). New Evidence on Uruk Expansion in the Central Mesopotamian Zagros Piedmont, *Paléorient* **43**(1).

Vallet, R. (ed.) (2015) Report on the First Season of Excavations at Girdi Qala and Logardan, Directorate of Antiquities of Suleymanieh, General Directorate of Antiquities of Kurdistan Regional Government.

Marty, P., Darras, L., Tabbagh, J., Benech, C., Simon, F.-X. and Thiesson, J. (2015) WuMapPy – an open-source software for geophysical prospection data processing, *Archaeologia Polona* **53**: 563-566.

Augmenting the interpretative potential of landscape-scale geophysical data - a case from the Stonehenge landscape

Philippe De Smedt[(1)], Henry Chapman[(2)] and Paul Garwood[(2)]

[(1)]Dept. of Soil Management, Ghent University, Ghent, Belgium; [(2)]Dept. of Classics, Ancient History and Archaeology, University of Birmingham, Birmingham, United Kingdom

Philippe.DeSmedt@UGent.be

Geophysical prospecting is increasingly being applied at a landscape scale. The motorized implementation of geophysical instruments has put theory into practice, and now enables mapping vast landscapes, surpassing a 1 km² area, at sub-meter resolution within a matter of weeks. Geophysics is on the verge of becoming a standard approach in landscape archaeology, alongside other non-invasive methods such as aerial remote sensing. Following these technological advances, a new challenge lies in fully, and robustly, exploiting the interpretative potential of landscape scale geophysical survey data.

Transferring geophysical data to a basis for archaeological research is challenging in any application, but the challenge in bridging the gap between survey data and archaeology increases along with the size of the survey area. Pedological variations or changes in deeper geology, for instance, can influence the geophysical response over surveyed areas, whereas differing survey conditions between field campaigns can affect the uniformity of detected contrasts. While often negligible in small-scale area surveys, increasing survey scale exacerbates such issues.

The combined use of different prospection methods can help facilitate subsequent interpretation, but we are nevertheless (partly) reliant on conjecture to transform our geophysical data to archaeological information. On an intra-site scale, comparing existing or newly gathered archaeological data with geophysical survey results is often practically feasible and implemented. In this respect, 'archaeological feedback' between excavator and geophysical surveyor has increased substantially over the past decades (consider, for instance, the rarity of such studies two decades ago, as discussed by Boucher (1996)). For landscape-wide geophysical survey, providing archaeological feedback is more burdensome. Selecting relevant features is a highly subjective task, based primarily on conjecture, and designing a sampling scheme that provides statistically relevant information is challenging when validating survey datasets that surpass 1 km² in scale.

Despite the challenges in gathering true ground information, such data remain pivotal in bridging the interpretational gap. It is, in many cases, the only possible way to increase the level of certainty of any geophysical data interpretation. When analysing geophysical datasets on a purely physical basis (e.g. through data inversion), their interpretive weaknesses are mostly well integrated into the discussion or presentation of derived results. The ill-posed and non-unique nature of inversion results is intrinsic to the process. It is (or at least should) therefore be standard practice to define and communicate uncertainties alongside such interpretations of geophysical data. Or, as Everett (2013, 9) states:

'It is well known that geophysical data are insufficient to uniquely determine the distribution of subsurface properties, to any level of precision. There are always ambiguities in the interpretation of geophysical data. A major challenge for the near-surface geophysicist is to decide how the uncertainty associated with a given subsurface image should be communicated to stakeholders.'

In archaeological applications, this uncertainty is often discussed only implicitly. However, the anthropological dimension that is added to the problem posed only amplifies the uncertainty of its geophysical solution. Here, the level of conjecture needed impacts the interpretative procedure heavily. For specific sites, landscapes, or even entire types of archaeological features, geophysical data can, however, be transformed to archaeological information with high degrees of certainty (e.g. -Neubauer *et al.* 2002). However, in many cases ready archaeological interpretation of detected features cannot be conducted based on the available geophysical data.

The electromagnetic induction (EMI) data gathered within the Stonehenge landscape (Fig. 1) are an example of the latter. Collected within a landscape that is dominated by traces of prehistoric activity, the current 2.5 km^2 dataset comprises a plethora of geological, pedological, and archaeological features that, through knowledge of the existing (geo-)archaeological framework can only be interpreted to a certain extent (De Smedt *et al.* 2013).

The EMI surveys were based on a number of research questions (De Smedt *et al.* 2013, De Smedt 2015), of which the two most general aimed to evaluate whether: sediments are present within the survey area that contain palaeo-environmental information relating to the formation of the landscape (e.g. palaeochannel deposits);

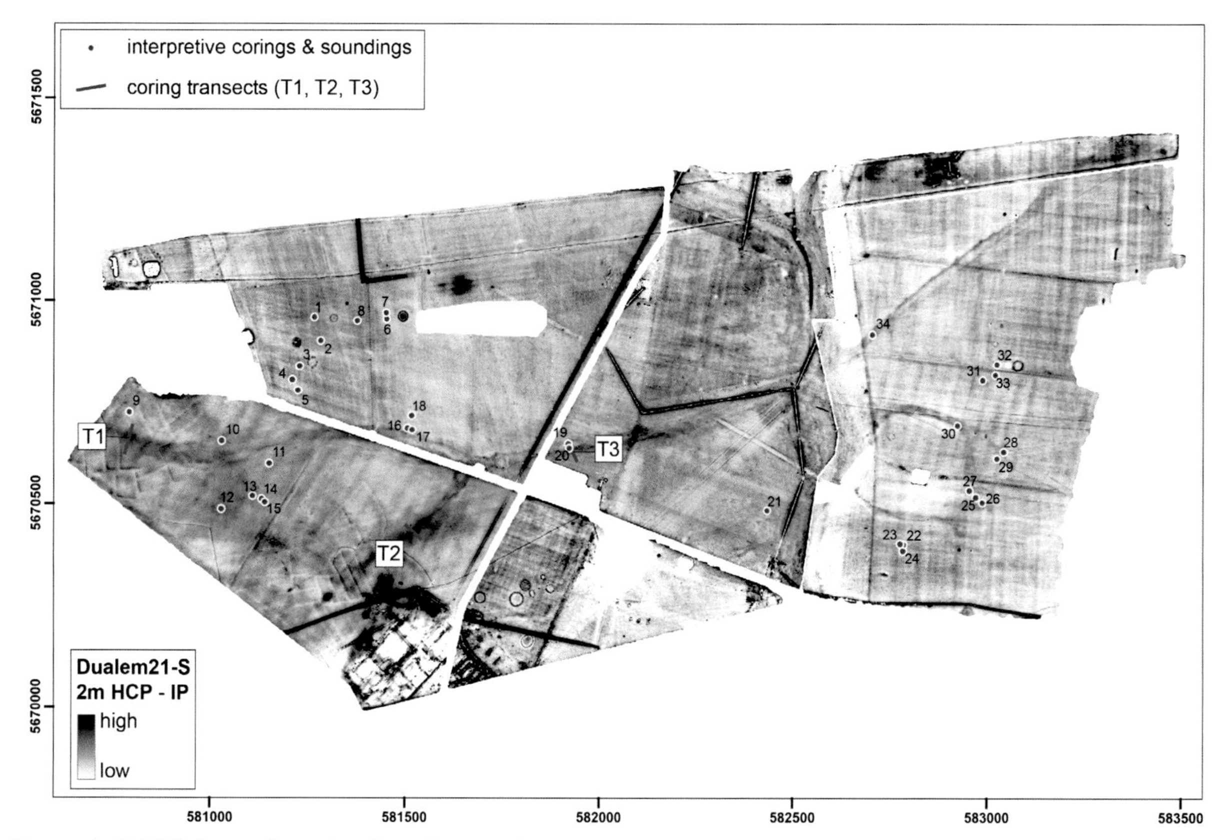

Figure 1: IP EMI data collected at Stonehenge, showing the location of 34 sampling locations alongside three coring transects that were evaluated in November 2016. (Coordinates in meters OSGB 1936).

and whether ephemeral prehistoric (i.e. non-monumental) traces within the landscape can be traced by means of EMI prospecting, potentially adding to the current state of the art on the earlier occupation of the landscape.

The EMI data alone, however, lack interpretative potential to resolve these issues.

In November 2016, we therefore conducted a one-week coring campaign throughout the entire survey area. Based on the quadrature-phase electrical conductivity (ECa) data and the in-phase magnetic susceptibility (IP) data, 45 sampling locations were selected. At each of these locations, one borehole was made for lithological descriptions and geo-archaeological soil sampling using a 5 cm diameter riverside corer. In addition, a second borehole was made with a 2 cm diameter gouge auger for subsequent magnetic susceptibility logging.

The selection of sampling locations was essentially a subjective process. Based on the ECa and IP data, distinctive feature types of presumed geological, archaeological or unidentified origin were selected for investigation. Particular emphasis was given to the identification of sequences and deposits containing potential palaeo-environmental information, and on confirming (or disproving) the archaeological origin of certain types of discrete anomalies.

The limited invasive surveying made it possible to characterise specific deposits and sediment types producing electrical and magnetic anomalies in the EMI data, hereby answering research questions relating to the composition of natural sediments in the area. The combination of lithological description and soil sampling made it possible to assess the palaeo-environmental potential of specific sequences (including viability for OSL dating, pollen analysis, and radiocarbon dating), while magnetic susceptibility logging provided the geophysical data needed to couple visual log descriptions to the IP data. Additional textural and sedimentological analyses are currently in progress to further consolidate correlation to the ECa data.

The results of the fieldwork show how even very limited invasive research can help bridge the gap between archaeological and geophysical information at a landscape scale. Designed specifically to corroborate the EMI survey data, this methodology does not in itself provide a basis for comprehensive archaeological evaluation of the surveyed area, but instead helps to resolve general research questions driving the EMI surveys, and increases the certainty of interpreting the data that these surveys generate.

Bibliography

Boucher, A. R. (1996) Archaeological feedback in geophysics. *Archaeological Prospection* **3**: 129-140.

De Smedt, P., Van Meirvenne, M., Saey, T., Baldwin, E., Gaffney, C. and Gaffney, V. (2014) Unveiling the prehistoric landscape at Stonehenge through multi-receiver EMI. *Journal of Archaeological Science* **50**: 16-23.

De Smedt, P. (2015) *Changing landscapes: an electromagnetic revision of prehistoric land-use at the Stonehenge-Avebury site complex*. Research Proposal – Research Foundation Flanders, Flanders.

Everett, M. E. (2013) *Near-Surface Applied Geophysics*. Cambridge: Cambridge University Press.

Neubauer, W., Eder-Hinterleitner, A., Seren, S. and Melichar, P. (2002) Georadar in the Roman Civil Town Carnuntum, Austria: an Approach for Archaeological Interpretation of GPR Data. *Archaeological Prospection* **9**: 135-156.

Mediterranean sites in archaeological prospection: the case study of Osor, Croatia

Nives Doneus[(1)], Petra Schneidhofer[(1)], Michael Doneus[(2,1)], Manuel Gabler[(3)], Hannes Schiel[(1)], Viktor Jansa[(1)] and Matthias Kucera[(1)]

[(1)]Ludwig Boltzmann Institute for Archaeological Prospection and Virtual Archaeology; [(2)]Department of Prehistoric and Historical Archaeology, University of Vienna; [(3)]Norwegian Institute for Cultural Heritage

nives.doneus@archpro.lbg.ac.at

Introduction

Mediterranean cities often show a long continuity of occupation, in some cases being populated for more than 2000 years. Naturally, these sites provide extensive archaeological archives and thus are favoured objects of investigations. Densely developed urban areas, however, often limit or even prevent detailed archaeological investigations. This applies to invasive excavations, but even more affects large- scale, non-invasive archaeological prospection techniques.

Outside developed areas, archaeological prospection is technically feasible. Nevertheless, the environmental settings prevailing in the Mediterranean landscapes can be challenging. Extensive and detailed cropmarks of archaeological structures, which are common in Central and Northern Europe are usually not present. Aerial archaeology is therefore mainly recording stone settings of hillforts, barrows, or submerged harbour sites and the like. The same also applies to airborne laser scanning (ALS), which experiences similar difficulties due to the presence of dense, low and mainly evergreen Mediterranean vegetation as well as rising sea levels, resulting in submerged former coastal structures. Dense vegetation, vineyards and olive plantations as well as uneven, highly eroded stony ground considerably limit the extent to which (motorised) geophysical prospection can be applied.

Because of these difficulties, archaeological investigations in the Mediterranean landscape of the Eastern Adriatic mainly relied on field walking as well as terrestrial and marine excavations. This has resulted so far in a fragmented picture of individual sites rather than in the reconstruction of detailed historical landscapes.

In 2012, a strategic project was initiated by the LBI ArchPro that focused on these challenging conditions. Its aim was to test the applicability of various non-invasive prospection techniques to achieve an integrative approach for documentation, investigation and interpretation of archaeological landscapes.

Case Study Area

Osor is located on a land bridge between the islands of Cres and Lošinj, which together with Krk and Rab form the most northern group of Croatian Adriatic islands. The case study area consists of carstic limestone densely covered by rigid, mostly evergreen shrubbery known as macchia. In some areas, abandoned and recently used olive tree plantations can be found, which are enclosed by dry stone walls.

Today, Osor accommodates only 60 inhabitants, but its initial extent evidences archaeological traces of several thousand years, including monuments from Iron/Roman Age to medieval times (Blečić Kavur 2015, Faber 1982, Bully *et al.* 2015). During the 16th century, the urban area was reduced in size due to the decreasing population (Sušanj Protić 2015). Consequently, the eastern part of the ancient city, which accommodates remains of several settlement phases, is today undeveloped and thus accessible for archaeological prospection techniques.

Methods

The case study at Osor used an integrated prospection approach combining airborne laser scanning (ALS), airborne laser bathymetry (ALB) and ground penetrating radar (GPR).

ALB utilizes green lasers resulting in a detailed DTM combining land-based with underwater topography to reconstruct former coastlines, identify sunken archaeological structures and locate potential former harbor sites (Doneus *et al.* 2013). To face the difficult environmental conditions mentioned above, a full-waveform laser scanning device and a data acquisition strategy, which would guarantee a dense point-cloud was chosen. The combined ALS/ALB data acquisition took place at the end of March 2012 covering amongst others Osor with parts of its hinterland at Punta Križa. The resulting DTM has a resolution of 0.5 metres.

GPR surveys concentrating on the archaeological area of Osor were conducted in 2014 and 2015. Two different high-resolution motorised GPR systems have been used: a six channel 500 MHz array (SPIDAR) with 25 cm cross-line spacing and a 16 channel 400 MHz MALÅ Imaging Radar Array (MIRA) with 8 cm cross-line spacing.

Results

Preliminary results from the integrated data interpretation are very encouraging. In the Hinterland of Osor, abundant information (dry-stone walls, lime-kilns, prehistoric hillforts, a historic settlement, barrows, terraces etc.) could be obtained on human

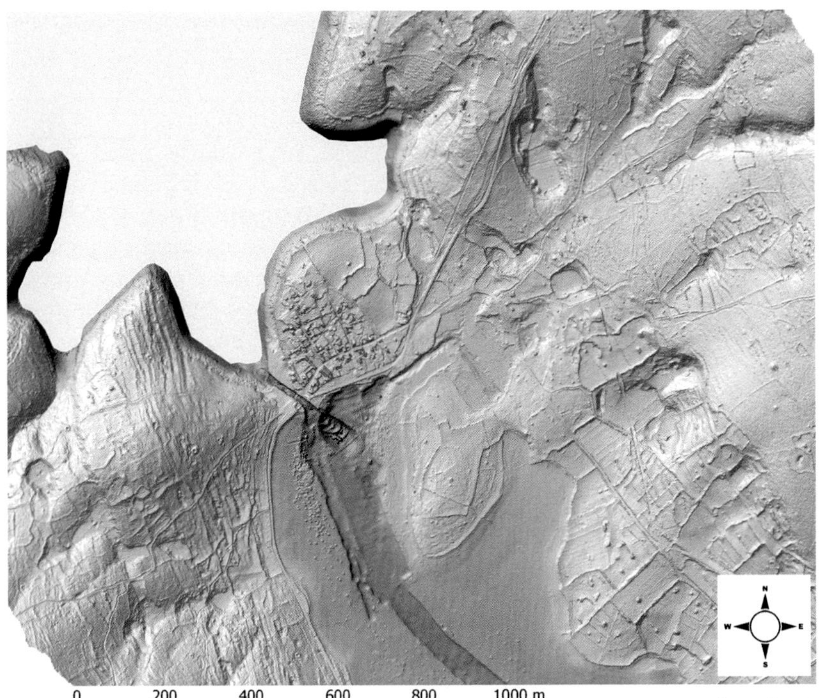

Figure 1: Osor – ALS/ALB-derived filtered digital terrain model. On land, abundant information on human occupation can be obtained despite the dense Mediterranean vegetation.

occupation in today's densely overgrown areas. The underwater topography reveals linear, deeper areas, which are interpreted as being dug out to maintain a navigable fairway (Fig. 1).

Both GPR surveys targeting the eastern part of Osor yielded traces of the ancient city. The use of high resolution GPR data allowed the documentation of settlement features like streets and buildings from Roman and post-Roman periods (Fig. 2).

Figure 2: The GPR interpretation shows the road network and stone buildings from Roman and late antiquity.

During the GPR data interpretation, however, differences in the visibility of archaeological features were observed. This has various reasons. Eroded surfaces allow better visibility of the city structures as they lie directly below the surface. This is the case in the northern survey areas, which are situated higher and closer to the shoreline and more exposed to the erosional forces of wind and waves. Also, differences in urban development in the eastern and western part of the city have affected the results. A better visibility can be observed in the western part of the ancient city, since settlement activities stopped here in the 16th century. And finally, the presence of calcareous subsoil located only 1.5 m underneath the topsoil might have led to poor visibility and/or the lack of wall foundations in parts of the survey area (Fig. 3).

The application of archaeological prospection methods provides – as in the case of Osor – new insights on old questions. The results demonstrate that the integrated use of airborne laser scanning (ALS/ALB) and ground penetrating radar (GPR) has brought meaningful results even in difficult environmental settings. The topography as well as archaeological remains hidden by the vegetation, seawater and soil were detected and mapped.

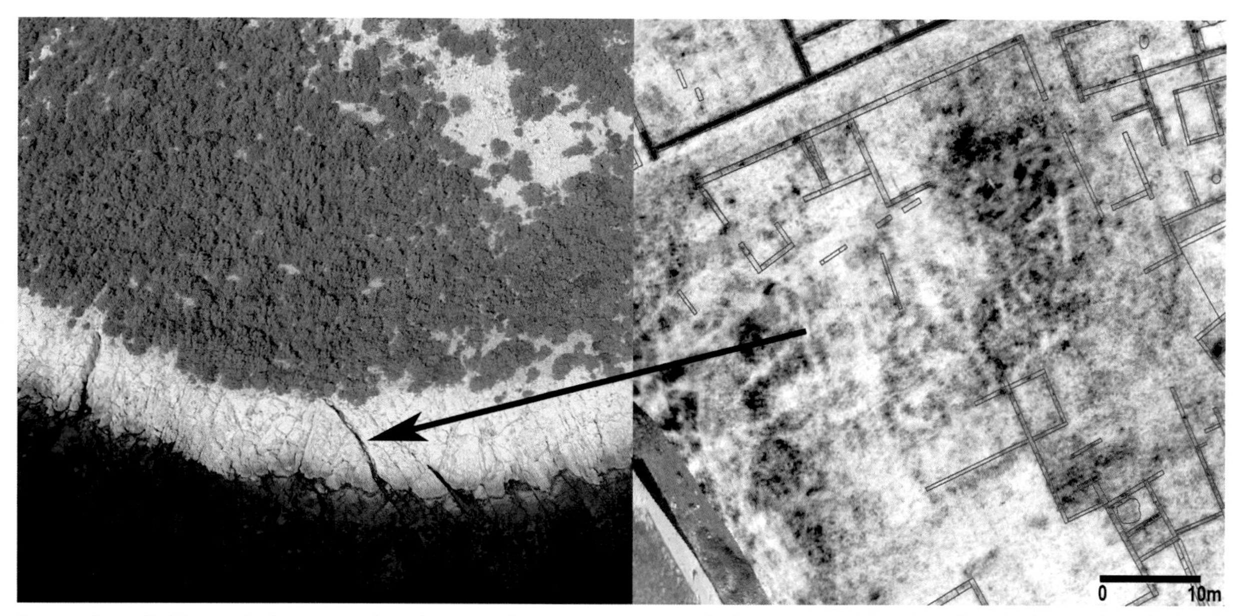

Figure 3: Left: Present-day example of exposed, uneven and carstic limestone at the shoreline in the region. Right: A similar situation might have contributed to the poor visibility of wall foundations in some parts of measured areas.

Bibliography

Blečić Kavur, M. (2015) *Povezanost perspektive. Osor u kulturnim kontaktima mlađeg željeznog doba. A coherence of perspective. Osor in cultural contacts during the Late Iron Age*. Koper, Lošinj: Založba Univerze na Primorskem.

Bully, S., Jurković, M., Čaušević-Bully, M. and Marić, I. (2015) Benediktinska opatija Sv. Petra u Osoru - arheološka istraživanja 2006-2013. *Izdanja Hrvatskog arheološkog društva* **30**: 103-127.

Doneus, M., Doneus, N., Briese, C., Pregesbauer, M., Mandlburger, G. and Verhoeven, G. (2013) Airborne Laser Bathymetry – detecting and recording submerged archaeological sites from the air. *Journal of Archaeological Science* **40**: 2136-2151.

Faber, A. (1982) Počeci urbanizacije na otocima sjevernog Jadrana. *Izdanja Hrvatskog arheološkog društva* **7**: 61-78.

Sušanj Protić, T. (2015) O urbanizmu Osora nakon 1450. godine. *Ars Adriatica* **5**: 95-114.

Acknowledgements

Archaeological research in Osor was done in collaboration with Lošinj Museum, Croatia. GPR measurements were carried out by the LBI ArchPro team: Manuel Gabler, Matthias Kucera, Hannes Schiel, Viktor Jansa, Michael Doneus and Nives Doneus.

Transforming the search for human origins using new digital technologies, low altitude imaging, and citizen science

Adrian Evans[(1)], Thomas Sparrow[(1)], Louise Leakey[(2)], Andrew Wilson[(1)], Randy Donahue[(1)]

[(1)]School of Archaeological Sciences, University of Bradford, United Kingdom; [(2)]Turkana Basin Institute, Nairobi, Kenya

a.a.evans@bradford.ac.uk

Introduction

In this paper we describe the augmentation of traditional fossil prospection using citizen science. It describes a project that has attempted to revolutionize how fossil discovery is conducted by providing a platform from which the public can be the eyes on the ground. To do this some technological developments had to be made. The technique required:

- a system that could image areas where fossil prospection takes place
- the images produced by the system to have a resolution suitable to enable the study of fragments as small as teeth
- the images to have metadata that would allow the physical relocation of the imaged surfaces to allow the collection/direct inspection of notable objects
- a basis in consumer grade equipment at as low a cost as possible

Hardware platform

At the inception of the project it was determined that a good resolution was required to allow images to be useful, having sufficient detail to resolve objects that can normally be seen while field walking. Human eye resolution is on the order of 0.02 degrees (Yanoff and Duker 2009) so at standing height this can be estimated to 0.6 mm on the ground. The project aimed to capture at double this resolution. Research and development in the area led to an evaluation of different flight platforms. Initial research concluded that a petrol powered remote piloted traditional single rotor helicopter would be the most suitable system. This class of craft are able to resist strong winds, can fly at very low speeds, can rapidly correct speed and altitude, and can fly for extended durations. The limiting factor for such a system is the size of the memory card available to use for the camera system. The disadvantages for such a system are the multiple single points of failure that would result in catastrophic system collapse, and the maintenance cost associated with working in dusty environments. Fixed wing craft (traditional airplanes) are the most efficient airfame, but cannot fly at the slow speeds required to capture images of a sufficient resolution without motion blur (given shutter speed limitations). Multi-rotor platforms have similar properties to traditional helicopters with the advantage of a less complicated mechanical system and ability to compensate for motor failure. The primary disadvantage to such a system is a reliance on battery power. Lithium ion batteries for multi-rotor copters of the size capable

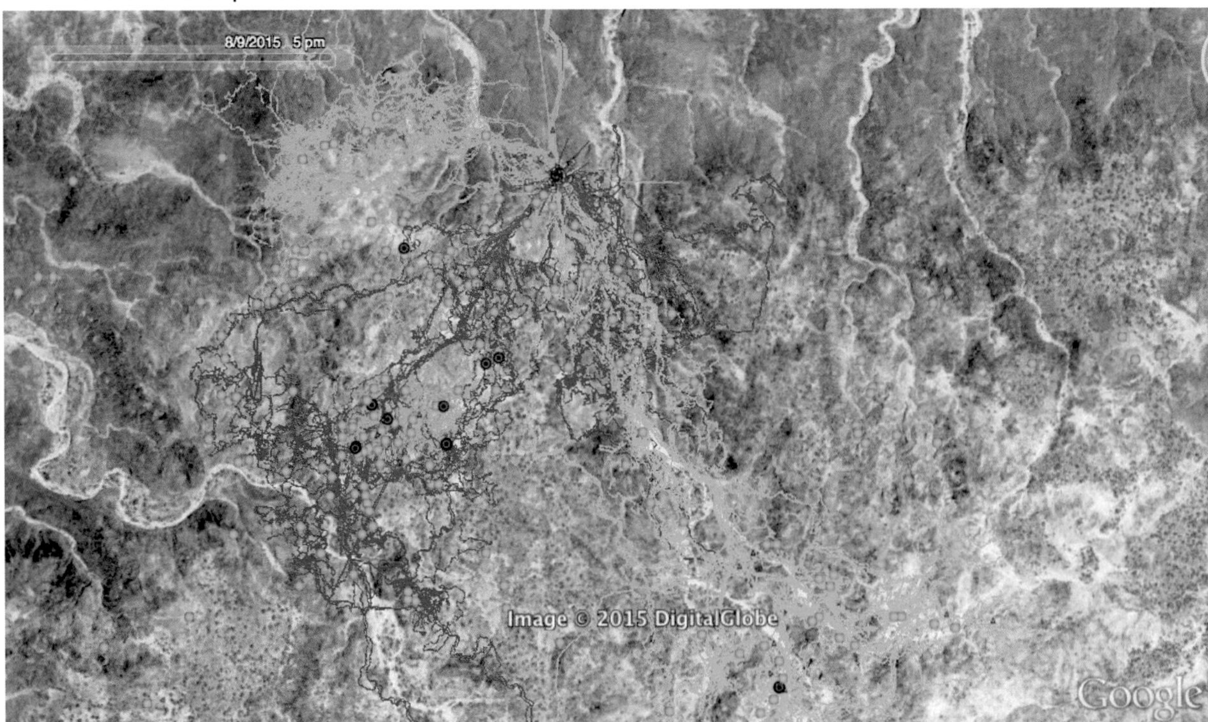

Figure 1: Satalite image of a survey region at lake Turkana overlaid with GPS traces from individual surveyors operating in the area.

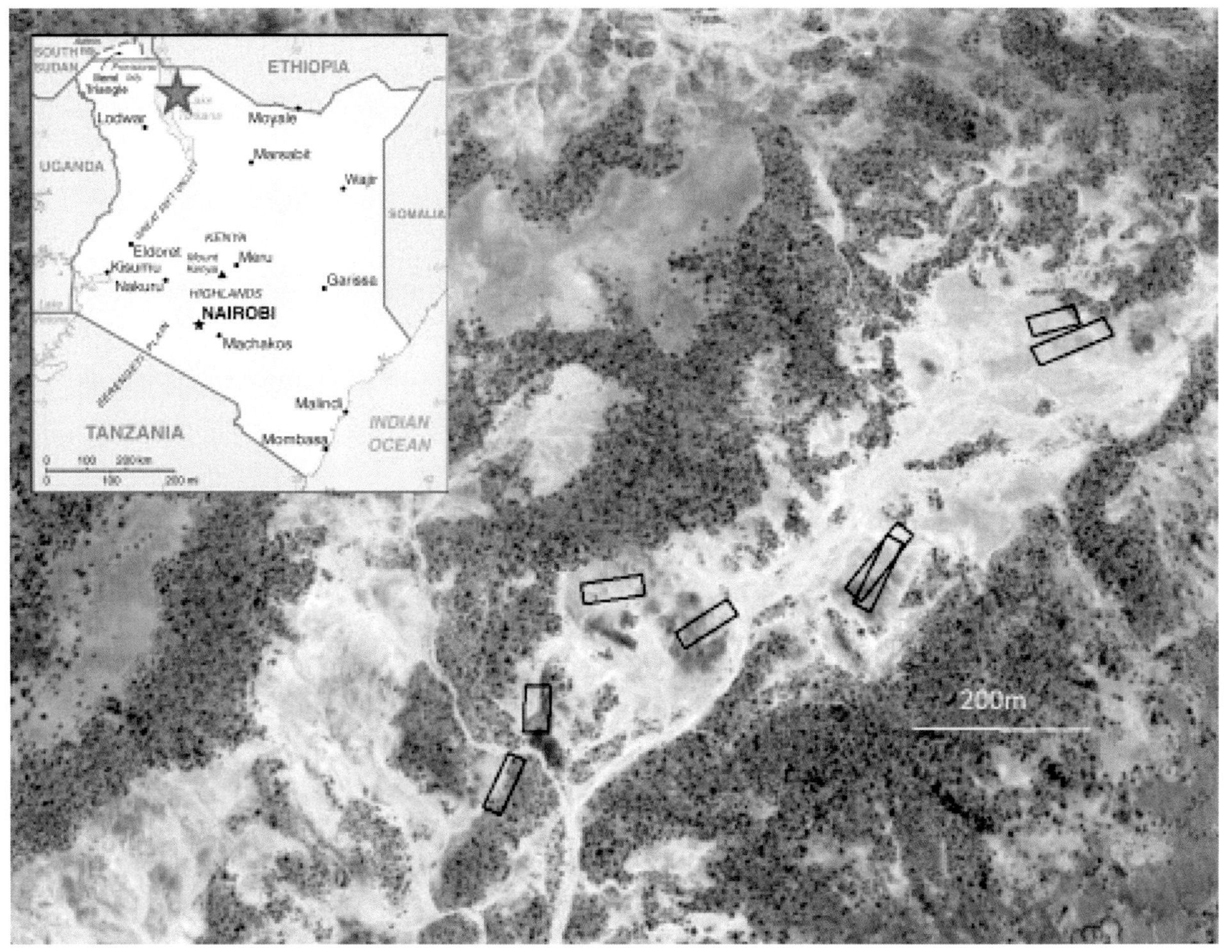

Figure 2: Locations of work in Kenya with outlines of areas imaged during season 1.

of lifting the imaging system required for this project are expensive and come with a need for generator powered in-field charging that can disrupt flight schedules. Low-cost development of a traditional helicopter auto-pilot system could not be achieved before fieldwork could begin. The current trend towards multi-rotor systems made adopting this technology unchallenging. Therefore, a decision was made to initiate the project using a multi-rotor copter system.

Deployment

The research area is defined by the east lake Turkana survey area surrounding the Turkana Institute's field station near Ileret village in northern Kenya. Over two field seasons, almost 1 million image 'subjects' were collected.

The first season of image data collection was conducted within area 8B and consisted of several low altitude flights following a raster sampling protocol. The aim of the strategy was to gather near 100% coverage from the selected locations. 6 locations were chosen for gathering this type of image data.

During this season the platform was also used to collect lower resolution imagery of the areas of interest to provide higher quality orthographic imagery of the areas and to help with the registration of image data collected at lower resolutions. This data could also be used to produce high resolution 3d topographic models of the surveyed areas.

The Online Panoptes system

The project has utilized the publicly accessible panoptes system produced by the zooniverse group and the university of Oxford headed by Christopher Linott. The project website can be found at https://www.fossilfinder.org.

Image Processing

In preparation for use on the online 'panoptes' citizen science platform each image needed to be prepared as a subject. It was determined by exploration of common screen size and the layout of the citizen science platform that an image less than 800 pixels wide was to be used. This resulted in a decision to split each image into a 7x7 frame of 49 images which would become individual subjects.

Launch and citizen science uptake. Forum discussion

At the first point of data testing 40 thousand images had been studied by a group of six thousand citizen

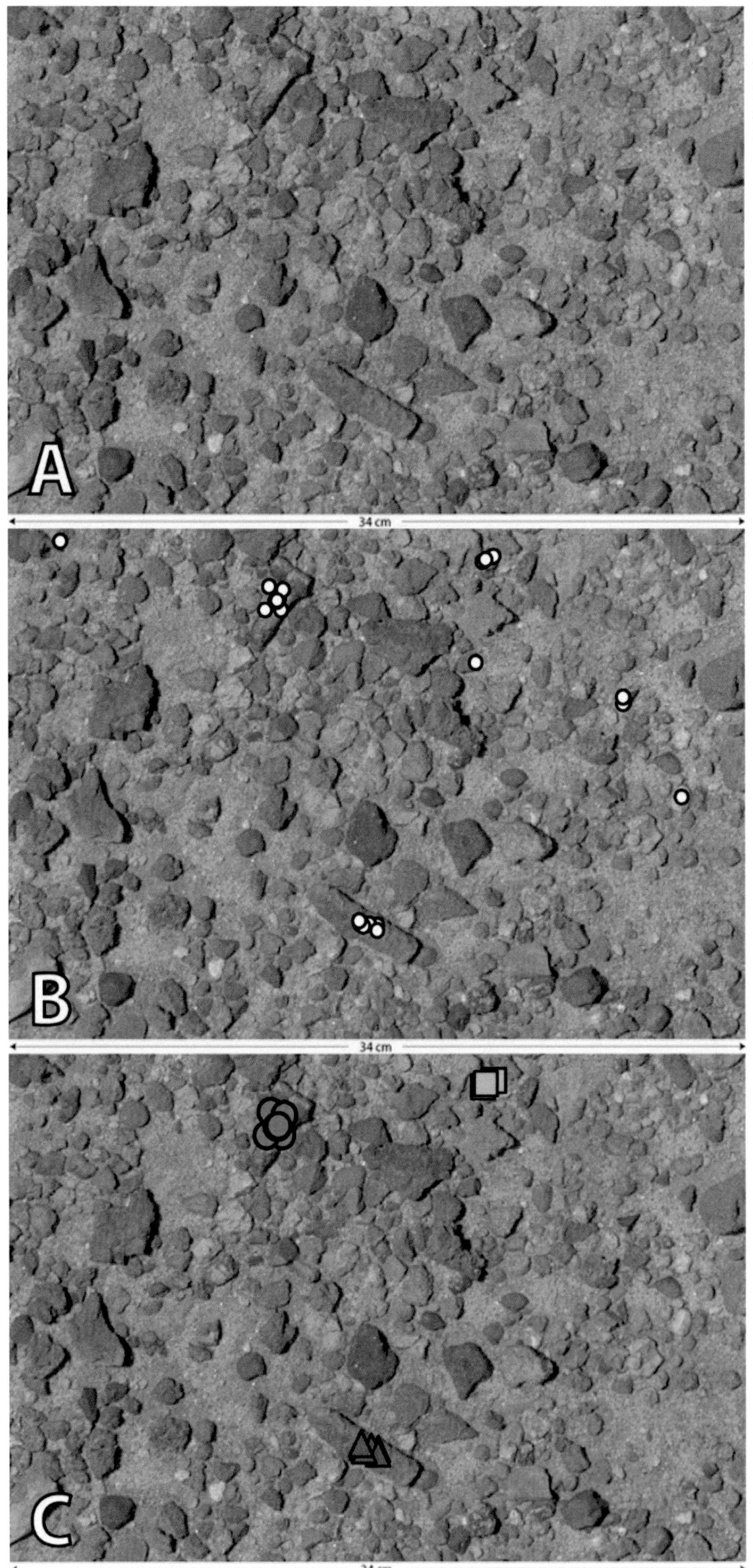

Figure 3: A. Subject 703837, B. All click data from citizens, C. Results of cluster analysis (blue circles= BONE, green squares=MBS, red triangles= MBS*).

scientists. Each subject studied had been labelled by up to 10 citizen scientists who labelled images based on the question set posed in the citizen science platform. These questions pertained to image quality and identification of features within each subject. Each subject was studied by 10 individuals to enable a blind group consensus of object classification. Consensus and identification of objects has been carried out with the DBSCAN density based clustering method (Ester *et al.*). Clusters with a membership of less than three (or with less than three contributors) were omitted. Consensus of identification of each object derived from clusters was identified using the modal label. If a 70% consensus in label was not reached a warning maker was placed on the label. Figure 1 shows an example of this process in which the location of a bone fragment was recognized by consensus.

Bibliography

Caliński, T. and Harabasz, J. (1974) A dendrite method for cluster analysis. *Communications in Statistics* **3**: 1-27.

Ester, M., Kriegel, H.-P., Sander, J. and Xu, X. (1996) A density-based algorithm for discovering clusters in large spatial databases with noise. In E. Simoudis, J. Han and U. Fayyad (Eds.) *KDD'96 Proceedings of the Second International Conference on Knowledge Discovery and Data Mining.* Oregon: AAAI Press. 226-231.

Leakey, M. D. and Harris, J. M. (1987) *Laetoli, a Pliocene site in northern Tanzania.* Oxford: Oxford University Press.

Lin, A. Y. M., Huynh, A., Lanckriet, G and Barrington, L. (2014) Crowdsourcing the Unknown: The Satellite Search for Genghis Khan. *Plos One* **9**(12): e114046.

Murphey, P. C., Knauss, G. E., Fisk, L. H., Démére, T. A., Reynolds, R. E., Trujillo, K. C., and Strauss, J.J. (2014) A foundation for best practices in mitigation paleontology. *Dakoterra* **6**: 243-285.

Njau, J. K. and Hlusko, L. J. (2010) Fine-tuning paleoanthropological reconnaissance with high-resolution satellite imagery: The discovery of 28 new sites in Tanzania. *Journal of human evolution* **59**: 680-684.

Roach, N. T., Hatala, K. G., Ostrofsky, K. R., Villmoare, B., Reeves, J. S., Du, A., Braun, D. R., Harris, J. W. K., Behrensmeyer, A. K. and Richmond, B. G. (2016) Pleistocene footprints show intensive use of lake margin habitats by Homo erectus groups. *Scientific reports* **6**: 26374.

Westphal, A. J., Butterworth, A. L., Snead, C. J., Craig, N., Anderson, D., Jones, S. M., Brownlee, D. E., Farnsworth, R. and Zolensky, M. E. (2005) Stardust@ home: a massively distributed public search for interstellar dust in the stardust interstellar dust collector. *Lunar and Planetary Science* **36**(21).

Yanoff, M. and Duker, J. (2009) *Ophthalmology.* Saunders: Elsevier.

Magnetometer prospection of Neo-Assyrian sites in the Peshdar Plain, Iraqi Kurdistan

Jörg W E Fassbinder[(1)], Andrei Asăndulesei[(2)], Karen Radner[(3)], Janoscha Kreppner[(3)] and Andrea Squiteri[(3)]

[(1)]Department of Earth and Environmental Sciences, Ludwig-Maxililians-Universität München, Theresienstrasse 41, 80333 München, Germany; [(2)]Interdisciplinary Research Department – Field Science, "Alexandru Ioan Cuza" University from Iași, Romania; [(3)]Faculty of History and the Arts Ancient History Ludwig-Maxililians-Universität München, Germany

fassbinder@geophysik.uni-muenchen.de

The Peshdar Plain in the Neo-Assyrian Period: the Border March of the Palace Herald

Our joint research project (Peshdar Plain Project Publications 2016) in north-western Iraqi-Kurdistan has shed new light on a hitherto little known eastern frontier region of the Assyrian Empire, specifically the Border March of the Palace Herald at the border with the kingdoms of Mannea and Ḫubuškia. Gird-i Bazar is the first unequivocally Neo-Assyrian site to be excavated in the region. The occupation layers beginning to be uncovered here offer the rare opportunity, firstly, to explore a decidedly non-elite settlement of the Neo-Assyrian period, secondly, to further our understanding of how the Assyrian Empire organized its frontier zone and, thirdly, to synchronize the Western Iranian pottery cultures (with the key sites Hasanlu, Godin Tepe, Nush-i Jan, and Baba Jan) with the Assyrian ceramic material of the 8th and 7th centuries BC. Karen Radner analysed the Neo-Assyrian textual sources (clay tablets) discovered on the site in 2013, which indicate that as part of the Border March of the Palace Herald, the Peshdar Plain was located directly on the empire's frontier with Mannea and Ḫubuškia (Fig. 1).

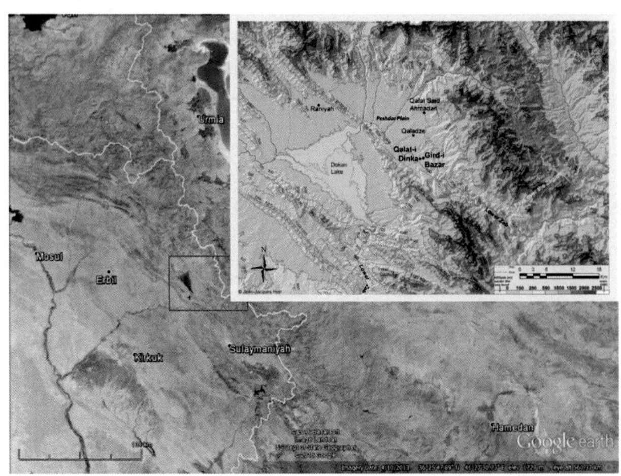

Figure 1: Location of the Peshdar Plain and its key sites (after Andrea Squiteri & Jean-Jacques Herr (Peshdar Plain Project Publications 2016)).

The geo-archaeological survey of Mark Altaweel and Anke Marsh provide a geo-archaeological assessment based on a large survey conducted in August 2015 by Jessica Giraud (Altaweel and Marsh 2016, Giraud 2016). Both studies strongly suggest that Gird-i Bazar and Qalat-i Dinka were part of one extended Neo-Assyrian settlement that we call the "Dinka settlement complex". Based on these studies, Jörg Fassbinder and Andrei Asăndulesei undertook in 2015 and 2016 the first large-scale high-resolution caesium-magnetometer surveys on selected sites of the Peshdar Plain (Fassbinder and Asăndulesei 2016, Fassbinder 2015).

Results of the Magnetometer Prospecting

The first test magnetometer measurements were undertaken in August 2015 on the western slope and on the eastern plateau of the fortification known as Qalat-i Dinka (Fassbinder and Asăndulesei 2016). The small tell Gird-i Bazar, previously assumed to have housed a small settlement, and which had been partly destroyed by the construction of a chicken farm and enclosed by a metal fence, seemed at first sight not very suitable for magnetometer surveying, but nevertheless revealed a detailed ground plan of a settlement.

The geological background on the slope of the Qalat, and also of the entire surrounding area, is dominated by a para-brown-earth developed on gravels and alluvia. The survey conditions were manifold: parts of the area were harvested, others where roughly ploughed, and others were simply not accessible due to thick plant (thistle) growth. Parts of the total survey area (ca. 1 km^2 in total) consist of uneven terrain with steep slopes and gullies. Accordingly, a handheld total field caesium-magnetometer was selected for our survey. The first tests were carried out on areas that had been ploughed in the year before. These areas were almost undisturbed and flat, so that we could expect the best conditions and almost no disturbances in the topsoil.

Results of the Qalat-i Dinka (Western Slope)

Qalat-i Dinka dominates the Peshdar Plain and the lower courses of the Zab River. The chosen survey area is situated on the south-western slope of the mound, and measured 120 x 120 m. The survey area is delimited by the steep slope of the Zab River, and enclosed by a modern field border and fence. In August 2015, the ground of our survey area was harvested but not ploughed, and hence was only little disturbed by plough furrows. The strong magnetic enhancement of the topsoil and archaeological soils compared to the weak magnetic susceptibility

Figure 2: Qalat-i Dinka. Magnetometer measurement of the survey area (ca. 120 x 120 m), on the slope of the fortification. Caesium total field magnetometer Scintrex, SMG-4 special in duo-sensor variometer configuration, total Earth's magnetic field at the Peshdar Plain 08/2015, 47.500 ±20 Nanotesla, sensitivity ±10 Picotesla, sampling density 25 x 50 cm, interpolated to 25 x 25 cm, dynamics in 256 grey scales, 40 m grid, high-pass filter overlay.

of the bedrock would have facilitated the detection and interpretation of the archaeological structures, but some strongly magnetized gravels obscured their detection beneath the ground. The resulting picture (Fig. 2) is dominated by archaeological activity, and few of the features are clearly visible. These features are concentrated in the upper part of the slope, while in the lower part they are clearly enclosed by the remains of the foundations of a wall. Little activity was detected outside of the delineation works. The magnetometer survey of the eastern plateau of the Qalat revealed a dense activity and many archaeological features, including some foundations, pits and very probably some fortification works.

Results of the Dinka Settlement Complex

The starting point for the survey in the plain was the former tell site that was already half-occupied by the modern farm house and chicken farm. On an area of 60 x 20 m we found, in 2015, the ground plan of a dense settlement. The adjacent field was not accessible for magnetometer prospecting in 2015, but in September 2016 it was harvested and cleared, and measurements were therefore be extended by an area of ca. 400 x 400 metres. The magnetometer data revealed structures that were more clear and distinct than was expected from such an area. The magnetogram revealed very precisely the foundations of a large settlement and, as it seems, at the moment, to very probably be an urban district of a large city. The buildings are separated by small streets, and the orientation of the complex tends to face the fortification of the Qalat (Fig. 3).

Summary

The geophysical results presented above reveal not only several archaeological features, but an overall clear ground map of an urban district of a city, including domestic buildings, workshops and production sites. The palace and most of the administrative buildings were probably close to, or on, the slope of the Qalat-i Dinka, and will hopefully be detected in the next prospection campaigns. The shapes, architecture and layout of the Neo-Assyrian buildings that we detected will, furthermore, provide the basic information that will likely be the groundwork for specific architectural studies that have for the moment no parallels in the specialised literature.

Figure 3: Dinka plane. Magnetometer measurement of the survey area (ca. 400 x 400 m), adjacent to tell Gird-i Bazar and Qalat-i Dinka. Caesium total field magnetometer Scintrex, SMG-4 special in duo-sensor variometer configuration, total Earth's magnetic field at the Peshdar Plain 09/2016, was 47.600 ±30 Nanotesla, sensitivity ±10 Picotesla, sampling density 25 x 50 cm, interpolated to 25×25 cm, dynamics in 256 grey scales, 10 x 10 high-pass filter, 40 m grid.

Bibliography

Altaweel, M. and Marsh, A. (2016) *Landscape and geoarchaeology of the Bora Plain*. Peshdar Plain Project Publications 1: 23-28.

Giraud, J. (2016). *Surface survey of the Dinka settlement complex, 2013-2015*. Peshdar Plain Project Publications 1: 29-35.

Fassbinder, J. W. E. and Asandulesei, A. (2016) *The magnetometer survey of Qalat-i Dinka and Gird-i Bazar, 2015*. Peshdar Plain Project Publications 1: 36-42.

Fassbinder, J. W. E. (2015) Seeing beneath the farmland, steppe and desert soils: magnetic prospecting and soil magnetism. *Journal of Archaeological Science* **56**: 85-95.

Peshdar Plain Project Publications (2016), Vol 1 Ed. Karen Radner. Exploring the Neo-Assyrian Frontier with Western Iran. The 2015 Season at Gird-i Bazar and Qalat-i Dinka, 129.

Integrated geophysical prospection in a Hittite Empire city (Šapinuwa)

Mahmut Göktuğ Drahor[1], Meriç Aziz Berge[1], Caner Öztük[2], Buket Ortan[2], Atilla Ongar[1], Aygül Süel[3], Sedef Ayyildiz[4], Önder Şeref Avsever[2] and Funda İçke[1]

[1]Dokuz Eylül University, Department of Geophysical Engineering, İzmir, Turkey; [2]GEOIM Engineering, Consulting, Software and Construction LTD, İzmir, Turkey; [3]Ankara University, Department of Hittitology, Ankara, Turkey; [4]Bitlis Eren University, Department of History, Bitlis, Turkey

goktug.drahor@deu.edu.tr

Introduction

Interest in integrated applications of geophysical prospection methods, which harbours a great potential to maximize information gained about archaeological structures buried in the shallow subsurface, has seen a rapid increase over the past decade in archaeological prospection (Diamanti *et al.* 2005; Drahor 2006, 2011; Papadopoulos *et al.* 2012; Drahor *et al.* 2015). The aim of the integrative prospection is to gain more reliable interpretations from the integrated geophysical data concerning complex soil and subsurface characteristics of archaeological sites. The combined application of geophysical techniques gives more useful results to determine the location, depth, dimension, and characterization of buried archaeological features. In the integrated geophysical investigations, the success of the used geophysical techniques also depends on the respective physical contrasts between the subsoil and the buried archaeological structures.

Šapinuwa was one of the kingdom cities in the Hittite Empire, and the other was the Hattuşa. The archaeological site of Šapinuwa is located in the Çorum region of Central Anatolia, Turkey. The goal of integrative surveys was to reveal the buried structures in the investigated site and to produce maps of the architectural plan of the ancient city by using a number of different geophysical techniques. To this purpose, magnetic gradiometry, ground-penetrating radar (GPR), electrical resistivity tomography (ERT), induced polarisation tomography (IPT), seismic refraction tomography (SRT) and multi-channel analysis of surface wave tomography (MASWT) have been applied in the city area of Šapinuwa since 2013.

Data Acquisition and Processing

The integrated geophysical investigations were carried out in the Tepelerarası site (ancient city area). Initially, geophysical survey grids and lines were established by using a highly precise GNSS system. Then the integrated data were collected in gridded areas. Magnetic gradiometry and GPR techniques were used on a large-scale area to determine the general settlement plan of ancient city, while the ERT, IPT, SRT and MASWT studies were applied in limited fields to test the identification capability of the mentioned techniques, which could be useful in the investigation of detailed subsurface characterisation in the area. The data acquisition parameters and processing stages of integrated investigations are presented in Table 1.

Results and Discussion

In this study, six different geophysical techniques were utilized to define possibly existing archaeological evidence (walls, pits, kiln, burnt materials, etc.) and to reveal the buried architectural

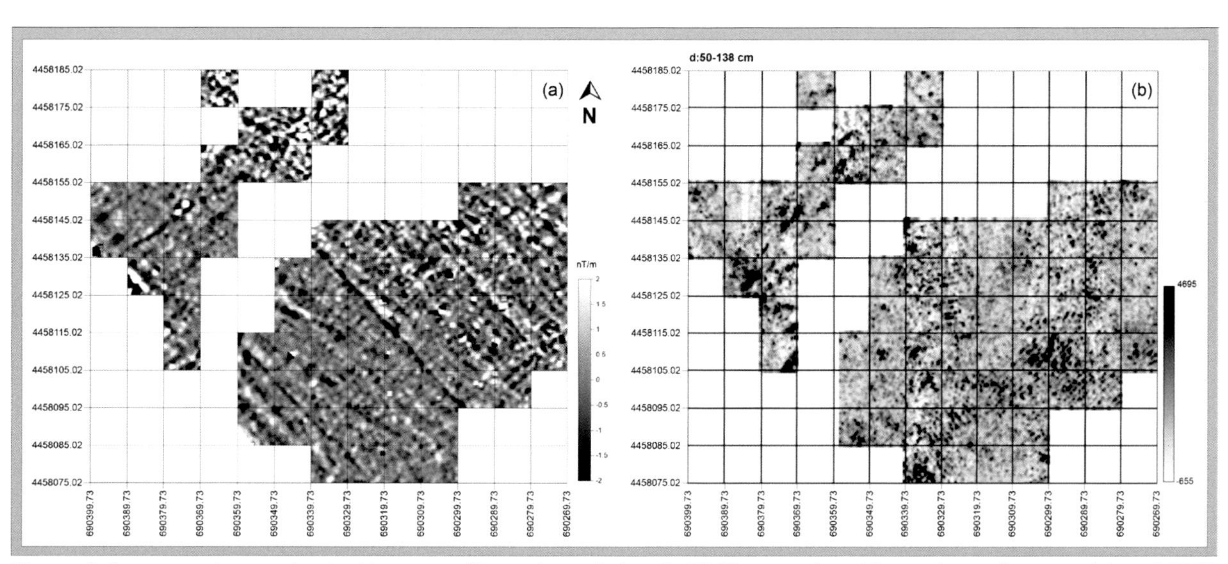

Figure 1: Large-scale geophysical images of investigated sites in 2016 campaign. Magnetic gradiometer (a) and GPR (b) grey-scale images of southern part of integrated investigation site of Šapinuwa.

Table 1: Data acquisition and processing steps of integrated surveys in Šapinuwa.

Method	Instrument	Measuring technique	Measuring and line intervals	Software / Data Processing
Magnetic Gradiometry (2012-2016)	Geoscan FM256 Fluxgate gradiometer	Parallel traverses	0.25 × 0.5 m	Geoplot zero mean traverse and grid corrections, despiking, interpolation, low-pass filtering
GPR (2013-2016)	GSSI SIR3000	400 MHz shielded antenna	0.125 × 0.5 m Time window: 75 ns Time sampling: 512 Trace increment: 0.125	GPR-Slice background removal, static correction, regain, boxcar smoothing, DC drift removal, bandpass filter, migration
ERT (2013-2016) IPT (2016)	AGI-SuperSting R1/IP resistivity meter with multicore cable	Wenner-Schlumberger array	1 × 1 m	Res2Dinv-Res3Dinv Robust (blocky) inversion using finite-element forward routine
SRT (2015)	Geometrics Geode (24 channel)	P-wave	2 × 1 m Record time:0.3s Sampling:0.25ms	SeisImager/2D 2D Refraction tomography
MASWT (2015)	Geometrics Geode (24 channel)	S-wave	2 × 1 m Record time:1s Sampling:0.5ms	SeisImager/SW 2D inversion
GPS (2012-2016)	Trimble			

plans of a Hittite Empire city. In addition, this study purposes to define the capabilities of integrated geophysical investigations in determination of subsurface archaeological relics within the context of the monumental structures located at the investigated area. The large-scale geophysical investigations consisted of GPR and magnetic gradiometry, and revealed the localisation of subsurface archaeological features and the city plan. In Figure 1, GPR and magnetic gradiometer images of the sites measured in the 2016 campaign are presented. The magnetic image reveals the general distribution of buried archaeological structures, which are oriented in NE-SW and NW-SE directions. The presence of buried structures is more distinctive in the northern and eastern part of the area. Negative anomalies corresponded to walls built of limestone, mud brick and rammed earth materials, while the positive anomalies are associated with the burnt mud brick materials and some volcanic stones used in wall construction. The general extensions of positive and negative magnetic anomalies reveal the presence of a regular city plan in the Šapinuwa archaeological site. The GPR depth slice obtained from 50-138 cm is clearly supported by the magnetic image in most locations. In this slice, the presence of important archaeological structures oriented along a NE-SW direction are evidently displayed in the northern and north-western part of the area. In other parts of investigated area, the very regular architectural distribution with a NE-SW orientation attracts the attention. In addition, some of them are slightly rotated from this direction. These structures display constructions that include small spaces.

In some areas of the Tepelerarası site of Šapinuwa, the integrated investigations consisted of magnetic, ERT, GPR, SRT and MASWT techniques carried out to reveal the characteristics of archaeological structures and covering soil. Therefore, we obtained detailed information of the subsurface characteristic including changes of resistivity, dielectricity, magnetic properties and seismic velocities (*Vp* and *Vs*). This study reveals the usefulness of the integrated approaches of magnetic gradiometry, GPR, ERT, SRT, and MASWT techniques in archaeological site investigation. By making use of tomographic techniques, we can obtain various images and depth slices that will prove to be useful for the selection of targeted excavation locations, rendering the excavations more cost and time efficient for the involves archaeologists. In Figure 2, the results of five geophysical techniques used in a selected area in the Tepelerarası site are given. In this area, the magnetic image shows a presence of regular magnetic anomalies oriented in NE-SW and NW-SE directions. This result reveals some archaeological vestiges that show a regular architectural plan. GPR results selected from different depth slices (0-0.5, 0.5-1.5 and 1.5-2.5 m) support the magnetic image. ERT investigation reveals the existence of archaeological structures found between 0 and 1.5 m depths. Archaeological structures have high resistivity values between 30-70 Ohm/m, while the subsoil values are 13-30

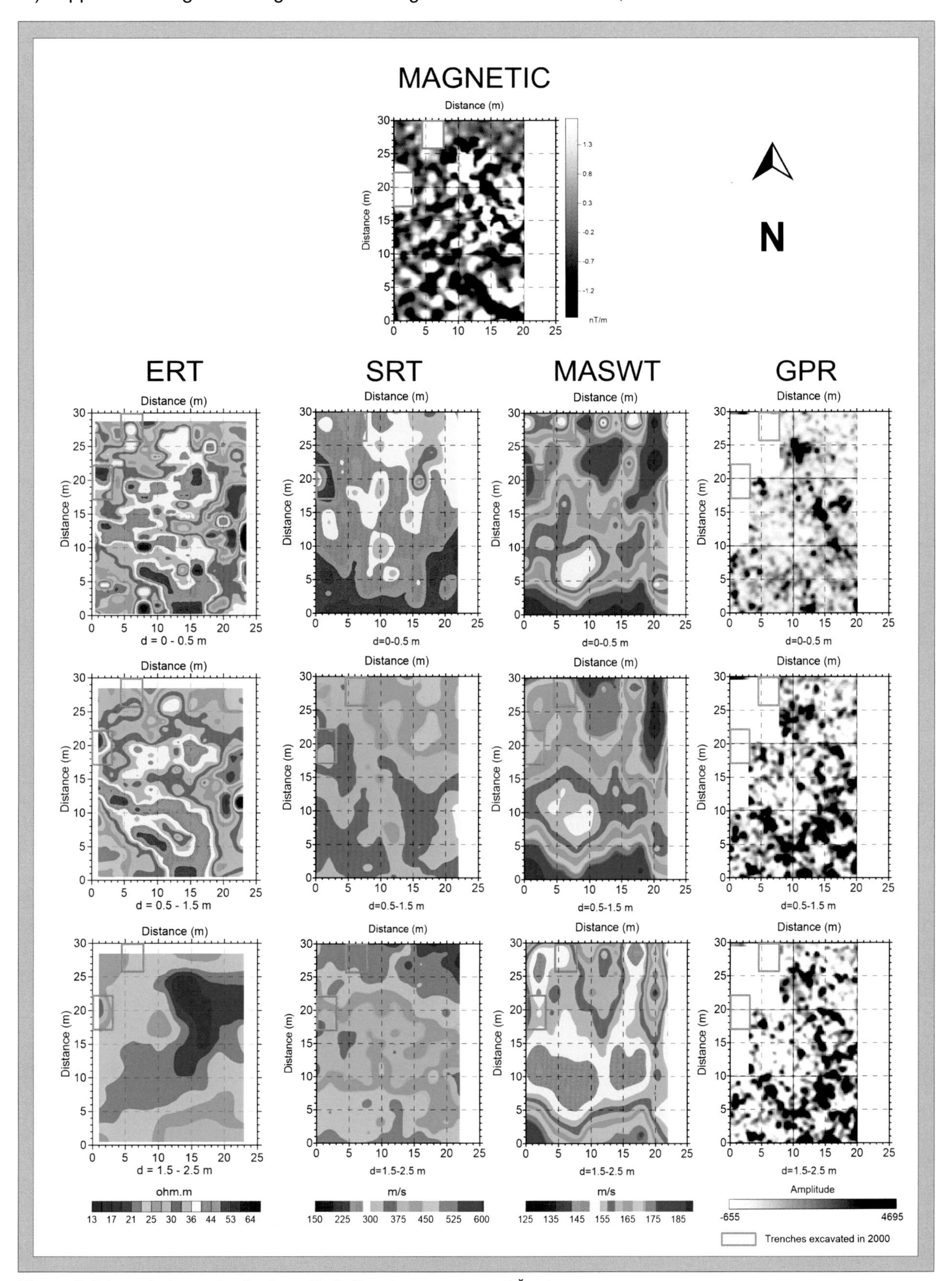

Figure 2: Integrated geophysical results in Tepelerarası area of Šapinuwa.

Ohm/m. The buried structures are oriented in NE-SW and NW-SE directions. After the archaeological structures, the resistivity values are suddenly decreased possibly due to the presence of clay materials, which might demonstrate the clay basement under the archaeological structures. SRT investigation results obtained from the same depth slices show the seismic P wave distribution in the area. In the first depth slice, the seismic P wave values are mostly very low (150-370 m/s). Particularly, the seismic P values are decreased against the southern part of the area. This feature reveals the dry subsurface soil conditions and the presence of high porosity and dispersed materials at shallow depth in this area. The P wave velocities in the second depth slice are slightly increased, and the P values are between 300-450 m/s. This increasing should be associated with the archaeological context in this depth. In the last depth slice, seismic P values are low (375-425 m/s) in the midpoint of the area, while the area enclosing the archaeological structures displayed high P wave velocities (450-600 m/s). MASWT proved useful and surprisingly successful in results. In the first depth slice of MASWT, the low S values are encountered with archaeological features clearly revealed by magnetic, ERT and GPR. The seismic S values of archaeological structures vary between 125 and 160 m/s, and the high S values (160-200 m/s) are enclosed in this area. In other depth slices, seismic S wave values and anomaly characteristics are generally similar to the first depth slice. This result revealed the sensitive and useful characteristics of this technique in archaeological site application.

Conclusions

The overall results addressed the usefulness of the integrated usage of geophysical techniques to determine the various characteristics of subsurface archaeological structures and context. The results highlighted the reliability of the integrated prospection approach, particularly during the interpretation stage involving various physical parameters, revealing different sensitivities and resolution characteristics regarding subsurface imaging. As a result, the geophysical prospection methods used enabled a better definition of the position, localisation, depth, thickness, extension, and physical characteristics of buried archaeological structures as well as the geological substratum at archaeological sites comparable to that of Šapinuwa.

Bibliography

Diamanti, N., Tsokas, G., Tsourlos, P. and Vafidis, A. (2005) Integrated interpretation of geophysical data in the archaeological site of Europos (northern Greece). *Archaeological Prospection* **12**: 79–91.

Drahor, M. G. (2006) Integrated geophysical studies in the upper part of Sardis archaeological site, Turkey. *Journal of Applied Geophysics* **59**: 205–223.

Drahor, M. G. (2011) A review of integrated geophysical investigations from archaeological and cultural sites under encroaching urbanisation in Izmir, Turkey. *Physics and Chemistry of the Earth* **36**: 1294–1309.

Drahor, M. G., Berge, M. A., Öztürk, C. and Ortan, B. (2015) Integrated geophysical investigations at a sacred Hittite Area in Central Anatolia, Turkey. *Near Surface Geophysics* **13**: 523-543.

Papadopoulos, N. G., Sarris, A., Salvi, M. C., Dederix, S., Soupios, P. and Dikmen, U. (2012) Rediscovering the small theatre and amphitheatre of ancient Ierapytna (SE Crete) by integrated geophysical methods. *Journal of Archaeological Science* **39**: 1960–1973.

Marine seismics along the Kane Peninsula

Annika Fediuk[1], Dennis Wilken[1], Tina Wunderlich [1] and Wolfgang Rabbel[1]

[1]Christian-Albrechts-University Kiel, Institue of Geosciences, Department Geophysics

annika@geophysik.uni-kiel.de

Introduction

The ancient harbour site Kane is situated 40 km to the ancient city of Pergamum, western Anatolia, which flourished during Hellenistic times (Radt 1999). Kane is located at the coastline of the Aegean sea on a promontory on modern Karadağ peninsula. The peninsula is shaped by the Karadağ mountains, volcanism and active seismicity (Brinkmann 1976). Historical sources prove that in 191/190 BC a Roman fleet overwintered in Kane (Pirson 2015, 2016). Therefore, Kane was considered to belong to Pergamum's harbour supply network of coastal settlements. In the research plan of the project Maritime Topography of the Ancient Kane Peninsula: A Micro-Regional Approach to the Impact of Harbours and Anchorages on Politics, Economy and Communication of a Western Anatolian Landscape, marine seismic prospection was performed on a bay flanking the 400 m by 150 m promontory in 2014. The Rome's Mediterranean Ports Advanced Grant project of the European Research Council funded the surveys. The objective was to investigate archaeological port structures, the navigability in the bay and to analyse the development of the ancient landscape.

Methodology

A marine seismic reflection system was used for the campaign. A rubber dinghy pulled a buoy carrying a piezoelectric transducer ("Pinger") as the seismic source, two hydrophones as receivers and a real time kinematic DGPS for positioning. Both seismic and DGPS Data revealed an accuracy within decimetre range. 143 profiles were recorded in 7 days, that is a 49km profile length covering an area of ~34 ha. The seismic data processing included band pass filtering, a signal deconvolution, a Normal Moveout correction and a 2D Stolt migration. The data enabled the production of a rough bathymetric map of the area by using the seafloor reflection, and the analysis of the stratigraphy of the Kane bay as well as the sedimentation rates since antiquity. Furthermore, a "breakwater", an elongated shoal in the north-western Kane bay was thoroughly imaged.

Results

From the bathymetric data, a possible box-shaped harbour basin with a size of 100 m x 50 m is evident. It is sheltered from wave motion by an eye-catching first breakwater in the north-western part of the bay. A second breakwater protects the harbour basin to the south. The first breakwater is formed by the local bedrock rising towards the coast (Fig. 1). Its bathymetry shows steep slopes towards the open sea and accumulations of rocks on its surface. At the edges of the breakwater there are also anthropogenic accumulations 3m below its surface. These data suggest that the breakwater is of geological origin and was reshaped by

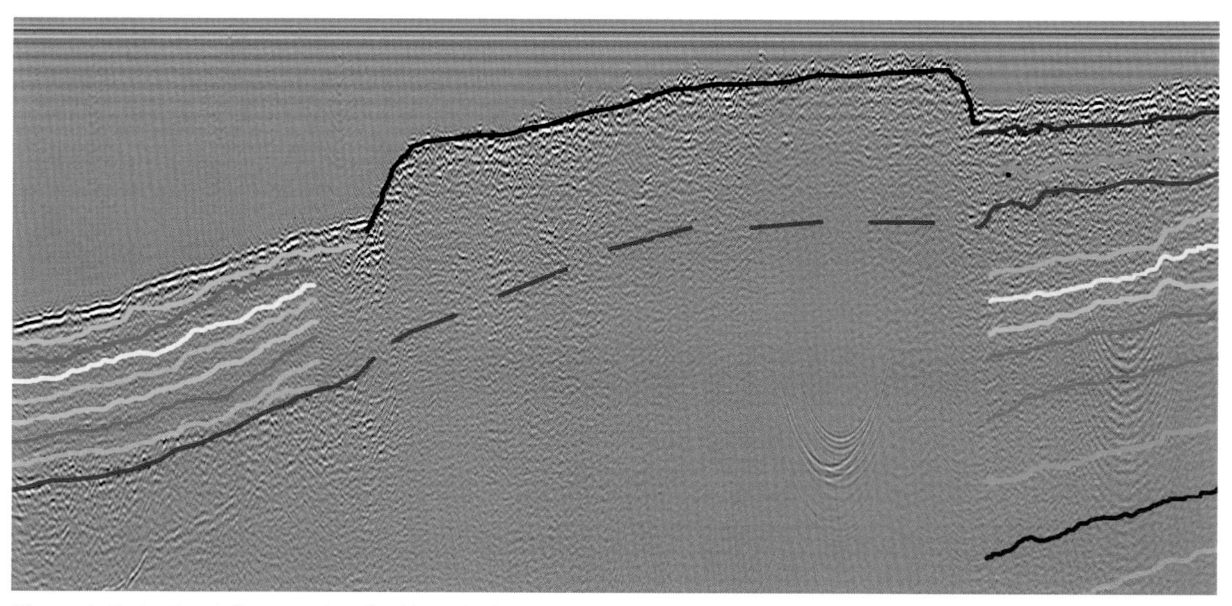

Figure 1: Seismic profile crossing the Kane harbour bay from West to East. Seismic horizons are highlighted in colour, the bedrock is coloured red. The stratigraphy of the deep basin and the harbour basin is evident. Due to the breakwater scattering most energy, its subsurface is transparent.

humans. In 2015, an ancient bollard was detected in the harbour basin. Based on reflection seismic interpretation, a minimum sedimentation rate of ca. 1 mm/year since antiquity was estimated close to the first breakwater.

The water depth of the Kane bay ranges from -0.6 to -13m NN. The seismic profiles show sedimentary layers also in the deeper parts of the Kane Bay (Fig. 1). The underlying bedrock is characterized by large-amplitude, chaotic reflection patterns. It is found in depths from -26 m NN to -3 m NN, dipping seaward with about 12°. The flexure of the basement can be regarded as evidence of faulting, possibly caused by volcanic uplift.

Conclusion

Bathymetric and seismic reflection measurements brought evidence that the shoal located in the Kane bay is likely to represent a breakwater of the ancient harbour of Kane. The base of breakwater is of geological origin and shows signs of anthropogenic reworking. Seismic reflection sections show the sedimentation history of the Kane Bay including faulting by small tectonic events. However, no sub-bottom constructions could be detected because the possible construction material does not show seismic contrasts to the surrounding boulders or bedrock.

Bibliography

Radt, W. (1999) *Pergamon - Geschichte und Bauten einer antiken Metropole*. Darmstadt: Wissenschaftliche Buchgesellschaft.

Brinkmann, R. (1976) *Geology of Turkey*. Stuttgart: Ferndinand Enke Verlag.

Pirson, F. (2015) Der neue Survey auf der Kane-Halbinsel („Kane Regional Harbour Survey"). *Archäologischer Anzeiger* **2015***(2):* 139-150.

Pirson, F. (2016) Der neue Survey auf der Kane-Halbinsel („Kane Regional Harbour Survey"). *Archäologischer Anzeiger* **2016***(2)*: in press.

Acknowledgements

The authors gratefully acknowledge funding from the European Research Council's project "Rome's Mediterranean Ports Advanced Grant", with Simon Keay (Southhampton) as principal investigator, and support from the Deutsches Archäologisches Institut (DAI) Istanbul, especially Prof. Dr. Felix Pirson (DAI) and the Kane survey team from 2014-2015. Special thanks go to Clemens Mohr and Detlef Schulte-Kortnack for constructing and maintaining the seismic acquisition systems.

Out of the blue: exploring Lost Frontiers in Doggerland

Simon Fitch[(1)]

[(1)]School of Archaeological Sciences, University of Bradford, BD7 1DP. UK

S.Fitch@Bradford.ac.uk

Introduction

The North Sea has long been known by archaeologists as an area of Mesolithic occupation, and in recent years there has been a growing appreciation of the archaeological potential of the coastal shelf. Yet, due to the submergence of this landscape the area remains difficult for archaeologists to explore, and the nature of its occupation remains tantalisingly elusive. However, due to its submergence, this region now contains one of the most detailed and comprehensive records of the Late Quaternary and Holocene, and its preserved sedimentary successions represent a mine of information that remains largely untapped by archaeologists. However, the lack of detailed data pertaining to this region results in our knowledge of the distribution of deposits of archaeological interest being at best patchy and limited to areas of good data coverage.

This paper will present some of the emerging results of the Europe's Lost Frontiers project (Gaffney *et al.* 2017), that demonstrate that the utilisation of spatially extensive oil industry data allows the recovery of information pertaining to the actual Mesolithic landscape of the North Sea and that extensive mapping of the region is now possible. This information not only reveals the diversity of this landscape but that it may have been longer lived than previously realised. Whilst the study of such landscapes is still at an early stage, the information provided by these studies offers the possibility of transforming how we interpret traditional terrestrial data and their relationship to the larger European Mesolithic.

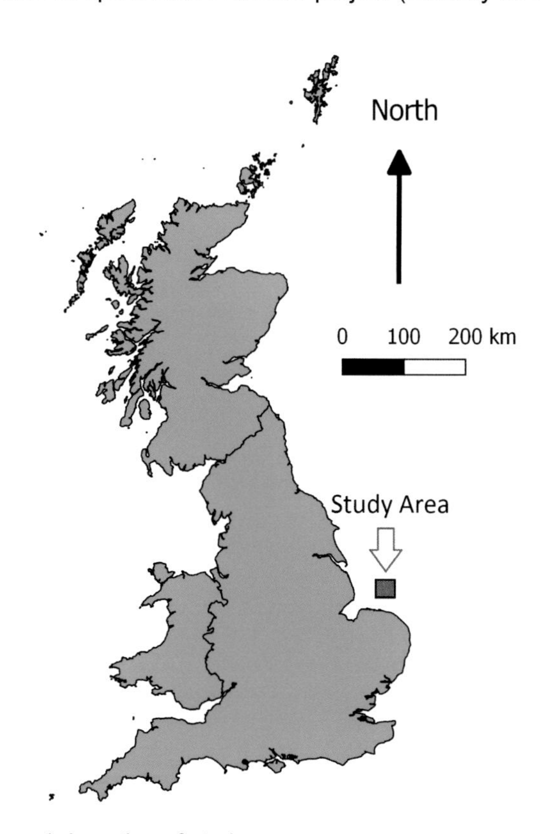

Figure 1: Location of study area

Methods

In this paper, a focus area of 20 x 30 km has been chosen to illustrate the work currently being undertaken within this region. The study is located 32km from the east coast of the UK landmass (Figure 1). The main source of data for this paper is a series of seismic lines and multibeam sonar data which was acquired prior to the project for the purposes of environmental characterisation.

The initial visualisation of this data was achieved with the IHS Kingdom suite, using the techniques illustrated in Fitch *et al.* 2011. Through the combination of seismic data and sonar (in the form of a surface) information was provisioned on the sedimentology and geomorphology of the landscape, and this enabled the location of features of interest by the project team. This analysis permitted a range of Holocene landscape features to be identified within the area and Figure 2 illustrates one of these; a fluvial feature that was active during the Mesolithic period. In conjunction with geophysical work, additional non-geophysical datasets (e.g. core databases or geological mapping) were integrated within the interpretation system. Cross correlation with these supported the interpretation of otherwise problematic features. Such data fusion assisted in producing a detailed understanding of the areas likely to yield traditional palaeoenvironmetal data or SedaDNA, which were subsequently targeted for sampling. In 2016 these targets were successfully sampled and the information derived was fed back into the interpretation to guide additional coring which is currently ongoing (2017).

Discussion

At a basic level, this research has shown that is possible to use a combination of datasets to enhance material recovery during sampling. This alone is a significant result when considering the cost and difficulty of recovering such material from marine environments.

The considerable detail provided by this work is very significant and provides the ability to recover

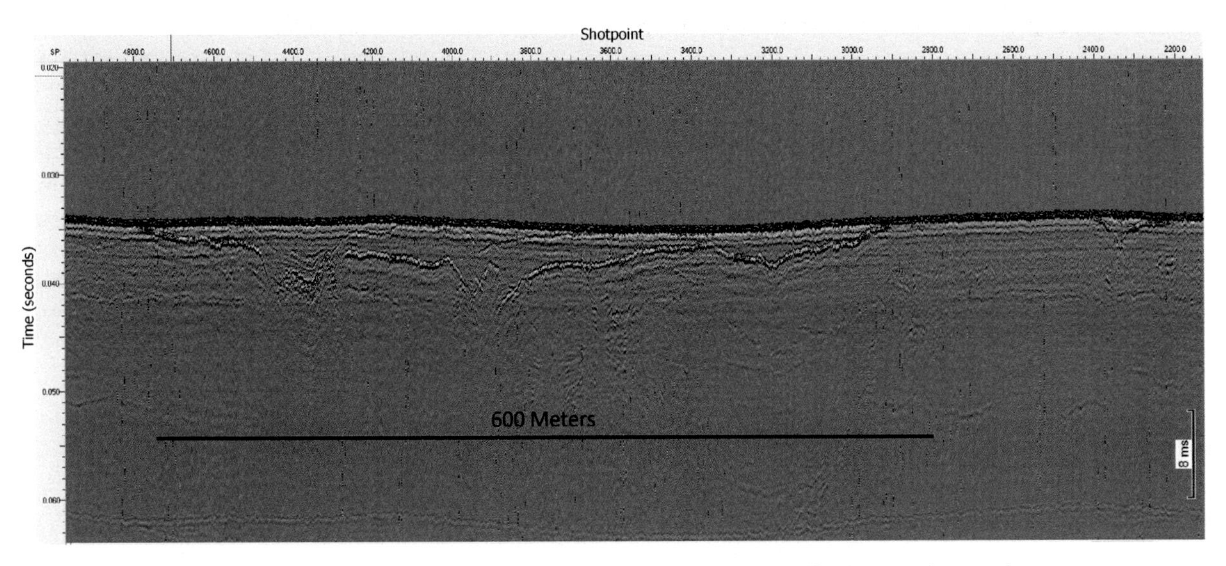

Figure 2: Fluvial feature identified in a Holocene landscape through analysis of seismins data and sonar.

and understand palaeoenvironmetal material in its landscape context. The data reveals a diverse landscape with many phases of activity during the Early Holocene as the landscape responded to changing climatic conditions. Additionally, the results also records evidence on the nature of the landscapes final submergence. Such information is important when increasing evidence shows that local sea level curves are critical in understanding landscape flooding and human activity in such a landscape (Tappin et al. 2011), and yet research suggests that regional sea level curves may be of limited utility in such situations (Kiden et al. 2002). However, the data provided here resolves this situation to some extent and allows us to determine the palaeoenvironmental potential of an otherwise inaccessible, and archaeologically unexplored, area

Ultimately, it is inevitable that researchers will wish to locate archaeological sites within this region to answer complex questions on topics such as migration, culture and population levels. To achieve this, the improvements outlined here offer the opportunity to provide data that will support modelling at a resolution that would have previously been unthinkable. This future work will represent a step change in the exploration of this smaller region and submerged landscapes worldwide.

References

Fitch, S., Gaffney, V., Gearey, B. and Ramsey, E. 2011. *Between the Lines- enhancing methodologies for the exploration of extensive inundated palaeolandscapes.* In: Cowley D. (ed.) Remote Sensing for Archaeological Heritage Management (EAC Occasional Paper No.5). EAC (Brussels).

Gaffney V., Allaby R., Bates R., Bates M., Ch'ng E., Fitch S., Garwood P., Momber G., Murgatroyd P., Pallen M., Ramsey E., Smith D., Smith O. (2017). Doggerland and the Lost Frontiers Project (2015-2020). In: Geoff Bailey G., Harff, J. and Sakellariou D. (eds.) *Under the Sea: Archaeology and Palaeolandscapes of the Continental Shelf.* Springer Press.

Kiden, P., Denys, L. and Johnston, P. 2002: Late Quaternary sea-level change and isostatic and tectonic land movements along the Belgian-Dutch North Sea coast: geological data and model results. *Journal of Quaternary Science*. 17, 535-46

Tappin, D. R., Pearce, B., Fitch, S., Dove, D., Gearey, B., Hill, J. M., Chambers, C., Bates, R., Pinnion, J., Diaz Doce, D., Green, M., Gallyot, J., Georgiou, L., Brutto, D., Marzialetti, S., Hopla, E., Ramsay, E. and Fielding, H. (2011) *The Humber Regional Environmental Characterisation. Marine Aggregate Levy Sustainability Fund*, 345pp. (OR/10/054)

Investigation and virtual visualisation of a probable burial mound and later motte-and-bailey castle from Lower Austria

Roland Filzwieser[(1)], Leopold Toriser[(1)], Juan Torrejón Valdelomar[(1)] and Wolfgang Neubauer[(1)]

[(1)]Ludwig Boltzmann Institute for Archaeological Prospection and Virtual Archaeology, Vienna, Austria

roland.filzwieser@archpro.lbg.ac.at

Research history and historical sources

The mound called *Runder Berg* in the county of Lower Austria, about 200m southeast of the Austrian-Czech border, was first investigated in the 1880s by Josef Szombathy, who is also known for his discovery of the Venus of Willendorf. In autumn 1887, a trench was dug through the mound by the owner, baron Suttner. In April 1888, Szombathy extended the trench, leaving a detailed documentation in his diary (Szombathy 1888). The trench is described as being 6m wide in the southeast and 4m wide in the northwest. The trench expands to a width of 9m in the centre of the 5m high, 42m diameter *tumulus*. The base of the mound consisted of two layers of compressed soil. On the surface of the lower layer, glazed sherds of two to three pots were found (Fig. 1, a). Medieval pottery was also collected in the surrounding field.

The structure is already marked as *large mound* on a historical map from 1711, which is drawn by Johann Jakob Marinoni – an 18th century Austrian astronomer and mathematician of the imperial court. Marinoni includes the mound in the south-eastern corner of an approximately 300m x 100m large enclosed area, in which he notes, "*Here the village of Grafenwasen" is said to have stood* (Fig. 2).

Also, the historian and topographer Franz Xaver Schweickhardt wrote in 1834 that at the foot of the mound traces of building foundations were repeatedly found. In this, he saw proof for a legend about a deserted village, destroyed by Swedish troops, that should have been located at that very spot. In contrast to Marinoni, however, the name of the settlement is not known to him anymore (Schweickhardt 1834).

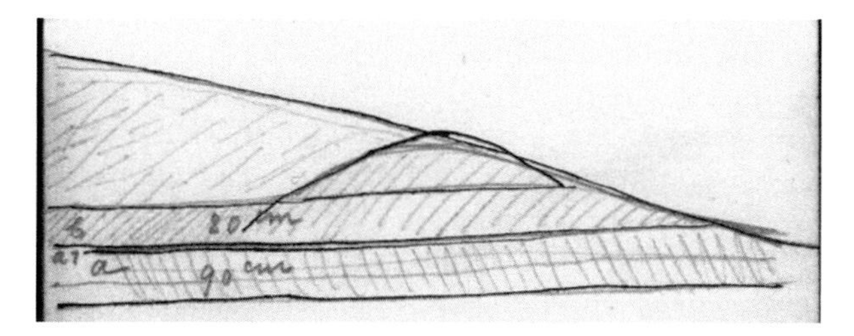

Figure 1: Section of the excavation from Szombathy´s diary (Szombathy 1888).

Figure 2: Historic map, drawn by Marinoni in 1711, depicting the mound and the position of the former village *Grafenwasen,* in superposition to a DTM.

In close vicinity to the mound, two further grave mounds are depicted on Marinoni´s map. One of those, the so-called *Schmalzberg*, was excavated in the 1980s. The archaeologists dated the probable Lombard grave mound to the 6th century AD. Its foundation was constructed by radially placed stones on which grass sods were situated. However, the grave itself had been robbed completely (Neugebauer and Neugebauer 1988).

Geophysical Prospection and Interpretation

In 2015 and 2016, the Ludwig Boltzmann Institute for Archaeological Prospection and Virtual Archaeology (LBI ArchPro) conducted a large-scale motorised geophysical prospection survey (Trinks *et al.* 2015) in this area close to the Czech border, where a wide variety of buried archaeological structures were discovered and mapped. In the course of this survey, the mound and its surrounding area were also measured through the application of magnetometry and ground-penetrating radar (GPR). In the magnetic data (Fig. 3), many large anomalies with a diameter up to 4m can be seen and can most likely be interpreted as the remains of prehistoric pit houses. Between the larger pits, many posthole-like anomalies were observed that could either also belong to the prehistoric settlement, or be of later origin. Most of these features are encircled by a slope on the western side, leading down to the river and a 3m wide segmented ditch to the north and east, with gaps every 20-50m – probably dating to the Neolithic period. Quite prominently, in the centre between this ditch and the settlement traces, a circular ditch can be observed with a width of about 4m surrounding the mound. The ditch is intersected in the northeast,

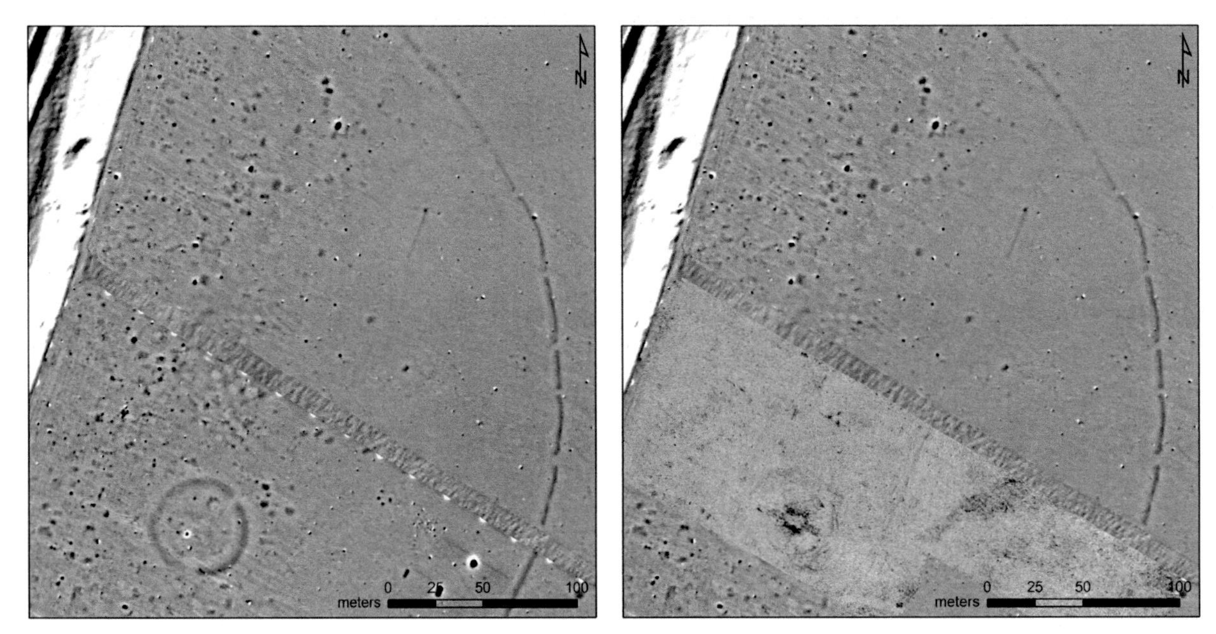

Figure 3: Left: magnetic data (-6 nT white clip-off to +6 nT black clip-off) with the circular ditch and a clear intersection at its northeast. Right: superimposed GPR depth-slice (0.40 m – 0.45 m), depicting the extent of the 1888 excavation trench as well as the southern part of the faintly visible village-boundary ditch.

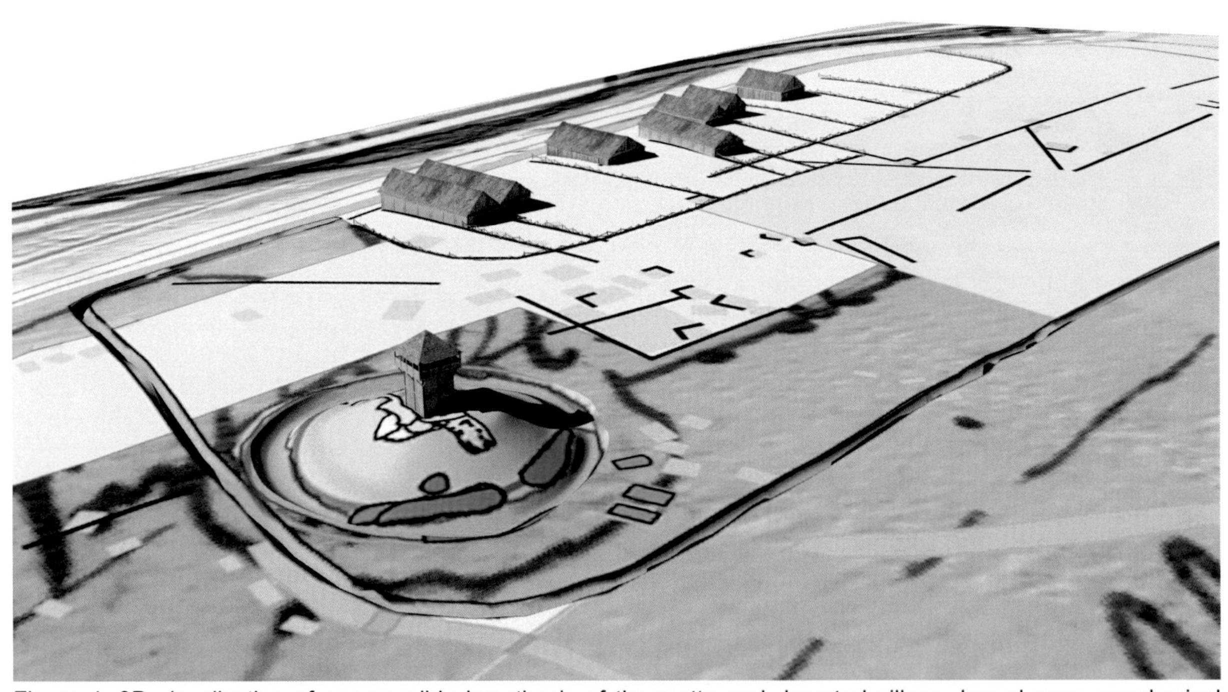

Figure 4: 3D visualisation of one possible hypothesis of the motte and deserted village, based upon geophysical prospection and historical sources. The 3D models are placed on top of a DTM, which is textured with a map showing both, historical map and interpretation of the geophysical data.

but no clear internal structures of the mound can be made out. However, another faint ditch can be seen in the data (although not over its entire length) that also surrounds the other magnetic anomalies, including the mound. This is strongly reminiscent of the course of the line drawn by Marinoni that marked the position of the deserted village.

With ground-penetrating radar, so far it was only possible to survey the southern field, where the mound is located. The main portion of the suspected village to the north could not be measured. Based on the existing data, however, the ditch can be followed further to the south and east of the mound. In addition to the circular ditch, several strongly reflective deposits can also be observed beneath the mound. The GPR data reveals the surrounding ditch, which shows some similarities with other known mottes surveyed with GPR (Verdonck 2012).

The radar also shows the internal structure of the mound, which corresponds very well with the 19th century excavation trench.

Discussion

The integrated interpretation of the geophysical data sets, together with the historical map and the written sources, lead the authors to believe that the fainter enclosing ditch and some of the postholes may derive from the village that also might have included a small motte-and-bailey castle, which would have been situated on the mound. Whether the mound was erected exclusively for the defensive structure or if it is rather to be seen as a reused prehistoric grave mound, could not be decided unambiguously. Szombathy only found a bronze needle as well as a blue glass bracelet, yet this does not seem to have convinced him that it was a *tumulus*. Furthermore, the term *tumulus* could also have referred to a motte during this time period (Schad´n 1950). However, the compressed soil *trampled compactly* with glazed sherds on its surface, indicates a later use as a motte. Another argument for this interpretation is the name *Grafenwasen* itself, as the word *wasen* is etymologically related to the German word *Rasen* and can also refer to a motte (Felgenhauer-Schmiedt 2007). It seems plausible that, while in 1711, roughly 70 years after the Swedes ravaged the country, some people still remembered the name *Grafenwasen*. While another 123 years later when Schweickhardt made inquiries, this knowledge was already lost. However, all this information together led to the here-presented visualisation of the settlement, as one possible interpretation of all the gathered data.

This virtual visualisation (Fig. 4) is based upon a rather limited set of sources and, consequently, cannot be taken as a final hypothesis, but instead more as a plausible depiction of the site. Nevertheless, it has helped the researchers to better understand the site and to reconsider the datasets provided by the geophysical surveys carried out in the area. Since all the available sources have been included and described, a fair level of scientific transparency concerning the virtual visualisation can be assured.

Bibliography

Felgenhauer-Schmiedt, S. (2007) Hausberge im Niederösterreichischen Weinviertel. *Beiträge zur Mittelalterarchäologie in Österreich*, **23**: 163–180.

Neugebauer, C. and Neugebauer, J. W. (1988) KG Neudorf. *Fundberichte aus Österreich (FÖ)*, **24/25**: 331–333.

Schad´n, H. (1950) *Die Hausberge und verwandten Wehranlagen in Niederösterreich - Ein Beitrag zur Geschichte des mittelalterlichen Befestigungswesens und seiner Entwicklung vom Ringwall bis zur Mauerburg und Stadtumwehrung. Mitteilungen der Anthropologischen Gesellschaft in Wien* (Vol. 80).

Schweickhardt, F. X. (1834) Mitterhof. In *Darstellung des Erzherzogthums Oesterreich unter der Ens. Bd.4 Viertel unterm Manhartsberg.*

Szombathy, J. (1888) Tagebuch von Josef Szombathy. Wien: Fundaktenarchiv der Prähistorischen Abteilung des Naturhistorischen Museums Wien.

Trinks, I., Neubauer, W., Doneus, M., Hinterleitner, A., Doneus, N., Verhoeven, G., Löcker, K., Kucera, M., Nau, E., Wallner, M. and Seren, S. (2015) Interdisciplinary archaeological prospection at unprecented scale and resolution. The first five years of the LBI ArchPro Research Initiative 2010-2015. *Archaeologia Polona*, **53**: 144–147.

Verdonck, L. (2012) *High-Resolution Ground-Penetrating Radar Prospection with a Modular Configuration. Potential for the Detailed Imaging of Buried Archaeological Remains.* PhD Ghent University.

Archaeologicla validation of geophysical data: risks of the archaeological interpretation

Ekhine Garcia-Garcia[1, 2], Antonietta Lerz[3], Roger Sala[4], Arantza Aranburu[2], Julian Hill[3] and Juantxo Agirre-Mauleon[1]

[1]ARANZADI Society of Science, Donostia, Basque Country; [2]Euskal Herriko Unibertsitatea (UPV/EHU), Mineralogia eta Petrologia, Leioa, Basque Country; [3]Museum of London Archaeology (MOLA), London, United Kingdom; [4]SOT Prospecció Arqueològica, Barcelona, Catalonia

ekhinegarcia@yahoo.com

Introduction

The archaeological interpretation of the anomalies detected by geophysical systems is one of the challenges of archaeological geophysics. Indeed, many scenarios can generate a similar geophysical response, and it is the experience of the geophysical team, together with the knowledge of the site's characteristics, which is used to make the final interpretation.

The validation through excavation leads to three possibilities: the interpretation was correct, it was wrong but one can explain why, or it was wrong and one cannot even explain why.

The objective of this presentation is to stimulate the interest in validating the archaeological interpretations based on geophysical data. Results of a Roman site in Navarre will be used to illustrate examples of the three possible scenarios mentioned above.

Previous Research

The Aranzadi Science Society has been engaged in a project to delimit and characterize a newly-discovered Roman city in Auritz/Burguete (Navarra). Since the discovery, the geophysical techniques have been applied as the main methodology to detect and describe the archaeological remains.

The magnetic method was used to make a first assessment and select the areas of interest. Data acquisition was performed using Bartington G-601-dual fluxgate gradiometer. Measurements were taken every 0.25 m along the traverses, 0.5 m apart, and reading in zigzag mode, using marked measurement tapes for positioning. The magnetic contrast was good enough to successfully describe the main remains of the central area, consisting of an urban occupation of circa 4.5 ha arranged along the road (Garcia-Garcia *et al.* 2016). Therefore, efforts focused on areas where the magnetic results were not satisfactory.

After some tests, GPR was chosen as the main complementary technique. Data acquisition was performed using an IDS Hi-Mod instrument with two multi-frequency antennas (200 MHz and 600 MHz). Measurements were taken every 2.5 cm along parallel traverses, 0.2 m apart, and in zigzag mode. Two of the surveyed areas are presented here, corresponding to the areas where excavations were later performed.

Area 1 was suspected to have a singular function within the settlement. Because of the weak magnetic contrast in some of the features (Fig. 1), the goal was to have complementary information. The concordance between the two techniques was

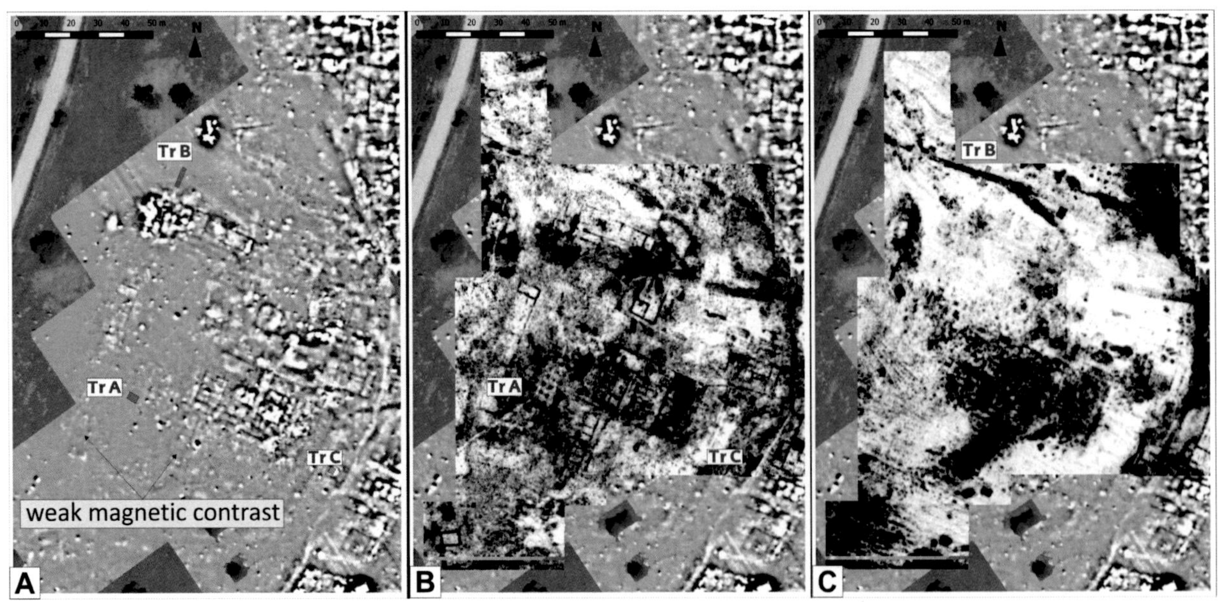

Figure 1: Geophysical results on area 1 and excavation trenches. A) Gradiometer response (-10nT blue, 10nT white). B) GPR amplitude map at 0.62-0.78m (v=8cm/ns, high amplitudes in black). C) GPR amplitude map at 1.11-1.27m (v=8cm/ns high amplitudes in black).

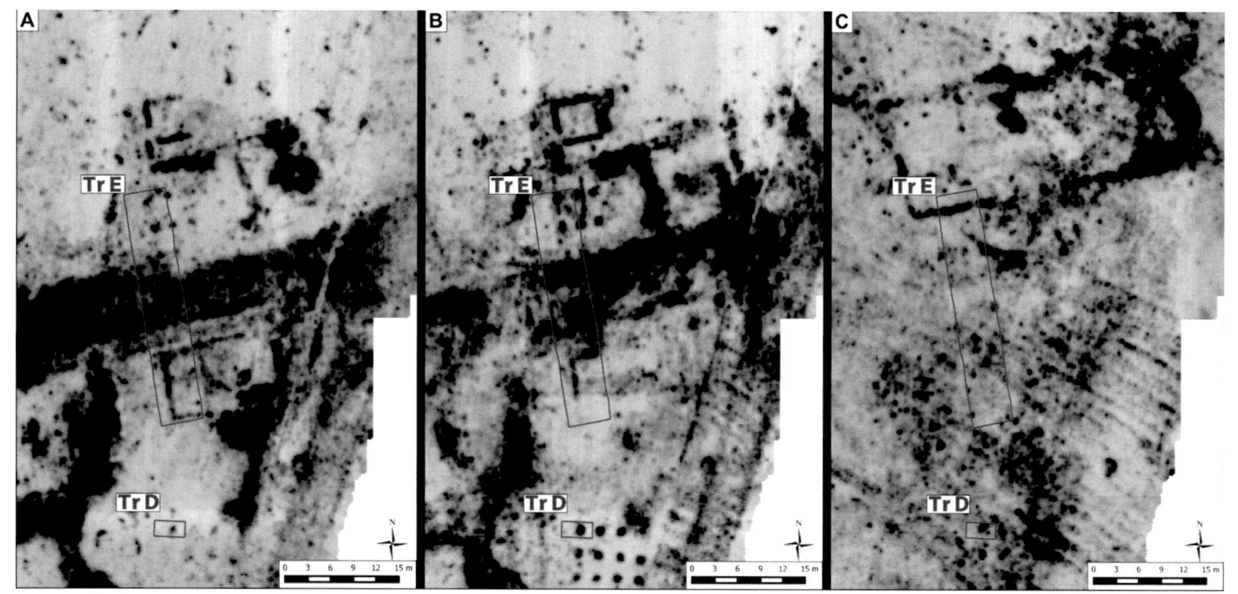

Figure 2: GPR results on area 2 and excavation trenches. A) GPR amplitude map at 0.28-0.40 m. B) GPR amplitude map at 0.48-0.59m. C) GPR amplitude map at 0.85-0.97 m. High amplitudes in black, v=8cm/ns.

high, but new features have been detected (Garcia-Garcia *et al.* 2015). In particular, the circular anomalies in the centre have been interpreted as the columns of a temple. Trench A was placed on one of those circular anomalies. In the south-east corner, results show a circular anomaly of positive magnetic contrast. The GPR results was in agreement with a circular feature filled by homogeneous sediments. Trench C was located over this anomaly. Trench B was located to verify a linear anomaly detected in the north, attributed to a water canalization.

In the area 2 the magnetic survey did not detect any significant anomaly. A GPR survey was performed to locate the Roman road which was known to be there. The road was clearly visible on amplitude maps and unexpected archaeological remains have been detected next to it (Fig. 2). Trench D was conceived to verify one of the circular anomalies. Trench E was positioned along the Roman road at a point where it is flanked by two different types of structures.

Excavation and Comparison

Excavations were performed in 2015 and 2016 by a team from Aranzadi Society of Science and from Museum of London Archaeology, MOLA (Harrison *et al.* 2015; Lerz *et al.* 2017).

Trenches A and D showed the circular anomalies as expected, but only the basement levels were conserved. Any clear information about the typology of the columns was not obtained, but it is suspected that they could be made in wood.

In trench B, a constructed water canalization was expected. Instead, the excavation revealed a superposition of different soils that filled a cut in the bedrock. The gravel level detected at the bottom was coincident with the reflective anomaly observed in the amplitude maps.

Excavation of trench C revealed a stone-walled structure in the location of the GPR anomaly and it soon became apparent that is was a furnace or kiln, rather than a well. The large quantities of metal-working waste, or slag, recovered from the associated deposits, suggest it was for iron working.

The excavation of trench E targeted the expected road and two of the buildings on either side of it. However, the linear feature predicted by the geophysical survey at the north end of the trench (see Figure 2, C) was not identified during the excavation.

Conclusion

The agreement between the geophysical data and excavation was high. In trenches A, D and E, the lack of magnetic contrast can be explained by the thin archaeological layers and the low susceptibility contrast of the remains. The archaeological expectation was however frustrated and only the basement of the columns was found.

In the case of the trench C, the weak magnetic contrast led to attribute this structure to a well filled with sediment. Instead, heat affected sediments had been found. The feature was not completely excavated but it could be for iron working. The magnetic contrast, however, is weaker than expected for such kind of feature.

In trench E the interpretation of the main anomalies was right. However, one of the expected walls was not found. The excavation did not identify the variations that caused the geophysical anomaly. Indeed, some stones appeared in the expected location, but they were not considered archaeologically relevant. Following rain during the night, a clear distinction in the deposits was visible in the section.

Summarizing, despite the good correspondence between geophysics and excavation we did not find what was expected. One of the reasons is that the interpretation was biased by our desires and perspective, and we did not use all of the information we had available to our interpretations. The unidentified wall of trench E illustrates that, in some cases, the variations causing the geophysical anomalies are not identified in excavation.

This demonstrates again the risks on the archaeological interpretation of the geophysical data, and the importance of the verification trenches.

Bibliography

Garcia-Garcia, E., Agirre-Mauleon, J., Aranburu, A., Arrazola, H., Hill, J., Etxegoien, J., Mtz. Txoperena, J. M., Rauxloh, P., Sala, R. and Zubiria, R. (2015) The Roman settlement at Auritz (Navarre): preliminary results of a multi-system approach to asses the functionality of a singular area. *Archaeologia Polona* **53**: 92-94.

Garcia-Garcia, E., Mtz. Txoperena, J. M., Sala, R., Aranburu, A. and Agirre-Mauleon, J. (2016) Magnetometer Survey at the Newly-discovered Roman City of Auritz/Burguete (Navarre). Results and Preliminary Archaeological Interpretation. *Archaeological Prospection* **23**(4): 243-256.

Harrison, D., Hill, J., Lerz, A. and Rauxloh, P. (2015) *Iturissa Roman Town. Report on an Archaeological Evaluation. December 2015.* London: Museum Of London Archaeology. Unpublished technical report.

Lerz, A.and Hill, J. (2017) *Iturissa Roman Town. Report on an Archaeological Evaluation. January 2017.* London: Museum Of London Archaeology. Unpublished technical report.

The planning of Daskyleion (Turkey), the Achaemenid capital of the Hellespontine Phrygia: report on three survey campaigns (2014-2016)

Sébastien Gondet[1]

[1]Univ Lyon, CNRS-Université Lyon 2, Maison de l'Orient et de la Méditerranée, UMR 5133 Archéorient

sebastien.gondet@mom.fr

Framework

The surveys at Daskyleion were carried out in the frame of a Turkish archaeological project jointly managed by Professor Dr. Kaan Iren (Muğla University) and by Doç. Dr. Sedef Çokay Kepçe (Istanbul University). These fieldworks relied upon the funding support of the Soudavar Memorial Foundation. The geophysical works were implemented in collaboration and thanks to the instrumental support of the Archéorient (UMR 5133, CNRS/Lyon 2 University) and of the Metis (UMR 7619, Sorbonne Université) teams.

Site Presentation

Daskyleion is located to the north-west of the Anatolian peninsula, 25km south of the Marmara Sea coast. It lies on the south shore of the shallow 160km² Manyas Lake. This city was founded at the eastern end of the large alluvial fan of the Kocaçay River that is the main stream of the lake drainage area. This alluvial plain is surrounded by low hills mostly sedimentary with intrusions of volcanic rocks like andesite. A part of the site is located on the top of one of these hills called Hisartepe 20 m above the plain and extending over 4 ha.

During the Achaemenid period (ca. 550-330 BC), Anatolia soon came under the dominion of the Persian Empire. The peninsula was divided into several administrative provinces, with the capitals located in still extant important cities. Daskyleion was already a significant centre during the first half of the 1st millennium BC and became the capital of the Hellespontic Phrygia province. The later periods corresponded with a decreasing importance of the site and the location of Daskyleion was forgotten for a while. Nevertheless, the classical authors have perpetuated its memory as a brilliant Persian capital. They emphasize it as a famous paradise, i.e. a large park connected to the nearby game-rich lake and encompassing the city and its palaces.

Since the rediscovery of the site and the beginning of the excavations during the 1950s, the successive Turkish archaeological teams bring to light its archaeological history (Bakır 2011). The excavations focus on the Hisartepe hill. They have proved that Achaemenid impact on the city was important. Recent results have suggested a deep reshaping of the urban planning during the Achaemenid period (Iren 2010). It has also been supposed that the Persian urbanized area extended far beyond Hisartepe towards the surrounding hillslopes, maybe as far as a cluster of funeral tumuli located 2.5km to the south.

Objectives

Our survey of Daskyleion is part of a wider project for a comparative archaeological study of the ancient Achaemenid capitals based on results obtained in southern Iran. In this region, the centre of the Empire, we have widened the scale of study in sites like Pasargadae and Persepolis thanks to the use of several complementary survey methods, especially geophysics. The results have enabled us to suggest that these Achaemenid cities corresponded to open cityscapes where green areas took a prominent place (Boucharlat *et al.* 2012). Taking into account

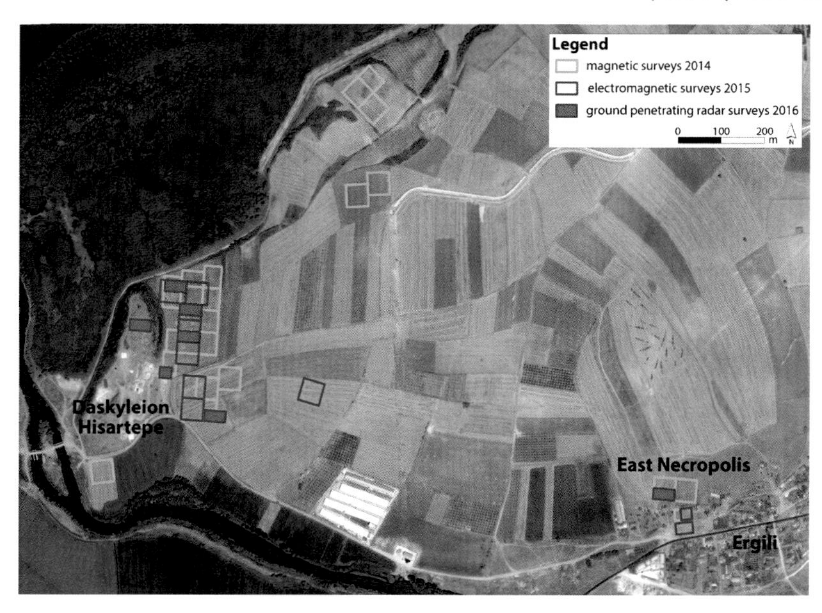

Figure 1: Location map of the surveyed areas over Daskyleion arranged by geophysical method (base satellite image: Google Earth).

Figure 2: Magnetogram east of the Hisartepe hill (dynamics -7/+7 nT/m).

the classical descriptions of Daskyleion, we might face a similar open urban layout showing a possible diffusion towards Anatolia of this new way to conceive city. Then to draw comparisons we have implemented a similar strategy in Daskyleion.

The main objective of our survey program in Daskyleion is to map the Hisartepe hill surroundings to reveal the extent as well as the general layout of the city. We also intend to approach the challenging question of revealing the garden areas. We have focused our efforts on the geophysical methods by implementing concurrent fieldwalking surveys. The first three campaigns (2012-2014) were mainly dedicated to test the efficiency of several geophysical methods for revealing the Daskyleion cityscape as well as the potential of different archaeological contexts on places strategically distributed over the site.

Results overview

During the survey campaigns we tested three different methods: magnetometry (2014), EM survey (2015), and GPR (2016). In this paper we will focus on results from the magnetic and EM surveys as GPR data are still under process (see Fig. 1 for the location of the surveyed areas).

The magnetic surveys were implemented by using a Geometrics G858 caesium gradiometer. We chose to start the survey tests with this method as it enables us to cover and evaluate large surfaces in the time-restricted frame of this first campaign. We focused a main part of our efforts on a small valley east of the Hisartepe hill. The resulting magnetogram shows disturbed results (Fig. 2). As a matter of fact, the andesite stone was intensively used for building the several constructions. The available data from previous excavations showed

Figure 3: On left: in-phase (apparent magnetic susceptibility) survey northeast of the Hisartepe hill in HCP configuration and 1.18 m coils distance (dynamics -1/2 ppt). On right: interpretative sketch map.

that the size of these andesite structures would have been imposing and, under a ploughed layer of almost 50 cm, quite well-preserved. This brought us to think that the magnetic method would have been able to reveal them. However, the scattered andesite fragments over the surveyed area, coming from the destruction of the buildings, induced important disturbances. The magnetograms have allowed us to outline several zones with high density of magnetic materials, certainly heaps of andesite blocks linked to ancient buildings.

Taking these results into account, we carried out surveys with the EM GF Instrument CMD Mini-Explorer. By focusing on the in-phase signal related to the soil magnetic susceptibility we hoped to reduce the effects of the widespread andesite blocks that have high remnant magnetism. In this respect the in-phase maps obtained in HCP configuration and for the maximum 1.18 m coils distance have provided us with promising results (Fig. 3). For the test area; the valley to the east, the map reveals the outlines of at least three partly preserved buildings at the same location where the magnetogram shows unshaped clusters of strong positive anomalies. In between these buildings sections of linear features may correspond to preserved parts of a drainage system. However, the most interesting result concerns the northwest part of the surveyed area. Here one of the revealed buildings is located in an area, triangular in shape, where the in-phase responses are lower than further east. An archaeological reading would be that the occupation was less dense and intensive in this area. The challenging question is to assess if it would be a hint for a real ancient zoning in the surveyed area and a division into, to the northwest, a green area encompassing an isolated building resulting in soils of lower magnetic susceptibility, and an active settled sector of the city to the east. This would be a promising result considering that the scope of our surveys in Daskyleion is to reveal a cityscape where gardens could have taken a prominent place.

Bibliography

Bakır, T. (2011) *Daskyleion*. Balıkesir: Balıkesir Valiliği.

Boucharlat, R., De Schacht, T. and Gondet, S. (2012) Surface Reconnaissance in the Persepolis Plain (2005-2008). New Data on the City Organisation and Landscape Management. In: G. P. Basello and A. Rossi (eds.) *Persepolis and his Settlements* (= Università di Napoli "L'Orientale" - Series Minor 77). Napoli: Università di Napoli "L'Orientale", 123-166.

Iren K. (2010) A New Discovery in Dascylium: the Persian Destruction Layer. In Matthiae P., Pinnock F., Nigro L., Marchetti N. (eds.), *Proceedings of the 6th International Congress on the Archaeology of the Ancient Near East - Vol. 2*. Weisbaden: Harrasowitz Verlag, 249-263.

Automation, automation, automation: a novel approach to improving the pre-excavation detection of inhumations

Ashely Green[(1)], Paul Cheetham[(1)] and Timothy Darvill[(1)]

[(1)]Department of Archaeology, Anthropology and Forensic Science, Faculty of Science and Technology, Bournemouth University, Poole, Dorset, UK BH12 5BB

agreen@bournemouth.ac.uk

As the use of mechanised systems and large area surveys continues to increase in commercial and research geophysics, there is still a markedly low ability to geophysically detect archaeological and, in some instances, forensic inhumations. This ongoing research project aims to produce automatic feature detection software with semi-automatic capabilities which will enhance grave-like responses in a range of datasets, thereby increasing the confidence of obtaining true positive results in both small-scale and landscape surveys. From this, it will also be possible to create an *ad hoc* workflow model to demonstrate appropriate survey methodologies, in addition to current guidelines (David *et al.* 2008, Schmidt *et al.* 2015, Bonsall *et al.* 2014), for achieving maximum potential to detect typical medieval inhumations.

The library for this software is primarily based on a sample of high resolution ground-penetrating radar (GPR) surveys. All data in the library are legacy data which have been ground-truthed in order to determine which responses of interest noted in the geophysical survey reports were true positives and which were false positives. Data included in the library are in both two-dimensional format (depth profiles, time-slices, etc.) and three-dimensional format (amplitude maps, isosurfaces, etc.) which have been improved but not processed. There are plans to include supplemental electromagnetic induction (low frequency electromagnetic methods), magnetic, and earth resistance survey data in a composite library to further enhance the scope of the project. Study sites focus on areas of known medieval settlement in Ireland and southwest England.

The software will employ established open source algorithms adapted to suit GPR data formats. Automatic object extraction and classification algorithms have been widely applied to spatial data, geological data, and aerial imagery (Arango *et al.* 2016, Eberle *et al.* 2015, Demir *et al.* 2011, Maulik and Chakraborty 2017) and more recently focused on GPR data (Jafrasteh and Fathianpour 2017). These fields have utilised a number of support vector machine and neural network approaches for supervised and semi-supervised machine

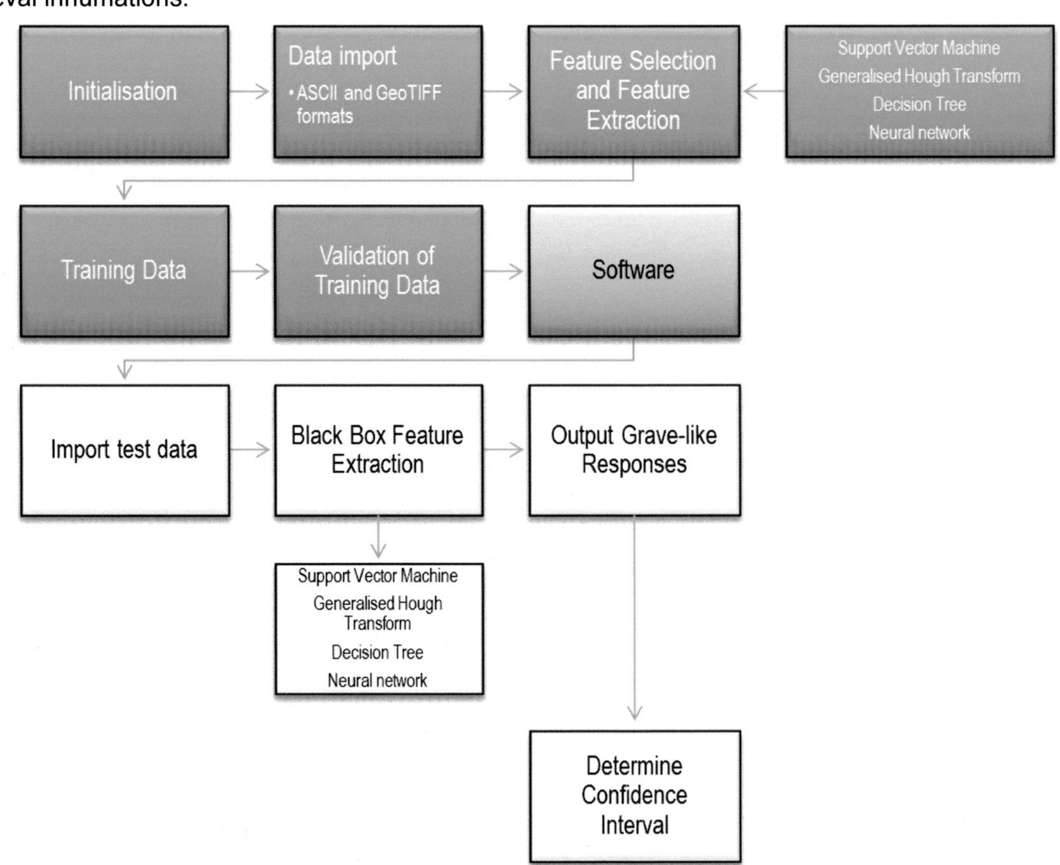

Figure 1: Simplified workflow for training dataset and prototype software methodology

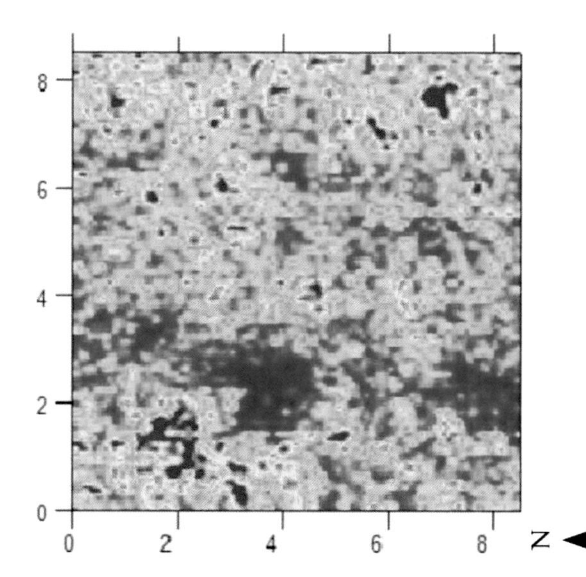

Figure 2: Sample of GPR data (0.10m traverse interval; 0.02m sampling interval) from the cloister garth at the Dominican Friary in Trim, Co Meath, Republic of Ireland, which will has interpreted manually and will undergo automatic 'interpretation' with the software in order to determine if the ease and accuracy significantly improve when automated.

learning with notable success. The software utilises a supervised machine learning approach, in which the prototypes employ support vector machine algorithms for computational geometry based pattern recognition (e.g. shape similarity), decision trees, and Generalised Hough Transforms. The real-world success rate of each is determined by comparing the automated 'interpretation' of collected data from study sites to excavation results in order to analyse the occurrence of true positive results. Isosurface extraction and analysis as well as remote sensing imagery overlay are to be incorporated where possible.

Recent progress of this ongoing doctoral research project is to be presented. With particular attention to the characteristics (amplitude, contrast between inhumation and background material, top- and side-view morphologies, conductivity, magnetic susceptibility, resistance, and resistivity) of true positive results. A selection of the study sites which are to be used to determine the success of the real-world application of the software, particularly the Dominican Friary in Trim, Co Meath, Republic of Ireland (O'Carroll 2014, Shine *et al.* 2016), and St. Brendan's Church in Birr, Co Offaly, Republic of Ireland, and their characteristics are examined.

As much of the GPR data collected during archaeological survey are similar to that shown in Figure 2, this software aims to improve data processing speed and reduce human error by objectively interpreting the data. By detecting human internments during the pre-excavation stage of an investigation, archaeologists are able to maximise the recovery potential of any human remains while also informing excavation and post-excavation strategies. In developing an *ad hoc* workflow model to determine the appropriate technical survey parameters while taking into account controllable and uncontrollable variables in any given survey, it is hoped to offer the highest probability of detecting burial(s) in a given survey with respect to environmental and anthropogenic factors.

Bibliography

Arango, R., Díaz, I., Campos, A., Canas, E., and Combarro, E. (2016) Automatic arable land detection with supervised machine learning, *Earth Sci Inform* **9**: 535-545.

Bonsall, J., Gaffney, C. and Armit, I. (2014) *Preparing for the Future: A reappraisal of archaeo-geophysical surveying on Irish National Road Schemes 2001-2010.* Bradford: School of Archaeological Sciences, University of Bradford.

David, A., Linford, N., and Linford, P. (2008) *Geophysical Survey in Archaeological Field Evaluation.* Swindon: English Heritage.

Demir, B., Persello, C. and Bruzzone, L. (2011) Batch-mode active learning methods for the interactive classification of remote sensing images. *IEEE Trans Geoscience Remote Sensing* **49**(3), 1014-1031.

Eberle, D., Hutchins, D., Das, S., Majumdar, A., and Paasche, H. (2015) *Automated pattern recognition to support geological mapping and exploration target generation – A case study from southern Namibia. Journal of African Earth Sciences* **106**, 60-74.

Jafrasteh, B. and Fathianpour, N. (2017) Automatic extraction of geometrical characteristics hidden in ground-penetrating radar sectional images using simultaneous perturbation artificial bee colony algorithm. *Geophysical Prospecting* **65**: 324-336.

Maulik, U. and Chakraborty, D. (2017) Remote sensing image classification – A survey of support-vector-machine-based advanced techniques. *IEEE Geoscience and Remote Sensing Magazine*, **March**: 33-52.

O'Carroll, F. (2014) *Archaeological Research Excavations at The Black Friary, Trim, Co Meath – Interim Report.* Retrieved from http://iafs.ie/wp-content/uploads/2014/06/Blackfriary-E4127-C240-Report-2014.pdf

Schmidt, A., Linford, P., Linford, N., David, A., Gaffney, C., Sarris, A. and Fassbinder, J. (2015) *EAC Guidelines for the Use of Geophysics in Archaeology: Questions to ask and points to consider.* Belgium: Europae Archaeologia Consilium (EAC), Association Internationale sans But Lucratif (AISBL).

Shine, D., Green, A., O'Carroll, F., Mandal, S. and Mullee, B. (2016) What Lies Beneath – Chasing the Trim Town Wall Circuit. *Archaeology Ireland* **30**(1), 34-38.

Non-invasive investigations at early medieval strongholds in Lubuskie province (western Poland)

Bartłomiej Gruszka[(1)] and Łukasz Pospieszny[(1)]

[(1)]Institute of Archaeology and Ethnology, Polish Academy of Sciences, ul. Rubież 46, 61-612 Poznań, Poland

bartek.gruszka@iaepan.poznan.pl, lukasz.pospieszny@iaepan.poznan.pl

Introduction

Interdisciplinary research of early medieval stronghold settlements in the area of the Middle Oder basin has been conducted since 2012. One of its components is geophysical prospection. So far magnetometer surveys have been carried out at four strongholds: Grodziszcze, Klenica, Przytok, and, partly, Świebodzin (Fig. 1.a).

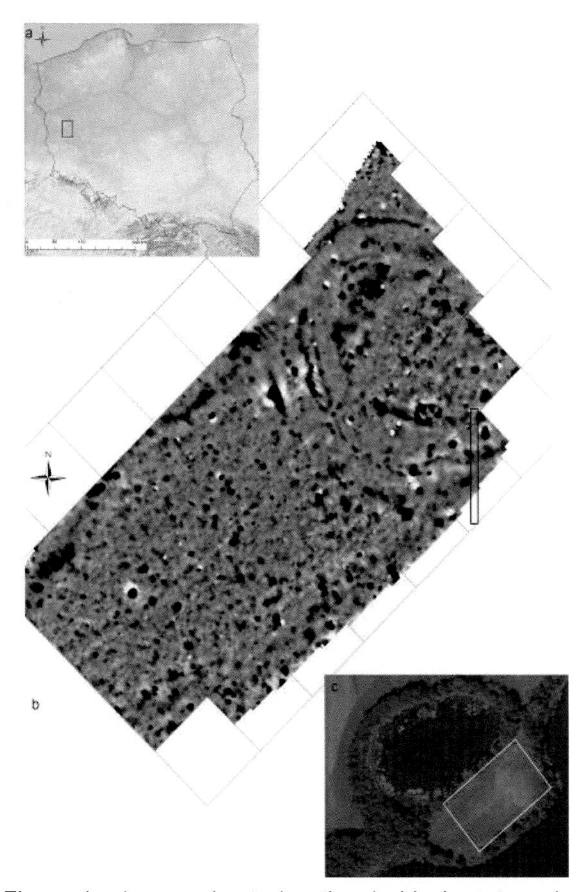

Figure 1: a) approximate location (a black rectangular frame) of the strongholds at: Grodziszcze, Klenica, Przytok, and Świebodzin. By A. Łuczak, B. Gruszka; b) Grodziszcze, Świebodzin commune, site 1. Geophysical image of examined area (gradiometer Bartington Grad 601-2; measurement grid 0.5 x 0.25 m, interpolated to 0.25 × 0.25 m, dynamics -10/+10 nT, white to black). A black frame marks the trench from 2015/2016. By Ł. Pospieszny; c) orthophotomap of the stronghold and its adjacent ancillary settlement at Grodziszcze. By B. Gruszka.

The survey areas at each site were subdivided into 20 x 20 m data grids using RTK GPS receiver. The surveys were undertaken using Bartington Grad 601-2 dual sensor fluxgate gradiometer. The data were recorded in a parallel mode, from south to north, at intervals of 0.25 m along lines spaced 0.5 m apart. Processing and image preparation were done using TerraSurveyorLite 3.0 and Surfer 9.0 software.

The Stronghold at Grodziszcze Site 1

The site is located approximately 1 km east of the town of Świebodzin, at the bottom of a wide glacial trough, currently filled in part with the waters of Lubinieckie and Zamecko lakes (Fig. 1.c). The results of previous investigations have shown that the stronghold, together with its immediately adjacent ancillary settlement, was erected on a peninsula surrounded on three sides by the waters of a lake. Today it is a narrow isthmus between Lubinieckie Lake and an unnamed water reservoir. The results of excavations carried out at the site in 1960s suggested that it was a ring-shaped stronghold with one line of ramparts. To better understand the results of the old excavations, the non-invasive investigations including gradiometer survey were initiated in 2015 (Fig. 1.b). The latter allowed for testing several hypotheses related to the function, size and spatial layout of the settlements. First of all, it revealed that the stronghold had been repeatedly extended by building the subsequent rings of ramparts on the west, at the expense of the adjacent ancillary settlement. In this way, the area of the stronghold was enlarged at least twice. The geophysical prospection has also allowed for re-evaluating earlier assumptions about the location of the gate. It showed that in the western side of the site, where scholars once located the gate and a road leading to the stronghold, the layout of magnetic anomalies indicates an undisturbed course of the rampart's remains. On the other hand, a clear discontinuity of these anomalies, which can point to the location of the main entrance to the stronghold, was identified in the south-western side of the site (Fig. 1.b).

In order to verify the results of geophysical investigations which provided the grounds for the hypothesis about two-phases of the stronghold, small scale excavations were carried out in 2015 and 2016. The locality selected for the study was characterised by the presence of anomalies that suggested the occurrence of wooden structures. A trench 30 x 2 m was opened to expose the course of two ramparts and a fragment of the stronghold's ward. The excavations confirmed the validity of the results reached through the magnetic survey. In the place where burnt relics of the internal wall had been expected, charred

Figure 2: Grodziszcze, Świebodzin commune, site 1: a) cross-section and b) plan view of the rampart built in AD 951. Charred elements of the rampart's construction were displayed as a strong magnetic anomaly. By B. Gruszka.

elements of a log-cabin construction were revealed, faced on the outside with stones bonded with clay (Fig. 2). On the basis of dendrochronological dating, it has been concluded that the rampart was built from oaks cut in the year AD 951. Probably, this was not the first phase of early medieval settlement in the area. In the layer deposited between the construction elements, numerous fragments of vessels have been discovered, dating to the second half of the 9th century AD and to the early 10th century AD. To the south of the rampart constructed in the middle of 10th century AD, in the place of further magnetic anomalies, the relics of boxes built using log construction were found. These were the elements of an unburnt, younger rampart. The magnetic anomalies were caused by the presence of boxes filled with charred material acquired from dismantled constructions from the year AD 951. On the basis of radiocarbon AMS dating it can be concluded that the rampart was probably built in the second half of the 12th century AD.

The Stronghold at Klenica Site 3

The early medieval settlement complex at Klenica, including a stronghold and its adjacent settlement is located in a broad valley of the Oder, on a floodplain. The stronghold is situated on an exposed dune, rising about 2 m above the surrounding flat area (fig. 3a). The site was discovered and excavated by Ernst Petersen in 1936 (Kieseler 2016). In

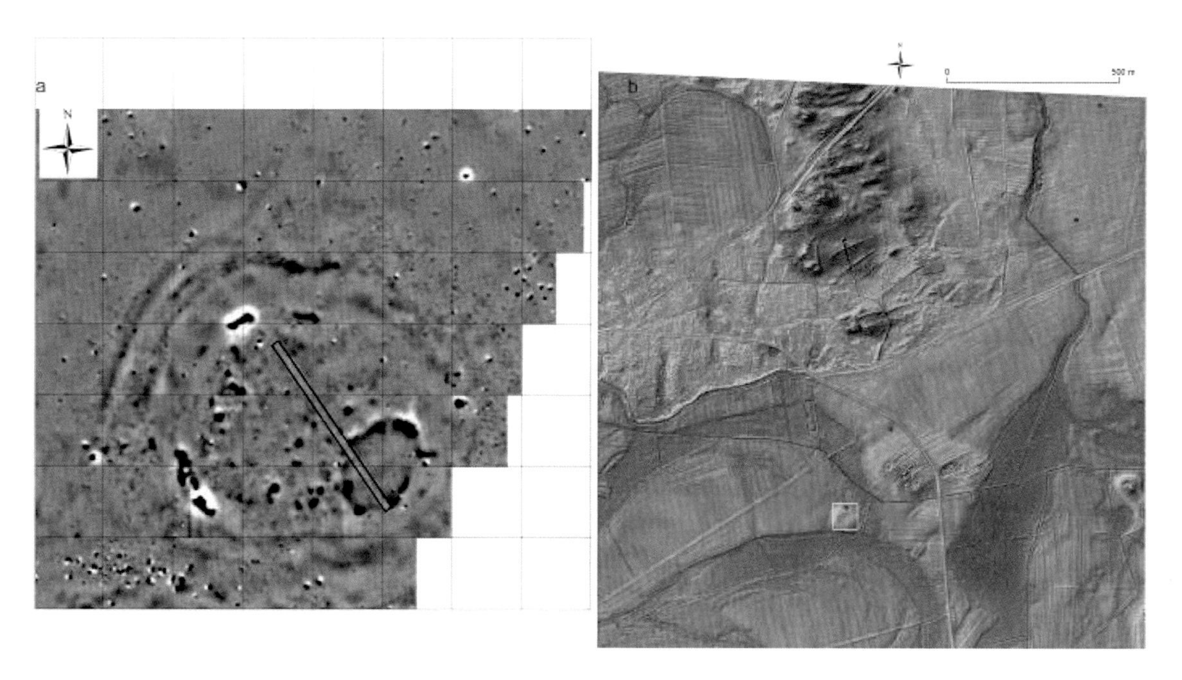

Figure 3: Klenica, Bojadła commune, site 3: a) geophysical image of examined area (gradiometer Bartington Grad 601-2; measurement grid 0.5 x 0.25 m, interpolated to 0.25 x 0.25 m, dynamics -10/+10 nT, white to black). A black frame marks the trench from 1936. By Ł. Pospieszny; b) digital terrain model of the surroundings of the settlement complex. Based on ALS/LiDAR data from project ISOK ("IT system of the country's protection against extreme hazards"). By A. Łuczak, B. Gruszka.

1960 Adam Kołodziejski discovered the adjacent settlement site, located at a distance of several dozen meters to the east (Gruszka 2016). Further research was conducted in 2007 under the direction of Felix Biermann. During the investigations, both in 1936 and 2007, in the highest, south-eastern part of the stronghold, features resembling earth mounds were uncovered. Petersen saw them as the relics of an internal rampart, while the researchers who carried out the fieldwork in 2007 suggested that these were the mounds of rubbish left by the inhabitants of houses identified nearby (Biermann *et al.* 2011, 330-333, Figs. 2, 3).

The magnetometer survey carried out in September 2016 (Fig. 3.b) allowed for testing both the abovementioned hypotheses. The investigations revealed the course of the trench opened in 1936, visible as a stripe of low magnetic susceptibility. Its southern part clearly cut an ellipse-shaped area (of 28 x 21 m in size, 60 m in diameter) of increased magnetism. The space limited by that anomaly appears to be an open area with no houses, in contrast to the ward. On the basis of the excavations it was possible to determine the size of mounds. They had a diameter and height reaching up to 1 m, and they were spaced up to 1.5 m apart. It can be assumed that there are at least 18-20 features of this type within the stronghold. The results of geophysical survey permit us to reject the hypotheses about both the internal rampart and rubbish mounds. However, the function of these mysterious features is still unknown.

In addition, the geophysical investigations have allowed for detailed recognition of the size of the stronghold and the organization of space surrounded by the ramparts. They have also indicated the location of the stronghold's gate and revealed the southern boundary of nearby settlement.

Bibliography

Biermann, F., Kieseler, A. and Nowakowski, D. (2011) Od ogniska do zniszczenia pożarem. Grodzisko w Klenicy, gm. Bojadła w świetle nowych wyników badań wykopaliskowych. In A. Jaszewska and A. Michalak (eds.) *Ogień – żywioł ujarzmiony i nieujarzmiony, VI Polsko-Niemieckie Spotkania Archeologiczne, Garbicz, 5-6 czerwca 2008.* SNAP: Zielona Góra, 329-348.

Gruszka, B. (2016) Ein Siedlung am slawischen Burgwall von Kleinitz (Klenica) – Aufarbeitung der Altgrabung von 1962. In F. Biermann, D. Nowakowski and A. Kieseler (eds.) *Burg, Herrschaft und Siedlung im mittelalterlischen Niederschlesien. Die slawische Ringwälle von Köben (Chobienia) und Kleinitz (Klenica) im Kontext der Frühgeschichte des mittleren Oderraums*, Studien zur Archäologie Europas, Bd. 27. Bonn: Verlag Dr. Rudolf Habelt, 495-532.

Kieseler, A. (2016) Der slawische Burgwall von Klenitz (Klenica) im nördlischen Niederschlesien. In F. Biermann, D. Nowakowski and A. Kieseler (eds.) *Burg, Herrschaft und Siedlung im mittelalterlischen Niederschlesien. Die slawische Ringwälle von Köben (Chobienia) und Kleinitz (Klenica) im Kontext der Frühgeschichte des mittleren Oderraums*, Studien zur Archäologie Europas, Bd. 27. Bonn: Verlag Dr. Rudolf Habelt, 211-466.

Acknowledgements

The excavations, dendrochronological dating and part of geophysical surveys were funded by National Science Centre within the project: *Early medieval strongholds of the part of the Middle Odra river territory to mid-11th century in archaeological research* (2015/16/S/HS3/00274).

The part of geophysical surveys was financed by the Ministry of Culture and National Heritage within priority 5 – Protection of archaeological artefacts: *A comprehensive, non-destructive recognition of early medieval settlement complex at Klenica in Lubuskie province.*

Geophysical and geochemical definition of a rural medieval churchyard at Furulund, Hedmark, Norway

Lars Gustavsen[2], Rebecca J S Cannell[1], Monica Kristiansen[2] and Erik Nau[2]

[1]Department of Archaeology, Anthropology and Forensic Science, Bournemouth University, Poole, BH12 5BB, UK; [2]Norwegian Institute for Cultural Heritage Research, P.O. Box 736 Sentrum, 0105 Oslo, Norway

rebecca.cannell@gmail.com

Background

In Norway, ca. 2000 churches are believed to have been in existence in the Middle Ages. Of these, 647 are still in use, and a further 614 sites are attested in historical sources, but now abandoned. This leaves a considerable number attested only via hints in toponymical and folkloric sources (Brendalsmo and Eriksson 2015). Although automatically protected by the Norwegian Cultural Heritage Act, their inexact locations render them inadequately maintained and threatened by continual natural processes, agricultural activities or acts of destruction.

Given the large number of potential sites, their mapping by way of intrusive methods is deemed costly thus unfeasible. An urgent need exists to develop alternative approaches so that these sites can be protected.

Small, rural, abandoned medieval church sites tend to lead a fairly anonymous existence. However, in 2014 the Norwegian Directorate for Cultural Heritage (NO: *Riksantikvaren*), and Hedmark County Council received alarming information regarding the church site at Furulund north of the town of Kongsvinger (Fig. 1). Human skeletal remains began surfacing as a result of ploughing, prompting concern that the graveyard was rapidly being decimated. The Directorate sought advice from the Norwegian Institute for Cultural Heritage Research (NIKU) on how to map the site using non-intrusive methods. Two methods were proposed and ultimately employed; preliminary geochemical sampling and analysis using portable Xray fluorescence (pXRF) followed by highresolution groundpenetrating radar (GPR) surveys.

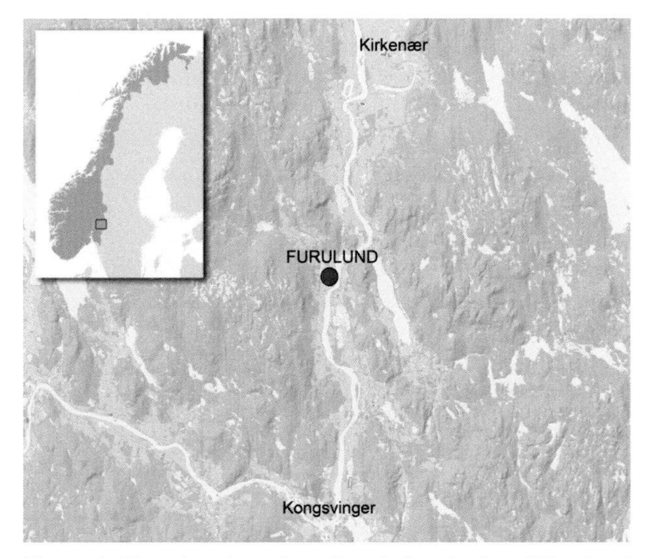

Figure 1: The abandoned medieval church site of Furulund is situated along the River Glomma, between the town of Kongsvinger and the village of Kirkenær in Hedmark County. Map source: The Norwegian Mapping Authority, 2016.

Portable XRF has been successfully applied to a variety of archaeological settlement and industrial sites (Hayes 2013, Gauss *et al.* 2013) but has never before seen use to delimit a mortuary site. Geochemistry was chosen on the assumption that the systematic mapping of certain elements across the church site would yield relatively enhanced values that would map differential land use and the presence of ploughed up burials, and thus delimit the cemetery. Portable XRF was used as it is flexible, rapid, cost effective and the instrumental resolution sufficient for the purpose.

The use of geophysical methods to detect and map graves, clandestine or otherwise, has a long and well-established history, and a considerable body of literature exists on the subject (e.g. Vaughan 1986, Bevan 1991, Davenport 2001, Cheetham 2005, Jones 2008). Due to its comparatively high spatial resolution and its capability to resolve relatively small targets whilst simultaneously providing depth information, GPR is generally considered the most suitable solution for mapping inhumation burials in graveyards and cemeteries (Conyers 2006, Jones 2008, Moffat 2015). Alternative geophysical methods have also seen some success, particularly when combined with other techniques (e.g. Davenport 2001, Nobes 1999, Linford 2004, Dalan *et al.* 2010).

Method

The probable graveyard area was estimated to be within a 50 x 50m area, encompassing both the area the farmer had set aside as the church location and the area where bones were found ploughed to the surface. To keep costs minimal, transects were used for geochemical sampling to delimit the graveyard. Five transects were established with a sample spacing of 5 m. In total, 61 samples were taken over the graveyard, with additional background samples taken in an area outside the graveyard (Fig. 3). Samples were taken with a push auger used to the base of the plough soil and the sample extracted.

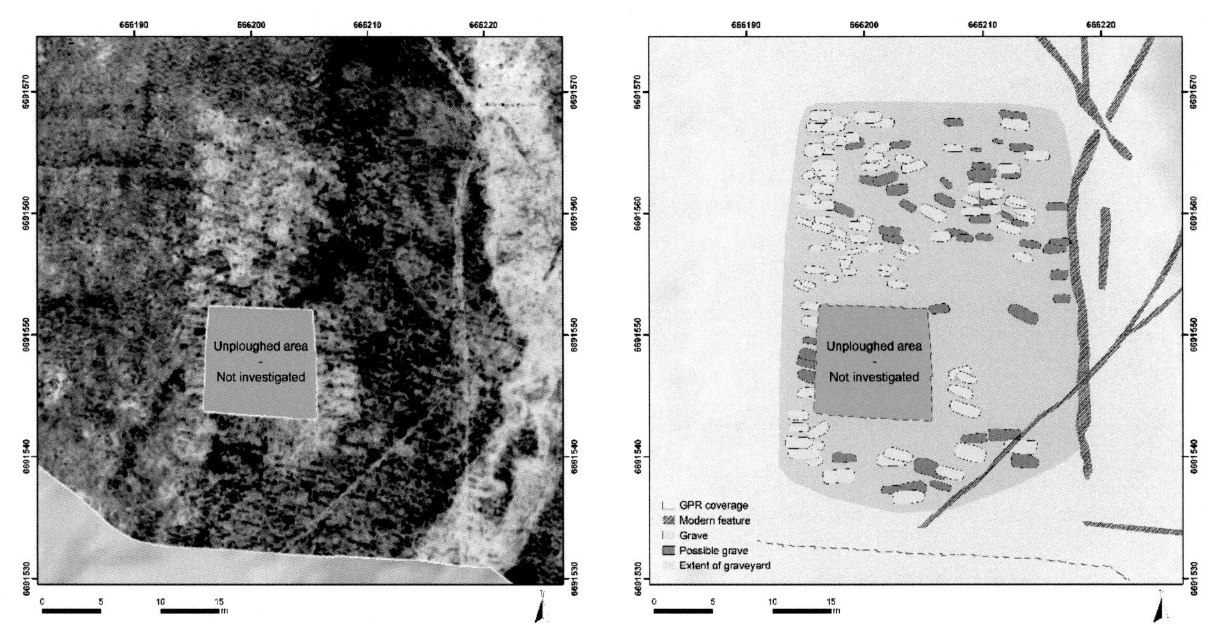

Figure 2: Left: GPR depth slice and, right: interpretation of the graveyard area. Map source: Norwegian mapping Authority 2016.

In the laboratory, samples were dried, crushed and homogenised prior to analysis using a Niton/Thermo Scientific XLt3 GOLDD+ portable XRF in mining mode. Standard reference materials were used for empirical calibration. Samples were analysed in cups with a 6µm polypropylene film, with the instrument in a field stand. The analytical time was 300 seconds between all filters, the longer duration necessary for lighter element detection (Z=<22) as helium purge was not available. The calibrated values for the selected elements were imported into the geographical information system *ESRI ArcGIS 10.2.2*. Using the *Geostatistical Analyst* extension, interpolated and gridded surfaces representing trends in the values were generated using ordinary kriging, which were then combined with other data sources for further analysis.

The GPR survey followed several weeks of cold (c. 0-12°C), but unusually dry (0-0.2mm) weather. A total of 1.8 hectares was surveyed using a motorized 16-channel, 400MHz *MALÅ Imaging Radar Array* (MIRA) from *MALÅ Geoscience*. Antenna spacing was set to 10.5 cm and the measurements time-triggered at a rate of 50Hz.

Once collected, the data were processed using the *ApRadar* software, developed by *ZAMG ArcheoProspections®/LBI ArchPro*, where trace interpolation, time-zero corrections, band-pass frequency filtering, spike removal, de-wow filters, average-trace-removal, amplitude-gain corrections, amplitude balancing and Hilbert transformations were applied. Time-to-depth conversion was set to a velocity of 10 cm/ns for the upper parts of the dataset, down to 10 ns, decreasing to 8 cm/ns at 20 ns and beyond. The conversion was based on hyperbola fitting carried out in *Sandmeier Scientific ReflexW*. The data were then resampled to a resolution of 8 x 8 cm, and subsequently interpolated into a 3D data block from which georeferenced depth slices were generated. In order to visualise, analyse and interpret the data, the depthslices, in the form of grey-scale TIFF images, were then imported into *ArcGIS*, where they were combined with other data sources, visualised and interpreted.

Results

The GPR survey identified a cluster of features, which is interpreted as graves belonging to the former church site. These features are largely E-W orientated, rectangular to sub-rectangular in plan and containing homogeneous, absorbing backfills. As a group, they are clearly defined against the natural subsoil, which has strongly reflecting properties. A total of 130 individual graves have been identified, 84 of which have been classified as "certain", the remaining 46 classified as "possible". Those features that can be positively and clearly identified as graves, measure between 80 – 250 cm in length, and 35 – 80 cm in width. Combined, the graves form a distinct clustering with a relatively clear outline and delineation (Fig. 2).

The data from elements commonly associated with human activity were visually compared to the GPR interpretations. Of these, Fe (iron), Ca (calcium), P (phosphorous), and Cu (copper) were clearly spatially associated with the graveyard. Ca was enhanced only where bones were visible on the ploughed surface, whereas Fe was connected to soil processes and the enhanced organic inputs. P was less defined, but enhanced by the

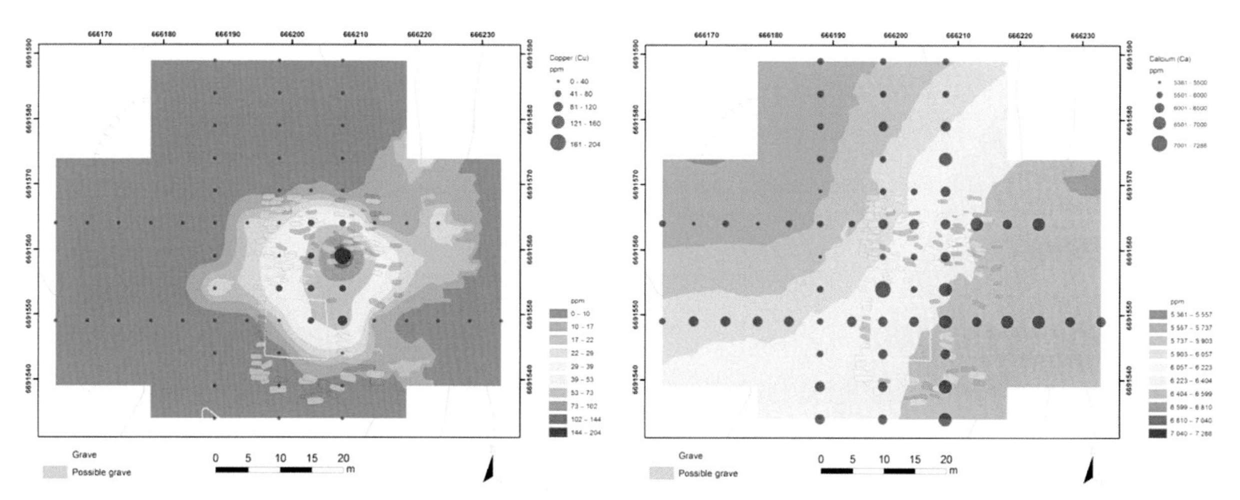

Figure 3: Ordinary kriging of elemental data for Cu (left) and Ca (right), with the GPR interpretations. All values in parts per million (ppm). Map source: Norwegian Mapping Authority, 2016.

cemetery. Surprisingly, the enhancement of Cu was concentrated in the area with the graveyard where graves are less abundant, and the concentration of Cu is tentatively interpreted as the church location (Fig. 3).

Conclusion

The church and associated graveyard were efficiently located and defined with the combination of non-destructive prospection methods, allowing for their future protection from further damage. There is great potential for the combined approach to define and thus protect the many other modest medieval rural graveyards in Norway, many of which are equally under threat from modern land use.

Bibliography

Bevan, B. W. (1991) The search for graves. *Geophysics*, **56**: 1310-1319.

Brendalsmo, J. & Eriksson, J. E. G. (2015) De middelalderske sakrale stedene. In: L. Johannessen and J.-E. G. Eriksson (eds.) *Faglig program for middelalderarkeologi: Byer, sakrale steder, befestninger og borger*. Oslo: Riksantikvaren.

Cheetham, P. (2005) Forensic Geophysical Survey. In: J. Hunter and M. Cox (eds.) *Forensic Archaeology: Advances in Theory and Practice*. London: Routledge.

Conyers, L. B. (2006) Ground-Penetrating Radar Techniques to Discover and Map Historic Graves. *Historical Archaeology* **40**: 64-73.

Dalan, R. A., De Vore, S. L. and Clay, R. B. (2010) Geophysical identification of unmarked historic graves. *Geoarchaeology* **25**: 572-601.

Davenport, G. C. (2001) Remote Sensing Applications in Forensic Investigations. *Historical Archaeology* **35**: 87-100.

Gauss, R. K., Bátora, J., Nowaczinski, E., Rassmann, K. and Schukraft, G. (2013) The Early Bronze Age settlement of Fidvár, Vráble (Slovakia): reconstructing prehistoric settlement patterns using portable XRF. *Journal of Archaeological Science* **40**: 2942-2960.

Hayes, K. (2013) Parameters in the Use of pXRF for Archaeological Site Prospection: A Case Study at the Reaume Fort Site, Central Minnesota. *Journal of Archaeological Science* **40**: 3193-3211.

Jones, G. (2008) Geophysical Mapping of Historic Cemeteries. *Technical Briefs in Historical Archaeology* **2008**: 25-38.

Linford, N. T. (2004) Magnetic Ghosts: Mineral Magnetic Measurements on Roman and Anglo-Saxon Graves. *Archaeological Prospection* **11**: 167-180.

Moffat, I. (2015) Locating Graves with Geophysics. In: A. Sarris (ed.) *Best Practices of Geoinformatic Technologies for the Mapping of Archaeolandscapes*. Oxford: Achaeopress, 45-53.

Nobes, D. C. (1999) Geophysical Surveys of Burial Sites: A Case Study of the Oaro Urupa. *Geophysics* **64**: 357-367.

Vaughan, C. J. (1986) Ground-penetrating radar surveys used in archaeological investigations. *Geophysics* **51**: 595-604.

Assessing the effect of modern ploughing practices on archaeological remains by combining geophysical surveys and systematic metal detecting

Lars Gustavsen[(1)], Monica Kristiansen[(1)], Erich Nau[(1)] and Bernt Egil Tafjord[(2)]

[(1)]The Norwegian Institute for Cultural Heritage Research, Oslo, Norway; [(2)]Buskerud County Municipality, Drammen, Norway

lars.gustavsen@niku.no

Introduction

It is a well-known fact that modern ploughing regimes can be highly detrimental to archaeological sites (Ammerman 1985, Haldenby and Richards 2010). Regular deep-ploughing in order to replenish the topsoil truncates the underlying archaeological features, whilst simultaneously bringing artefacts to the surface, artefacts that are commonly found by metal detectorists. Without further information about their original context, however, the scientific value of the finds is often limited to their physical and aesthetic properties rather than their cultural significance. Through the example below, we argue that by using geophysical methods in conjunction with systematic metal detecting, we may not only increase our knowledge of the site itself, but also be able to further our understanding of how, and to what extent, archaeological sites are damaged by the plough.

Background

In 2014 a group of experienced metal detectorists recovered a number of unusual artefacts near the farm Sem in Øvre Eiker municipality in the southeast of Norway (Fig. 1), a municipality best known in archaeological terms for the spectacular 9th century *Hoen hoard*, unearthed in 1834 near Hokksund some three kilometres to the north (Fuglesang and Wilson 2006).

Amongst the 323 artefacts recovered in 2014 were lead weights, fittings in the insular style and Arabic coin fragments, as well as ingots, crucible fragments and slag, suggestive of a site engaged both in local fine smithing and international trade. Such sites are a rarity in the Norwegian archaeological record, and in order to explore the finds' origins, a geophysical survey was carried out the following year. This employed a 400 MHz 16-channel MALÅ Imaging Radar Array GPR system (Trinks et al. 2010), with a high spatial sampling resolution of 4 x 10,5 cm - resampled to 10 x 10 cm. 11 hectares were surveyed over the course of three days, revealing traces of a production and trading site, a mound cemetery and an extensive settlement site with a

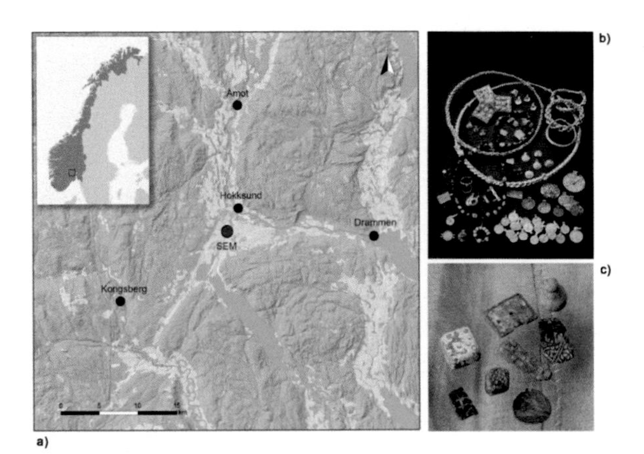

Figure 1: a) The location of Sem in Øvre Eiker, Buskerud, Norway (Base map: The Norwegian Mapping Authority, 2017); b) The 9th century hoard found at Hoen near Hokksund, three kilometres north of Sem (© 2017 Museum of Cultural History, University of Oslo/CC BY-SA 4.0); c) A selection of artefacts recovered during the metal detecting campaigns of 2014 and 2016 (© Bjørn Johnsen, Buskerud County Administration).

broad time span. A second, more systematic metal detecting campaign was then mounted in 2016, in order to extract more information from the site. Although still ongoing, the analyses of the metal detecting finds have already revealed interesting distribution patterns that we believe warrant further investigation.

Preliminary Results

The GPR surveys revealed a complex settlement extending across the survey area (Fig. 2). In the northern part, a post-built structure and sundry postholes with no clear organisation can be seen in connection with the Medieval and post-Medieval activity on the site. Of particular interest is a large, angular anomaly which continues into the fields to the north, representing gardens of a 17th century royal estate. In the eastern part of the survey area anomalies forming at least 10 circular structures, measuring some 7 – 18 m in diameter, clearly representing ring ditches of ploughed out burial mounds. In the southern part of the field the GPR surveys yielded evidence for a well-defined area consisting of numerous pits of varying size and shape. The larger of these are thought to represent pit-houses used in connection with fine smithing, whereas the smaller pit-like features may represent hearths or refuse pits associated with production.

Combining the information from the metal detecting surveys with the GPR data shows that the distribution of metal finds corresponds well with the position of the archaeological features detected by GPR, with distinct clusters over the production and trading site as well as the mound cemetery (Fig. 3). In this preliminary study we have specifically analysed the spatial distribution of Iron Age metal objects related

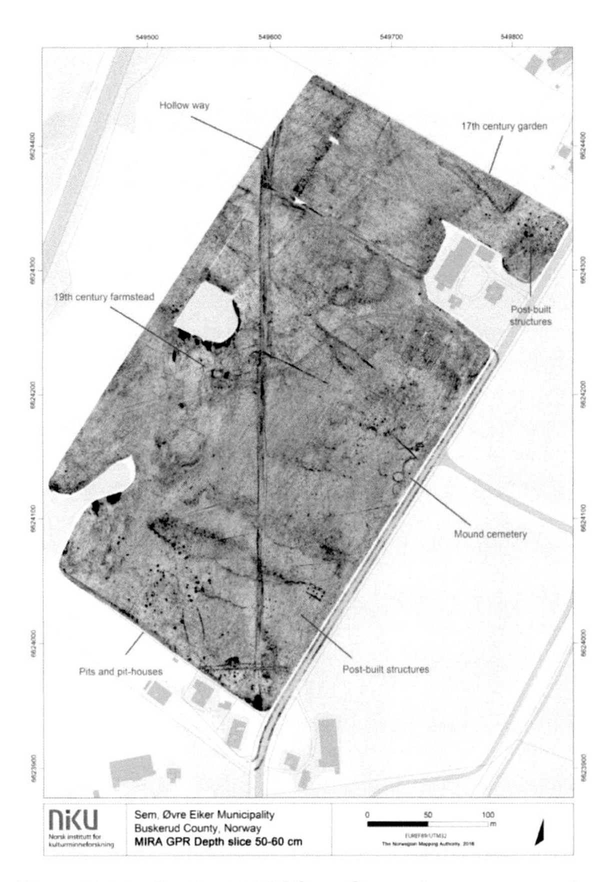

Figure 2: Depth slice 50-60 from Sem, showing a complex settlement area comprising post built houses, production pits, pit houses and several burial mounds (Base map: The Norwegian Mapping Authority, 2017).

Figure 3: Distribution of finds recovered by systematic metal detecting combined with interpretations of the GPR datasets (LiDAR data: The Norwegian Mapping Authority, 2017).

to trade and production, in relation to features interpreted as contemporaneous working pits and waste pits. Assuming that there is a correlation between the metal objects and the features detected by GPR, there seems to be considerable lateral movement of objects in the plough soil. By measuring the distance between every single feature and the nearest find we can assert that, at least in the southernmost activity zone, the majority of finds is located 4 – 7 m away from the nearest feature. This is surprising given that the metal detecting was carried out shortly after the autumn ploughing, and clearly indicates that even a single ploughing event has caused significant damage to the underlying archaeology. Further preliminary analyses of the data also indicate distinct distribution patterns in the type and material of the metal finds. For example, there is a clear correlation between the distribution of lead based finds and the mound cemetery, whereas copper alloy finds are largely concentrated around the production and trading zone. The ongoing analysis and studies of the finds will hopefully produce important information as to the original function, date and composition of each find, allowing the objects to be sourced to their original contexts. This, we believe, will make it possible to assess the spatial movement of the finds and to potentially quantify the effect of the current agricultural regime on the archaeological features, so that preventive measures may be introduced. Further analyses of the finds will of course also enhance the dating of the site, and will refine its chronological, functional and spatial development, while simultaneously providing suitable targets for future, limited excavations and geophysical surveys.

Bibliography

Ammerman, A. J. (1985) Plow-Zone Experiments in Calabria, Italy. *Journal of Field Archaeology* **12**(1): 33-40.

Fuglesang, S. H. and Wilson, D. M. (2006) *The Hoen hoard : a Viking gold treasure of the ninth century.* Norske oldfunn. Vol. 20. Oslo: Museum of Cultural History, University of Oslo.

Haldenby, D. and Richards, J. D. (2010) Charting the effects of plough damage using metal-detected assemblages. *Antiquity* **84**(326): 1151-62.

Trinks, I., Johansson, B., Gustafsson, J., Emilsson, J., Friborg, J., Gustafsson, C., Nissen, J. and Hinterleitner, A. (2010) Efficient, large-scale archaeological prospection using a true three-dimensional ground-penetrating Radar Array system. *Archaeological Prospection* **17**(3): 175-86.

Nebelivka, Ukraine: geophysical survey of a complete Trypillia mega-site

Duncan Hale(1), John Chapman(1), Mikhail Videiko(2), Bisserka Gaydarska(1), Natalia Burdo(2), Richie Villis(1), Natalie Swann(1), Patricia Voke(1), Nathan Thomas(1), Andrew Blair(1), Ashley Bryant(1), Marco Nebbia(1), Andrew Millard(1) and Vitalij Rud(2)

(1)Department of Archaeology, Durham University, Durham, UK; (2)Institute of Archaeology, Kyiv, Ukraine

d.n.hale@durham.ac.uk

The Trypillia (Russian *Tripolye*) mega-sites in the Kirovograd and Cherkassy regions of Ukraine constitute the largest settlements in 4th millennium BC Europe. Discovered in the 1970s, the five largest mega-sites comprise Taljanky (350ha), Maydanetskoe (270ha), Nebelivka and Dobrovody (each 250ha) and Tomashivka (220ha). A geophysical plan of the mega-site at Nebelivka has recently been completed as part of a joint project between Durham University (UK) and the Kyiv Institute of Archaeology (Ukraine). This site plan informs the strategy for intrusive investigation and sampling, to establish internal site phasing, and provides a platform for understanding social space at Nebelivka, its role in regional settlement structures and the possibility of state-level societies of the same date as Uruk in Iraq.

Between 2009 and 2013 a team of archaeological geophysicists from Durham University completed the first detailed magnetic survey over an entire Trypillia mega-site (Chapman *et al.* 2014). The Durham team used Bartington Grad 601-2 dual sensor fluxgate gradiometers to cover a total area of 286ha. The vast majority of anomalies detected relate to burnt, partly burnt or unburnt structures, as well as soil-filled pits, ditches and palaeochannels. Over 1,350 structures have been identified in the geophysical survey, the vast majority of which are assumed to be dwelling houses. Two-thirds of the structures appear to have been burnt at the end of their 'use-lives', with one-third unburnt. Two new important classes of feature were also discovered during these surveys. The first was a type of structure much larger than the usual dwelling house, provisionally termed 'Assembly House'; the second was a perimeter boundary ditch. The latter was significant in that it demonstrated that some Trypillia communities were concerned to define the limits of their settlements, to distinguish 'inside' from 'outside', an idea previously discussed by Harding *et al.* (2006).

Figure 1: Magnetic survey at Nebelivka.

Figure 2: Stylistic interpretation of site plan.

In plan the site is broadly oval, principally comprising two concentric circuits of houses enclosed by a perimeter ditch. The outer ditch measures approximately 5.9km in length and encloses an area of 238ha. It appears to have defined the site's extent rather than served a defensive function as there are several, often wide, causeways across it (reminiscent of Western European causewayed enclosures) and on excavation it was found to be relatively shallow; the ditch measured up to 4m in width and up to 2m in depth. The layout of the two house circuits shows both segmentation and irregularity, ranging from groups of closely-set structures which perhaps shared walls, through small groups of parallel structures, to widely spaced structures with perhaps more individual than group identity. The composition of the streets shows as much variability as that of the circuits, including at least three arrangements which resemble small open 'squares'. Further houses were identified inside the inner circuit, typically arranged along almost 50 radially oriented streets, though the large central area of the site (ca. 65ha) appears completely devoid of any structures.

The area between the house circuits was generally clear of structures, though those structures that are present are typically larger than the houses in the circuits. Occasional buildings were also detected outside the outer house circuit, but within the perimeter boundary ditch; these buildings are also often larger than the circuit houses. Many of these larger buildings have been identified in pairs, often with one structure located between the circuits and the other located outside the outer circuit. There are also several instances of single large structures. They are believed to have served as public places, and have been provisionally termed 'Assembly Houses' (Chapman *et al.* 2016).

The locations of Assembly Houses are considered to be of critical importance for the spatial division of the house circuits and inner radial streets into a

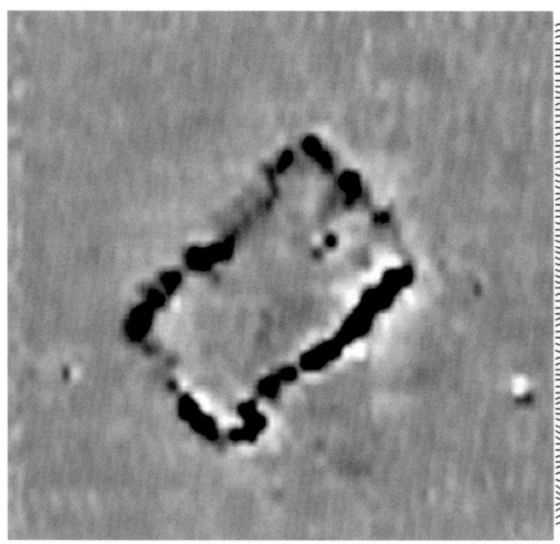

Figure 3: Greyscale and trace plot of a typical burnt 'assembly house' (24m x 13.5m).

level of spatial order termed 'Quarters'. Eight criteria have so far been used to partition the Nebelivka mega-site into fourteen Quarters, each typically having one or one pair of Assembly Houses. Quarters appear to have developed in markedly different ways, perhaps indicating a localised, bottom-up decision-making process, within overall planning constraints.

The largest structure detected at Nebelivka measures approximately 60m x 20m and occupies a gap in the eastern side of the inner house circuit. This is the largest structure yet to be found on any Trypillia site. The variation in the strength of the magnetic anomalies largely reflects the varying amounts of burnt daub and fired ceramics recorded during subsequent excavation. The eastern part of the large rectilinear anomaly therefore reflects an open, featureless, enclosed yard or entrance area, whilst the central part of the anomaly reflects a series of rooms, possibly roofed, with another open area and probable further rooms to the west.

The 'mega-structure' is aligned east-west along what appears to be a broad band of relatively near-surface granite rockhead; this may have been more evident as a ridge before the loess was deposited. Indeed, four of the five larger buildings in this area sit on top of this geological feature. Given the absence of overlapping house-floors, the principal data processing issue was the removal of the strong magnetic signal from the Ukrainian 'granite shield'. This was achieved by the application of a high pass filter to the data, which has preserved high frequency small-scale spatial detail whilst suppressing the low frequency large-scale detail.

One of the biggest research challenges in investigating such a large settlement is how to build an internal chronological sequence for a site whose occupation falls entirely within a single ceramic stage in the relative chronological scheme. Approximately 80 samples of either animal bone or plant macrofossil remains have been subject to for AMS dating, with subsequent Bayesian modelling of the dates. However, it has proved impossible to statistically distinguish separate phases of activity or house-building at the site, which appears to have been very short-lived, perhaps only occupied for 100-200 years, between 3970 BC and 3800 BC. The aim remains to estimate how many structures were occupied at the same time. Only when this estimate is made can we make a further estimate of possible population levels at different phases of the Nebelivka mega-site's use.

The surveys were undertaken as part of the AHRC-funded 'Trypillia Mega-Sites Project: Early urbanism in prehistoric Europe?' directed by John Chapman (Durham University, UK) and Mikhail Videiko (Institute of Archaeology, Kyiv, Ukraine).

Bibliography

Chapman, J., Videiko, M., Gaydarska, B., Burdo, N., Hale, D., Villis, R., Swann, N., Thomas, N., Edwards, P., Blair, A., Hayes, A., Nebbia, M. and Rud, V., (2014). The Planning of the Earliest European Proto-Towns: A New Geophysical Plan of the Trypillia Mega-Site of Nebelivka, Kirovograd Domain, Ukraine. *Antiquity Project Gallery* **88**(339): published online at http://antiquity.ac.uk.ezphost.dur.ac.uk/projgall/chapman339/

Chapman, J., Gaydarska, B. and Hale, D. (2016) Nebelivka: Assembly Houses, Ditches and Social Structure. In J. Müller, K. Rassmann, and M. Videiko (eds.) *Trypillia Mega-Sites and European Prehistory 4100-3400 BCE.* Abingdon: Routledge. Themes in Contemporary Archaeology Volume 2: 117-131

Harding, A., Sievers, S. and Venclová, N. (eds.) (2006) *Enclosing the Past: Inside and Outside in Prehistory.* Sheffield Archaeological Monographs 15. Sheffield: J R Collis Publications.

Geophysical prospection in the Natal landscape of the Buddha, southern Nepal

Duncan Hale[1], Robin Coningham[1], Kosh Prasad Acharya[2], Mark Manuel[1], Chris Davis[1] and Patricia Voke[1]

[1]Department of Archaeology, Durham University, Durham, UK; [2]Department of Archaeology, Government of Nepal, Kathmandu, Nepal

d.n.hale@durham.ac.uk

Introduction

These surveys form part of an ongoing conservation and management project for Lumbini and other sites associated with the early life of the Buddha in the Nepal Terai, directed by Robin Coningham and Kosh Prasad Acharya. The project is largely funded by UNESCO Japanese Funds-in-Trust, and supported by the Lumbini Development Trust (LDT), the Department of Archaeology, Government of Nepal (DoA) and Durham University. The Buddha was born in Lumbini, a village in southern Nepal near the border with India. Recent excavations and scientific dating have indicated probable sixth century BC dates for the life of the Buddha (Coningham *et al.* 2013).

Approximately 25km west of Lumbini is the village of Tilaurakot, which is believed to have been 'Kapilavastu', the former capital of the ancient Shakya republic where the Buddha spent his early, princely, life before leaving to seek enlightenment. The site includes the remains of a citadel, measuring ca. 500m x 400m, and a series of external monuments. Pilot geophysical evaluations within the citadel in 1997 and 1999 identified a number of probable wall footings and other features (Schmidt *et al.* 2011). Between 2014 and 2016 a team from Archaeological Services Durham University conducted more extensive geomagnetic surveys at Tilaurakot, covering the entire area within the Kapilavastu ramparts and several areas beyond the ramparts in each direction. This is believed to be the first complete geophysical plan of an early historic town in South Asia.

During the same period, surveys were also conducted for the DoA at several other early historic sites in Buddha's natal landscape, including Kudan, Araurakot, Sisaniya, Dohani, Chitradei and Sagrahawa. The principal aim of each survey was to assess the nature and extent of any sub-surface features of potential archaeological significance, which would in turn facilitate research and inform management and conservation issues, particularly with regard to any future development. The sites surveyed for this project are protected by the Nepali government and are defined on the ground by concrete maker posts or fences. The protected areas typically include visible standing remains or earthen mounds or platforms, but the recent surveys have shown that the archaeological sites often extend for considerable distances beyond the protected areas.

Tilaurakot- Kapilavastu

A grid system of roads has been identified within the citadel, aligned broadly north-south and east-west. Numerous buildings and other probable wall remains have been detected as well as open, undeveloped areas perhaps used for markets, recreation or ceremonies. Two broad categories of building have been detected, most comprising brick footings for a timber and thatch super-structure, with no associated rubble spreads, and some apparently more substantial structures, which may have had tiled roofs or brick floors, indicated by rubble spreads. A large walled complex has been identified in the central part of the citadel, with gates where two roads pass through and several probable buildings. It is likely that this complex may have been a palace.

One large rectangular structure close to the possible palace has been confirmed by excavation as a substantial brick-lined water tank. Further north a rectangular concentration of probable building debris probably reflects the remains of another substantial building complex, perhaps associated with a courtyard or temple. A small Hindu temple sits on this area of slightly raised ground.

The possible remains of further buildings and roads have been detected to the north and west of the fortified area, whilst to the south a broad in-filled moat was detected together with some possible furnace bases along the northern side of a mound of metal-working debris. To the east of the citadel, survey around the Eastern Stupa has revealed a rectilinear complex of walls, buildings and brick-lined water tanks extending over several hectares; this appears to be a major religious complex.

Figure 1: Geomagnetic survey near the Eastern Stupa, Tilaurakot.

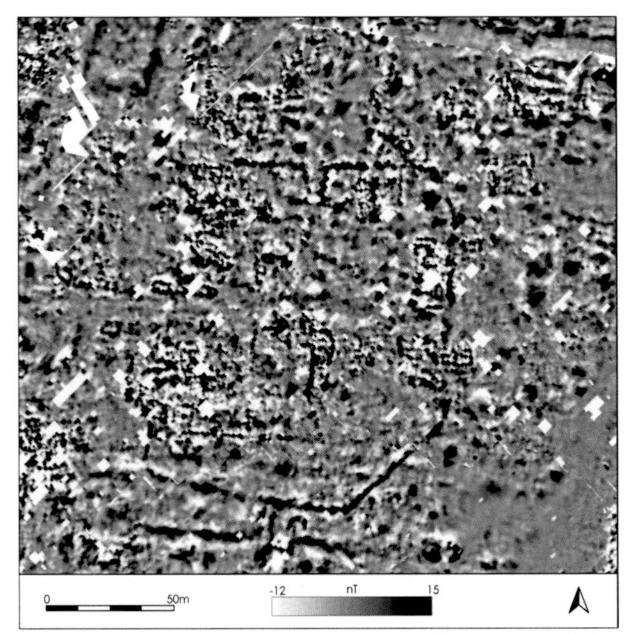

Figure 2: Sample of geomagnetic survey from the citadel of Kapilavastu, Tilaurakot

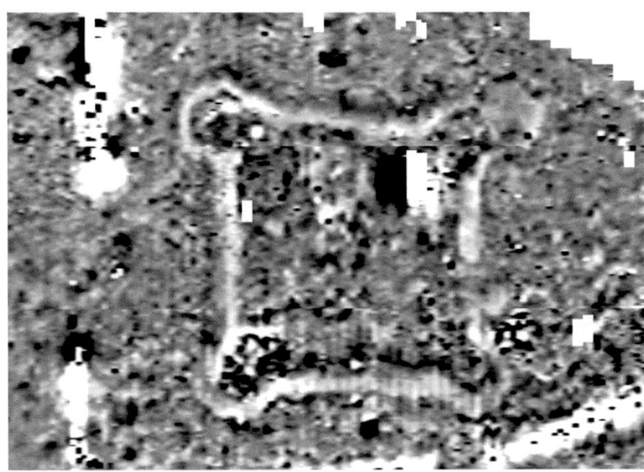

Figure 3: Survey results from Dohani (displayed white -12nT to black +15nT).

Kudan

The survey covered an early historic religious site comprising two substantial brick temples with a large water tank to the east. Many geomagnetic anomalies were detected, some of which almost certainly reflect significant archaeological remains including possible brick stupas, circumambulatory paths and occasional wall remains. Survey over a large mound in the south-east of the site revealed the remains of a probable third large temple.

Araurakot

This is an early historic defended site, currently used for grazing and cricket. Strong geomagnetic anomalies along the eastern and southern sides of the site show that the ramparts there were constructed of fired brick, whereas the ramparts to the north and west were predominantly constructed of local subsoils. Five stirrup bastions were also detected. The remains of at least two probable buildings were detected in the south-east of the interior, one within a walled compound.

Sisaniya

Survey of an extensive flat mound produced the plan of a former settlement, with the remains of buildings, roads and small open areas, all broadly aligned north-south and east-west. A substantial brick-lined tank was detected to the north of the settlement together with two rows of probable brick kilns. The principal road through the settlement could be part of a much more extensive north-south routeway with various other early historic sites along its course, such as Dohani, Araurakot, Niglihawa and Sagrahawa.

Dohani

The main survey area contained a broad low mound previously thought to be the remains of a temple. Geomagnetic survey revealed the plan of a small fort measuring approximately 50m across with a stirrup bastion at each corner and possible external ditch.

Chitradei

The protected zone here comprises a central mound (former temple) covered in jungle. Survey around the mound detected the remains of further brick-built structures and areas of probable brick rubble.

Sagrahawa

The survey detected probable remains of brick structures close to previously excavated areas. Some of these almost certainly reflect large buildings and enclosures while others may reflect small buildings, platforms or stupas. The detected features at Sagrahawa are broadly aligned with the four cardinal points, as has been the case with all the sites surveyed so far in and around Tilaurakot.

Bibliography

Coningham, R. A. E., Acharya, K. P., Strickland, K. M., Davis, C. E., Manuel, M. J., Simpson, I. A., Gilliland, K., Tremblay, J., Kinnaird, T. C. and Sanderson, D. C. W. (2013) The Earliest Buddhist Shrine: Excavating the Birthplace of the Buddha, Lumbini (Nepal). *Antiquity* **87**(338): 1104-1123.

Schmidt, A., Coningham, R. A. E., Strickland, K. M. and Davis, C. E. (2011) A Pilot Geophysical Evaluation of the Site of Tilaurakot, Nepal. *Ancient Nepal* **177**: 1-16.

A largescale simultaneous magnetometer and electromagnetic induction survey at Stična Hillfort, Slovenia

Chrys Harris[1,2], Ian Armit[2], Finnegan Pope-Carter[1], Graeme Attwood[1], Lindsey Büster[2] and Chris Gaffney[2]

[1]Magnitude Surveys Ltd, Bradford, United Kingdom; [2]University of Bradford, Bradford, United Kingdom

c.harris@magnitudesurveys.co.uk

The Iron Age of Slovenia was a period of dynamic cultural exchange and development (Armit and Büster 2016, Armit *et al.* 2014). The expansive site of Stična hillfort in central Slovenia constitutes one of the largest Iron Age settlements in the region. A simultaneous ca. 12ha fluxgate gradiometer and electromagnetic induction (EM) survey was undertaken across the site as part of the Encounters and Transformations in Iron Age Europe (ENTRANS) project. Increasingly, LiDAR survey had been effective for mapping such scales of sites in Slovenia; geophysical investigations have been hampered by the extensive forest area covering the country. The work at Stična is therefore unprecedented in scale for ground-based geophysical hillfort studies in Slovenia. The geophysical investigations sought to comprehensively map the majority of the site and aimed to more accurately integrate the work of previous archaeological investigations.

Figure 1: Fluxgate gradiometer results at Sticna hillfort (white to black: -1 to 2 nT).

The magnetic and EM data were collected using a bespoke cart-based system that the operator pulled by hand. The configuration of the sensors provided a line separation of 1.0m for the magnetic data and 2.0m for the EM data. Data at a sampling rate of 10Hz and 5Hz for the magnetic and EM data respectively, using an RTK GPS for positioning. As the survey at Stična Hillfort was conducted at the end of summer, it presented a challenging survey environment. To minimise sensor drift and inaccuracies, data collection was limited to the early morning and dusk hours, due to intense heat and sun. Tie-in lines for calibrating instrument drift have not been found successful with the CMD Mini Explorer EM instrument used for the survey; minor instrument drift was found to be effectively corrected with a rolling median following data collection. The CMD Mini Explorer collected quadrature-phase and in-phase measurements over three soil volumes. From relatively shallow to deepest, these areas: C1, C2 and C3, for the quadrature-phase, and I1, I2 and I3, for the in-phase. Combined with the fluxgate gradiometer results, seven unique datasets revealed both broad-scale and fine resolution of the settlement at Stična.

The results of the geophysical surveys at Stična have confirmed an organised dense internal settlement within the bounds of the hillfort. The detailed spatial resolution provided by the fluxgate gradiometer survey delineates a range of different types of features, which represent intensive settlement activity. Several distinct areas of activity have been delineated in the magnetic results. The southern half of the site contains a potentially earlier settlement area with evidence for limited discrete areas of potential industrial activity. To the north of this is potential earlier settlement, is a structured settlement area characterised by a more regular grid patterning, which includes enclosures, pits, ditches and possible roads or tracks. In contrast, the northern half of the site is markedly different in appearance than the southern half, with a less structured pattern for settlement activity.

The EM results do not have as fine a resolution as the magnetic results, owing in part to the differences in sampling density. Compared to the magnetic results, the EM results are more useful for understanding broad trends within the sites. Both the quadrature-phase and in-phase results reveal the rampart features, which have been detected as three concentric sets of curvilinear

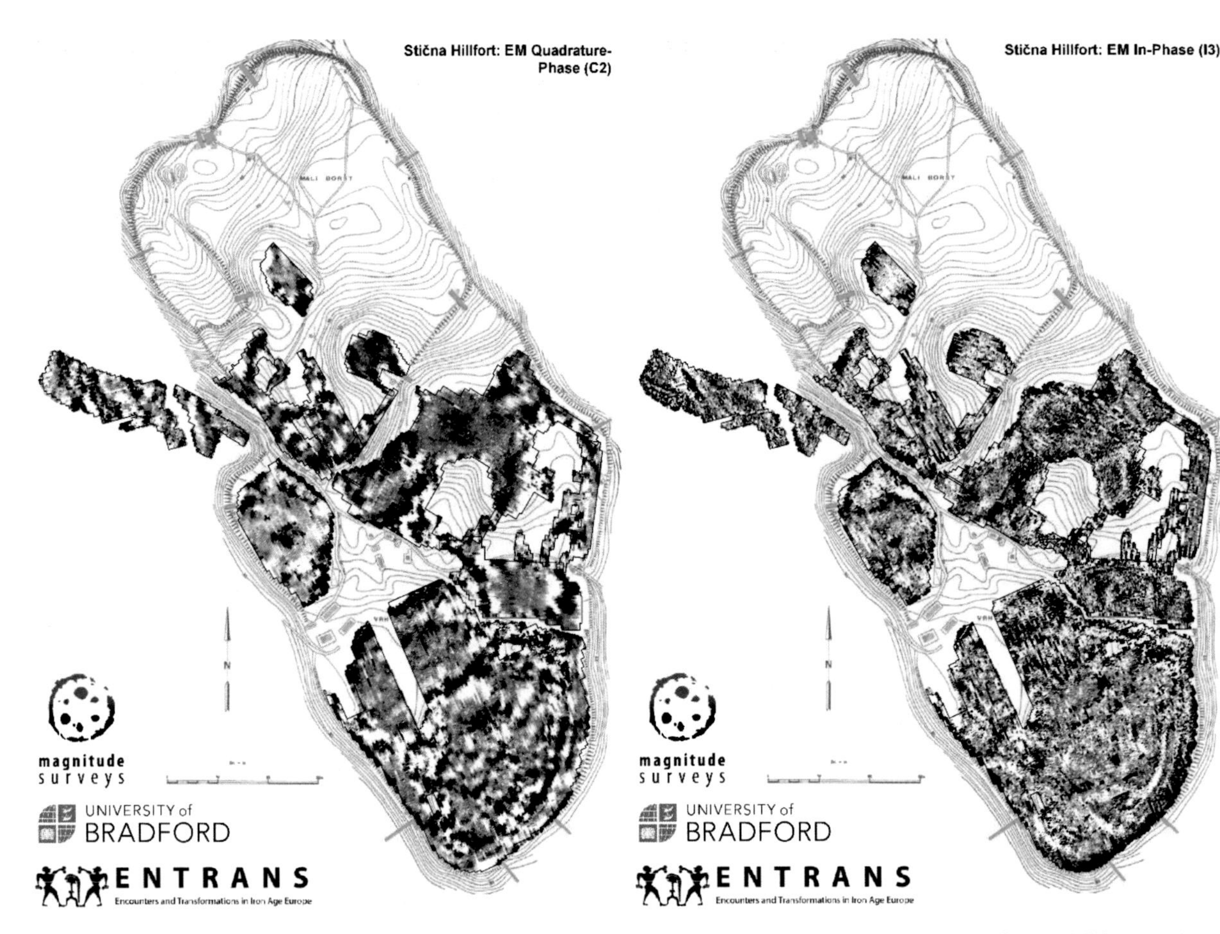

Figure 2: EM quadrature-phase results at Sticna hillfort (white to black: 5 to 95 percentile).

Figure 3: EM in-phase results at Sticna hillfort (white to black: 5 to 95 percentile).

anomalies extending across the eastern end of the site. The correlation of discrete pit-like anomalies in the magnetic and EM results support the interpretation for fired or kiln-like features. As well as the correlation between detected anomalies, the magnetic and EM results showed good correlations for areas that appeared devoid of any anomalies.

Alongside other, more limited programmes of geophysical survey at other sites in Slovenia and Croatia, supported in the latter case by targeted excavation, the work has begun to demonstrate that Iron Age settlements in the region were densely occupied with highly organised interiors. In this respect, they resemble settlement interiors in the Mediterranean world to the south rather than those of the 'barbarian' north (Armit *et al.* 2012, 2014). This contributes to the overall results of the ENTRANS Project in demonstrating the deep social impact of cultural interactions between societies in this region during the Iron Age.

References

Armit, I. and Büster, L. (2016) Meet the barbarians: taking a fresh look at the Iron Age of south-east Europe. *Current World Archaeology* **80**, 38-42.

Armit, I., Gaffney, C. and Hayes, A. (2012) Space and movement in an Iron Age oppidum: integrating geophysical and topographic survey at Entremont, Provence. *Antiquity* **86**, 191-206.

Armit, I., Potrebica, H., Črešnar, M., Mason, P. and Büster, L. (2014) Encounters and transformations in Iron Age Europe: the ENTRANS Project. *Antiquity* **88**: 342.

Geophysical surveying in Egypt and Sudan periodical report for 2015-2016

Tomasz Herbich[(1)]

[(1)]Institute of Archaeology and Ethnology, Polish Academy of Sciences, Warsaw, Poland

herbich@iaepan.edu.pl

Political issues in Egypt were the reason for fewer projects being carried out in the field and this translated itself into a growing interest of researchers in the territory of Nubia and northern Sudan. With the number of projects in Sudan surpassing those in Egypt, the author has revised the formula of his biannual reports to include from now on a presentation of work on archaeological sites in Sudan. In Egypt, geophysical surveys were conducted at Heliopolis, Karnak, Tanis, Kom Bahig and Kom el-Dahab (the latter two are new projects, the others a continuation of earlier research). The prospection in Sudan encompassed the Affad basin, Doukki Gel, Selib, Argi and Naga.

The electrical resistivity method was applied at one site (Heliopolis). The magnetic method was predominant, using fluxgate gradiometers FM256 with a sampling density of 0.5m x 0.25m, in grids 20 x 20m, in parallel mode (zigzag mode only in Selib). An ADA05 apparatus by Elmes was used for the resitivity survey. The fieldwork was carried out by Robert Ryndziewicz, Krzysztof Kiersnowski, Dawid Święch and Jakub Ordutowski.

EGYPT

Heliopolis/El-Matariya *(Old Kingdom and Late period; project of Leipzig University, project director Dietrich Raue)*

The electrical resistivity survey aimed to trace the remains of the temple of Re-Horakhty and Atum preserved under 2m of alluvial mud of modern date (for earlier work, see Herbich 2013, 241–242 and Herbich 2015, 208–209). Profiling with an asymmetric Schlumbereger array AM=7m and MN=1m covered an area of 0.4ha in the south-western corner of the site, still not obscured by modern rubbish dumped over the past ten years. Elongated and parallel structures of higher resistivity recorded by the survey could be a reflection of stone architecture.

Karnak *(New Kingdom, Centre Franco-Égyptien d'Étude des Temples de Karnak, Christoph Thiers)*

Work carried out in Karnak in 2006 had shown the efficacy of the magnetic method for recording remains of architecture in the temple of Amun (Herbich 2007, 189). Current prospection in the northern part of the precinct, on the southern side of the sacred lake, around the Ptah temple and between VIII and IX pylons, covered an area of 2.3ha and mapped the remains of mud-brick architecture of unidentified nature.

Kom Bahig *(Late Period and Ptolemaic-Roman; Centre for Alexandrian Studies, Marie-Dominique Nenna)*

The site on the north-western fringes of the Nile Delta covers 17ha to judge by the pottery scatter. Remains of mud-brick and stone architecture were observed in places, especially on two mounds, 10 and 15m high respectively, covering a combined surface of about 3.8ha, of which 2.8ha were surveyed. Architecture forming a rectangular block, 65 x 130m, was traced on the western mound (Fig. 1). The detailed magnetic image permits a recreation of the street network and the outline of particular buildings. The block of architecture does not correspond to the mound, suggesting the natural origin of the hill. The architecture traced on the other mound was irregular in layout.

Kom el-Dahab *(Ptolemaic-Roman; Chicago Oriental Institute, Gregory Marouard)*

The site is situated on an island, about 40 ha in area, lying about 2km from the western edge of Lake Menzaleh. Relics of unexplored architecture that are visible in satellite images are for the most part regular, following a strictly orthogonal grid inspired by the Hellenistic Hippodamian plan; a preliminary interpretation of these structures has been carried out on this basis (Marouard 2014). A magnetic field survey was carried out in six areas (combined total of 3.6ha), identifying a theatre (Fig. 2), large public buildings (presumed temples), storerooms and a

Figure 1: Kom Bahig. Magnetic map.

Figure 2: Kom el-Dahab. Magnetic map.

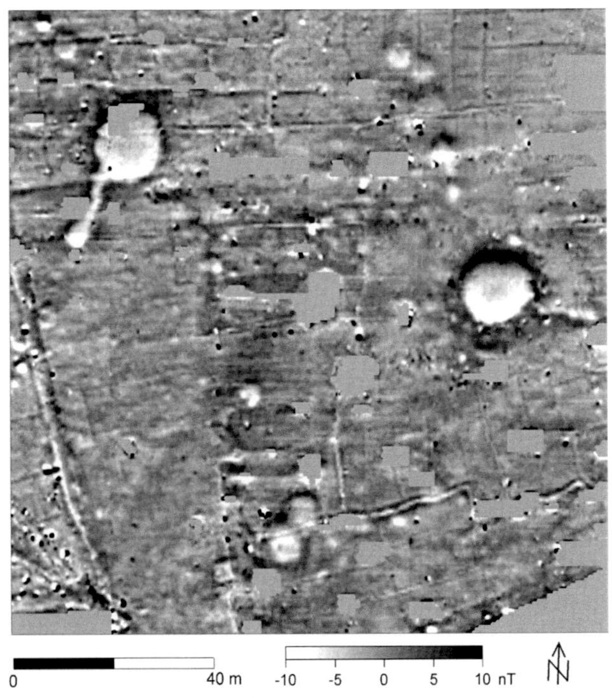

Figure 3: Doukki Gel. Magnetic map of Roman theatre.

lighthouse. The magnetic image helped to clarify building plans traced from satellite photos.

Tanis *(Late and Hellenistic-Roman, École pratique des hautes études, François Leclére)*

Work continued southward of the areas mapped in 2014–2016 (Herbich 2015, 210, Leclere 2016). Mapping 10ha (altogether 45ha since 2014) revealed city architecture from the Late period and a number of industrial features of presumably Hellenistic–Roman date. A section of the northern part of the temenos of the temple of Horus and remains of a collonade known from *Desription de l'Égypte* were traced.

SUDAN

Affad Basin *(Paleolithic and Neolithic, Polsih National Science Centre and Institute of Archaeology and Ethnology, Polish Academy of Sciences, Marta Osypińska)*

The survey carried out within an area of recorded settlement traces aimed to reconstruct the late Pleistocene and early Holocene paleolandscape. Magnetic measurements of 3.4ha gave a precise image of the changing Nile bed.

Doukki Gel *(New Kingdom, Napatan and Meroitic, University of Geneve, Charles Bonnet)*

A town established by pharaohs of the 18th Dynasty of Egypt, located 1km north of Kerma. Research was carried out in two areas: directly next to a complex of temple and palatial architecture and in a palm grove outside the archaeological reserve. Inside the reserve, the survey noted remains of old excavations along with pits and archaeological dumps. Outside the reserve, the prospection led to the discovery of a number of circular features from 4m to more than 20m in diameter. Buttresses or towers were traced on the external side of the two largest features (Fig. 3). These were interpreted as temples by analogy to earlier discoveries (Honegger 2009). The temple layout and the smaller circular features suggest a settlement that could be connected to Nilotic or Central African cultures.

Selib *(Meroitic and Christian, Polish Center of Mediterranean Archaeology - PCMA, Bogdan Żurawski)*

The survey of the Meroitic settlement mapped houses built of mud brick and traced the extent of the settlement. The prospection also established the changing bed of the Nile. Combined with earlier surveying, the magnetic map now covers an area of 5.6ha.

Argi *(Meroitic, PCMA, Bogdan Żurawski)*

The cemetery was discovered when a few mud-brick tombs were recorded on the surface, exciting the attention of illicit diggers. The magnetic survey preceded salvage research on the site. The number of tombs was established thanks to prospecting of an area of 1.4ha and the extent of the burial ground (covering 0.8ha) was traced.

Naga *(Meroitic; Egyptian Museum in Munich, Sylvia Schoske, Karla Kröper, Dietrich Wildung)*

The survey covered 0.8ha around the Lion Temple, mapping the course of stone walls, which were earlier partly investigated by clearing the tops. The prospection mapped anomalies typical of furnaces and hearths. Regularly spaced anomalies along the temple enclosure wall may correspond to remains of granite and basalt statues.

Bibliography

Herbich, T. (2007) Geophysical surveying in Egypt: periodic report for 2005-2006. *Študijné zvesti Archeologického ústavu* **41**: 188-191.

Herbich, T. (2013) Geophysical surveying in Egypt: periodic report for 2011-2012. In: W. Neubauer, I. Trink, R. B. Salisbury and C. Einwögerer (eds.), *Archaeological Prospection. Proceedings of the 10th International Conference on Archaeological Prospection, Vienna*, 241-245.

Herbich T (2015) Geophysical surveying in Egypt: periodical report for 2013-2015. *Archaeolgia Polona* **53**: 206-212.

Honegger M, Bonnet C & collab. (2009) Archaeological excavations at Kerma (Sudan), *Documents de la mission archéologique suisse au Soudan* **1.**

Leclére F, Payraudeau F, Herbich T (2016) Nouvelles recherches sur Tell Sân el-Hagar (Tanis), *Égypte, Afrique & Orient* **81**: 39-52.

Marouard G (2014) Kom el-Dahab interpreted. *Egyptian Archaeology* **45**: 25-27.

Not-so good vibrations: removing measurement induced noise from motorized multi-sensor magnetometry data

Alois Hinterleitner[1,2], Immo Trinks[2], Klaus Löcker[1,2], Jakob Kainz[3], Ralf Totschnig[1], Matthias Kucera[2] and Wolfgang Neubauer[2,3]

[1]Zentralanstalt für Meteorologie und Geodynamik (ZAMG), Vienna, Austria; [2]Ludwig Bolzmann Institute for Archaeological Prospection and Virtual Archaeology (LBI ArchPro), Vienna, Austria; [3] University of Vienna, Vienna, Austria

alois.hinterleitner@zamg.ac.at

Introduction

Today motorized multi-sensor magnetic prospection systems are widely used for detailed efficient large-area surveys. The magnetometer systems are operated at relatively high speed on all suited terrains. The drawbacks of such speedy data acquisition are the necessary required high sampling rates (≥ 50 Hz) and considerable involved perturbations caused by vibrations and oscillations of the sensor cart. The motorized fluxgate type magnetometer prospection systems set up and used at the LBI ArchPro together with ZAMG *Archeo Prospections®* utilize eight gradiometer sensors mounted in parallel with a horizontal distance of 25 cm, towed by quad-bikes at speeds in general between 15 and 25 km/h (Fig. 1).

Noise observed in the prospection data with characteristic frequencies of 16.7 Hz and 50 Hz, which is due to electrical currents in the ground, and typically affects data recorded at high sampling rates, can easily be filtered out because of the known and constant frequency. However, measurement induced noise caused by vibrations and oscillations of the entire sensors system is spatially variable in regard to strength and frequency, depending on the surface conditions, the resonance frequency of the cart, the vibrations produced by the tow-vehicle, the geographic direction of driving, and the velocity of the entire system. These disturbances are therefore difficult to remove from the data, especially because they occur in the same amplitude and spatial range as the magnetic response caused by archaeological features of interest, such as pits, postholes and ditches. While the special construction of the sensor carts of the LBI ArchPro prospection systems attempts to mitigate the negative effects as much as possible, they cannot be avoided completely under all environmental conditions encountered.

Figure 1: LBI ArchPro motorized fluxgate magnetometer system with 10 sensors (Foerster Ferex CON650) mounted in an array on a cart with 25 cm horizontal distance.

The vibrations and oscillations of the fluxgate type gradiometer probes – measuring the vertical component of the magnetic field – and sensor system produce perturbations in the magnetic data that have three different causes:

- variations of the sensor orientation in the magnetic field,
- oscillations and directional variations of the entire magnetometer system, including tow vehicle, in the magnetic field, and
- variations of the distance of the gradiometer sensors to the ground (and contained archaeological features)

We present two different methods to remove the magnetic disturbances produced by these variations, which attempt to leave the magnetic effects that are due of the buried archaeological features unaffected.

Method 1

The first method removes vibrations that result a sinusoidal signal component within a predefined frequency range. To this purpose, the parameters of a single sinus wave, that is its frequency, its amplitude and its phase shift, are determined by application of a best-fit algorithm for all locations, within the high-pass filtered data of each magnetometer sensor. The high-pass filtering step removes frequency components from the data below a lower threshold value of the predefined frequency range, in order to remove the influence of strongly magnetic data values and spatially large-scale background variations in the measured magnetic field. All the individual best-fitting sinusoidal wave trains are compiled and subsequently subtracted from the unfiltered magnetic data. The advantage of this algorithm, compared to a simple frequency filtering, is that the frequency can vary in-between a predefined range and that only the local most dominant disturbing frequencies are removed. This algorithm is applied to each sensor separately. It removes mainly perturbations of type 1 and 2 that for our fluxgate prospection systems occur in the frequency range of 4 to 8 Hz.

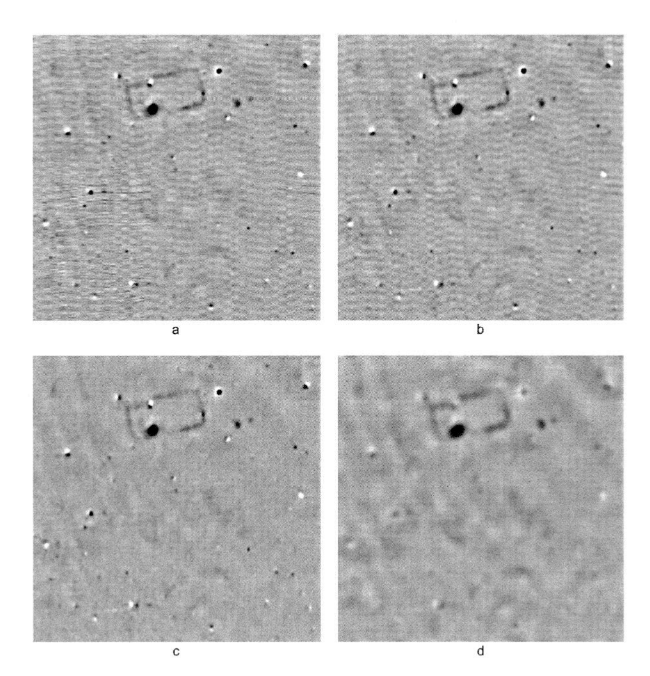

Figure 2: Detail of a magnetic survey in Krummnussbaum/ Austria conducted by Klaus Löcker and Ralf Totschnig, using a motorized LBI ArchPro fluxgate magnetometer system with eight sensors. a) raw data; b) after removal of a fitted sine wave within a frequency range of 4-8 Hz using Method 1; c) same as b with additional application of filter Method 2; d) low-pass filtered visualization of the raw data using a Gaussian filter with 2.0 m filter radius. Each image area measures 68×66 m. Greyscale visualization: 254 values from black [3 nT] to white [-2 nT].

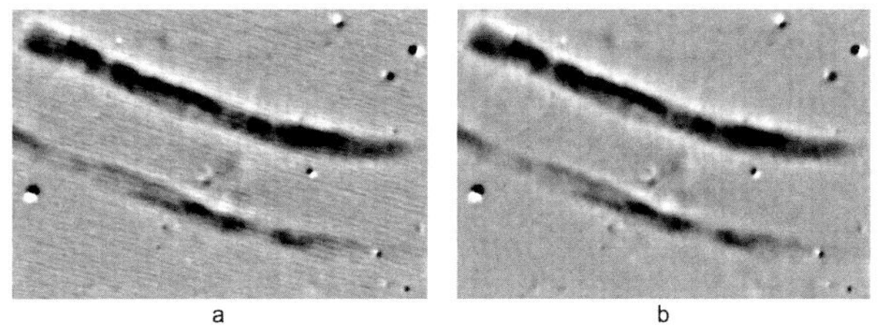

Figure 3: Detail of a magnetic survey conducted in Westphalia/Germany, conducted by Jakob Kainz using a motorized LBI ArchPro fluxgate magnetometer system with eight sensors. a) raw data; b) after removal of a fitted sine wave within a frequency range of 4-8 Hz using Method 1 and Method 2. The image areas measure 56×39 m. Greyscale visualization: 254 values from black [6 nT] to white [-4 nT].

Method 2

The second method has a completely different approach: it tries to determine magnetic influences caused by variations of type 2 and 3 that simultaneously affect all sensors on the cart. As the sensor cart moves not only up and down while being driven across the field, but is also tilting from one side to the other, we determine the influence of the left half and right half of the sensor array separately, in the direction of motion. This method also requires a predefined time-range within which to estimate the influence on the sensor array. It is limited to variations within an amplitude range of ±1 nT. In contrast to the first method, this approach is not restricted to a certain wave form or a certain frequency range. In order to leave magnetic values of the archaeological features unchanged, the data values of the sensor array are analyzed statistically: high standard deviations across the array width indicate important magnetic anomalies. The corresponding magnetic data values are subsequently excluded from the estimation of the noise component. This method works very well for low frequency changes in the range below 3 Hz, and for measurement induced signal variations that do not have a sinusoidal wave form.

Results

Figure 2 illustrates the application of Method 1 (Fig. 2b) and Method 1 and Method 2 (Fig. 2c) to a detail in a larger magnetic prospection survey (Fig. 2a) conducted in Krummnussbaum/Austria. The data were recorded using a motorized LBI ArchPro magnetic system with eight fluxgate sensors (Förster FEREX CON650) mounted in an array with 25 cm horizontal distance. Even though the influences produced by mechanic vibrations have amplitudes smaller than 1.0 nT, these effects are recognizable in the data visualizations. This effect is even more pronounced when the archaeological features also display very low amplitudes, such as those observed on Neolithic sites in loess soils. Both methods permit data visualizations with significantly less blurring of the archaeologically interesting anomalies compared to conventional filtering methods (Fig. 2.d)

This proposed approach is able to reconstruct the correct shape of anomalies as it can be seen in Figure 3 in case of the ditch anomaly.

Electrostatic and GPR survey: case study of the Neuville-aux-Bois church (Loiret, France)

Guillaume Hulin[(1)], François Capron[(2)], Sébastien Flageul[(3)], François-Xavier Simon[(1)] and Alain Tabbagh[(3)]

[(1)]Inrap, Paris, France; [(2)]Inrap, Saint-Cyr-en-Val, France; [(3)]Université Pierre et Marie Curie, UMR Metis, Paris, France

guillaume.hulin@inrap.fr

Archaeological investigations in urban contexts and/or inside buildings are often a difficult issue for archaeologists. Excavation areas are usually limited by practical considerations. In these cases, geophysics can be applied even if this type of context remains a challenge. The main difficulties are caused by magnetic and electromagnetic disturbances (which prevent magnetic and often low-frequency EM methods) but are also due to the nature of the ground surface which forbids galvanic contact with electrodes in case of DC resistivity survey.

For this type of study, GPR is often the only geophysical method used. Nevertheless, another method, the electrostatic method, can be very efficient although nowadays it is under used. This fact comes from the lack of off-the-shelf devices compared to the number of available GPR instruments. With the electrostatic method, the injection of the current is achieved by an open capacitor and the voltage generated on the ground surface by the resulting current distribution is measured using another open capacitor(s). This method which is also called 'capacitive resistivity' or 'capacitive coupled resistivity' corresponds to a generalization of the DC electrical method and the interpretation can be made using the same inversion processes. As the frequency cannot be lower than several kHz, its use is limited to shallow depths of investigation (less than 20 m in general) by induction effects, but this does not restrict its relevance in archaeological prospection. Thus, in contexts where the galvanic contact between the electrodes and the ground can be problematic, the electrostatic method allows the determination of the ground resistivity (Tabbagh *et al.* 1993; Dabas *et al.* 2000; Perez-Garcia *et al.* 2009) and is a perfect complement to GPR reflectors characterization.

This combination is especially useful on sites where soils have a high clay content as in the case of the Saint-Symphorien church at Neuville-aux-Bois (Loiret, France). Restoration of the building led to a preventive excavation which was performed by INRAP (Institut National de Recherches Archéologiques Préventives) in 2015-2016. Some test trenches were dug around pillars and outside the church. The church dates back to the end of the 15th century / beginning of 16th century but two older building phases (11-12th century and 13-15th century) were recognised during the excavation. A large ditch (more than 4 m wide and 2 m deep – filled by clay) was also excavated outside the church. This could be linked to the second phase of the church (13-15th century) and was dug as a fortification during the Hundred Years War.

Geophysics was employed to complete the excavation results which were obviously limited by practical aspects and the preventive archaeological context. Two techniques were implemented. A GSSI UtilityScan with a dual-frequency antenna (300/800MHz) was used for the GPR survey. A 'MPG' prototype from UMR Metis (Université Pierre et Marie Curie, Paris-Sorbonne) was used for the electrostatic survey (Fig. 1). The 'MPG' multipole, in a V-shaped geometry, consists of a transmitting dipole A-B (1.1 m apart) followed at 1.2 m by a first receiving dipole M1-N1 (0.9 m apart, channel 1) and at 2 m a second receiving dipole M2-N2 (1.85 m apart, channel 2). The corresponding depths of investigation are 1 m and 2 m.

Figure 1: Installation of the 'MPG' multipole on the site.

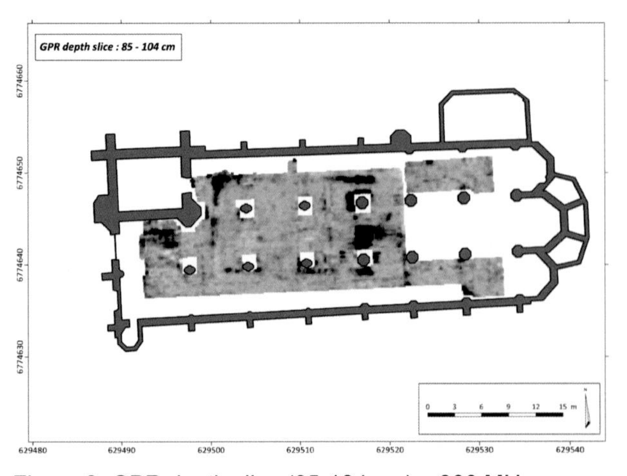

Figure 2: GPR depth slice (85-104 cm) – 800 MHz.

GPR results clearly reveal walls of the second phase of the nave (10x18 m) with an orientation slightly different from the present day church (Fig. 2). Electrostatic results are smoother than GPR due to the large volume of soil measured but give some useful information (Fig. 3). The smallest dipole indicates that the first meter of soil is resistive (more than 100 ohm/m). The larger dipole shows that a large part of the church has a lower electrical resistivity (less than 70 ohm/m) demonstrating the presence of more deeply conductive materials. Consequently, this affects the GPR reliability as it is well reputed that the presence of a conductive soil prevents propagation of high-frequency EM waves. From an archaeological point of view, the large conductive anomaly located at the south-west of the survey (larger dipole) could correspond to the ditch linked to the second phase of the church as observed during the excavation.

Once again, a combination of several methods and measurements of different physical parameters prove to be a key point for a better understanding of geophysical results and lead to a greater archaeological interpretation. In urban contexts and/or indoors, the range of geophysical methods is rather limited especially where the soil is conductive. In this case, the electrostatic method is an efficient method and should be used more often to complete and improve the reliability of GPR results.

Bibliography

Dabas, M., Camerlynck, C. and Freixas i Camps, P. (2000) Simultaneous use of electrostatic quadrupole and GPR in urban context: Investigation of the basement of the Cathedral of Girona (Catalunya, Spain). *Geophysics* **65**: 526-532.

Perez-Gracia, V., Caselles, J. O., Clapes, J., Osorio, R., Martınez, G. and Canas, J. A. (2009) Integrated near-surface geophysical survey of the Cathedral of Mallorca. *Journal of Archaeological Science* **36**: 1289-1299.

Tabbagh, A., Hesse, A. and Grard, R. (1993) Determination of electrical properties of the ground at shallow depth with an electrostatic quadrupole: field trials on archaeological sites. *Geophysical Prospecting* **41**(5): 579-597.

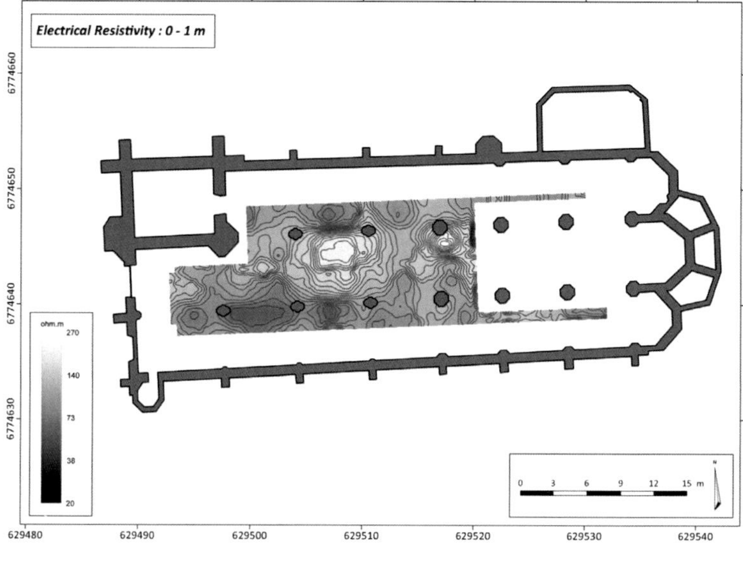

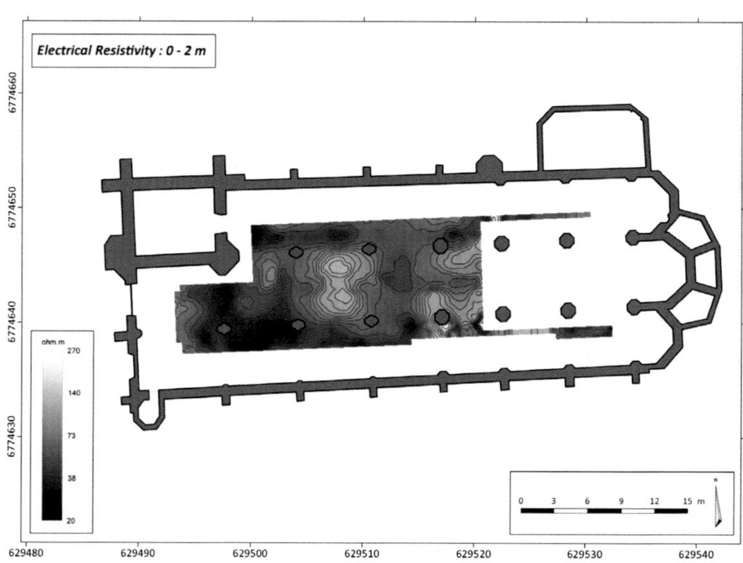

Figure 3: Resistivity results obtained with the 'MPG' Multipole (depths of investigation: top, 0 -1 m; bottom, 0 - 2 m).

Medieval monks seen through a modern landscape

Freya Horsfield[1]

[1]Durham University, Durham, UK

freya.horsfield@durham.ac.uk

Integration and interpretation of archaeological prospection are key aspects of a project which aims to understand the landscapes associated with the first Cistercian monastery in the North of England, Rievaulx Abbey.

The physical impact of monasteries on the wider landscape has not generally been well understood, despite the important role played by Medieval monasteries in the creation of European cultures (Bond 2004). New approaches have transformative potential for research into the monastic order whose contested reputation has been especially entwined with practical matters in around 900 years of teaching and scholarship. The Cistercian Order has been characterised as one of the last great pan-European phenomena before the rise of the nation state (Tobin 1995), was one of the most successful and spatially extensive of the coenobitic orders which flourished in the High Middle Ages, and has attracted a significant proportion of the scholarship into European monastic orders (Burton and Kerr 2011).

The landscape record associated with monasticism is challenging to understand. The physical remains may be buried from unaided human view, religious expression can obscure interpretation of utilitarian activity, and the approaches required to understand landscape lie at the intersection of multiple disciplines. Monastic archaeology has tended to be been dominated by study of discrete sites and standing buildings, particularly those at the monastic core, and the discipline has considerable scope for transformation through the critical integration of techniques such as GIS (Gilchrist 2014).

Figure 1: Rievaulx Abbey, the first Cistercian house in the North of England. Photo credit: Clarissa DiSantis Humphreys.

The Rievaulx Landscapes project aims to develop an integrative workflow for monastic landscape research. The project uses the physical landscape record available through extensive LiDAR coverage, as a basis to interpret multiple other lines of evidence. These approaches are intended to help understand how Rievaulx Abbey managed its substantial land interests. The abbey acquired interests across the North of England, from the Tees through to South Yorkshire, with its heartland in the Western North York Moors and the adjoining Vale of Pickering. The conventual core of the abbey is considered a heritage icon, was taken into national guardianship in 1917, and is now on the National Heritage List for England (Historic England 2017: entries 1175724 and 1012065).

At the abbey's conventual core in the Rye Valley, our understanding of when and how the monastic precinct was laid out depends on a theory devised more than 100 years ago. This theory was devised in collaboration using documentary references and field observations (Atkinson 1899, Rye 1905). Atkinson and Rye suggested that the River Rye had flowed down the centre of the valley when the monastery was established, and that Rievaulx carried out major diversions to the river thus increasing the land within the monastic precinct. Modern measurement estimates that these postulated diversions affected around 60% of the land within the precinct as defined in the National Heritage List. Atkinson and Rye also suggested that Rievaulx created canals which facilitated the transport of building stone from quarries upstream on the River Rye. Several archaeological field observers have however since expressed reservations about aspects of the Atkinson-Rye theory, which they suggest placed too much reliance on an interpretation of the historic documents which is physically unfeasible. Lidar analysis (begun at the University of Birmingham and now at the University of Durham) has further suggested that the evidence for watercourse management to the north of the abbey may be more complex and extensive than previously thought. In 2015, the University of Bradford began an ongoing programme of geophysical survey into these locations.

The Cistercian Order had an historic reputation for the transformation of marginal land, yet Donkin suggested that there were relatively few examples of Cistercian estates which were founded on primeval land. Donkin further suggested that one of the few examples of Cistercian occupation of primeval "waste" was land in the Vale of Pickering given to the abbey by King Henry II (Donkin 1978). Research at Durham University is however suggesting a more complex and interesting narrative for Rievaulx's estates in the Vale of Pickering. Henry II instructed

Figure 2: Rievaulx Abbey: looking beyond the monastery. Photo credit: Clarissa DiSantis Humphreys.

his officials to declare "waste" before giving it to Rievaulx Abbey (Jamroziak 2005), and the strategic location of this land suggests a complex motivation by Henry II for the gift of this land to a Cistercian house. As defined by watercourses named in the Rievaulx cartulary, the Pickering "waste" commanded a substantial frontage along the River Derwent. Place-name evidence suggests that the marsh may have been occupied by the time of the Domesday survey in the 11th century (Darby and Versey 1975), so it is questionable whether this land was "primeval waste" at the time of Cistercian settlement as suggested by Donkin.

By the time of the Dissolution of the monastery, Rievaulx had developed multiple granges in the Vale of Pickering. These outlying agricultural units were important to its substantial sheep farming activities (Jamroziak 2005). The extent of one of these granges within the land given by Henry II in the Vale of Pickering has been traced using 19th century tithe documentation. Certain Cistercian newly-cultivated land was exempt from payment of church tithes and this exemption could pass to successor landowners, which was a valuable exemption sometimes recorded in tithe documentation (Platt 1969). Analysis of tithe documentation in the Vale of Pickering has revealed the extent of Deerholme Grange, which was one Rievaulx's granges listed in the Dissolution accounts for the abbey. Such documentary evidence is facilitating spatial targeting of archaeological prospection. The Vale of Pickering is flat, and has been intensively farmed and drained for hundreds of years. Surface expression of archaeology therefore may be extremely slight and challenging to discern. Within the tithe-exempt extent of Deerholme Grange, features suggestive of former water channels are being delineated using LiDAR data. Once the LiDAR feature mapping is complete, sedimentary analysis of features within the extent of Deerholme Grange could offer empirical evidence for drainage history. This Cistercian grange has a clear documentary description as "waste" before Cistercian occupation. Deerholme Grange therefore offers a location at which it may be possible to compare physical evidence against the Cistercian reputation for landscape transformation.

Aside from the future elements of this research, however, it is already clear that the former marshland owned by Rievaulx Abbey in the Vale of Pickering is less well understood and consequently less protected by law and practice than the charismatic standing buildings at the abbey's conventual core. The historic landscape in both study areas is complex, but the relationship between the challenges faced by Medieval and modern occupants of this land is noteworthy. The project workflow is therefore intended to provide information useful to landscape managers and to modern communities which need to balance multiple challenges as part of living sustainably in this flood-prone landscape.

Bibliography

Atkinson, J. C. (1889) *Cartularium abbathiæ de Rievalle, Ordinis cisterciensis fundatæ anno MCXXXII*. Durham: Surtees Society.

Bond, J. (2004). *Monastic landscapes*. Stroud: Tempus.

Burton, J. E. and Kerr, J. (2011). *The Cistercians in the Middle Ages*. Woodbridge : Boydell Press.

Darby, H. C. and Versey, G.R. (1975). *Domesday Gazetteer*. Cambridge: Cambridge University Press.

Donkin, R. A. (1978) *The Cistercians: studies in the geography of medieval England and Wales, Studies and texts / Pontifical Institute of Mediaeval Studies*. Toronto: Pontifical Institute of Mediaeval Studies.

Gilchrist, R. (2014). Monastic and Church Archaeology. *Annual Review of Anthropology* **43**: 235–250.

Historic England (2017) *The National Heritage List for England*. https://historicengland.org.uk/listing/ Accessed 12.5.2017

Jamroziak, E. (2005) *Rievaulx Abbey and Its Social Context, 1132-1300: Memory, Locality, and Networks*. Turnhout: Brepols Publishers.

Platt, C. P. S. (1969). *The monastic grange in medieval England: a reassessment*. London: Macmillan.

Rye, A. (1900). Rievaulx Abbey, its Canals and Building Stones. *The Archaeological Journal* **57**: 69–77.

Tobin, S. (1995). *Cistercians: Monks and Monasteries of Europe*. London: Herbert Press.

Changing faces: archaeological interpretations and the multi-stage archaeological prospection of the Roman town of Aregenua

Karine Jardel[(1)], Armin Schmidt[(2)], Michel Dabas[(3)] and Roger Sala[(4)]

[(1)]Service Archéologie – Département du Calvados, Caen, France; [(2)]Dr Schmidt - GeodataWIZ, Remagen, Germany; [(3)]École Normale Supérieure, Paris, France; [(3)]SOT Archaeological Prospection, Barcelona, Spain

A.Schmidt@GeodataWIZ.com

Introduction

The process of archaeological interpretation is dynamic. At any point in time it depends on the data available, their previous analysis and the human agents interacting with them. The relevant data may either be direct material manifestations (e.g. archaeological objects or excavation results) or indirect scientific results (e.g. aerial photographs, geophysical anomalies) and are in most instances incomplete. For example, excavation results may be limited due to taphonomic processes or contexts may have been indistinguishable during excavation. Geophysical data are limited by the techniques used, the methodology applied, and prevailing soil and environmental conditions. It is therefore necessary to change archaeological interpretations as new data are becoming available, which may lead to revisions or augmentations.

The Roman town of *Aregenua* (today's Vieux-la-Romaine, near Caen, Normandy) has been investigated for centuries through excavations, airphoto analysis and geophysical surveys. Each stage of this process led to updated archaeological interpretations of the site, which is exemplified here using geophysical results from the champ des crêtes.

Background

The Roman town of *Aregenua* was occupied from the 1st to 5th century AD and these phases were detected in the recent open-area excavation (1500 m^2) in the eastern forum, revealing a succession of public buildings. A program of geophysical investigations was started to provide a broader overview of the town's layout. In 2005 an earth resistance survey with a three-level ARP instrument (spatial resolution approximately 1 m) showed substantial structural remains buried on the site, with considerable depth extent. The area was subsequently re-investigated with a multi-coil FDEM system (DualEM421S) and a spatial resolution of approximately 0.25 m x 0.75 m that identified additional anomalies. In 2016 a 600 MHz GPR survey (IDS 5-antenna system) covered the majority of the area at a spatial resolution of 0.02 m x 0.20 m and provided detailed data, also with high depth resolution. The archaeological understanding of the site evolved accordingly and each new data set led to the augmentation and refinement of the archaeological model.

Archaeological Interpretation

Three areas proved particularly interesting in terms of the archaeological interpretation of the geophysical results and are highlighted in the subsequent diagrams. Figure 1 shows ARP apparent earth resistivity results (ARP 2 m; Fig. 1.a) and FDEM apparent conductivity data (coil separation 6.2 m HCP; high-pass filtered and unfiltered; Figures 1b&c). Figure 2 presents the same ARP data (Fig. 2.a) alongside the GPR data at a pseudo-depth of 1.2 m (Fig. 2.b). Especially in Area 3 the foundation walls of the buried structures show a high apparent resistivity and, surprisingly, also a high apparent conductivity. The latter is attributed to the large FDEM coil separation (6.2 m) in comparison to the burial depth. In HCP mode this can lead to an inversion of the signal's polarity. The

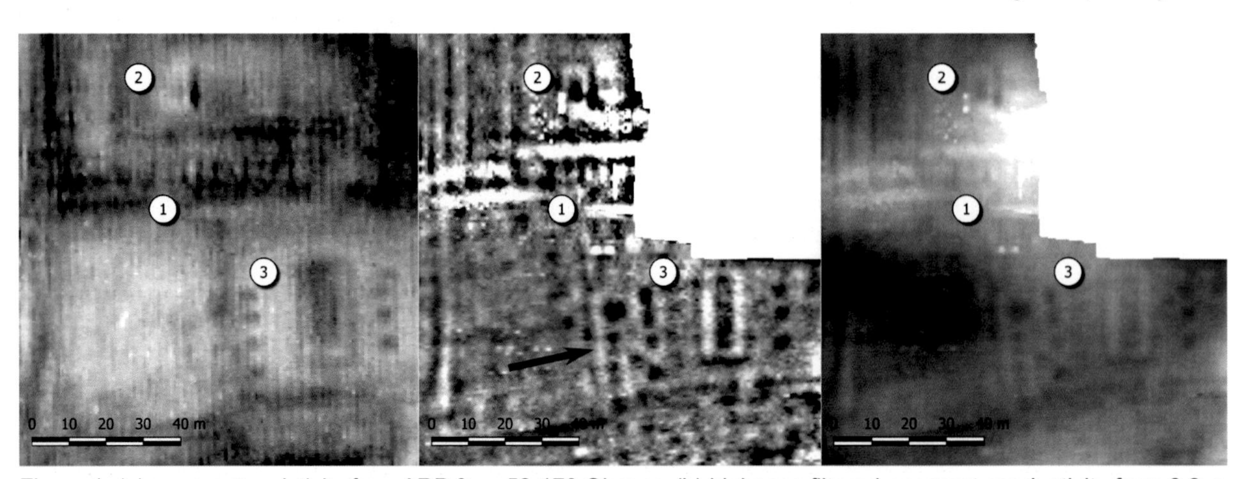

Figure 1: (a) apparent resistivity from ARP 2 m, 52-170 Ohm m; (b) highpass filtered apparent conductivity from 6.2 m HCP FDEM, ±100 mSi m-1; (c) unfiltered, 5-19 mSi m-1; black is high for all.

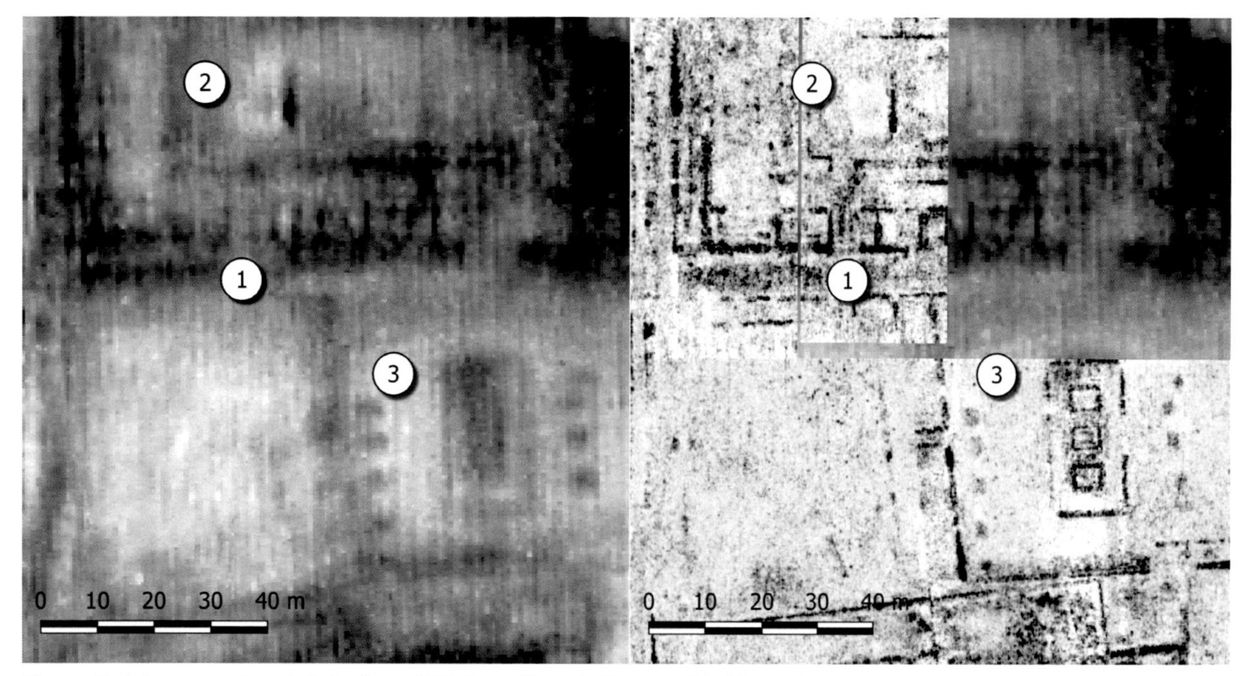

Figure 2: (a) apparent resistivity from ARP 2 m, 52-170 Ohm m; (b) GPR reflectivity; black is high for all.

interpretative clarity of the GPR data is also a result of their high spatial resolution.

1 Southern Entrance to the Forum

The earth resistance data indicated a gap in the gallery that forms the southern enclosure of the forum. However, the breaks in the northern and southern anomaly of the gallery appeared poorly defined and could have been caused by the removal of stones in more recent times. The FDEM data showed that the gap is flanked on its western and eastern side by linear structures, but a band of low conductivity anomalies obscured the rest of the gallery so that it remained unclear even, with these additional data, whether this was indeed the location of an entrance. The GPR results demonstrated clearly that there is a deliberate opening in the well-defined southern gallery, possibly with small structures inside the entrance way (possibly drainage or paving).

2 Possible Location of the Imperial Temple in the Forum

It is conceivable that, like in many other Roman towns, the temple for the Imperial cult was located in *Aregenua's* forum. In Area 2 the earth resistance data showed a rectangular region of low resistance (10 m x 15 m), bordered in the east by a narrow elongate high resistance anomaly. An excavation trench was placed over this anomaly and revealed a marble threshold with a worked face on its western side. If this was originally part of the temple, the low resistance region to its west could have been a platform in front of the temple's stairs. However, after the FDEM data were inspected the highpass filtered results (Fig. 1.b) seemed to show anomalies that enclosed the rectangular region and a further anomaly inside it. At the south-western corner of the region a 5 m long anomaly of pronounced low conductivity was visible. This led to the assumption that this was not, after all, an empty space but some structure, possibly related to the Imperial temple. However, the subsequently acquired GPR data characterised the rectangular region again as an area of uniform low reflectivity (down to a depth of 1.5 m), similar to the earth resistance data. A further detailed examination of the unprocessed FDEM data (Fig. 1.c) revealed that the low-conductivity anomaly in the south-western corner actually consisted of two individual readings along the survey transect. It is likely that a single small object near the surface (e.g. some modern metal debris) created this typical 'double-peaking' response and the subsequent highpass filtering and greyscale display merged the two peaks into a seemingly elongated anomaly. The linear FDEM anomalies on the western and eastern edge of the rectangular space are exactly aligned with the survey transects and continue, albeit weaker, to the north and south (similar to the long 'stripes' visible further west). It is hence possible that they are artefacts of data acquisition and interpolation. The archaeological interpretation, based on the geophysical data so far is that only a single threshold with substantial foundation remains from the building/temple, and that the area to its west was the platform in front of it. On other sites such platforms were found to be paved with pebbles. That there was a central feature on this platform, as indicated by the FDEM data, cannot be ruled out.

3 Ritual complex

The earth resistance and FDEM data south of the forum both show several wall-like anomalies that form an outer and inner enclosure around a rectangular central structure. Together with eight small rectangular features just inside the outer enclosure this area was interpreted as a ritual complex. After the FDEM data were acquired it became clear that the western section of the outer enclosure was not just a wall, but consisted of several cells that may have been shops or rooms (arrow in Fig. 1.b). Inspection of the GPR data then revealed that the central rectangular anomaly actually consisted of three separate entities. However, the cells in the western section of the enclosure were not visible in the GPR data.

Conclusion

The archaeological interpretation of geophysical data from the site changed as new data became available. The three different data sets were partly complementary but sometimes contradictory. Only their comparison alerted to possible non-archaeological causes for some of the apparently visible anomalies. The archaeological understanding of the site evolved with the acquisition of new and different data sets and highlights the interpretative nature of geophysical data analysis.

From large- to medium- to small- scale geophysical prospection

Jakob Kainz[1, 2]

[1]LWL - Archäologie für Westfalen, Münster, Germany; [2]University of Vienna, Vienna, Austria

jakob.kainz@lwl.org, jakob.kainz@univie.ac.at

Introduction

The paper focuses on large and small scale geophysical prospection and the results that such surveys can contribute to archaeological understanding. Examples shown are from surveys undertaken for the collaboration between the Ludwig Boltzmann Institute for Archaeological Prospection and the Landesverband Westfalen-Lippe - Archäologie für Westfalen and an excavation by the University of Vienna. Large scale surveys can contribute as much information towards the archaeological understanding of a landscape as well as about single features. Often such surveys, however, raise as many questions as they can answer. Thus, it often requires smaller, more detailed geophysical surveys and samples for laboratory studies to obtain a better insight into anomalies. Depending on the feature and the question to be investigated, different survey sizes, measuring raster, and methodologies can be applied. Examples of such can be seen in the collected magnetometry results, were a ditch is visible as a positive anomaly but is then a negative anomaly, graves which only appear after these have been excavated and back filled or a palisade which is hardly visible in the magnetometry but visible during excavation. Therefore, different methodologies and surveys were conducted during the excavation of a middle Neolithic ditched enclosure (Kreisgrabenanlage) in Lower Austria.

Firstly, a large-scale survey was conducted on the surrounding landscape. Secondly, surveys with different geophysical methods (magnetometry, ground penetrating radar and magnetic susceptibility) were carried out prior to and after the top soil removal of the excavation. Thirdly, small scale geophysical surveys (magnetometry and magnetic susceptibility) in different sizes and measuring rasters, ranging from 60 m x 20 m to 1.5 m x 0.9 m in area and 0.5 m x 0.25 m to 0.05 m x 0.05 m in measuring raster, were collected. Fourthly, samples were taken from these features or areas in order to undertake simple (mass magnetic susceptibility and frequency dependency calculations) to more complicated measurements (magnetic susceptibility spectroscopy measurements and magnetic grain size calculations) in the laboratory. Not only did this provide an opportunity to have a more detailed geophysical insight of different anomalies and features from the large-scale survey, but also permitted testing different measuring rasters. This permitted the improvement of methodologies for investigating features of varying sizes and geophysical contrast. The results lead to a better understanding of the responses observed from the different geophysical methods used, as well as furthering the archaeological interpretation.

Background

Buried archaeological remains are precious commodities of our cultural heritage, often under the threat of destruction. Their investigation and management should, therefore, involve minimal damage to or destruction of them. Archaeology, therefore, has a need for fast and accurate recording of such buried remains. Allowing for these to be preserved, managed or evaluated in the most cost-effective form of excavation or investigation. Thus, numerous non-invasive archaeological prospection methods, such as geophysical prospection and remote-sensing techniques, permit the investigation of such sites and structures in detail. The success of these measurements (the results and interpretation) largely depend on the interaction between the applied method, the subsoil and the archaeological feature. Often a deeper understanding is needed of these and can be achieved by small scale investigations.

Mostly buried archaeological structures are visible in archaeological prospection methods because these differentiate themselves from the surrounding matrix in which they are buried. This is due to their physical and chemical differences which contrast with the surrounding subsoil. Such anomalies can be of anthropogenic (pits, ditches etc.), geological (bedrock), geomorphological (palaeochannels) or biological nature (animal dens). The various prospection methods respond to different chemical and physical properties. Magnetometry results reflect the magnetic properties of the subsurface whereas ground penetrating radar measurements relate to the relative electrical permittivity properties of the soil (Gaffney and Gater 2003). Archaeological structures can be identified in aerial photographs because of the varying physical and chemical properties visible as soil, moisture or crop marks (Hejcman *et al.* 2013). Investigating the physical properties of anomalies can bring further insights into these, as such are often not easy to understand in prospection data. Better knowledge of an anomaly's physical properties, as the makeup of such, can improve this (Fassbinder 2015). This can help identify reasons some features are detected by certain methods and not by others.

In the 21st century there have been major incentives within archaeological prospection. One of these has been moving the focal point from the local to a landscape approach. This has been the case to understand the individual archaeological features detected in terms of the 'site' they form, as well as the 'site' in terms of its surrounding landscape, both in its archaeological and natural background. Technological advancement has permitted large-scale surveys to be carried out over entire buried cities and landscapes (Gaffney *et al.* 2000, Gaffney *et al.* 2012, Powlesland *et al.* 2006, Neubauer *et al.* 2012). These result in large datasets over varying subsoils and archaeological features, which differently manifest themselves within these. Another incentive has been a multiple method approach as archaeological structures can manifest themselves differently in the various techniques (Gaffney *et al.* 2000, Neubauer 2001, Kvamme 2007, Keay *et al.* 2009, Kainz 2017). This is due to the variations in the physical properties of different archaeological features. Therefore such an approach can be carried out over a small area to collect confirmatory, complementary and new information on archaeological and non-archaeological anomalies observed in the large scale datasets.

Methods

During the excavation magnetometry and ground penetrating radar measurements were carried out prior to the excavation and after topsoil removal. Magnetometry surveys were additionally carried out over the various infills of the southern innermost ditch in the entrance passage. This allowed identifying different features which were not visible in the original surface survey and assessing the contribution of the various layers of the innermost ditch to the original surface magnetogram.

To compliment the magnetometry data magnetic susceptibility measurements at various scales and measuring raster were carried out over horizontal and vertical areas of different archaeological features, such as the ditch section and over the palisade. Different measuring rasters were applied for the various objectives needed, which allowed for an assessment of useful measuring raster sizes for the intended purposes. These ranged in size from the complete excavation trench (20 m x 60 m) and a 0.5 m x 0.25 m measuring raster to smaller areas (1.5 m x 0.9 m) with high resolution measurements (0.05 m x 0.05 m).

This data corroborated the magnetometry data but also allowed for a better insight of different measuring raster for different purposes. Furthermore, 299 samples were collected from a variety of features. These were subjected to magnetic susceptibility measurements in the laboratory, ranging from simple low and high frequency measurements and the calculation of their frequency dependency as well as examining the effects of different measuring times on the results. Magnetic susceptibility spectroscopy measurements were carried out on a number of samples from the innermost ditch. This involved a sample's magnetic susceptibility being measured over a wide spectrum (150 – 63 000 Hz) allowing an assessment of a samples grain size distribution to be made. These measurements allowed for a comparison to be made with the more common laboratory measurements mentioned above as well as a deeper understanding of the samples submitted to the spectroscopy measurements.

Outcome

This allowed correlation of archaeological anomalies from the various prospection methods made above ground with measurements undertaken during the excavation. This provided complementary surveys collected prior and during the excavation. This permitted collecting more in situ measurements which helped understanding the initial measurements as well as providing more insight into these as well as answering archaeological questions concerning these.

Bibliography

Fassbinder, J. (2015) Seeing beneath the farmland, steppe and desert soil: magnetic prospecting and soil magnetism. *Journal of Archaeological Science* **56**: 85-95.

Gaffney, C., Gater, J. A., Linford, P., Gaffney, V., and White, R. (2000) Large-Scale Systematic Fluxgate Gradiometry at The Roman City of Wroxeter. *Archaeological Prospection* **7**: 81–99.

Gaffney, C. and Gater, J. (2003) *Revealing the Buried Past: Geophysics for Archaeologists*. Stroud: Tempus Publishing.

Gaffney, C., Gaffney, V., Neubauer, W., Baldwin, E., Chapman, H., Garwood, P., Moulden, H., Sparrow, T., Bates, R., Löcker, K., Hinterleitner, A., Trinks, I., Nau, E., Zitz, T., Floery, S., Verhoeven, G., and Doneus, M. (2012) The Stonehenge Hidden Landscapes Project, *Archaeological Prospection* **19**: 147-155.

Kainz, J. (2017) An Integrated Archaeological Prospection and Excavation Approach at a Middle Neolithic Circular Ditch Enclosure in Austria. In M. Forte and S. Campana (eds.) *Digital Methods and Remote Sensing in Archaeology - Archaeology in the Age of Sensing*. Cham: Springer, 371 – 404.

Keay, S., Earl, G., Hay, S., Kay, S., Ogden, J., and Strutt, K. D. (2009) The Role of Integrated Geophysical Survey Methods in the Assessment of Archaeological Landscapes: the Case of Portus. *Archaeological Prospection* **16**: 154-166.

Kvamme K.L., (2007) Integrating Multiple Geophysical Datasets. In J. Wiseman and F. El-Baz (Eds.). *Remote Sensing in Archaeology*. Cham: Springer, 345-374.

Neubauer, W., Doneus, M., Trinks, I., Verhoeven, G.,

Hinterleitner, A., Seren, S. and Löcker, K. (2012) Long-term integrated archaeological prospection at the Roman town of Carnuntum/Austria. In P. Johnson and M. Millett (Eds.). *Archaeological survey and the city Vol. 2*. Oxford: Oxbow Books, 202-221.

Neubauer, W. (2001) Images of the invisible - prospection methods for the documentation of threatened archaeological sites, *Naturwissenschaften* **88**:13-24.

Powlesland, D., Lyall, J., Hopkinson, G., Donoghue, D., Beck, M., Harte, A. and Stott, D. (2006) Beneath the Sand Remote Sensing - Archaeology, Aggregates and Sustainability: a Case Study from Heslerton, the Vale of Pickering, North Yorkshire, UK. *Archaeological Prospection* **13**: 291 – 299.

The auxiliary castrum at Inlăceni (Énlaka), Romania: results of the geomagnetic survey 2016

Rainer Komp[1] and Ingo Petri[1]

[1]*German Archaeological Institute, Berlin, Germany*

rainer.komp@dainst.de

A remote sensing campaign using a large-scale magnetometer system was conducted by the German Archaeological Institute in cooperation with the University of Pécs to investigate the roman auxiliary castrum at Inlăceni (hungarian: Énlaka), community of Atid in the district of Harghita in Transylvania (Romania). The survey area is located east of the actual village, at a level of about 700 m a.s.l. in the Transylvanian highlands at the western side of the Carpathian mountains. These volcanic, late tertiary mountains arose through unfolding caused by the drift of the African massif. The deposits consist of sandstone, calcareous slate, marl clay, shale and volcanic tuff with layers up to 5000 m thick. The ridge of the site is a watershed. Only a thin layer of top soil covers the bedrock.

The site known by the field name "Vir" occupies an east-west slope with terraces 5 to 10 m in width and steps of between 0.20 m and 1 m. The castrum, an object of archaeological research since the mid 19th century, is hitherto known by thorough examination and excavation as a well as geomagnetic survey.

The Roman auxiliary camp 'Augusta Praetoria', according to archaeological finds, was founded in the time of the emperors Traian and Hadrian and existed until the middle of the 3rd century AD as part of the Dacian Limes. Described for the first time by F. Müller in 1858, excavations were carried out in 1947 and 1950, and a summarizing study given by N. Gudea (1979). In 2010 a joint project including the Roman-Germanic Commission of the German Archaeological Institute, the Romanian Academy, the University of Kiel and the University of Chişinău performed geomagnetic surveys using a hand driven carrier with 6 probes at a distance of 0.5 m, thus covering a width of 3 m at a time (Popa 2010). Unfortunately, due to the partial inaccessibility of the fields, they could only cover parts of the castrum, missing completely the west wall and the principia. In summer 2016, an excavation was carried out by Zs. Visy, opening a trench in the zone of the principia.

What is clear to date, is that an original construction from wood and earth was replaced by a work of stone in the *opus incertum* technique. The general size, ditches, wall, four gates with towers and the principia as well as other buildings are known. Also, the nearby baths, 60 m southwest of the castrum, were investigated intensively. Furthermore, a vicus is assumed either west or south of the castrum. The limes road is identified running along the north of the site in east-west direction, still identical with the modern road (National Road 136B).

The general purpose was to investigate, in a non-destructive manner, historic features in the soil related to the known remains of the Roman fort in order to enhance knowledge of the surroundings and to guide future excavations. Furthermore, the idea now was to cover the vicus and the thermae.

The German Archaeological Institute owns a multi-sensor geomagnetic survey system, a Magneto MX v2 manufactured by SENSYS GmbH. The vehicle towed frame carries 16 magnetometers on an array width of 4 m and is perfectly suited for large scale landscape prospection. The sensors of gradiometer type FGM 650/3, two single-axis vertical fluxgate magnetometers, that are aligned vertically to each other in a distance of 650 mm, feature a standard measurement range of ±3.000 nT. The horizontal distance of the probes is 25 cm; sampling frequency at 20 Hz (the sampling distance depends on driving speed - typically between 10 and 20 cm). For towing the magnetometer system an all-terrain capable vehicle was kindly provided by the project partner. RTK DGPS ensured the precise geolocation of sensor tracks, for which a local geodetic reference station was set up and transmitted adjustment signals to the moving sensors (rover) by radio, resulting in an accuracy within centimetres.

Sensor and position data are logged in live stream so that tracks could already be seen on the fly at the control monitor. Retrieved data were exported to an open sensor archive format and prepared by a specific tool chain to be integrated directly into GIS. After applying filter and rasterizing mechanisms, visualisation and analysis were performed. The detailed GPS-guided survey also allowed for the creation of a DEM.

Thanks to the now largely free accessibility of the site (an area of more than 11 ha has been surveyed)

Figure 1: Magnetometer system in action at Inlăceni (Énlaka) (photo: Zs. Visy).

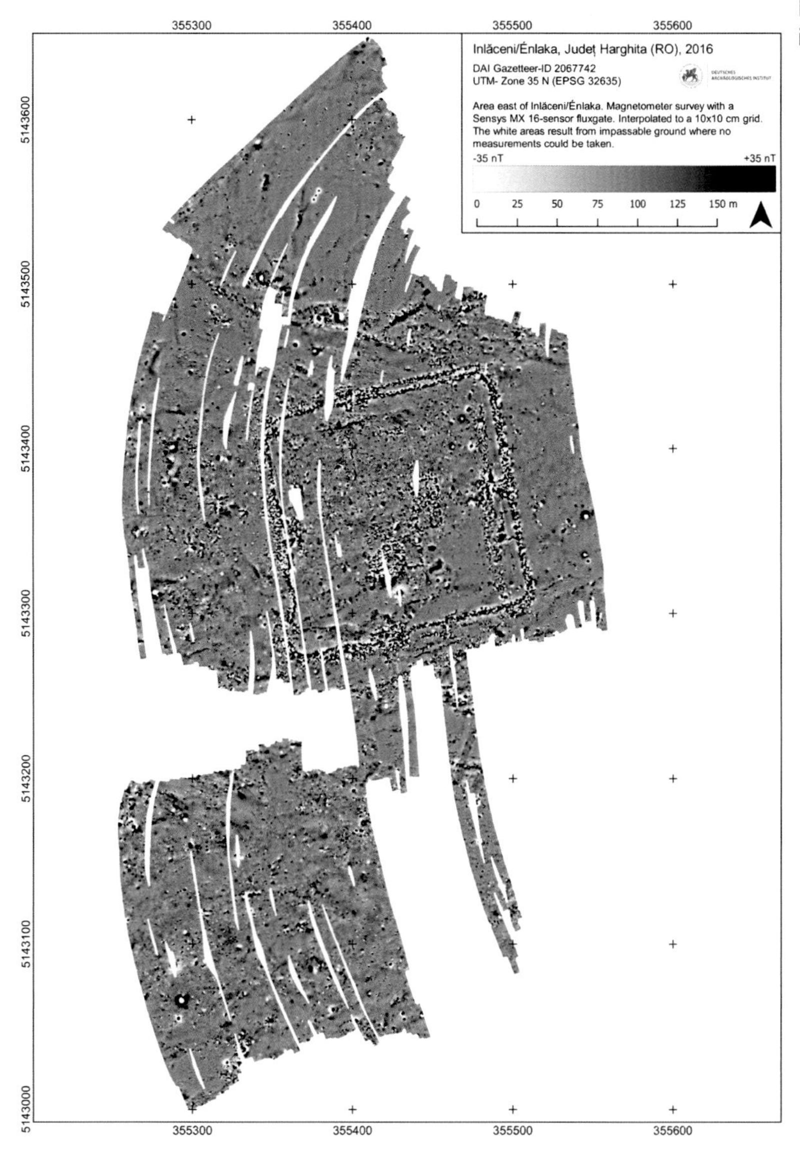

Figure 2: The castrum at Inlăceni (Énlaka) in the magnetogram.

and crucial advancements in technological development, the analysis resulted in significant enhancements concerning our knowledge of the castrum compared to earlier excavations and surveys. Due to the magnetic volcanic stone material used for construction, archaeological features can be difficult to differentiate, so that at first glance the image depicts a castrum with the general outlines only. In fact, we do now have an exact layout of the walls and important features. The castrum's history and development can be read through the results of geomagnetic survey, giving clear evidence for some hypothetical reconstructions. Most probably up to four phases of wooden and stone fortification walls, and the corresponding construction of gates with towers, can be identified. This corresponds to repairs and extensive rebuilding, known from reused stones with inscriptions dated at the end of the 2nd century and the middle of the 3rd century AD (Gudea 1997). The complete new development of the walls each time suggests that the castrum has been severely damaged or even destroyed if not abandoned and reclaimed several times during its function as an auxiliary castle for the limes. The construction phases show a development concerning dimensions and orientation, the sequence of which, unfortunately, cannot be derived from the magnetics. Detailed views of the interior allow the identification of more buildings than previously known.

References

Gudea, N. (1979) Castrul roman de la Inlăceni. *Acta Mus. Porolissensis* **3**: 149-273.

Gudea, N. (1997) Der dakische Limes. Materialien zu seiner Geschichte. *Jahrb. RGZM* **44**(2): 497-609 (59-60).

Popa, A., Cociș, S., Klein, Ch., Gaiu, C., and Man, N. (2010) Geophysikalische Prospektionen in Ostsiebenbürgen. Ein deutsch-rumänisch-moldauisches Forschungsprojekt an der Ostgrenze der römischen Provinz Dakia. *Ephemeris Napocensis* **20**: 101-128.

The results of magnetometer prospection as an indicator of the extent and intensity of soil erosion of archaeological sites

Roman Krivanek[(1)]

[(1)]Institute of Archaeology of the Czech Academy of Sciences, Prague, Czech Republic

krivanek@arup.cas.cz

Introduction

Between 80 and 90% of all archaeological sites in the Czech Republic are situated in agricultural lowland. The condition of these sites is connected with many anthropogenic factors and activities. Many factors associated with the expansion of modern settlements, industrial areas, the extraction of raw materials and traffic communications have a fatal and irreversible impact on the quality of preservation of sub-surface archaeological situations. Compared to the aforementioned impacts, agricultural activities seem to be much less dangerous forms of landscape use and sub-soil alteration. But that is a misconception. The intensity of erosion of sub-surface archaeological layers depends on the relief and climate, but mainly on the extent and manner of land cultivation. The agricultural land use changes in the Czech Republic during the last 70 years are not only connected with socialist agriculture on large field units, but also with important changes in the quality and thickness of the soil, changes in the ground water regime, climate, etc. All of these factors work against the good preservation of sub-surface archaeological remains of sites and affect the process of soil erosion. For this paper, three selected examples represent different stages in the condition and soil erosion of surveyed archaeological sites.

Example of the Local Erosion of an Archaeological Site

The polycultural prehistoric enclosed settlement area (named "Okrouhlík") of Bylany, near Český Brod in the Kolín district (Central Bohemia), is located on a locally sloped promontory of a flat plateau above the eastern bank of Bylanka Stream. Based on archaeological evidence (mainly from archaeological finds; the documentation from an earlier archaeological investigation was not preserved), the Eneolithic site is preliminarily dated

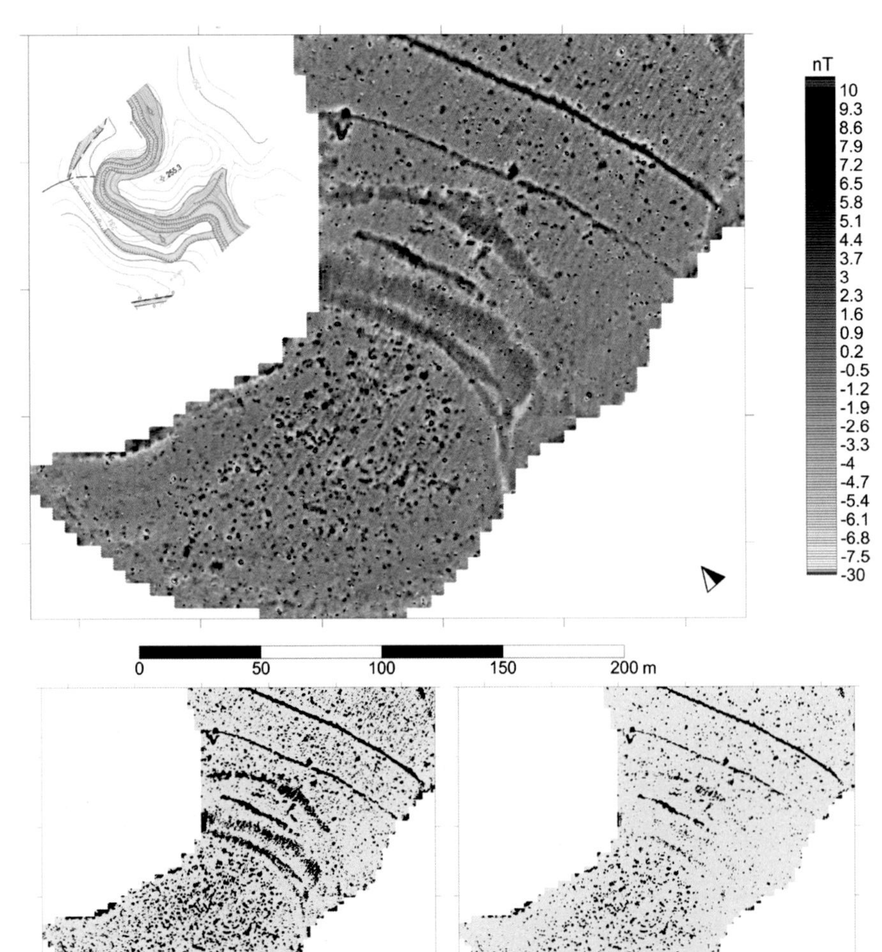

Figure 1: Bylany near Český Brod, Kolín district. The result of magnetometer survey of locally eroded polycultural prehistoric enclosed settlement with visible changes of amplitudes of magnetic anomalies (surveyed area: approx. 3.8 ha; survey: Křivánek 2016).

to between 3900 and 2800 BC (various Eneolithic cultures), with additional settlement activity in the early medieval periods. The extent of the whole site is not clear, and the enclosed settlement area by a ditch (ditches) was presumed only from aerial photographs on public websites.

The results of the magnetometer prospection of the arable field of the promontory confirmed the remains of very intensive and concentrated settlement with a large number of small oval features (pits). This intensive settlement covers nearly the whole inner area of the promontory and is probably enclosed by at least six various ditches. Different phases of Eneolithic activity are very probable at the site. On sloped terrains, some parts of these ditches and inner sunken features are not visible or have very low amplitudes (between +1.5 and +3.5 nT) of magnetic anomalies (Fig. 1). The central flat area with higher amplitudes of anomalies (between +3.5 and +10 nT) seems to be even better preserved than the outer sloped areas of the site. A less intensive settlement also continues in the eastern area of the large flat plateau outside of the enclosed area.

Example of the Deep Erosion of an Archaeological Site

The early medieval stronghold of Vlastislav, in the Litoměřice district (North Bohemia), is located on a strategic sloped promontory above the eastern bank of Modla Stream in the Central Bohemian Uplands north of the present village. Based on the results of former systematic archaeological investigations, the Slavic stronghold is dated to the period between the end of the 9th century and the 10th century AD. Archaeological evidence suggests the size of the fortified area is approximately 3.2 ha, with three areas (acropolis and two baileys) divided by a rampart with outer ditch. Most of the actual terrain covers fields, while an inaccessible part is covered with grass in private terraced gardens.

Geophysical measurements of the Slavic site were carried out during a project of regional cooperation (Non-destructive geophysical surveys of significant and endangered archaeological sites in the Ústí Region, Czech Republic; R300021421; Křivánek 2014-2016). Magnetometer prospection of all agricultural areas inside the stronghold and some smaller segments of the outer area confirmed the very poor state of the sub-surface preservation

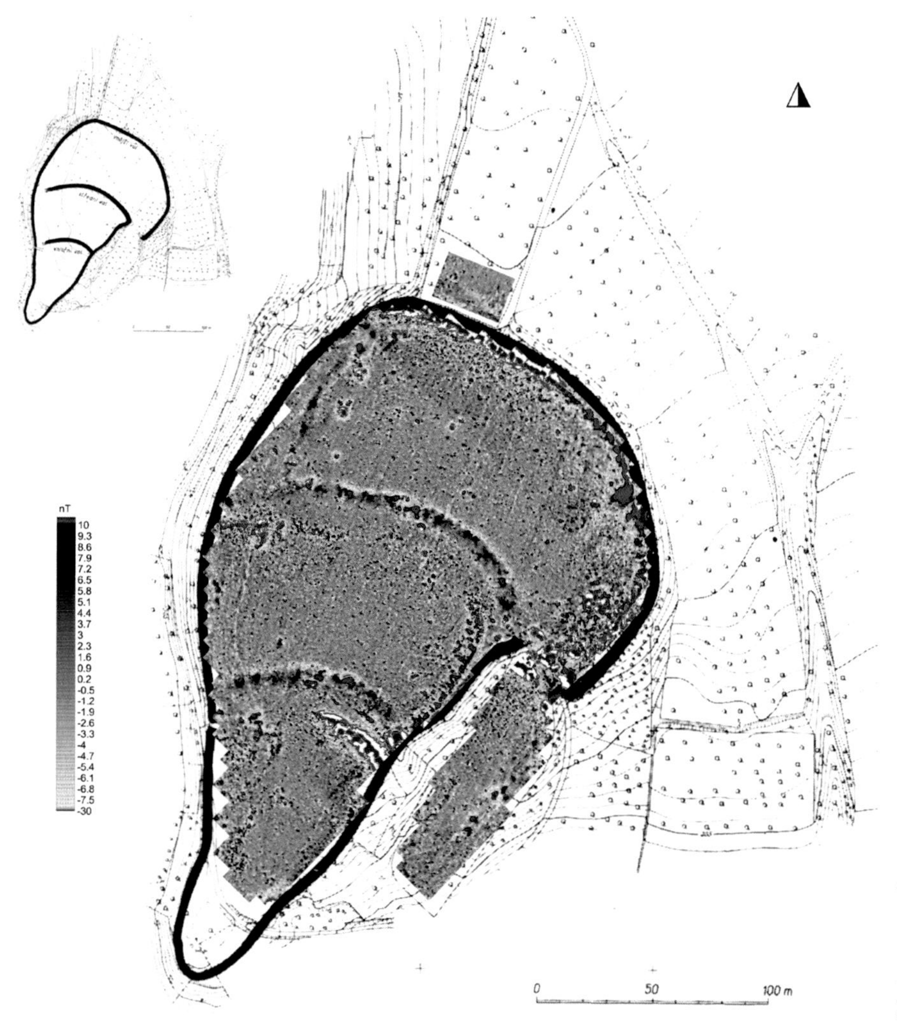

Figure 2: Vlastislav, Litoměřice district. The result of magnetometer survey of deeply ploughed-out early medieval stronghold compared with plan of the site from the time of archaeological investigations (Váňa 1953-1960; surveyed area: approx. 3.2 ha; survey: Křivánek 2016).

of archaeological situations (Fig. 2). Bows of burned internal ramparts known from previous archaeological excavations were inside ploughed-out sloped fields. The remains of ramparts are visible only in the shallow ploughed terrain near the edges of fields. Burned material from destroyed ramparts is probably also visible inside the filling of outer ditches and in several other places of the stronghold. Unfortunately, the results of the new magnetometer survey are also missing places or concentrations of sunken settlement features known from old excavations. Intensive ploughing of the inner areas of the stronghold after more than sixty years probably changed the state of most sub-soil layers at the site. Additional geoelectric resistivity measurement of a segment of the inner rampart and outer ditch confirmed the completely ploughed-out stone walls uncovered during earlier excavations.

Example of the Destruction of an Archaeological Site

The sub-surface remains of an abandoned La Téne quadrangular ditch enclosure of Stožice in the Strakonice district (South Bohemia) are located on slightly elevated sloped terrain near the abandoned springs of local streams south of the modern village. The remains of the quadrangular ditch enclosure ("Viereckschanzen") were detected on aerial photographs from public websites and subsequently verified by surface artefact collection in the field (Hlásek et al.). The situation was then verified in the field by magnetometer prospection in connection with other field prospection activities of the Museum of Prácheň in Písek (Hlásek) and the Institute of Archaeology at the University of South Bohemia in České Budějovice (John).

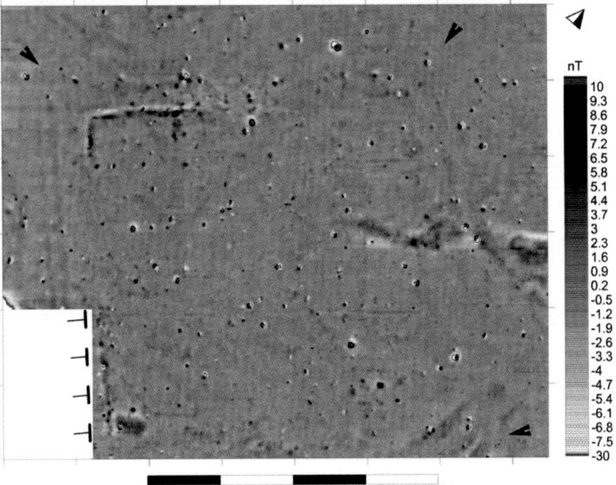

Figure 3: Stožice, Strakonice district. The result of magnetometer survey of nearly destroyed abandoned La Téne quadrangular ditch enclosure with aerial photography from public website from 2003 (source: www.mapy.cz; surveyed area: approx. 1.7 ha; survey: Křivánek 2016).

Unfortunately, the results of the magnetometer prospection of arable fields confirmed very poor sub-surface preservation of archaeological situations (Fig. 3). The quadrangular ditch enclosure was nearly completely ploughed out and deeply eroded. From the perimeter ditch we can see only some discontinuous low magnetic body. In the inner area of the destroyed quadrangular enclosure we can separate remains of a sunken corner construction, and in another corner one larger sub-rectangular sunken feature. The rest of the whole surveyed area seems to be without archaeological features; local anomalies indicate only sub-surface eroded and ploughed remains of the old field boundaries or an abandoned road and occurrence of recent metal intrusions in the soil layer. Additional studies of old aerial photographs also indicated deep erosion and modern remodelling of the landscape of the field at the site with a system of local ameliorative trenches in sloped areas.

Conclusion

Large-scale geophysical, and especially magnetometer, prospection could be a good tool for the rapid monitoring of the state of archaeological sites (archaeological landscape) in agricultural areas. It is clear that agricultural activity is the most important factor of soil erosion. The archaeological information in sub-soil layers, together with the quality of soils, is not unchangeable. This was also confirmed by results from the Czech Republic. Sometimes these changes can be identified in time, and this a success for archaeological heritage. Sometimes we can find what was lost in the last decades from the previous archaeological investigations and we can initiate rapid changes in the protection of clearly endangered archaeological terrain. But sometimes it is probably too late for some deeply eroded sites, and then the result of our work is quite discouraging. The change of the state of archaeological sites in ploughed agricultural land is only question of time.

When geology plays a major role in the results of archaeological prospection - case studies from Bohemia

Roman Krivanek[(1)]

[(1)] Institute of Archaeology of the Czech Academy of Sciences, Prague, Czech Republic

krivanek@arup.cas.cz

Introduction

Geological composition in different countries and regions is very diverse: in some places it is simple, elsewhere complicated. From the point of view of archaeology, this often indicates which non-destructive geophysical methods are (can be) used in the region efficiently. In the case of Central Europe and the Czech Republic, geological conditions are quite complicated, with regional and also local changes. Most of the country was formed by the Czech Massif and Proterozoic rocks, but the internal tectonic structure was formed later during the Paleozoic Variscan Orogeny with different degrees of metamorphic rocks. The other parts of country were also formed by Mesozoic (or Tertiary) sediments. Some mountain or upland regions were also formed during the Mesozoic Alpine Orogeny (and later) with a variety of neo-volcanic rocks. The final remodeling of the Czech landscape and the redistribution of sediments (including bedrock stones) was connected with Quaternary movements (and the melting) of the continental ice sheet. The mineralogical and geochemical composition of the constituent sedimentary, igneous or metamorphic formations is then very diverse and variable. The results of archaeological prospection in various Czech regions could not bring the same quality of results to archaeology. The specific geological conditions of some regions affected the choice of suitable geophysical methods. Possibilities of particular methods and the interpretation of results could in some cases also be connected with the type of surveyed archaeological situation. In this paper, selected examples will be discussed to demonstrate less typical results of the identification of archaeological sites in very complicated geological conditions.

Ctiněves (LT)

The sub-surface remains of the prehistoric burial mound in Ctiněves (North Bohemia) in the Litoměřice district, are located near the edge of a flat terrace with ploughed fields north-east of the modern village. The superposition of two right-angled linear anomolies was detected by aerial prospection. The area with the probable ditch enclosure of a destroyed burial mound was (after geophysical prospection and field artefact collections) investigated during the large research programme “Neglected Archaeology” of the University of West Bohemia in Plzeň (Gojda-Vařeka et al. 2005-2010; MSM 4977751314).

Compared to the clear results from aerial photographs, the results of geophysical measurements of this particular situation were significantly affected by the geological composition of sand-gravel terraces with a significant proportion of neo-volcanic stones (Fig. 1). But the other blocks of high magnetic neo-volcanic stones (sodalitic nephelinite, brought from Říp Hill, approximately 2km west) formed the original elevated construction of the burial mound. This stony elevation was destroyed in the 19th century (written sources: Krolmus 1855). The square shape of a perimeter ditch enclosure (25x25m) could be identified in the results of magnetometer measurements thanks only to the presence of neo-volcanic blocks in the filling of the ditch (Křivánek 2011). Additional geoelectric resistivity measurements of the situation (together with magnetometry) confirmed only the ploughed-out inner stony construction of the origin burial mound. The archaeological investigation of the site confirmed some remains of the central deep sunken pit (chamber?) with destroyed filling layers (Gojda and Trefný 2011). However, the final interpretation of the situation remains ambiguous due to polycultural activity at the site (mostly finds from the Late Bronze Age, but also Neolithic, Eneolithic and the Roman Period in places).

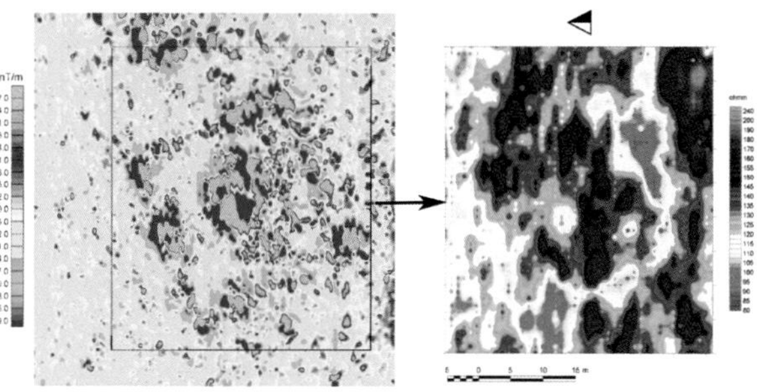

Figure 1: Ctiněves, Litoměřice district. The result of magnetometer and geoelectric resistivity surveys of sub-surface remains of the prehistoric burial mound compared with aerial photography (source: M. Gojda; surveyed area: M: 57x55m, R: 42x45m; survey: Křivánek 2008).

Starosedlský Hrádek (PB)

The sub-surface remains of the abandoned La Téne quadrangular ditch enclosure of Starosedský Hrádek in the Příbram district (Central Bohemia) are located on the southern slope above Kundratec Pond south of the medieval castle and village. The quadrangular ditch enclosure ("Viereckschanzen") was detected on aerial photographs from public websites (Korený-Krušinová). The situation was then verified in the field by geophysical prospection together with other field artefact collection conducted by the Mining Museum in Příbram.

For geophysical surveys of these types of features, we used (for the verification of shape and inner areas of the enclosure) magnetometry successfully in this region. Geological maps from this region did not indicate problems with the use of this geophysical method. Unfortunately, in the case of the enclosure near Starosedlský Hrádek, the results of magnetometer prospection include the magnetically very un-homogenous background of the site with many high magnetic (and also non-magnetic) linear anomalies. These anomalies are also variable in orientation and are related to changes of geology (granite with veins of granite porphyry and diorite porphyry). In the results of magnetometer measurements, we could distinguish only the perimeter trapezoidal shape of the enclosure due to low magnetic filling (soil without stones from bedrock) of the ditch. We also observed similar negative (low magnetic) linear anomalies of ditches with soil fillings in cases of prospection in areas of sand-gravel sediments with a high content of high magnetic neo-volcanic stone-gravel sediments (lowland areas along rivers from the neo-volcanic Doupov Mountains or Central Bohemian Uplands in NW Bohemia). Additional geoelectric resistivity measurements above two corners of the enclosure confirmed different fillings of the ditch and different material (more soils) inside the quadrangular ditch enclosure.

Figure 2: Starosedlský Hrádek, Příbram district. The results of magnetometer survey of the abandoned La Téne quadrangular ditch enclosure in complicated geological conditions with aerial photography from public website from 2012 (source: www.mapy.cz; surveyed area: approx. 1.25 ha; survey: Křivánek 2015).

Most (MO)

The forgotten sub-surface remains of abandoned relics of air defence units and other wartime activities from World War II of Most, in the Most district (North Bohemia), are located on the upper plateau of the large "Šibeník" urban park inside the town. Remains of artillery defences were identified as individual crop marks in aerial photographs from 1947 and 1953 (Čech). A part of the identified situation was then verified by geophysical prospection together with the field documentation conducted by P. Čech from the Institute of Archaeology in Prague.

Magnetometer prospection of a large wooded park was chosen as an experiment because the whole hill top of the town park was composed of high magnetic Tertiary neo-volcanic phonolite. It was clear that this type of magnetic bedrock must affect the legibility of results (expected +/-X0 nT). In the results of magnetometer prospection, we can see the very high variability of different high magnetic anomalies. Most of the highest dipole magnetic anomalies in the area (between +/-50 nT and +/- 200 nT) confirmed the presence of various metal components (including the remains of military equipment or constructions). After a comparison with old aerial photographs, we could confirm a nearly complete pentagonal ground plan with probable sunken emplacement of the original anti-aircraft cannons with a ferro-concrete base below the park surface. In other areas inside the pentagon, we can also expect sub-surface metal remains of other ancillary military facilities (may be also blasting pits or bombardment places). Due to extremely high magnetic anomalies from the WWII remains, the result showed the real possibility of identifying similar military relics in this area without limitations by magnetic bedrock.

Conclusion

The geological structure of the bedrock influences the composition of surface sedimentary deposits, soils, but also the possibility of prospecting archaeological situations. Knowledge of the geology of studied archaeological sites has a great impact on the selection of appropriate methods for their archaeological prospection. In Czech

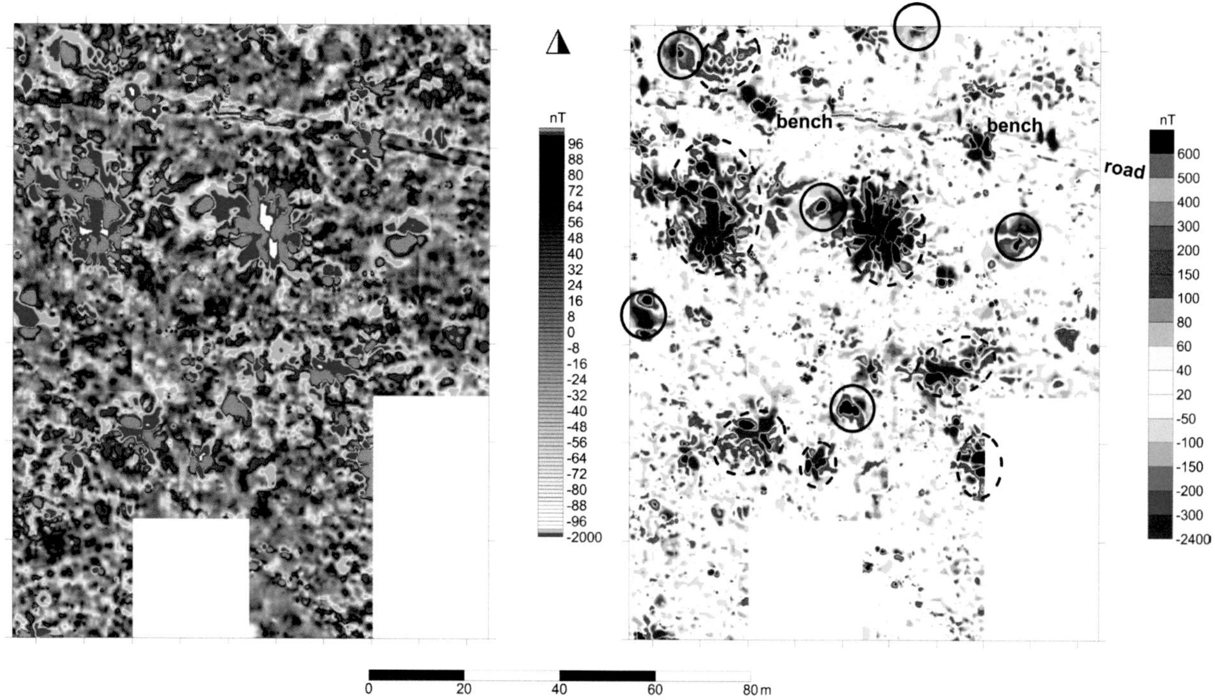

Figure 3: Most, Most district. The results of magnetometer survey of abandoned relics of air defence units and other wartime activities from World War II on neo-volcanic bedrock (surveyed area: approx. 1 ha; survey: Křivánek 2016).

archaeology we can find more examples in which geophysicists do not have sufficiently detailed information about the observed archaeological site and geology. Beneficial in these cases are test measurements in the field using more geophysical methods. The results of measurements sometimes showed unexpected geophysical results, anomalies and/or problems for survey or limits to archaeo-information in the data. The other results could then also discourage an archaeologist from continuing prospection in proved geologically inhomogeneous areas, or to think about different types of field verification of the site.

Bibliography

Gojda, M. and Trefný. (2011) *Archeologie krajiny pod Řípem – Archaeology in the landscape around the Hill of* Říp, Department of Archaeology, University of West Bohemia in Pilsen: Plzeň.

Křivánek, R. (2011) Kapitola 4. Využití archeogeofyzikálních měření při výzkumu Podřipska v letech 2005-2010 – Chapter 4. The application of geophysical survey in the Archaeology in the Landscape around the Říp Hill project. In: M. Gojda and M. Trefný (eds.) *Archeologie krajiny pod Řípem – Archaeology in the landscape around the Hill of* Říp, Department of Archaeology, University of West Bohemia in Pilsen: Plzeň.

Geophysical insights and problem solving at Chief Looking's Village, North Dakota, USA

Kenneth L Kvamme[(1)]

[(1)]Archeo-Imaging Lab & Department of Anthropology, University of Arkansas, Fayetteville, USA

kkvamme@uark.edu

Over the past decade, geophysical surveys coupled with small excavations, coring, historical aerial imagery, LiDAR, and most recently drone flights, have led to an increased understanding of Chief Looking's Village (CLV), a fortified ancestral site of the Mandan tribe located on a terrace above the Missouri River. Dating to the mid-sixteenth century, the site is significant because preservation is good and it was occupied during a period of transition. CLV contains long rectangular houses (about 8 x18 m) characteristic of earlier eras as well as rounded square to circular four-post houses (about 14-18 m in diameter), referred to as *earthlodges*, which ultimately came to dominate the region. All were built of wooden frames, with the former having gabled roofs (possibly hide-covered) while the latter were hemispherical and earth-covered. Both forms possessed long linear entryways which, in earlier times, always faced southwest. The transition from rectangular to four-post lodges probably reflects new ideas from interaction and a possible in-flux of peoples from further south (Mitchell 2013).

CLV is located within a city park of Bismarck that offers an excellent view of the region. Consequently, it has been moderately disturbed because it is a favorite picnic spot, and before it was a protected park vehicles would freely drive across the site leaving numerous trackways still visible topographically and in the geophysical data. Moreover, thousands of metallic items litter the site and impact geophysical results. In portions of the site a number of surface depressions are visible that point to the locations of prehistoric houses, a defensive ditch, and bastions. Visualization of these forms was initially improved through LiDAR, although with a coarse post spacing of 1.5 m; subsequent UAV flights conducted by Arlo McKee (University of Texas, Dallas) have yielded centimeter-level aerial imagery and digital surface models with spatial resolutions below 10 cm. Yet, enhanced computer visualizations were unable to reveal strong evidence of potential house locations elsewhere in the village.

Geophysical investigations were carried out at CLV to address multiple specific problems and issues: (1) locate houses hidden on the surface, (2) yield evidence about house form to identify locations of the two house types, (3) generate a map of the village showing all structures and defenses, (4) relocate reconstructed earthlodges built by the National Park Service (NPS) in the 1930s for tourism (later destroyed and abandoned), and (5) move away from reliance on magnetic gradiometry surveys (due to the plethora of metallic litter) by placing greater focus on GPR and electrical resistivity.

In the Northern Plains magnetometry is typically all that is required to reveal principal features of archaeological interest (see Kvamme 2003, 2008). At CLV not only are there decades of discarded metallic rubbish, but walkways were lined with wooden rails anchored with steel rebar. The resultant robust anomalies almost entirely obscured subtler ones arising from the prehistoric occupation. Although computer processing was partially successful in reducing the effects of the large dipolar forms, recent modernization of the walkways removed the rebar. Resurvey of the village core was therefore undertaken to produce a much cleaner magnetic data set that improved identification of prehistoric features, including hearths, storage pits, and house outlines (Fig. 1). Many of these findings were corroborated by down-hole magnetic susceptibility studies carried out by Rinita Dalan (see Dalan 2008). Moreover, clear evidence of the NPS lodges was revealed because their constructions included steel fittings and they were burned when abandoned (Fig. 1).

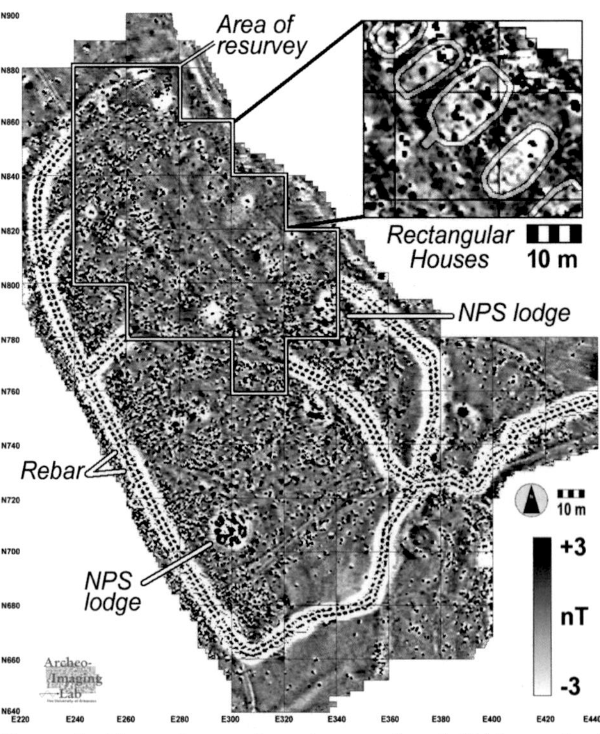

Figure 1: Magnetic gradiometry results at CLV acquired with a Bartington 601 dual fluxgate gradiometer. The outlined central resurvey area is after removal of rebar. The inset shows rectangular house outlines with anomalies representing hearths and corn storage pits.

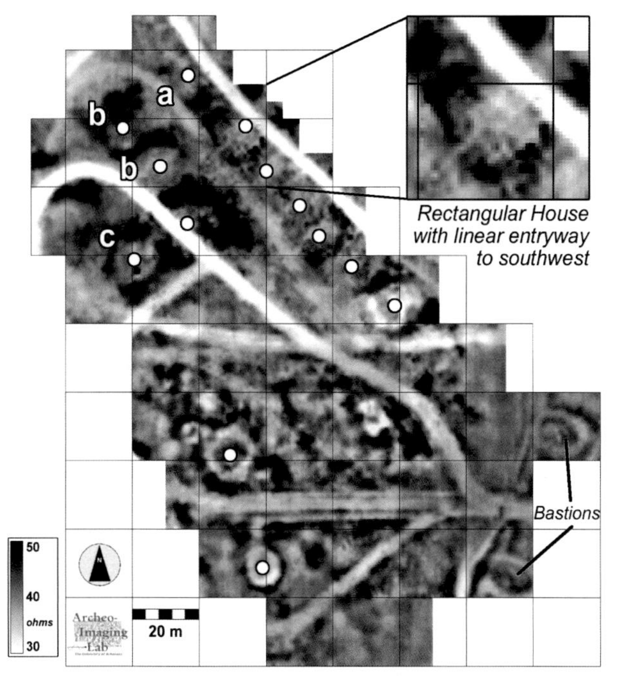

Figure 2: Results of electrical resistance survey made with a Geoscan Research RM-15 using four twin probe arrays in parallel, with 50 cm electrode spacing. Identified houses shown with white circles; letters refer to houses in Figure 3. Trails and former trails of the city park are well indicated.

To learn more about content and structure throughout the village required reliance on other geophysical methods. An electrical resistance survey revealed more about the distribution of houses and house shapes (Fig. 2), but smaller archaeological features were not indicated. Focus therefore turned to GPR, but with some reluctance. Nearly two decades of GPR explorations in Northern Plains villages did not recommend it because of the extensive rodent damage that commonly exists in these sites in the native prairie (see Kvamme 2008 for examples). The many anomalies caused by numerous burrows and dens make recognition of archaeological patterns and targets difficult using GPR (see exceptions in Kvamme 2003, 2008). CLV's location in a manicured city park therefore proved advantageous. Rodent damage is nearly absent and has been so for decades, permitting some of the cleanest GPR data sets seen in Northern Plains villages (Fig. 3).

The clarity of GPR results was also due to an unusual circumstance of climate. The Northern Plains are normally dry with an average annual rainfall in Bismarck of only 43 cm. Yet, in the six weeks prior to fieldwork in mid-June of 2015, Bismarck received an unusual 28.5 cm of rainfall compared to a reported average of about 10.2 cm for the same period. These wet conditions led to markedly improved results. Surrounding each house, even when completely buried, is a raised berm of sediments eroded from earthen roofs and sides that act much like a mulch in periods of great moisture, increasing GPR reflectivity. The result is a large contrast between house floors and their perimeters, giving excellent definition to house forms that enabled identification of square-to-circular four-post houses as well as rectangular forms (Fig. 3). Of particular significance was the revelation of superimposed houses, with circular forms above rectangular ones (Fig. 3.c). Nineteenth century ethnography suggests family "ownership" of parcels within a village, and this super-positioning may point to a transition of house forms within the same family plots. Ironically, repeat GPR surveys of the same areas almost exactly one year later with 16.7 cm transect separation (compared to the previous 50 cm), yielded uninformative results with poor contrasts owing to only 11.8 cm of precipitation in the six weeks prior to fieldwork.

Bibliography

Dalan, R. A. (2008) A review of the role of magnetic susceptibility in archaeogeophysical studies in the United States: recent developments and prospects. *Archaeological Prospection* **15**: 1-31.

Kvamme, K. L. (2003) Multidimensional prospecting in North American Great Plains village sites. *Archaeological Prospection* **10**: 131-142.

Kvamme, K. L. (2008) Archaeological prospecting at the Double Ditch State Historic Site, North Dakota, USA. *Archaeological Prospection* **15**: 62-79.

Mitchell, M. D. (2013) *Crafting History in the Northern Plains: A Political Economy of the Heart River Region, 1400-1750*. Tucson: University of Arizona Press.

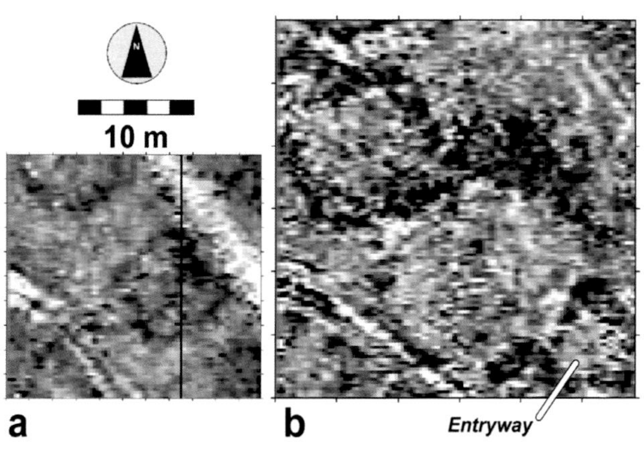

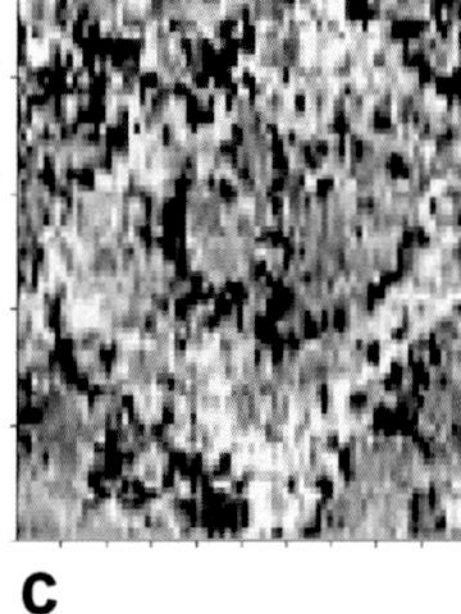

Figure 3: GPR depth slices 20-50 cm below surface obtained with a Geophysical Survey Systems Inc. SIR-2000 and 400 MHz antenna: a) rectangular house, b) two circular houses, lower one with prominent entryway, c) circular house centered over rectangular house.

The iron-age burial mounds of Epe-Niersen, the netherlands: results from magnetometry in the range of ±1.0 nT

Lena Lambers(1), Jörg W E Fassbinder(2), Karsten Lambers(1) and Quentin Bourgeois(1)

(1)Faculty of Archaeology, Leiden University, The Netherlands; (2)Geophysics Department of Earth and Environmental Sciences, LMU University, Munich, Germany

l.s.l.kuhne@arch.leidenuniv.nl

The archaeological site of Epe-Niersen in the central Netherlands consists of a 6km long linear arrangement of barrows dating to the 3rd and 2nd Millennium BC (Bakker 1976; Bourgeois 2013). In total, no less than 50 burial mounds of this alignment are still preserved within the modern-day landscape. At the start of the 20th century the curator of prehistory at the National Museum of Antiquities in the Netherlands (*Rijksmuseum van Oudheden*) investigated several of these mounds and discovered peculiar burials and grave gifts in them (Bourgeois *et al.* 2009). The research at the time, however, focussed solely on the burial mounds and gave little attention to the surroundings of these monuments. What lies beyond these monuments? Can we find evidence for activities contemporaneous with the construction of these mounds?

The current day terrain-use and the extent of the alignment does not lend itself to classic (and destructive) excavation methods. Therefore, it was decided to do a magnetometer survey, connected to radar and seismic research conducted by the Faculty of Civil Engineering and Geosciences of the TU Delft.

For the magnetometer survey we used the Scintrex SM4G-Special Caesium magnetometer in a duo-sensor and total field configuration, which we carried ca. 30 cm above the uneven ground at a sampling rate of 25 x 50 cm. The application of this magnetometer with a sensitivity of ±10 pT in the uncompensated total field configuration allowed us the maximum utilisation of the magnetic anomalies (Fassbinder 2016). The total Earth's magnetic field at Epe-Nierson in June 2016 was ca. 49100.00 ±10 nT. The sandy and partly acid soils in the Netherlands constitute unfavourable geoarchaeological conditions for magnetometer prospecting (Kattenberg 2008). Considering this, the survey yielded surprisingly good results. Displaying the data in the narrow range of ±1 nT has the effect that even a tiny variation in sensor height above ground, unavoidable on an uneven surface, shows up as a "mistake" of the surveyor. Moreover, the area is densely criss-crossed by pathways from all time periods. Nonetheless we could clearly trace the archaeological features in the ground.

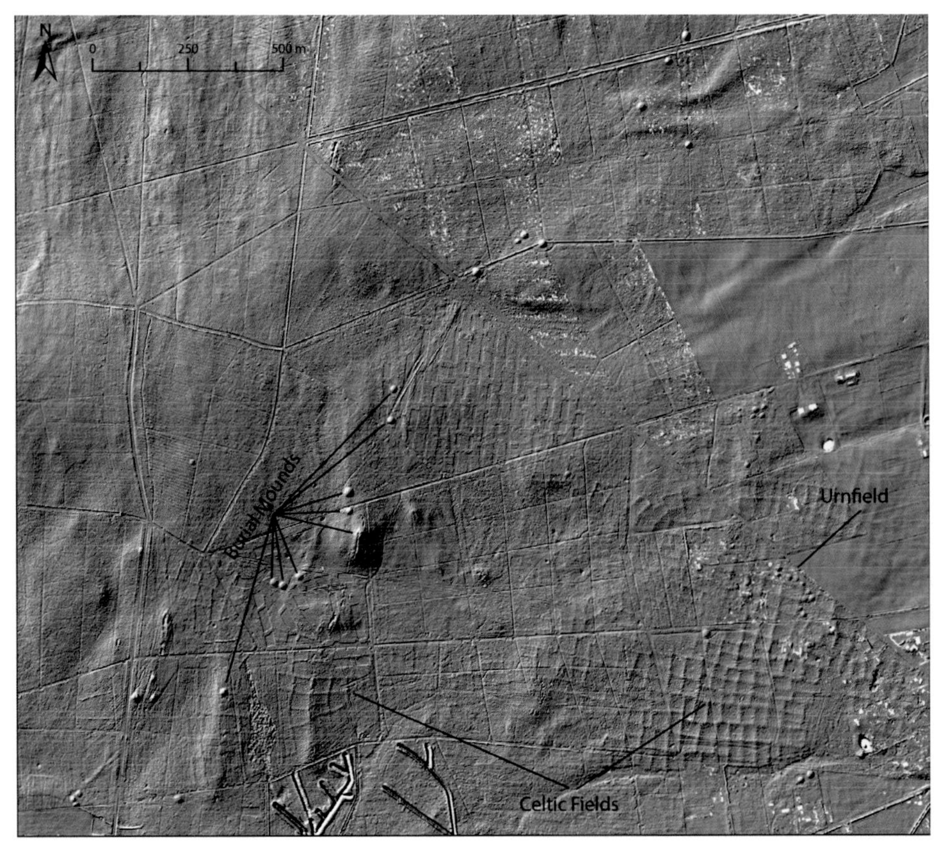

Figure 1: a hillshade-map of the Epe-Niersen barrow alignment, next to the barrows from the alignment a large field system dating to the Iron Age can also be seen (created from the *Actueel Hoogtebestand Nederland 2* or *AHN2*, www.ahn.nl).

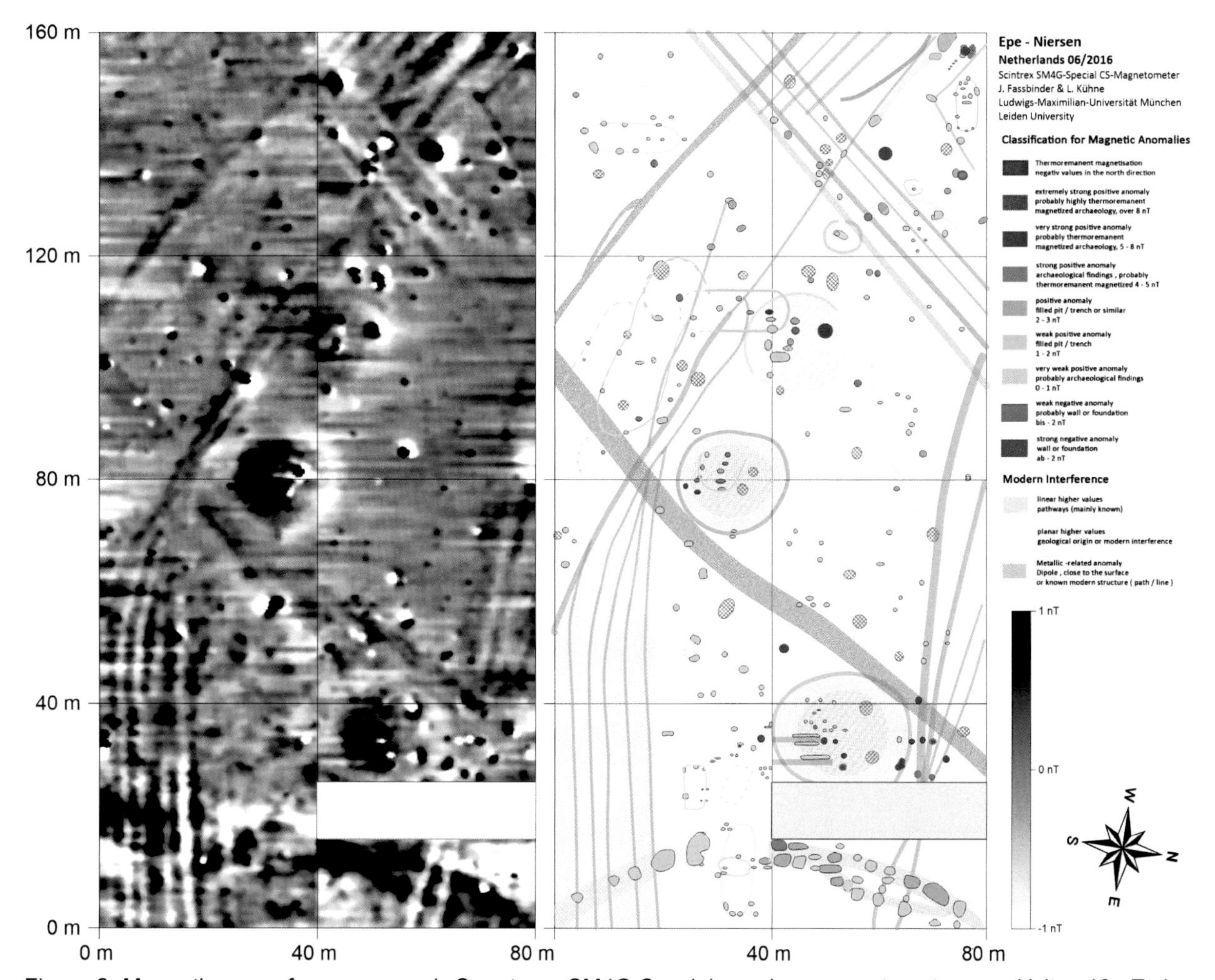

Figure 2: Magnetic map of survey area 1. Smartmag SM4G-Special caesium-magnetometer, sensitivity ±10 pT, duo-sensor configuration, sampling rate 25 x 50 cm, interpolated to 25 x 25 cm.

Apart from three known mounds the magnetogram shows two more burials, which also may have been small barrows, now erased. Additionally, within the known mounds we found a half-round structure in the range of 1 - 2 nT. Through archaeological fieldwork and core drilling this feature was later verified as an earlier construction phase of the barrow. The soil samples in this part differ clearly from the remaining mound.

We also detected thermoremanent magnetization in the range of ±5 - 8 nT. This value is nearly the strongest in our measurements corresponding to archaeology, and was probably caused by high temperature impact, which we interpret here as intentionally burned stones inside and outside the mounds. Such pits containing burned stones are known from contemporaneous burial sites in the wider region.

The main reason to undertake geophysical measurements at these sites, however, was to find out if there were any other structures, such as settlements or ritual places, adjacent to the mounds, similar to those found at the foot of Scythian Kurgans in Russia (Fassbinder *et al.* 2016). Indeed, we found up to ten ground plans of houses from different time periods. While they were almost all below 1 nT, they can still be detected in the magnetic image, if you look closely in the data. One of these ground plans is located beneath a topographically visible mound. We conclude that it must be older than 4000 years and hence belongs to the Neolithic period. Further investigations like trench excavation are planned to verify the results and provide us with more reliable dating of the ground map of the house and further structures.

Interestingly, in the eastern part of the surveyed terrain a possible circular pit-row showed up. Unfortunately, that part is heavily disturbed by the pathways leading through it, so we will have to wait for the excavation results to see if it is worth continuing our survey towards the east.

Another measurement was undertaken a few hundred metres to the southeast at a mound which belongs to the same burial alignment. The mound is notably higher than the previous ones and located next to a modern road. This may explain why we see a slightly higher range in the magnetometer data, but still no magnetic anomalies caused

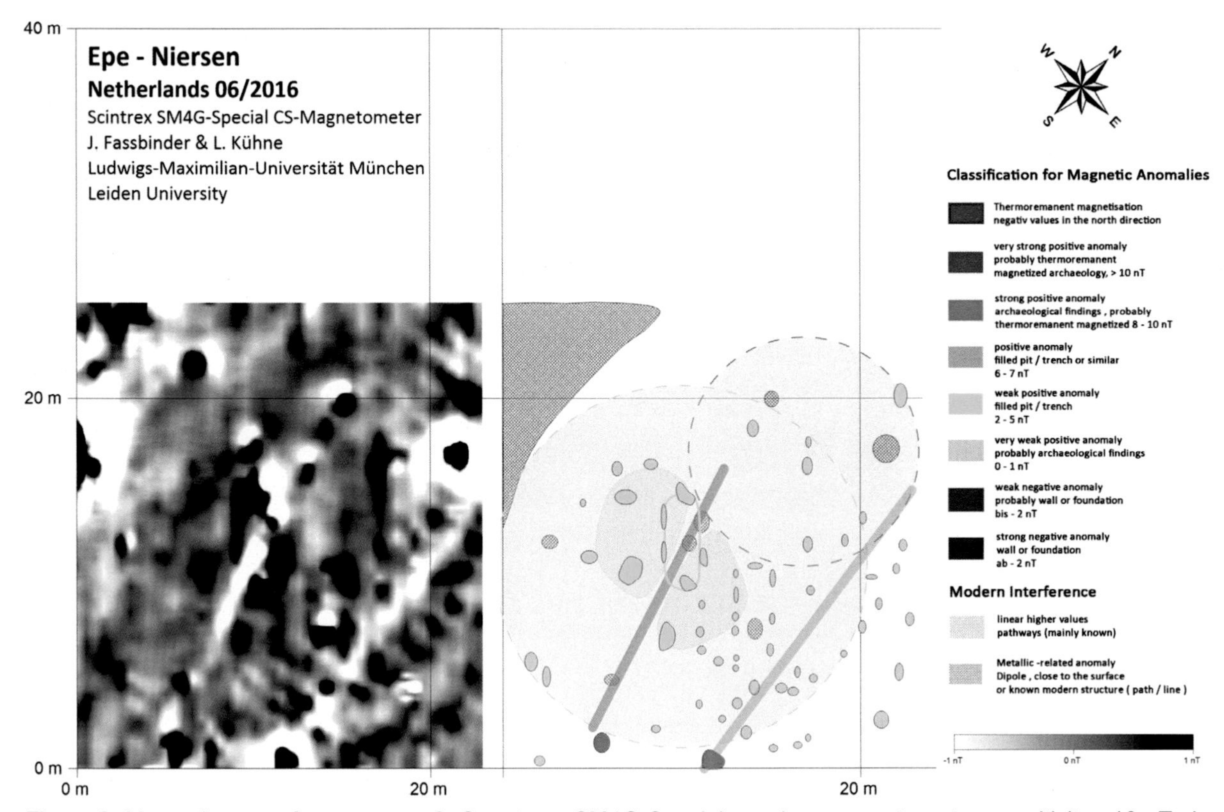

Figure 3: Magnetic map of survey area 2. Smartmag SM4G-Special caesium-magnetometer, sensitivity ±10 pT, duo-sensor configuration, sampling rate 25 x 50 cm, interpolated to 25 x 25 cm, total field fused by high-pass-filtered magnetogram.

by archaeology above ±5.0 nT. The two linear structures running through the mound from north to south were already visible in the LiDAR data and correspond to pathways probably dating to the medieval period. Apart from some pits and pit alignments we can recognise a burial in the centre that seems to be surrounded by a palisaded ditch, a feature known from other mounds (Bourgeois 2013).

What was not recognised before is that the mound is in fact composed of two distinct phases, suggesting that here as well, the large barrow actually consists of multiple construction phases.

Bibliography

Bakker, J. A. (1976) On the possibility of reconstructing roads from the TRB period, *Berichten van de Rijksdienst voor Oudheidkundig Bodemonderzoek* **26**: 63-91.

Bourgeois, Q. (2013) *Monuments on the Horizon: The formation of the Barrow Landscape throughout the 3rd and 2nd Millennium BC.* Leiden: Sidestone Press.

Bourgeois, Q. P.J., Amkreutz, L. and Panhuyzen, R. (2009) The Niersen Beaker Burial. a renewed study of a century old excavation, *Journal of Archaeology in the Low Countries* **1**(2): 83-105.

Fassbinder, J. W. E. (2016) Magnetometry for Archaeology, *Encyclopedia of Geoarchaeology*, Encyclopedia of Earth Sciences Series. Dordrech: Springer, 499-514

Kattenberg, A. E. (2008) *The Application of Magnetic Methods for Dutch Archaeological Resource Management.* Amsterdam: Geoarchaeological and Bioarchaeological Studies, Vol.9, PhD thesis.

Meninx – geophysical prospection of a Roman town in Jerba, Tunisia

Lena Lambers[2, 3], Jörg W E Fassbinder[1], Stefan Ritter[2] and Sami Ben Tahar[4]

[1]Geophysics Department and [2]Institute of Classical Archaeology, Ludwig-Maximilians University, Munich, Germany; [3]Faculty of Archaeology, Leiden University, The Netherlands; [4]Institut National du Patrimoine, Houmt Souk, Tunisia

l.s.l.kuhne@arch.leidenuniv.nl

Meninx, located in the southern part of the island of Jerba, was an important Roman harbour and one of the largest production sites of purple dye in antiquity. Existing since Punic times, the city experienced its cultural and economic heyday in the 2nd and 3rd centuries AD. The site extends ca. 1.7 km along the coast and ca. 200 to 600 m inland. However, it is still unclear how far it extends under water.

Since the 19th century, sporadic excavations near the supposed forum and outside the city were undertaken. From 1996 to 2006 Meninx was included in the research project "An Island through Time: Jerba Studies", during which comparatively small areas in the centre of the site were excavated (Drine *et al.* 2009). These excavations were accompanied by limited magnetometer measurements in the centre around the basilica and the *macellum*.

In the framework of the new project "The urban structure of the ancient town of Meninx" by the Ludwig-Maximilians University (LMU) Munich, a large-scale magnetometer prospection was conducted in 2015. The results provided the first coherent picture of the extensive core area of the ancient city that extends along the coast. For measurement we applied the Scintrex SM4G-Special Caesium magnetometer in a handheld duo-sensor configuration with a sensitivity of ±10 pT, and a sampling rate of 25 x 50 cm, interpolated to 25 x 25 cm. The configuration allowed us to cover almost the entire surface regardless of the extremely uneven topography, ruins, piles of rubble and sand as well as bushes and little trees which cover the area.

Results

Contrary to earlier assumptions based on some of the excavation results, the magnetogram shows no orthogonal street system in the old city centre of Meninx. While the streets themselves are not always clearly visible, the alignment of the buildings shows their layout throughout the city. Also, the position of the forum is not as clear as earlier research suggested.

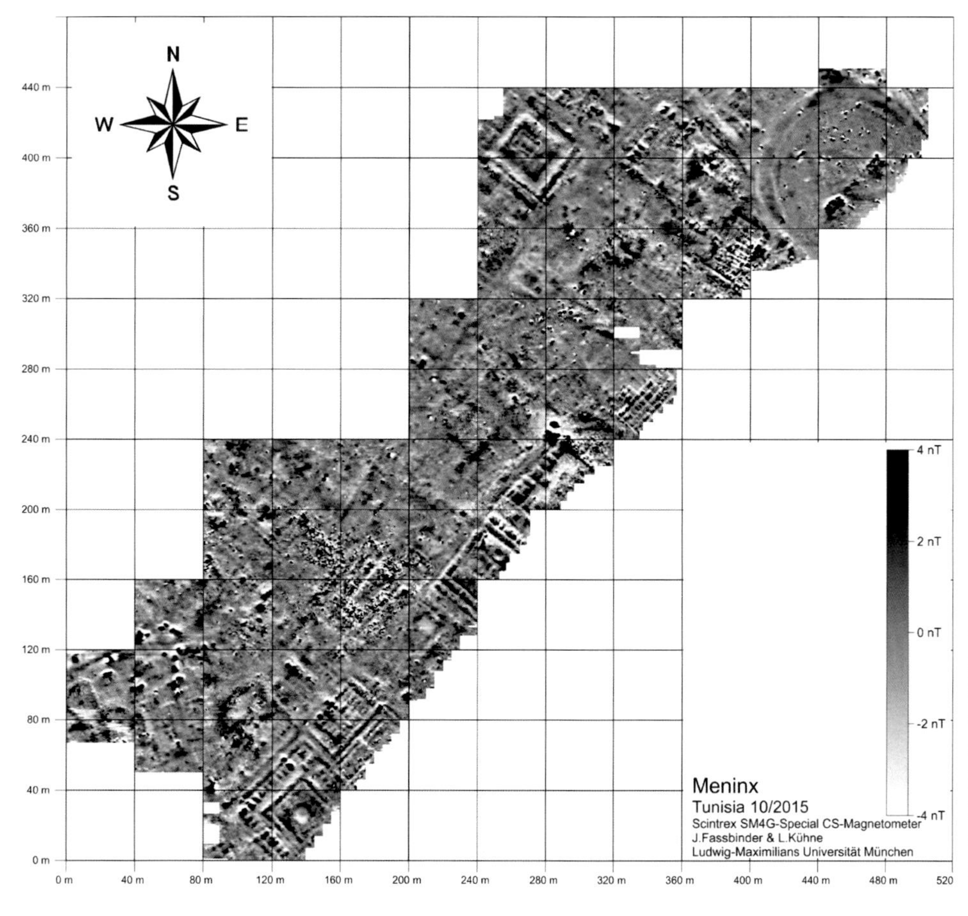

Figure 1: Magnetic map of the survey area. Smartmag SM4G-Special caesium-magnetometer, sensitivity ±10 pT, duo-sensor configuration, sampling rate 25 x 50 cm, interpolated to 25 x 25 cm.

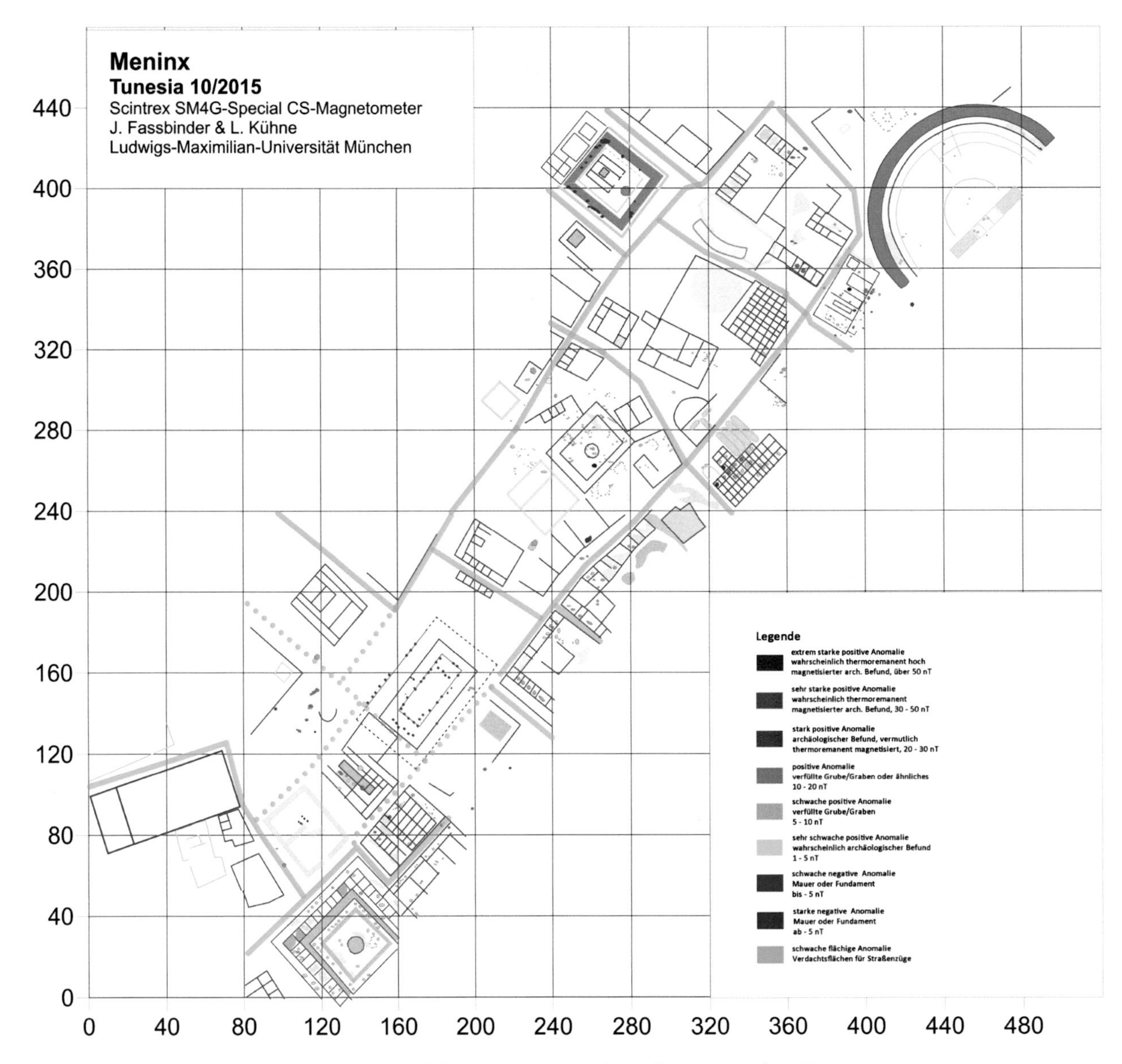

Figure 2: Classification and interpretation of the magnetogram based on magnetic value.

Although the partly excavated basilica is covered with pieces of highly magnetic rocks, it was possible to connect linear anomalies with column bases. Some columns are still visible above ground while others are not. For interpretation of data it was important to note that the columns and some of the representative buildings were made of marble and thus feature "negative" magnetic susceptibility. As a result, some large column fragments lying on the surface in the westernmost part of the site with a diameter of 80 cm and a length of up to 5 m long clearly show as a strong dipole anomaly as if they were pieces of iron.

The theatre is still visible on the surface through its huge debris and sand layers. In the magnetogram the substructures made of marble and other stones are clearly visible. Even parts of the stage are preserved, probably partly made of burned bricks. An enclosed area marked by piles behind the *cavea* seems to belong to the theatre setting. Connected to the theatre on its south-eastern side are several storehouses, joined by several cisterns a little further away, of which four were previously excavated.

The building south-east of the basilica seems to feature *tabernae* along the street, while the magnetic values from ±5 - ±20 nT seem to indicate organic material in the context of these rooms. Similar *tabernae* surround the *macellum* in the south. Some of them even have values up to ±50 nT, indicating either floor heating or cooking in these rooms. North-east of the *macellum* is another storehouse.

In a south-eastern direction our survey was limited by the coastline. If the city centre continued towards the south-west and if a former harbour was located there, its location will require an extended sonar survey.

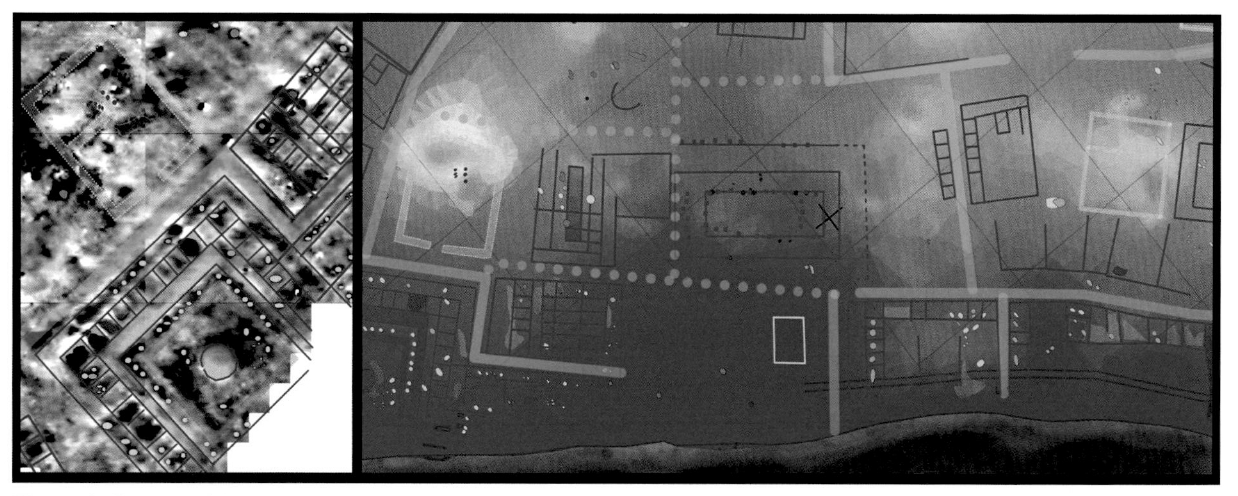

Figure 3: Details of the interpretation: *macellum* with high magnetic values in some *tabernae* (left); area of the basilica with two buildings showing as positive magnetic values (right).

Thanks to the interpretation based on the magnetic value, certain details of the site plan can be depicted very clearly, such as the different ranges of magnetization of the shops in the *macellum* mentioned above, which indicate different uses of the rooms. At first glance, positive values of two buildings in the city centre seemed to indicate an earlier construction phase with burned bricks or rocks with higher magnetic susceptibility than those of limestone. But the elevation map revealed that the higher values in the magnetogram are caused by the surrounding material. The thick debris layers contain a lot of diamagnetic shell waste material from the purple dye production during late antiquity. Thus, these buildings are covered by less magnetic material than the stone buildings themselves.

Bibliography

Drine, A., Fentress, E. and Holod R. (2009) *An Island through Time: Jerba Studies 1. The Punic and Roman Periods* (JRA Suppl. 71). Portsmouth: Journal of Roman Archaeology.

The application of semi-automated vector identification to large scale archaeological data sets considering anomaly morphology

Neil Linford[(1)] and Paul Linford[(1)]

[(1)]Historic England, Portsmouth, United Kingdom

neil.linford@historicengland.org.uk

Large area, high sample density data acquisition using both vehicle towed magnetometer arrays and multi-channel GPR systems provide highly detailed coverage of near-surface, archaeological remains at a landscape scale. This produces a considerable volume of data, which can present a considerable challenge for the subsequent identification and interpretation of significant archaeological anomalies. In the case of GPR surveys the problem is compounded as the data are often represented in the form of multiple amplitude time or depth slices of the buried ground surface. This presentation examines the application of semi-automated analysis of GPR and magnetic data sets to identify georeferenced vector objects from the original raster data.

Edge detection algorithms can rapidly abstract closed polygons from a user defined amplitude threshold and have been demonstrated to be very effective when applied to GPR data sets (Schmidt and Tsetskhladze 2013; Verdonck 2016; Leckebusch *et al.* 2008). The potential archaeological significance of the vector objects is then determined through a consideration of the morphology and continuity between multiple data sets. A particular application is made to the location of pit-type anomalies, which are often highly numerous across a landscape, yet can be time consuming to interpret through manual extraction of each individual response from multiple layers within a data set (cf. Linford 2005; Trier and Pilø 2012; Schneider *et al.* 2015).

Defining specific constraints, in terms of the accepted size of a potential pit-type feature, likely depth and whether it presents a circular outline can assist in the classification of the anomalies. In the case of the GPR data considered here the initial georeferenced raster amplitude time slices were first converted to a binary image, based on a user defined amplitude threshold (Linford *et al.* 2007: Fig. 1.A), which is then processed to extract connected objects based on size. This can be used to remove either very small objects, more likely to represent noise, or those too large to be representative of a pit-type feature. Further binary image processing is applied to close gaps within the objects based on an appropriate structuring element, for example a disc shape for detecting pit-type responses and rectangular shaped element for linear walls. Exterior boundaries of the binary objects are then traced to abstract a set of closed vector polygons.

Properties of the vector polygons, including the area enclosed and size of the perimeter, can be used as a further basis for the classification of circular, pit-type anomalies. A simple approach to determining whether an anomaly is approximately circular or not can be achieved through estimating the apparent radius of the circle based first upon the area enclosed by each vector polygon, and then by the perimeter. If the ratio of the two estimates for the radius approaches unity, then the anomaly is more likely to be circular, although in practise some relaxation of the morphological classification may be necessary to obtain the best results from real data (cf. Trier *et al.* 2009). Combining the morphological classification of vector polygons with other appropriate constraints, in terms of the size and depth of potential pit-type anomalies can be used to assist the selection (Fig. 1.A). A similar approach can be applied to target other anomaly types, for example by considering the rectilinear morphology and orientation of the vector polygons to identify Roman structural remains (Fig. 1.B), and with GPR data to consider the continuity of selections made between amplitude time slices.

For magnetometer data slightly different initial processing is required and for the current application

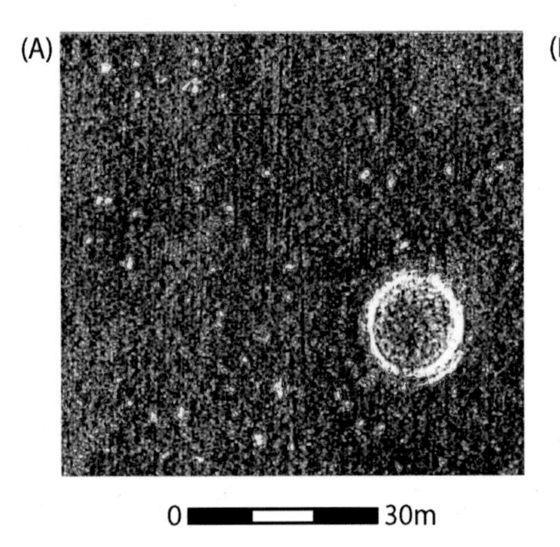

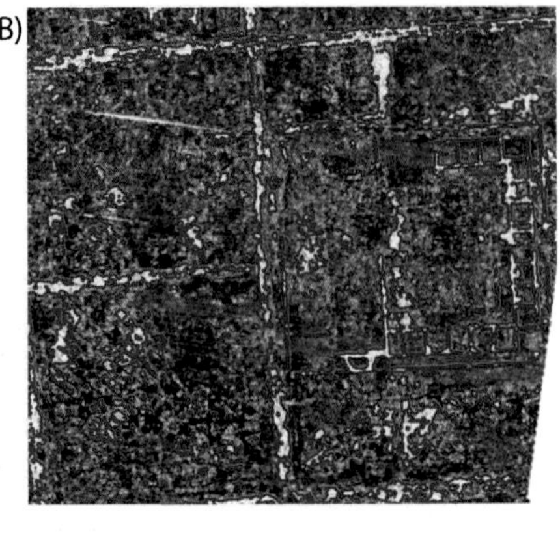

Figure 1: (A) Identification of approximately circular, pit-type anomalies (red vector outlines) superimposed over a greytone image of the GPR time slice, prior to size classification, and (B) identification of Roman structural remains without any morphological constraints.

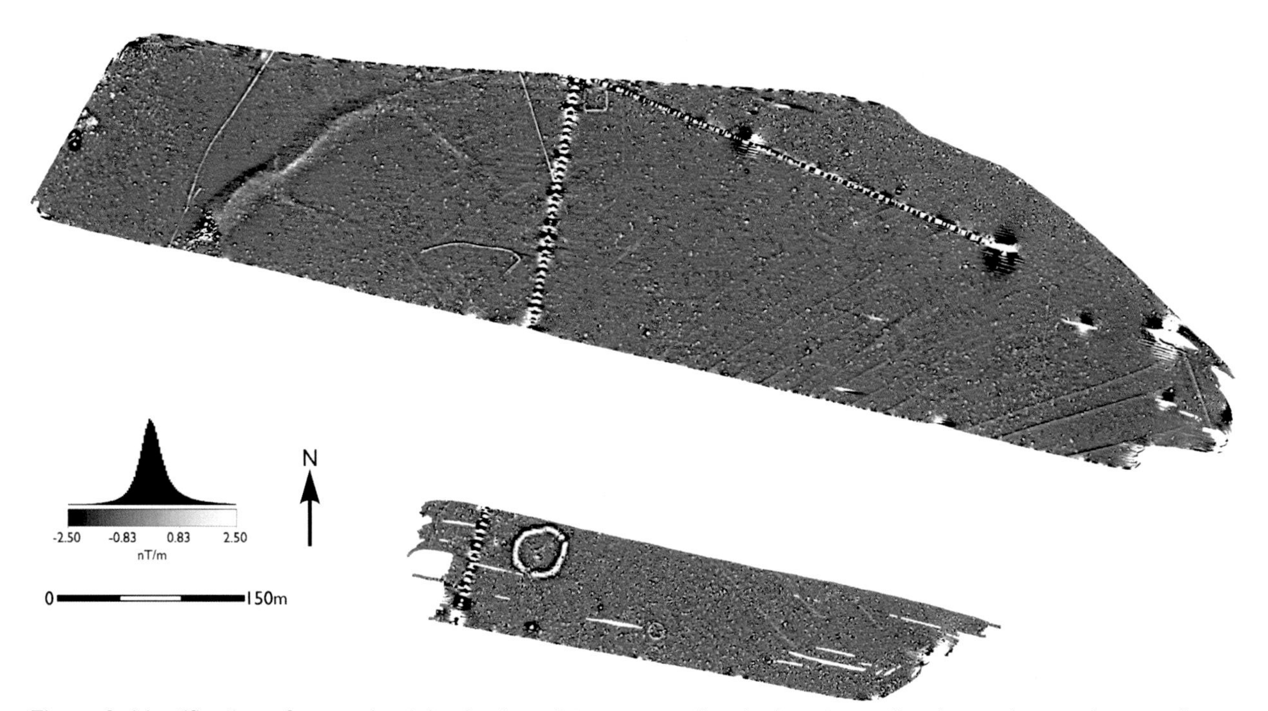

Figure 2: Identification of approximately circular, pit-type anomalies (red vector outlines) superimposed over a linear greytone image of a magnetic data set.

a refinement of the method suggested by Blakely and Simpson (1986) was chosen. In this approach the data are first transformed in the Fourier domain to the equivalent pseudo-gravity anomaly then the magnitude of the horizontal gradient at each point is calculated. Local maxima in the latter dataset will be positioned over the edges of any causative magnetic feature and their positions form a set of x, y points marking its perimeter. Clusters of nearby points can then be grouped into line segments based on their proximity and similarity in magnetic gradient strength at their location to provide vector anomaly outlines. Once anomalies are described in terms of line segments (effectively groups points) marking their perimeters, they can be classified using the same morphological analysis of vector polygon properties as for the GPR data (e.g. round line segments with a diameter less than ~3 m might be classified as denoting pit-type anomalies). Figure 2 shows the identification of pit-type anomalies over a large area magnetic survey where the algorithm has successfully discriminated possible pit anomalies from numerous responses due to near surface iron litter.

Where coverage with both magnetic and GPR survey is available, further refinement may be possible through comparing the identification of significant anomalies between the two techniques. However, this requires a diagnostic response to a particular feature type, such as a pit, to occur in both data sets.

Bibliography

Blakely, R. J. and Simpson, R. W. (1986) Approximating edges of source bodies from magnetic or gravity anomalies. *Geophysics* **51** (7): 1494-1498.

Leckebusch J., Weibel, A. and Bühler, F. (2008) Semi-automatic feature extraction from GPR data for archaeology. *Near Surface Geophysics* **6** (2): 75-84.

Linford, N., Linford, P., Martin, L. and Payne, A. (2007) Geophysical Evidence for Assessing Plough Damage. Študijné zvesti Archeologického ústavu SAV **41**: 212-213.

Linford, P. (2005) An Automated Approach to the Analysis of the Arrangement of Post-pits at Stanton Drew. Archaeological *Prospection* **12** (3): 137-150.

Schmidt, A. and Tsetskhladze, G. (2013) Raster was yesterday: using vector engines to process geophysical data. *Archaeological Prospection* **20** (1): 59-65.

Schneider, A., Takla, M., Nicolay, A., Raab, A. and Raab, T. (2015) A Template-matching Approach Combining Morphometric Variables for Automated Mapping of Charcoal Kiln Sites. *Archaeological Prospection* **22** (1): 45-62.

Trier, Ø., Larsen, S. and Solberg, R. (2009) Automatic detection of circular structures in high-resolution satellite images of agricultural land. *Archaeological Prospection* **16** (1), 1-15.

Trier, Ø. and Pilø, L. (2012) Automatic Detection of Pit Structures in Airborne Laser Scanning Data. *Archaeological Prospection* **19** (2): 103–121.

Verdonck, L. (2016) Detection of Buried Roman Wall Remains in Ground-penetrating Radar Data using Template Matching. *Archaeological Prospection* **23** (4): 257–272.

'Over head and ears in shells' recent examples of geophysical survey of historic designed landscapes and gardens

Neil Linford[(1)], Paul Linford[(1)] and Andrew Payne[(1)]

[(1)]Historic England, Portsmouth, United Kingdom

neil.linford@historicengland.org.uk

Introduction

Geophysical survey has long been applied to the investigation of garden archaeology, often concentrating on specific elements of formal planting within a much larger designed landscape (Cole *et al.* 1997; Keevil and Linford 1998). This approach has, perhaps, been dictated by the use of earth resistance as the most fruitful technique to apply, and a focus within specific projects to inform the accurate design and replanting of isolated formal elements of the wider garden layout. Recent developments in new instrumentation, for example array based Ground Penetrating Radar (GPR) systems, have improved both the sample density and speed of data acquisition (eg. Linford and Payne 2011), allowing a much greater appreciation of the designed landscapes complementing the setting of historic buildings, including the formal gardens and wider parkland vistas. This paper draws on recent work by Historic England, mainly in support of projects initiated by the English Heritage Trust, to enhance the investigation of designed landscapes through comparison with a range of historical sources. These sources, whilst not always completely reliable, are a testament to a continued fascination with the designed landscape and include the voices of many notable historical figures drawn to these gardens and even include evidence from an early popular music video.

Methodology

Vehicle towed magnetic and multi-channel GPR arrays were initially deployed, followed by targeted earth resistance survey to complement and extend the coverage where access was more difficult (Linford *et al.* 2010; Linford *et al.* 2015a). The magnetic survey was conducted with Geometrics G862 caesium vapour sensors at a sample interval of ~0.1m x 0.5m and the GPR coverage with a 3d-radar MkIV GeoScope and DXG1820 ground coupled antenna array at a sample density of 0.075m x 0.075m. Earth resistance measurements were made using a Geoscan RM15 resistance meter and a MPX15 multiplexer, to allow data sets with mobile electrode separations of 0.5m and 1.0m to be collected simultaneously at sample densities of 0.5m x 1.0m and 1.0m x 1.0m respectively.

Sites and Results

The Palladian villa, pleasure grounds and gardens built for Henrietta Howard, Countess of Suffolk, at Marble Hill House in Twickenham, cover an area of almost 27 hectares in the densely populated outskirts of London. Landscaping of the park began in 1724 as the house was constructed, and a neighbour, the poet Alexander Pope, suggested a planting scheme and garden features to the Countess and encouraged her decoration of two shell lined grottos. The grounds were altered in 1786 for the Earl of Buckinghamshire and in 1850 by Jonathan Peel, before being made open to the public in 1903 for recreational activities and use as allotments during the second world war. Despite potentially unfavourable site conditions, the GPR performed better than anticipated over the wet London Clay geology, detecting a wealth of superimposed anomalies, including a late 18th century Italianate garden design, reflecting the changing land use of the park through time (Fig. 1 and Linford *et al.* 2016).

While magnetic survey was of limited use at Marble Hill, it proved far more effective when applied to the parkland surrounding Belsay Hall and Castle in Northumberland. Created by the Middleton family, who were first recorded at Belsay in 1270 and lived there until 1962, the 14th century fortified tower-house was built as a statement of family pride and as a response to the unrest in the borders between England and Scotland. It sits within a wider medieval landscape, including the original medieval village and extensive patterns of ridge and furrow (Fig. 2). In the relative peace of the early 17th century, a mansion wing with elaborate walled gardens was added to the castle which has been elucidated by a combination of magnetic, GPR and earth resistance survey that questions the fidelity

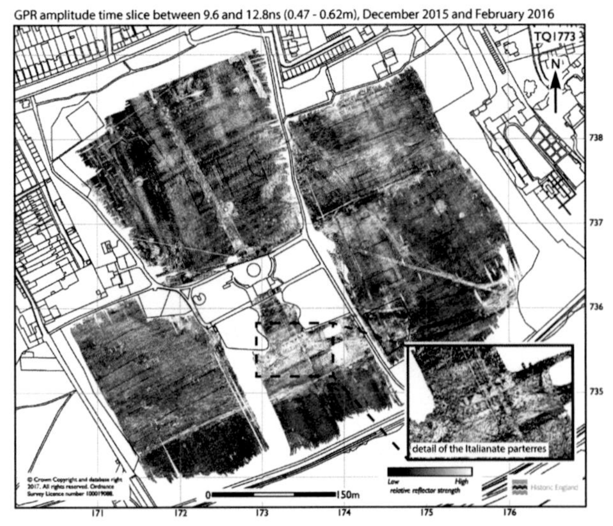

Figure 1: Large scale GPR survey at Marble Hill House showing inset detail of the Italianate formal gardens.

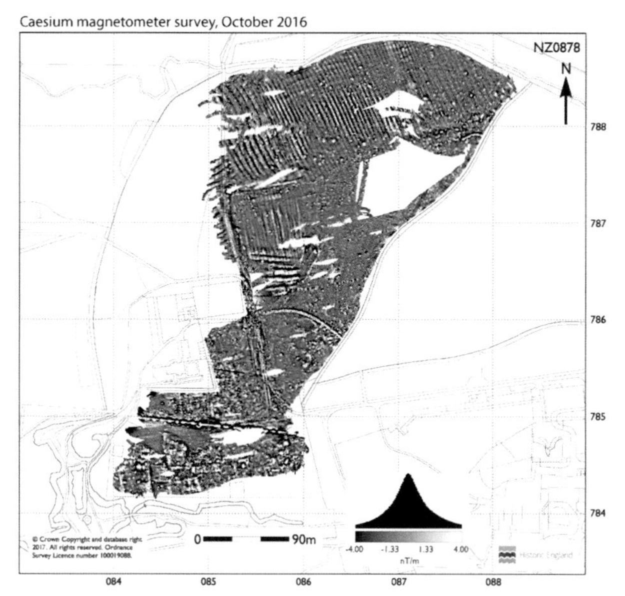

Figure 2: Magnetic survey over the medieval village and designed landscape surrounding Belsay Castle.

of a contemporary engraving of the site. Other elements of the site, including the remains of the original chapel and an ornate series of fishponds, have been revealed through geophysical survey in the vicinity of the new mansion built in 1817 when the Middletons moved from the castle.

Contemporary depictions of the gardens at Clandon Park House, Berkshire and Witley Court, Worcestershire proved to be more accurate and allowed a direct correlation with the geophysical data. Both sites have suffered considerable damage to the main historic buildings, and this was particularly devastating at Clandon Park House following the fire that engulfed the building in April 2015. Geophysical survey in advance of bringing heavy equipment on to site following the fire, helped prevent damage to the garden remains depicted in an early 18th century painting by Knyff and to the wider designed landscape laid out by Capability

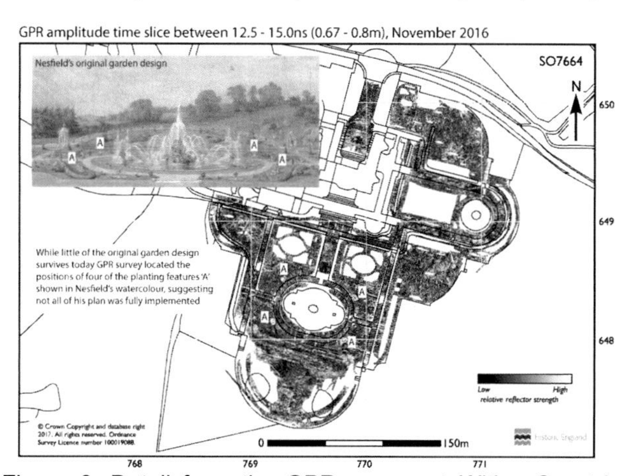

Figure 3: Detail from the GPR survey at Witley Court in the vicinity of the Perseus and Andromeda fountain. Inset shows a watercolour of Nesfield's design for the site.

Brown (Linford *et al.* 2015b). The more recent, highly elaborate 19th century garden designs and magnificent fountains at Witley Court, are recorded in their prime by contemporary photographs capturing many lost details, together with prominent visitors to the house. But some intriguing questions regarding the original water supply required for the Perseus and Andromeda fountain can, potentially, be found in comparison between a detailed GPR survey of the site and images from a promotional film made in the 1960s by Procol Harum (Fig. 3). This features views of the empty fountain bowl, including visible pipe work, before the partial restoration of the gardens and water features by English Heritage in the 1990s.

Bibliography

Cole, M. A., David, A. E. U., Linford, N. T., Linford, P. K. and Payne, A. W. (1997) Non-destructive techniques in English gardens: geophysical prospecting. *Journal of Garden History* **17**: 26-39.

Keevil, G. D. and Linford, N. (1998) Landscape with gardens: aerial, topographic and geophysical survey at Hamstead Marshall, Berkshire. In Pattison, P. (editor) *There by Design, Field archaeology in parks and gardens. British Archaeological Report, Vol. 267.* Oxford: Royal Commission on the Historical Monuments of England. 13-22.

Linford, N., Linford, P., Martin, L. and Payne, A. (2010) Stepped-frequency GPR survey with a multi-element array antenna: Results from field application on archaeological sites. *Archaeological Prospection* **17** (3): 187-198.

Linford, N., Linford, P. and Payne, A. (2015a) Chasing aeroplanes: developing a vehicle-towed caesium magnetometer array to complement aerial photography over three recently surveyed sites in the UK. *Near Surface Geophysics* **13** (6): 623-631.

Linford, N., Linford, P. and Payne, A. (2015b) *Clandon Park House, West Clandon, Surrey: Report On Geophysical Surveys, May 2015. [Research Reports Series]. London: Historic England. 83/2015. http://research.historicengland.org.uk/Report.aspx?i=15414*

Linford, N., Linford, P., Payne, A. and Pearce, C. (2016) *Marble Hill Park, Twickenham, London: Report On Geophysical Surveys, December 2015 And February 2016. [Research Reports Series].* London: Historic England. 19/2016. http://research.historicengland.org.uk/Report.aspx?i=15414

Linford, N. and Payne, A. (2011) Audley End House and Park, Saffron Walden, Essex: Report on Geophysical Surveys, 2009-2010. [Research Department Reports]. London: English Heritage. 6/2011.

Geophysical survey at bronze age sites in southwestern Slovakia: case studies of fortified settlement in Hoste and burial ground in Majcichov

Zuzana Litviaková[(1)], Roman Pašteka[(2)], David Kušnirák[(2)], Michal Felcan[(3)] and Martin Krajňák[(2)]

[(1)]Department of Archaeology, Comenius University, Bratislava, Slovakia; [(2)]Department of applied and environmental geophysics, Comenius University, Bratislava, Slovakia [(3)]Institute of Archaeology, Slovak Academy of Sciences, Nitra, Slovakia

litviakova2@uniba.sk

Geophysical methods can be used in more widely for research and monitoring of changes of different archaeological terrains and types of archaeological situations. This poster presents results of the geomagnetic prospection at two Bronze Age sites in southwestern Slovakia that are well-known in the archaeological literature since the most important finds, discovered during rescue excavations in the second half of the 20th century, have been published. Nevertheless, archaeological evidence from the settlement at Hoste and its adjacent burial ground in Majcichov can be still described as fragmented – composed of fragmentary data without any complex information.

Therefore, our aim was to obtain a large-scale conclusive picture of the sub-surface, both in the settlement and in the burial area (Fig. 1) using geophysical prospection methods, as well as to monitor influences of current agricultural activities on changing the archaeological features. The detailed geomagnetic surveys were successful in identifying anthropogenic anomalies at both sites. Some of the data from Hoste were acquired using TM-4 caesium magnetometer and later, and more extensively, just using Sensys Magneto MXPDA fluxgate magnetometer (Fig. 2), which enables comparison of sensitivity and accuracy of both methods used (Fig. 3).

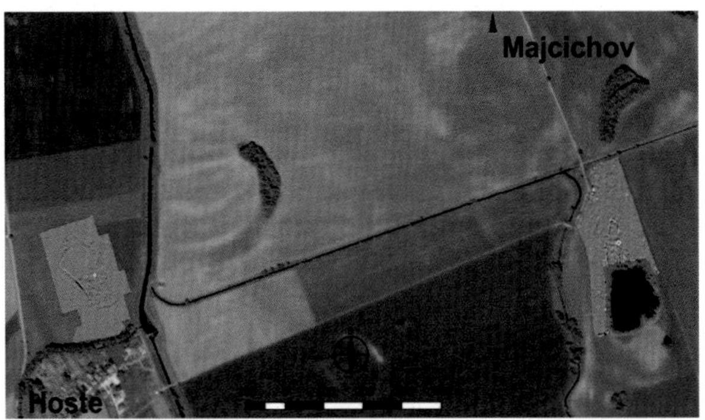

Figure 1: Maps of anomalous magnetic fields at archaeological sites in Hoste and Majcichov.

An area of more than 4.5 ha was measured in general (Fig. 2). As clearly visible in Figure 3 (right-hand situated 2 maps), resolution of directly measured gradiometric data is higher than the numerically evaluated version from the caesium magnetometer (this is partly given by a different separation distance among acquisition lines).

Geophysical techniques provide raw data which must then be processed and evaluated.The purpose of archaeological evaluation of obtained results was to determine the presence or absence, extent, quantity and quality of archaeological deposits within the area. In addition to empirical archaeological interpretation of the data, GRASS plugins of QGIS were used to extract structures of certain nT values and diameters. These values vary for different deposits and contexts. Prehistoric features with significant contrasts include pits with organic content, ditches filled with various layers, fire hearths, burned houses, etc.

Initially, geophysical prospection was focused on the Early Bronze Age fortified settlement in

Figure 2: Comparison of first measured area (color scale -5/+5 nT) and later measured (larger) area (grey scale -5/+5 nT).

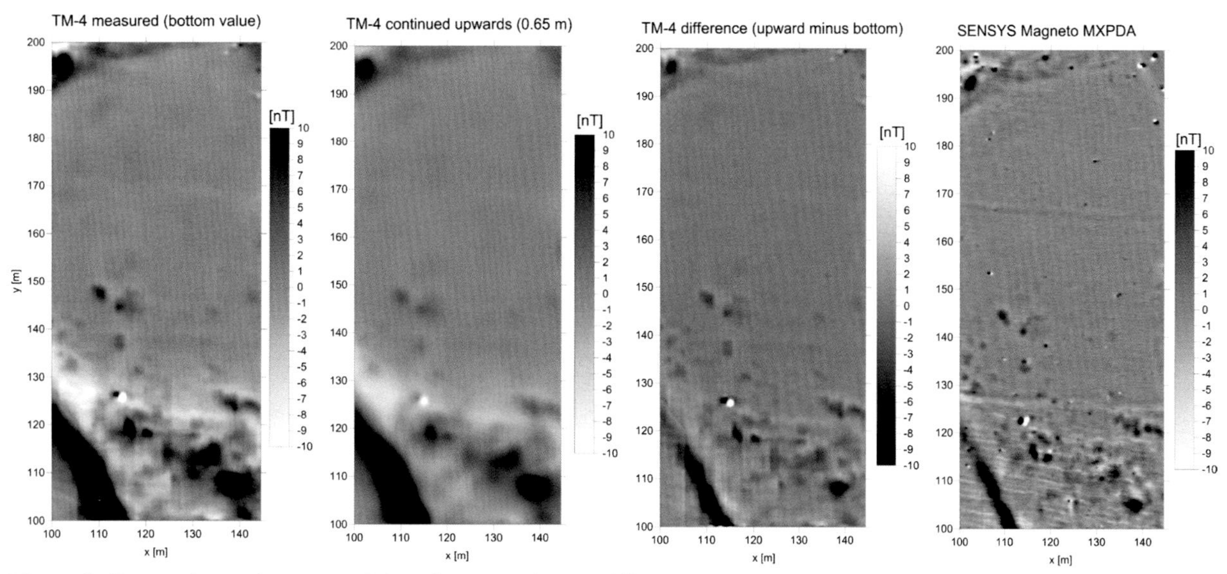

Figure 3: Comparison of measured data from caesium and fluxgate magnetometer.

Hoste (Fig. 1), which was situated in a favourable topographic and environmental position on a loess dune surrounded by fertile soil.

A very important advantage of the obtained results is the possibility to reconstruct the extent of previous rescue archaeological excavations in the 20th century. There is missing information about the exact uncovered area, and geomagnetic data helped us to get a more precise overview. As visible on the geophysical map (here the majority of magnetic anomalies are missing) (Fig. 2) excavations were focused on the northern part of the site which was most dominant in the past compared to surrounding flat land. Also, there were presumptions that fortification of the settlement in Hoste enclosed only this highest part following hypotheses that in most cases site location and establishment of fortified settlements were adapted to the landscape. They used to be situated on elevated positions and dominant places – in the hilly areas as well as in low lands. Described terrain characteristics were usually improved by enclosing structures and defensive structures such as walls, ramparts, etc.

However, magnetometric results helped to identify the fortification system of the settlement in Hoste as a whole. It consisted of a large and deep enclosure ditch with a system of smaller ditches on the northern part. The unique shape of the fortification (in terms of Central European Bronze Age research), together with its size and the fact that the settlement is not completely destroyed confirmed that the geomagnetic research is valuable for our purpose.

No features that could be traditionally interpreted like houses or residential structures have been excavated in Hoste so far. On the other hand, many settlement pits were excavated that are visible on the results from geophysical prospection as well. However, our results show an area of a large quantity of smaller rectangular features which are presumably the remains of houses. They are concentrated mostly in the southern part of the site (Fig. 2), which may be an effect of the fact that this part was not uncovered during excavation. Similarly, it is less influenced by the erosion and agricultural impact on damaging of sub-surface cultural layers. Furthermore, the results indicate that the settlement was concentrated inside the fortification (Fig. 2), without any marked extensions or re-buildings of the settled area.

The second analysed site is located in Majcichov, where a burial ground dated to the transitional period from Early Bronze Age to Middle Bronze Age was uncovered in the 20th century. There were excavated inhumations as well as cremation graves. It is located only 1200 m from the fortified settlement in Hoste (Fig. 1) and dated to the same period. Therefore, these two sites were probably in close relationship. Our aim was to discover if there are any potential graves still existing and to identify the extent of the burial site. An area of more than 4 ha was geomagnetically measured.

Archaeological interpretation of data from Majcichov is more complicated, as the prospection revealed a significant number of dipole anomalies and contemporary influences related to modern use of the site (for sand exploitation, agriculture and fishing). Nevertheless, there are some concentrations of features that could potentially be graves (intact or re-opened) according to their orientation and size.

From magnetic SQUID prospection to excavation – investigations at Fossa Carolina, Germany

S Linzen[(1)], M Schneider[(1,2)], S Berg-Hobohm[(3)], L Werther[(4)], P Ettel[(4)], C Zielhofer[(5)], J Schmidt[(5)], J W E Faßbinder[(3)], D Wilken[(6)], A Fediuk[(6)], S Dunkel[(1)], R Stolz[(1)], H-G Meyer[(1)] and C S Sommer[(3)]

[(1)]Leibniz Institute of Photonic Technology (IPHT), Jena, Germany; [(2)]Institute of Biomedical Engineering and Informatics, Ilmenau University of Technology, Germany; [(3)]Bavarian State Department of Monuments and Sites, Munich, Germany; [(4)] Seminar for Prehistory and Early History, Friedrich Schiller University Jena, Germany; [(5)]Institute of Geography, Leipzig University, Germany; [(6)]Institute of Geosciences, Christian Albrechts University of Kiel, Germany

sven.linzen@leibniz-ipht.de

In 2013, large scale magnetic prospection was carried out in Franconia, Germany, to reveal remains of a canal construction and corresponding infrastructure built by order of *Charlemagne* in the Early Middle Ages. The extended canal structures of the so-called *Fossa Carolina* were expected within a part of the main European watershed between *Altmühl* and *Rezat* – tributary streams of *Rhine/Main* and *Danube*. The construction of a first navigable continuous waterway between the latter seems to be forced by *Charlemagne's* military or economic strategy. Tremendous feats of engineering as well as enormous human and material resources were required.

The realized geophysical investigations are part of an ongoing multidisciplinary geo-archaeological research project embedded into the German Research Foundation (DFG) Priority Programme *Harbours from the Roman Period to the Middle Ages* (SPP 1630). The *Fossa Carolina* project combines geophysics, physical geography, archaeology, history and archaeometry to contribute to fundamental questions like: What was the entire extension and hydro-engineering concept of the canal? Was it been finished and put into operation? Which structures in the hinterland belong to the building?

The site has been prospected magnetically on a large scale by means of a SQUID (Superconducting Quantum Interference Device) measurement system (Linzen *et al.* 2007). The motorized, fast and high-sensitive system was used at a very early stage of the project to get data sets of the entire archaeological site as the basis for subsequent geophysical and geo-archaeological investigation. Thus, an area of more than 120 hectares was mapped in the vicinity of the last visible remains of

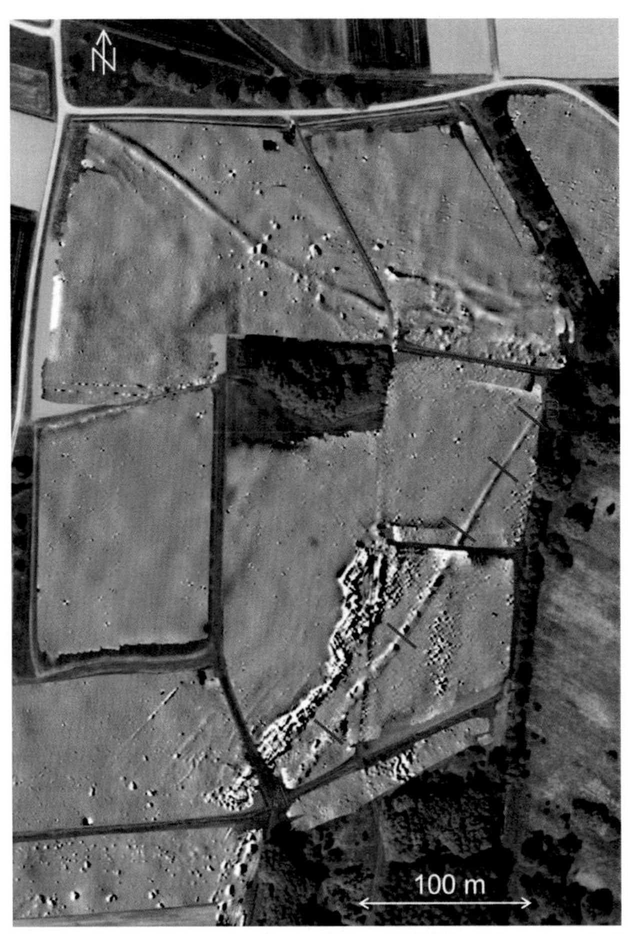

Figure 1: Northern section of the SQUID data recorded close to the current *Rezat* course near *Weißenburg*. The magnetogram shows a pronounced linear magnetic anomaly which has been interpreted as canal course and analysed on five different positions (red cuts). The positions A and B are 240 m apart. The magnetogram grey scale represents +-10 nT/m.

the *Fossa Carolina* as well as at a greater distance. Most of the area was measured deep in winter of February 2013. The frozen and snow-covered surface soil enabled the access to the intensively used farmland. The precisely georeferenced magnetic data were used in combination with topographic information and sediment analysis of well-positioned drillings (Zielhofer *et al.* 2014) to determine a promising position for a first excavation in summer 2013. This work led to the remarkable finding of a 5.3 m wide canal structure flanked by axed oak piles which were dated precisely to the year 793 AD (Werther *et al.* 2015). The canal bottom was found in a depth of 3 m referenced to the present-day surface.

The further analysis of the profuse SQUID data - which show a variety of magnetic signatures (Berg-Hobohm *et al.* 2015, Linzen and Schneider 2014) – was focussed on an extended linear anomaly corresponding to the canal course to the north, i.e. to the confluence with the *Rezat* river, see Figure 1. Calculations of the magnetic source layers on

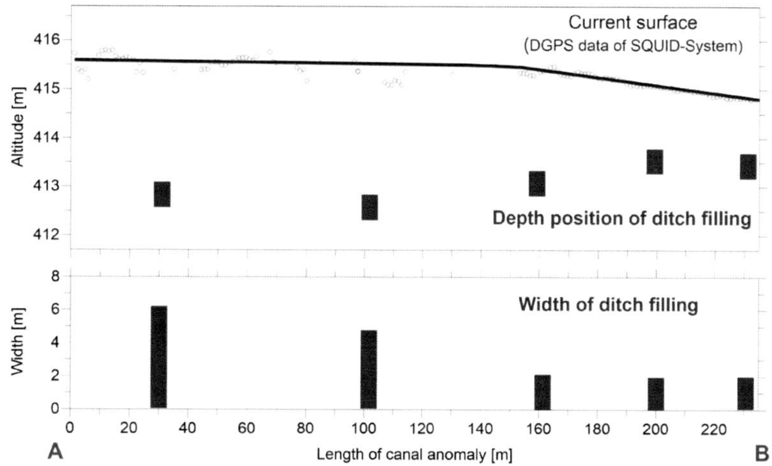

Figure 2: Result of magnetic inversion on the positions marked in Figure 1. Note the significant reduction of the canal filling width from 5-6 m to 2-3 m between the anomaly metre 100 and 160.

the base of the magnetic information from three different SQUID gradiometers and a polyhedral inversion algorithm (Schneider *et al.* 2014) revealed a significant change of the canal base geometry (Figure 2) hundreds of meters apart from the first excavation position. The width of the canal sole decreases by a factor of two within a distance of about 60 meters. This discovery was the base of a second excavation campaign carried out in summer 2016, see Figure 3. The calculated sole geometries were used to select two ideal excavation positions and to estimate the required time and digging effort.

The geo-archaeological results which have fully verified the magnetic inversion models will be presented and discussed. The excavations revealed canal sole layers with a high organic proportion (Figure 3) and an increased magnetic susceptibility which has been predicted by the inversion of the SQUID magnetic data. The calculated canal cross sections will be compared further with data of linear geophysical measurements like seismic SH-wave refraction and electrical resistivity tomography.

Figure 3: Northern most section of the canal excavated in 2016, located at meter 230 in Figure 2 and at position B in Figure 1 (L. Werther).

Bibliography

Berg-Hobohm, S., Linzen, S. and Faßbinder, J. W. E. (2014) Neue Forschungsergebnisse am Karlsgraben durch geophysikalische Messmethoden, *Denkmalpflege Informationen* **157**: 25-27.

Linzen, S., Chwala, A., Schultze, V., Schulz, M., Schüler, T., Stolz, R., Bondarenko, N. and Meyer, H.-G. (2007) A LTS-SQUID system for archaeological prospection and its practical test in Peru. *IEEE Transactions on Applied Superconductivity* **17**:750-755.

Linzen, S. and Schneider, M. (2014) Der Karlsgraben im Fokus der Geophysik, In Ettel P, Daim F, Berg-Hobohm S, Werther L, Zielhofer C (Eds.) *Großbaustelle 793 – Das Kanalprojekt Karls des Großen zwischen Rhein und Donau.* Mainz: Verlag RGZM. 29-32.

Schneider, M., Linzen, S., Schiffler, M., Pohl, E., Ahrens, B., Dunkel, S., Stolz, R., Bemmann, J., Meyer, H.-G. and Baumgarten, D. (2014) Inversion of Geo-Magnetic SQUID Gradiometer Prospection Data Using Polyhedral Model Interpretation of Elongated Anomalies, *IEEE Transactions on Magnetics* **50**(11): 6000704.

Werther, L., Leitholdt, E., Berg-Hobohm, S., Dunkel, S., Ettel, P., Herzig, F., Kirchner, A., Linzen, S., Schneider, M. and Zielhofer, C. (2015) Häfen verbinden. Neue Befunde zu Verlauf, wasserbaulichem Konzept und zur Verlandung des Karlsgrabens. In: T. Schmidts and M. Vucetic (eds.): *Häfen im 1. Millennium AD - Bauliche Konzepte, herrschaftliche und religiöse Einflüsse.* Mainz: Verlag RGZM, 151–185.

Zielhofer, C., Leitholdt, E., Werther, L., Stele, A., Bussmann, J., Linzen, S., Schneider, M., Meyer, C., Berg-Hobohm, S. and Ettel, P. (2014) Charlemagne's summit canal: an early medieval hydro-engineering project for passing the Central European Watershed. *PLOS ONE* **9**(9): e108194.

Acknowledgements

SL, MS, SD, RS and HGM would like to thank the Bavarian State Department of Monuments and Sites for financial and scientific support to realize the unique magnetic prospection. All authors thank further the German Research Foundation for financial support within the SPP 1630.

Multi-method prospection of an assumed early medieval harbour site and settlement in Goting, island of Föhr (Germany)

Bente Sven Majchczack[(1)], Steffen Schneider[(1)], Dennis Wilken[(2)] and Tina Wunderlich[(2)]

[(1)]Lower Saxony Institut for Historical Coastal Research, Wilhelmshaven, Germany; [(2)]Institute of Geosciences, Dept. of Geophysics, Christian-Albrechts-University of Kiel, Kiel, Germany

majchczack@nihk.de

Introduction

During the early medieval period, a network of trade routes connected settlement sites and specialized trading ports in the North Sea area, facilitating long-distance trade activities. On the southern North Sea coast, specialized settlements developed to act as harbours, linking the hinterland with the coastal shipping routes and housing traders and craftsmen. The characteristics of these trading ports are dependent on the natural landscape and topography, especially on natural waterways between the settlement areas and the open sea (Siegmüller and Jöns 2012). The project "Trading terps and Geest boundary harbours – medieval trading ports on the German North Sea coast" focuses on locating such harbour sites and investigates the archaeological structures of both harbour facilities and settlements, as well as the paleo-geographical conditions by coupling geophysical, geoarchaeological and archaeological methods. This paper presents results of prospections of the settlement 'Goting' on the island of Föhr (county Nordfriesland, Germany) which reveal a settlement with trade activities and maritime connections. One of the main research questions is whether the environs of the settlement might have served as a natural harbour.

Approach and Methodology

The archaeological site 'Goting' is located on the southern coast of the island of Föhr between the 'Goting'-cliff in the south and the 'Bruk'-basin, a tidally influenced wetland of c. 8.5 ha, in the north. The site was discovered decades ago, when a part of the settlement area was repeatedly eroded by the coastal cliff, which eventually led to the exposure of an abundance of archaeological finds. The initial archaeological analysis of the settlement's location and dating was gained from analysis of the archived find material. It reveals a multiphase settlement dating to the Younger Roman Iron Age and Migration Period (3rd to 6th century AD), as well

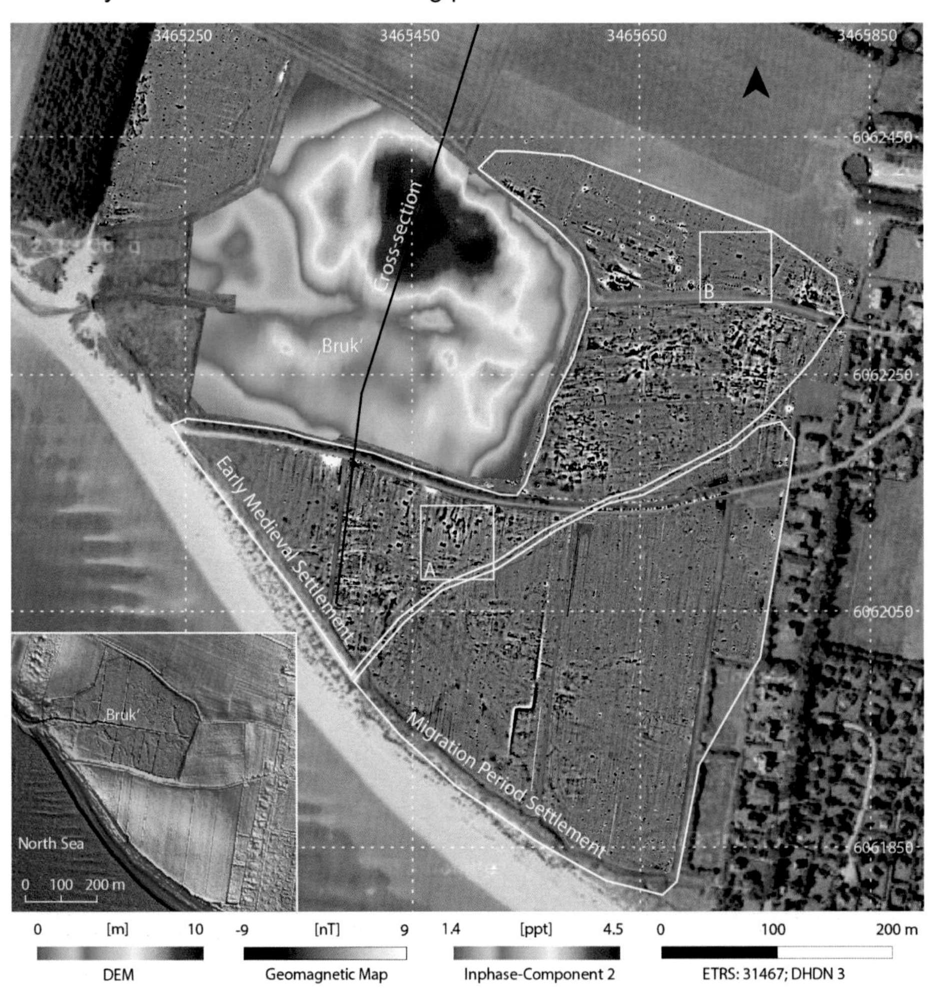

Figure 1: Map of the 'Goting'-site with geomagnetic map, electromagnetic induction map, outline of the settlement areas and the geoarchaeological transect. A and B: see Figure 2. Left corner: Topography of the 'Goting'-site in the Digital Elevation Model. Aerial Picture: ©Google. DEM: ©GeoBasis-DE/LVermGeo SH.

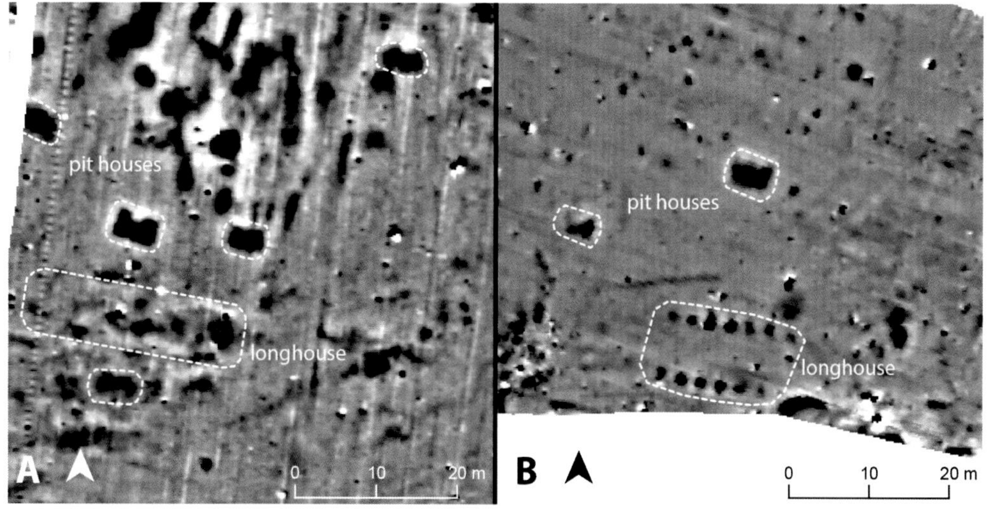

Figure 2: Details of the geomagnetic map (Fig. 1). A) Area with several rectangular pit house anomalies and one possible longhouse. B) Area with two pit house anomalies and one very detailed longhouse, where each wall-bearing post hole is indicated by the anomalies. Geomagnetic map: ±9 nT.

as the Early Medieval Period (8th to 11th century AD). The early medieval finds include imported goods from the Rhineland, the southern North Sea coast area and Scandinavia. The site location on the boundary between Pleistocene hinterland and open sea, and the evident trade connections emphasize a function as a trading port (Majchczack 2015).

The settlement site and the areas surrounding the 'Bruk' were prospected with magnetometry to establish the extension of the settlement and its layout (Fig. 1). An archaeological test excavation was carried out to determine function, age and the state of preservation of select structures.

Geoarchaeological investigations aim to reconstruct the paleo-environmental conditions of the 'Bruk'-basin by coupling geomorphological, sedimentological, and pedological analyses. Thirteen sediment cores were obtained by manual drilling with a diameter of 2 cm. Two sediment cores were obtained by means of an Atlas Copco Cobra vibracoring device with a diameter of 5 cm. Sediments and soils were documented and analysed following the guidelines of the German manual of soil mapping (Ad-hoc-AG Boden 2005). Volume magnetic susceptibility (κ) was measured by using the Bartington Instruments MS2 meter and the MS2H borehole probe. Electromagnetic induction (EMI) measurements were performed with the CMD MiniExplorer to support the geoarchaeological investigation.

Results

The geomagnetic prospections reveal the settlement layout. The southern part between cliff and Bruk-basin is occupied by anomalies typical for a rural settlement of the Younger Roman Iron Age and Migration Period, containing enclosed farmsteads and longhouses. The western corner yields characteristic rectangular anomalies of early medieval pit houses (Fig. 2.A; Majchczack 2015). These results are confirmed by the distribution of the archived find material from the cliff. The area east of the Bruk houses settlement with dense anomalies, containing pit houses and structures of parcelled out plots. These structures are reminiscent of a settlement layout of a market site, as it is known from early medieval trade emporia like Ribe, Southern Denmark (Feveile 2009). First excavations in this area proved the rectangular anomalies to be early medieval pit houses, adducing evidence for craft activities such as textile production, amber working and iron smithing.

The sedimentological-pedological cross-section across the 'Bruk'-basin allows first insights into landscape development and paleo-topography (Fig. 3). Although Pleistocene sands form the geological underground of the whole cross-section, the overlying soil and sediment sequences differ considerably between the northern slope (A), the Bruk basin (B) and the southern slope (C). Therefore, each of them must be regarded as an independent landscape element representing specific processes. (A) The northern slope of the basin dips gently southwards and transitions smoothly into the basin. It is covered by mostly undisturbed cambisols (A – Bw – C sequence) on glacial sands. (B) The almost level Bruk basin is built from a sequence of glacial sands, fossil soil, salt marsh sediments and top soil (base to top). The depth of the fossil soil and the thickness of the marsh sediments are greater in the northern part of the Basin (B-1), where also peat layers occur. In the southern part (B-2), the fossil soil gently ascends to the south. This asymmetry of the basin fill is confirmed by the EMI inphase map, where section B-1 has significantly higher values than section B-2 (Fig. 1) However, the values presented in the inphase map are integral values down to a depth of approximately 1m. To get more precise depth information, the data needs to be inverted. (C) The southern slope dips gently to the north and is separated from the basin by an

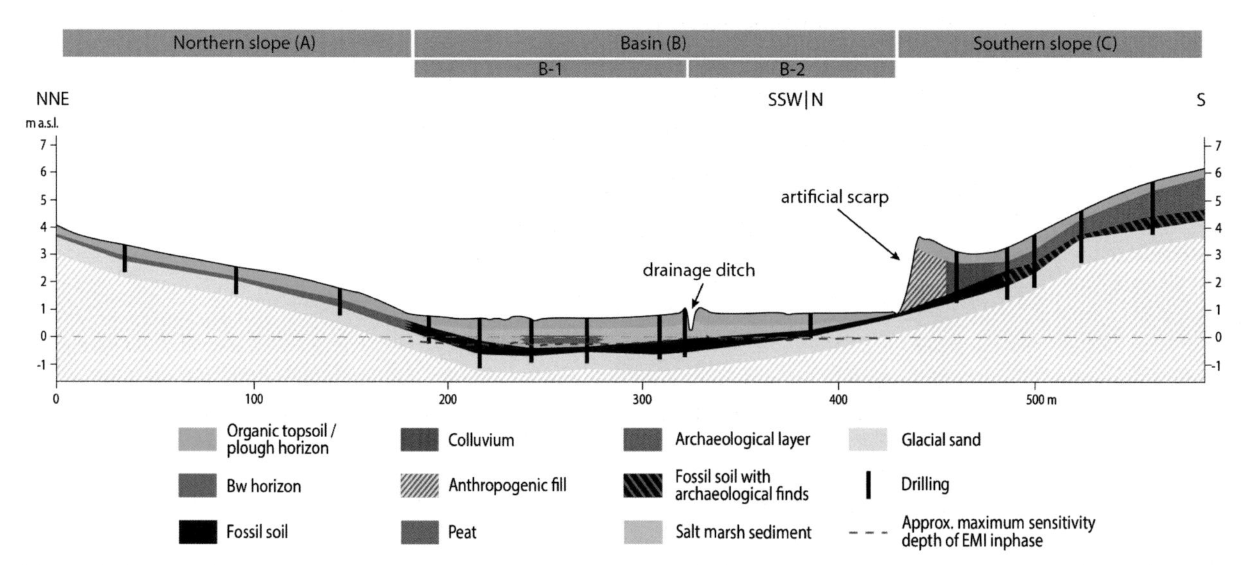

Figure 3: Sedimentologic-pedological cross section of the 'Bruk' with the approximate maximum sensitivity depth of the EMI inphase map (Fig.1).

accentuated artificial scarp. The drillings reveal fossil soil horizons, partly containing archaeological finds, which are covered by archaeological layers and findings up to a thickness of 2 m. Colluvials have formed at the footslope.

Conclusions

The geomagnetic prospections provide a very detailed insight into the settlement structures and a reliable basis for archaeological excavations. A test excavation proved a high consistency between geomagnetic anomalies and archaeological features. The early medieval settlement consists of numerous pit houses used for craft activities. An area close to the 'Bruk'-basin might be a marketplace for trade, connected to maritime trade routes.

As long as no absolute dating for soils and sediments are available, reliable conclusions on the landscape development and its chronological order are limited. However, first conclusive statements are possible. The sedimentological-pedological results from the southern slope (C) confirm the archaeological results and identify thickness (up to 1.5 m) and extent (at least 100 m in N-S direction) of the archaeological layers. The depth of an old surface could be located between 0.4 and 1.5 m below the present surface. The chronological order of archaeological site and old surface, but also artificial scarp, has yet to be clarified. (B) The sediment-soil sequence within the Bruk-basin gives evidence of (i) post-Pleistocene terrestrial soil formation and (ii) the onset of salt marsh sedimentation and peat formation, most likely due to the Holocene sea-level rise. Because of the high correlation between the pedologic-sedimentologic results and the EMI inphase map it seems promising that by coupling the two methods the fill of the Bruk-basin as well as the paleo-surface can be reconstructed in detail. Whether or not the 'Bruk'-basin might have served as a natural harbour in the early medieval period could not, as yet, be conclusively clarified. However, the existence and thickness of the salt marsh sediments within the basin give evidence of a watercourse with direct connection to the North Sea.

Bibliography

Ad-Hoc-AG Boden (2005) *Bodenkundliche Kartieranleitung*, Stuttgart: Schweizerbart.

Feveile, C. (2009) Ribe on the north side of the river, 8th-12th century – overview and interpretation. In C. Feveile (ed.), *Ribe Studier. Det ældste Ribe. Udgravningen på nordsiden af Ribe Å 1984–2000. 1.1.* Århus: Jysk Arkåologisk Selskabs skifter 51, 279–312.

Majchczack, B. (2015) Neues vom Goting-Kliff auf Föhr. Eine Siedlung von der Jüngeren Römischen Kaiserzeit bis ins Frühmittelalter im Spiegel alter Sammlungen und aktueller Prospektion. *Archäologie in Schleswig / Arkæologi i Slesvig* **15**: 139-152.

Siegmüller, A. and Jöns, H. (2012) Ufermärkte, Wurten, Geestrandburgen. Herausbildung differenter Siedlungstypen im Küstengebiet in Abhängigkeit von der Paläotopographie im 1. Jahrtausend. *Archäologisches Korrespondenzblatt* **42**: 573-590.

Acknowledgements

This work was funded by the German Research Foundation (Deutsche Forschungsgemeinschaft) in the projects RA 496/6-2, JO 304/8-2 and SE 2254/1-2 within the frame of the Priority Program 1630 "Harbours from the Roman Period to the Middle Ages".

Ripples in the sand: locating a complete aircraft in the inter-tidal zone

Peter Masters[(1)]

[(1)]Cranfield Forensic Institute, Cranfield University, Defence Academy of the United Kingdom, Shrivenham, SN6 8LA

p.masters@cranfield.ac.uk

A total of 8 aircraft crash sites dating from the Second World War have been recorded using an integrated approach since 2013. These have ranged from a Spitfire at Upavon on Salisbury Plain to a Messerschmitt MF110 at Lulworth Ranges, Dorset, and more recently the Spitfire in the Cambridgeshire Fen in October 2015.

This paper will demonstrate how important it is to record the remains methodically using the traditional and scientific methods employed in archaeology. The strategy and methodology used in these investigations showed how effective and important it is to recover as much of the remains as possible to place them into a meaningful context in order to understand the reasoning behind why they crashed. Techniques included metal detecting, fieldwalking, geophysics, excavation and post-excavation analysis of the aircraft parts.

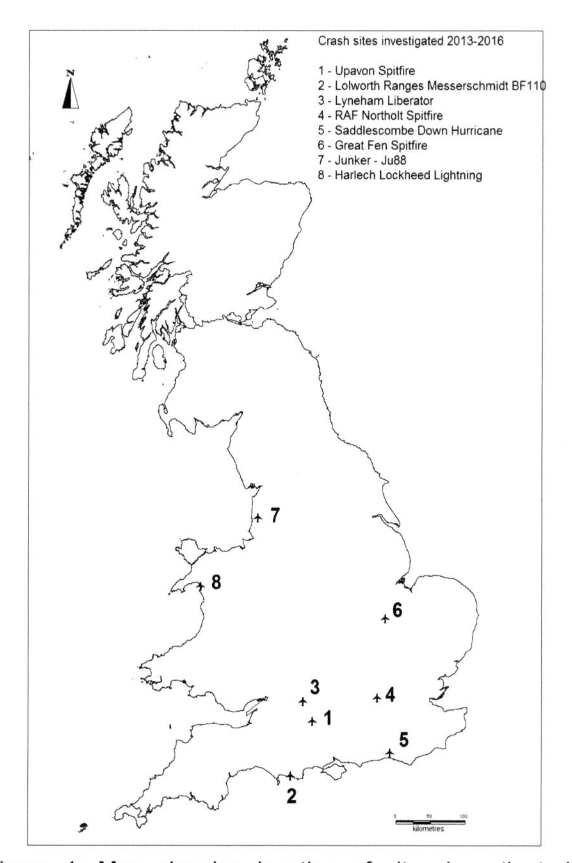

Figure 1: Map showing location of sites investigated since 2013.

Figure 2: Gradiometer survey of the crash site

Figure 3: EM38 survey of the crash site

Since 1986, all Military Aircraft Crash Sites are protected under the Protection of Military Remains Act (PMPA) and a licence is required from the Ministry of Defence to excavate or recover a military aircraft (UK Legislation 1986).

The Military Aircraft Crash Sites guidelines issued by Historic England (formerly English Heritage) in 2002 are currently being revised, which will encompass the work undertaken by the author in this paper.

These crash sites have shown similar results with some exceptions such as a complete aircraft off the coast of North Wales that is buried in shifting sands and is submerged at high tide.

The survey employed three techniques to record the exact location and associated remains: magnetometer, GPR and EM38 (Figures 1 and 2). This paper will illustrate the nature of how the survey was conducted under extreme conditions at this location off the coast of North Wales. Three techniques were employed: gradiometer, GPR and EM38. The results were extremely successful at locating the remains of the aircraft that is now buried in the shifting sands off the coast of North Wales.

In 1942, a Lockheed Lightning P38 crash landed off the coast of North Wales after the pilot realised that he had little fuel on board. The aircraft was successfully crash landed on the sand at low tide in a channel. The Lockheed P-38F-1 Lightning aircraft, given the name "Maid of Harlech" by The International Group for Historic Aircraft Recovery (TIGHAR), was landed just short of the beach at Morfa Harlech, Gwynedd, North Wales, on 27th September 1942 by the pilot, 2nd Lt Robert F Elliot. Since this time, the aircraft has been buried in the sand, except for 2007 and 2014 (TIGHAR 2007; Cundy *et al.* 2014; Masters 2016) when storms exposed substantial sections of aircraft remains.

The Harlech P-38F-1 is unique, in that it survives as a virtually complete and unaltered aircraft compared to the many thousands that crashed during WW2. The P-38F-1 is highly recognisable in its appearance and design, with its twin boom and central nacelle design. This is the only surviving 8th Air Force combat veteran P-38F-1 and probably the oldest surviving of the 8th Air Force combat aircraft.

The P-38F-1 was part of Operation Bolero, the wartime operational build-up of US Air Forces in the UK. The significance of this aircraft in terms of its completeness is rare.

This is quite a dynamic coastline with shifting sands and eddy currents which move markers or bury them in the sand as can be seen in Figure 1.

This paper will set out some of the issues of undertaking such surveys in the inter-tidal zones around the UK coast. It will also outline the standards required to achieve the best results.

Bibliography

Cundy, I. and Turner, W. (2014) *Survey and Recording of the Lockheed P-38F Lightning "Maid of Harlech"*. Report Ref: MADU-P38-2014.

Masters, P. and Osgood, R. (forthcoming) Towards an Integrated Approach to Recording Military Aircraft Crash Sites. *Antiquity.*

Historic England (2008) *Geophysical Survey in Archaeological Field Evaluation.* London, English Heritage: Research & Professional Guidelines No.1.

TIGHR (2007) *An Archaeological Survey and Preliminary Assessment for Recovery of a P38 WWII Airplane, Harlech, Gwynedd County, Wales, UK.* The International Group for Historic Aircraft Recovery (TIGHR).

Built to last: building a magnetometer cart - advantages and disadvantages in the construction of a bespoke system

Peter Masters[1] and Gary Cooper[2]

[1]Cranfield Forensic Institute, Cranfield University, Defence Academy of the United Kingdom, Shrivenham, SN6 8LA; [2]Cranfield University, Defence Academy of the United Kingdom, Shrivenham, SN6 8LA

p.masters@cranfield.ac.uk

Cranfield Forensic Institute has built its own multi-sensor cart-system for mounting 4 Grad-601 fluxgate gradiometers to carry out large scale surveys, which is a relatively recent innovation in the case of fluxgate gradiometers compared to alkali-vapour systems; it is also becoming more of a mainstream requirement in the field of commercial archaeology to cover large tracts of land for development (Historic England 2008; Schmidt *et al.* 2016). The benefits of using cart–systems means that the random noise can be reduced to a level that has minimal effect on the data recorded and the area covered per day can be up to 3 times more than compared to the traditional hand-held fluxgate gradiometers.

It is also being utilised in research based surveys, in particular for First World War landscapes. This paper will outline the successes and failures of building a bespoke cart system and outline some of the limitations.

The cart's construction has been made from non-magnetic material such as carbon-fibre tubing; wheels have been utilised from an MRI non-magnetic wheelchair. The connecting components have been designed using CAD software and manufactured using a 3D printer. These are then secured to the carbon fibre tubing at specific points using brass nuts and bolts. It was essential to make sure that the construction of the cart was robust enough to withstand uneven terrain; also, it was designed with minimal use of connector types and modular for ease of assembly in the field. During the testing on ground at Cranfield University, the cart worked well without any issues. However, during a large-scale survey, issues arose with the cart and it was found not to be robust enough for rough ground conditions.

Phase 1a allowed us to rectify some of the issues that arose from the Phase 1 build. These included bracing the sensor boom to aid stiffness and prevent excessive vibrations being detected in the captured data. Secondly, the mounting components were strengthened to prevent fatigue occurring when tightening the nuts and bolts to keep it secure whilst surveying across such terrain. This is essential to having a stable system without compromising the data quality and overburdening the cart. The whole concept of the cart system was to make sure it was robust enough for all survey conditions but at the same time to be light enough for the operator to push/pull the cart across uneven ground.

The integrated system using a DGPS to position each point is vitally important to record each data point along the traverse line. However, there are limitations to using DGPS in terms of accuracy of position especially along field boundaries where leaves and branches of trees will interfere and degrade the signal. Therefore, a good signal would be prevented from being received by the GPS antenna from the satellites. RTK would be ideal for preventing loss of signal and accuracy of each point. The problem with this is cost and how effective would it be. During previous surveys, the author has found that RTK signals can do the same thing depending on where the base station is positioned in relation to the area being surveyed. Also, where the satellites are positioned at the time of the survey can have an effect on the quality of the positioning data.

Lat/long vs UTM data is important to consider when capturing the coordinate positions. UTM is more reliable to provide accurate coordinates as this is based on a grid system compared to Lat/Long that will give you a coordinate of where you are in degrees but not necessarily the exact points on the ground.

RTK GPS would be the best for centimetre accuracy compared to DGPS which is sub-metre accuracy (up to 10cm) and is less susceptible to losing signal during a survey.

Phase 2 build will include a towing mechanism to allow the cart to be towed by a 4 x 4 vehicle such as a quad bike or Kawasaki Mule. This is currently being designed and built.

Figure 1: Cranfield Forensic Institute Multi-Sensor Cart System.

The Bartington Sensors designed for the Grad-601 have been used in this multi- sensor cart system. However, the issue with these sensors is that they are directional and still need to be set up in the same way as the dual hand-held system. These cannot at present be set up automatically compared to the Foerster Ferex 4.032 4-channel fluxgate system. Geophysical Surveys would be comparatively easier if this is the case with the Bartington system set up in a similar way.

GEOMAR software, MLgrad601 is used to capture the sensor data and GPS positions whilst Multi-Grad 601 v1.05 is designed to post-process the data collected in the field. This allows you to convert the data to an ASCII text file format and to correct created XYZ files for the delay (lag) caused by the system time constant. However, it has limitations in terms of data size and time taken to record it. This will be highlighted in the presentation of the paper.

Overall, the quality of the data recorded has been good during trials and using a cart is less likely to be susceptible to random noise or operator error.

Bibliography

Historic England *(2008) Geophysical Survey in Archaeological Field Evaluation.* London, English Heritage: Research & Professional Guidelines No.1.

Schmidt, A., Linford, P., Linford, N., David, A., Gaffney, C., Sarris, A. and Fassbinder, J. (2015) *EAC Guidelines for the Use of Geophysics in Archaeology: Questions to Ask and Points to Consider.* EAC Guidelines 2. Europae Archaeologia Consilium (EAC), Association Internationale sans But Lucratif (AISBL),Siége social: Namur, Belgium.

How to make sense out of incomplete geophysical data sets - cases from archaeological sites in north-eastern Croatia

Cornelius Meyer[(1)]

[(1)]Eastern Atlas GmbH & Co. KG, Berliner Str. 69, 13189 Berlin, Germany

cornelius@eastern-atlas.com

In autumn 2014 and 2016 two geophysical prospection campaigns on several archaeological sites in Eastern Croatia were completed. In 2014, arbitrarily distributed field strips at the Late Bronze Age / Early Iron Age site of Dolina (Nova Gradiška) were subject to magnetic prospection. In the 2016 campaign, 10 archaeological sites were investigated in Eastern Croatia in the plains between Drava, Sava and Danube rivers in a 12-day campaign. Data and interpretation from the Hallstatt site of Bistričak (Varaždin), from the oppidum of Markušica (Vinkovci) and from the multi-phase site of Dren (Vinkovci) are presented and discussed.

All data sets can be considered as incomplete, since the investigations faced a number of serious constraints. Firstly, those undertakings are often characterised by relatively low budgets. Secondly, intensive agricultural use, small property sizes and weather result in a limited accessibility of the areas to be investigated. On no account a complete coverage of the site's cores was possible. Moreover, the limited budgets also implicate a methodological narrowness: an archaeologist, a geophysicist, a van, a magnetic gradiometer array, occasionally a GPR, and GPS equipment are used, and nothing else matters.

For the data collection, the multi-gradiometer array LEA MAX (Meyer *et al.* 2015) was used, consisting of 5 to 7 Förster CON650 fluxgate sensors. The GPR data were collected with the SIR-3000 from GSSI and a 270-MHz antenna. Data positioning was realised be means of a RTK system of two GNSS receivers NovAtel SMART V1.

Do these incomplete and narrow data sets really contribute to increase our archaeological knowledge? From a pure scientific point of view, a total coverage not only of the "site", but also of the surrounding "landscape" is self-evident. A wide methodological spectrum and the registration of time series may complete this very reasonable approach. But in the reality of archaeological fieldwork and research in today's "austerity societies" this remains mere wishful thinking. Yet, could we not consider this kind of low-budget hit-and-run prospection campaigns a a full-value tool in our archaeological research kit?

Due to the wide availability of multi-channel equipment, data collection has become a standard procedure during the last 15 years. By contrast, the archaeological interpretation of the data is a matter of permanent epistemological development. The combination of the mentioned incomplete data sets and poor documentation of archaeological information on historical excavations and surveys present a challenge for both archaeologists and geophysicists.

The examples presented prove that substantial archaeological information can be gathered even from very limited data sets. Taking into account geological, geomorphological and archaeological information, however sketchy it may be, prevents us from both merely describing geophysical data, as well as from over-interpreting. It is self-evident that all information is assembled and presentable in GIS, so that continuous reworking of the data is possible for all involved parties.

Magnetic data from the site of Dolina exhibit lined-up features that resemble remains of houses, made of wooden posts and mud (Fig. 1). The strong dipole anomalies suggest the existence of numerous fire places. Subsequent archaeological excavations unveiled thick layers of burnt daub, several pavements of pottery fragments and even post holes but no real indications for the existence of houses. The excavated structures rather show the character of a work place, used only over a short period. So, the archaeological interpretation of the magnetic data gradually changed from "settlement" to "zone of human activity of unclear character". Another type of structure shows less clearly in the magnetic data but resulted in an unambiguous interpretation: the almost completely flattened remains of Late Bronze Age tumuli (Ložnjak Dizdar and Gavranović 2014). In this case, shape, dimension and context tell their own story, even though the accessible parts of the necropolis area again were rather limited.

The results from the Bistričak site display an insight into one of the most prominent Iron Age tumuli in Central Europe, the Gomila burial mound (Šimek and Kovačević 2014). The combination of GPR and magnetic results indicate a rectangular burial chamber, built of limestone and calcareous sandstone blocks. The data may even hint to looters' disturbances long ago. South of the tumulus a necropolis with two different types of burials was cut by the magnetic measurements. Now, further research on temporal and spatial correlation of the assumed burials is due.

The site of Dren is located near Vinkovci, north of the Bosut river, a tributary of the Sava river. Finds from the site are extremely numerous and date from Bronze Age to early medieval period. The site

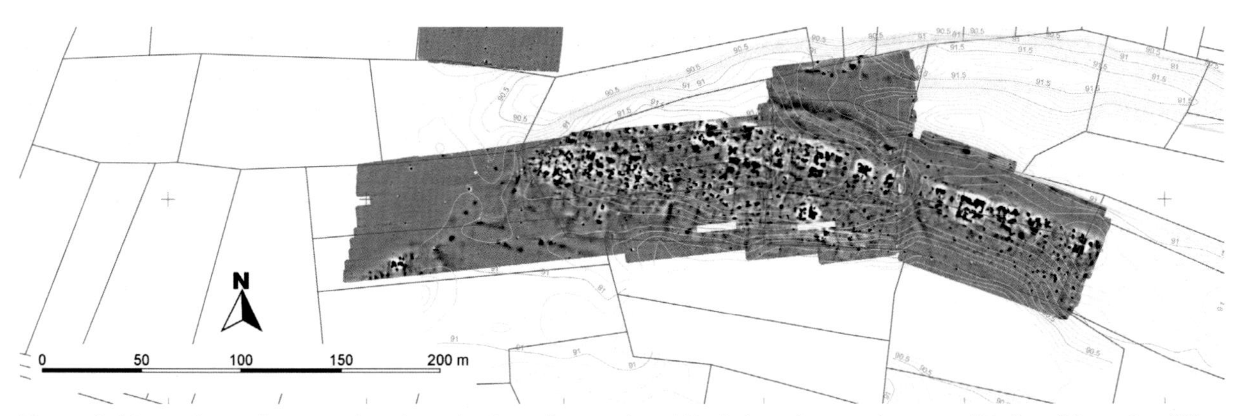

Figure 1: Magnetic gradiometry data from the Late Bronze Age / Early Iron Age settlement of Dolina (Nova Gradiška, Croatia), profile distance 50 cm, in-line point distance 10 cm, greyscale dynamics: -10 nT (white) to +10 nT (black).

shows signs of metalwork production from La Tène and Roman period, visible in bronze slag, unfinished fibulas and bronze casting extras. Corresponding to the finds, the magnetic data are characterised by likewise numerous indications of metal working places and ditches. Since the site seems to be much larger than the surface covered by magnetic measurements, a meaningful interpretation has to take into account the archaeological survey results and topographic features. Yet, the results can only be considered as intermediate ones and as a basis of further research on this complex site.

The data set from the oppidum Markušica steps out of line compared to the other examples. The settlement area was almost completely covered by magnetic measurements. It reveals a kind of textbook example, and the question is rather, which information is actually missing here?

It can be concluded that even these notoriously incomplete data sets hide archaeological treasures. Thorough interpretation and consideration of all other likewise incomplete information resources can contribute to important knowledge, both on a local and a wider level.

The geophysical prospection campaigns were initiated, funded and strongly supported by the Institut za Arheologiju Zagreb (Dolina, Bistričak), the OREA – Institut für Orientalische und Europäische Archäologie, Vienna (Dolina) and the Gradski muzej of Vinkovci (Markušica, Dren).

Bibliography

Ložnjak Dizdar, D. and Gavranović, M. (2014) Across the River. The Cemetery in Dolina and New Aspects of the Late Urnfeld Culture in Croatian Posavina and Northern Bosnia. *Archaeologia Austriaca,* **97–98**: 13–32.

Meyer, C., Zöllner, H., Pilz, D., Horejs, B. and Matthaei, A. (2015) LEA MAX – multi-purpose gradiometer array in the fields of the Kaikos valley (Bergama, Turkey). *Archaeologia Polona* **53**: 229-232.

Šimek, M. and Kovačević, S. (2014) Jalžabet – Bistričak: On the eve of the new research. *Prilozi Instituta za arheologiju u Zagrebu*, **31**(1): 231-238.

The story of two ceramic vessels: geophysical prospection and excavation in the premises of Volkswagen Slovakia

Peter Milo[1], Tomáš Tencer[1] and František Žák Matyasowszky[2]

[1]Department of Archaeology and Museology, Masaryk University, Brno, Czech Republic; [2] Archeologická Agentúra s.r.o., Bratislava, Slovak Republic

milop@post.sk, tomastencer@seznam.cz, info@archeologickaagentura.sk

The combination of large scale geophysical prospection and contract archaeology is still a new phenomenon in the area of central-east Europe. Only recently, places endangered by the construction activities started to be prospected on a wider scale. This paradigm shift in post- socialist archaeology is an outcome of several small changes.

Better quality and availability of the instrumentation at the archaeological institutions and universities, together with gradual generation change leads to new ideas and expectations in contract archaeology. The subsequential changes in the heritage management measures resulted in the necessity of geophysical prospection of the large areas affected by construction activities. Archaeological prospectors gained the trust of contract archaeologists as well as decision makers. Both archaeologists and investors start to benefit from the possibility to assess the archaeological potential of an area and to plan earthworks and logistics for the construction site in advance.

Within the last years, our Department of Archaeology and Museology at Masaryk University made several advance geophysical prospection surveys in the Czech and Slovak Republics. Surveys carried out in cooperation with Archeologická Agentúra s.r.o in 2015 and 2016 demonstrate the advantages of archaeogeophysical prospection conducted prior to the excavation itself. It also points out the weaknesses of the traditional excavation method used in preventive archaeology.

The objectives of the surveys presented here were to identify the archaeological potential of two adjoined fields (10 and 11 hectares) intended for an expansion of the car factory in Devinská Nová Ves (Slovakia), thus help to prepare and organize subsequent archaeological excavations. Two set fluxgate magnetometers were applied: Ferex (Förster) and LEA MAX (Eastern Atlas) with FEREX CON 650 probes.

The surveyed area is flat and previously arable farmed land, in some parts with traces of heavy field machinery. The area sits on the alluvium of the river Morava. The geological background of the area is composed of gravel, sandy gravel and gravel residual undifferentiated accumulation of the upper terraces.

The survey didn't detect any of the typical set of anomalies characteristic for the settlement area, burial ground or production places. Besides clearly recent features (pipelines, irrigation system and common metallic contamination), we detected several anomalies which, as we assumed, were of geological origin. Only 35 mostly magnetic positive anomalies were identified and interpreted as potential archaeological features. These were scattered over the area without any significant concentrations. Magnetic survey significantly improves the efficiency of the archaeological excavation, however, the nature of the many of the identified structures remains a mystery. Not even a half of the magnetic features were recognized and recorded by the subsequent excavations as archaeological features. The ones which were recognized were lacking archaeological material. Often they contained pebble stones, thus the excavation team concluded: these features were of

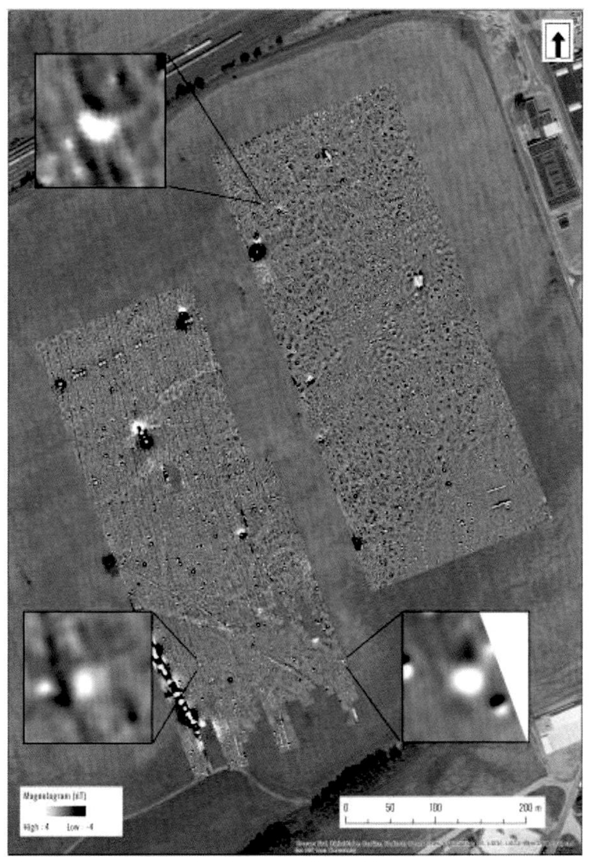

Figure 1: Devínska Nová Ves. Magnetogram of the surveyed area (21 ha), dynamic range [-4,4] nT -> [white , black].

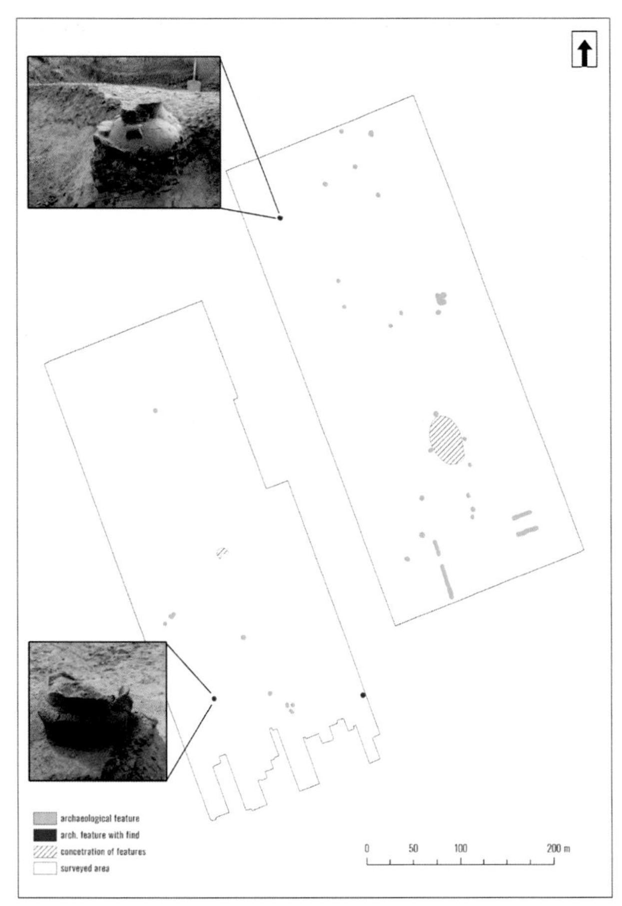

Figure 2: Devínska Nová Ves. Archaeological interpretation of the magnetic survey.

geological and non-anthropological origin and are somehow related to the activity of the river Morava.

The breakthrough came during the first season when at the bottom of one of the pits a completely preserved ceramic vessel was found. The pit was not different from the others and besides the vessel (early Bronze Age – c. 2000 BC), it was free of any anthropological evidence (pottery, stone or bone artefacts, metal etc.). Nevertheless, the next season the situation was repeated. The excavated features didn't contain any archaeological material. Only two features were different; one contained a complete ceramic vessel (early Bronze Age – c. 2000 BC), the other one a lithic flake again lying at the bottom of the pit.

These solitary finds on the 20-hectare area raised more questions than they answered. What kind of archaeological site are we actually dealing with? We are lacking possible analogies to this kind of archaeological site in the region. There was no evidence for funeral activities. Not a single bone was found and there were no bone fragments or ash. For the settlement, we are missing the usual finds such as charcoal, ash, sherds, daub etc.

However, we are not even convinced that the excavated features are of anthropological origin. They can easily be a result of the environmental activities. Moreover, the whole area is situated in the alluvium of the river and, as such, is not a suitable place neither for the long-term settlement nor for the funeral activities. Nevertheless, this area may have played a role in animal husbandry, fishing, and wood exploitation or served as a communication corridor.

The two ceramic vessels at the bottom of the otherwise empty pits may also indicate the specific character of the place. Further analysis of the ceramic vessels' fillings and application of additional scientific methods in future archaeological excavations of the nearby area may help to answer the question whether these two ceramic vessels were votive gifts to the gods or jugs left by a shepherd or a woodcutter in the pond.

Only one thing is certain, without the geophysical prospection the traditional excavation methods would not be able to identify any of these features. They would be lost and shredded in the mass of the topsoil removed by the heavy machinery and the site would be declared as free of archaeological potential.

It is questionable how many similar non-typical areas are spread out in the landscape in between "common" settlement, burial or production places, or more likely how many of them were overlooked in the past and irretrievably destroyed.

Geophysical survey for understanding Dousaku-Kofun structure

Chisako Miyamae[1], Yuki Itabashi[1] and Hiroyuki Kamei[1]

[1]Tokyo Institute of Technology, Tokyo, Japan

miyamae.c.aa@m.titech.ac.jp

Introduction

This article shows the results of archaeological prospection for Dousaku No.1 -Kofun tumuli which is 40 km north-west from Tokyo, Japan. It is said that Dousaku No.1-Kofun was built in the late 6th century AD. Kofun are ancient graves in Japan and were constructed between the early 3rd century AD and the early 7th century AD as massive mounds. Kofun have buried coffins or chambers as a main component. In many cases, numerous clay figures and potteries were arranged on the mound, and bronze mirrors and iron products were buried in the coffins or chambers. They are categorized by the shapes of the mound and the structures of the buried chamber.

Dousaku No. 1

Dousaku No.1-Kofun is categorized as 'Keyhole-shaped' tumuli that has square-shape mound at the front and round-shape mound at the back by its shape. The dimensions of this Kofun are:

- Whole length: 46m
- Maximum width of square area: 33m
- Diameter of rounded area: 25m
- Height: 4m

Although an area around this Kofun an area was excavated until 2012, this mound itself was unexcavated (Ui 2007; Kita 2011; Higure 2014), this Kofun was a designated cultural object, and was thus not allowed to be excavated.

Therefore, the authors tried to investigate this Kofun using non-destructive methods. The purpose of this study is to understand the structure of underground features.

Geophysical Survey

Ground penetrating radar (GPR) and magnetic survey have been carried out on this site from the 16th to 19th September 2016. Metal detector, G-858G Mag Mapper (Geometrics Inc., USA) and Pulse EKKO PRO GPR system with 250MHz antennas (Sensor and Software Inc., Canada) were used for this survey. Figure 1 shows the areas of this survey. Data were collected along the survey lines at 0.5m intervals.

GPR survey was carried out on these three areas:

- On the top of the buried mound: 8m x 28m (X lines, Y lines)
- Northern slope: 20x9~14m (Y lines)
- Southern slope: 21x10~15m (Y lines)

Magnetic survey was carried out on these two areas:

- On the top of the buried mound: 8 x 28m (X lines)
- Southern slope: 11x12m (X lines)

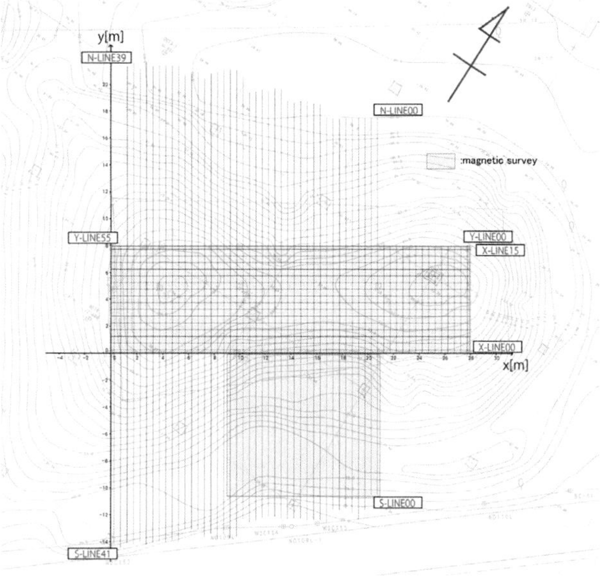

Figure 1: Survey areas.

Results

Results from Metal Detector and Magnetic Survey

The metal detector was used before undertaking magnetic survey. A gilt earing was found on the surface of the square mound by the detector. These products were usually put inside the chamber. Therefore, it is highly possible that the chamber had been disturbed by grave robbers. Characteristic anomalies from magnetic survey data were not detected. It is possible to say that numerous iron products that are often excavated from this type of Kofun do not exist underground now.

Analysis Method for GPR Data on Slopes

Several analysis methods for GPR data have been developed (Conyers and Goodman 1997). The standard visualization techniques today commonly produce amplitude slice maps from two-dimensional reflection profiles (Conyers 2015).

Survey lines on the ground are often considered as horizontal planes, when reflection profiles are created. But, this site has steep slopes. It is difficult to interpret underground structures accurately. The authors tried to visualize GPR data on slopes for more highly accurate interpretation (see Figure 2 (Y-LINE10)). On this study, steep slopes are separated every 0.2m on the contour map, and intervals are approximated linearly. Then amplitudes of received signals are plotted on lines vertical to the slope for data analysis. The electric wave velocity of the ground can be deduced when the velocity is adjusted to overlap the different signals on the same reflection object. It is calculated that the velocity of the ground was 0.06m/ns.

Figure 2 (N-LINE 32) shows a visualization example applied on this analysis. This is a reflection profile from the northern slope area (also see Figure 3). We can find the underground structure shown as a red line. It is assumed that this mound has two flat parts that are at different heights. These flat parts are detected in both areas (southern slope and northern slope, also see Fig. 3). Therefore, it is supposed that this Kofun was constructed with a double-terraced slope of the mound.

GPR Data Interpretation for Top of the Mound

GPR data for the top of the mound are analysed without topographical data correction, as it can be considered that the ground is horizontal. Spot B and Spot C on Figure 3 are anomalies from reflection traces data. These anomalies exist on the central axis. Therefore, it can be considered that these are the remains of the main buried coffins.

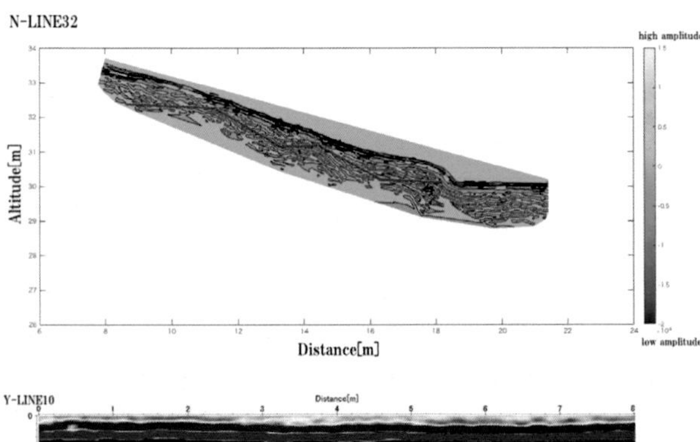

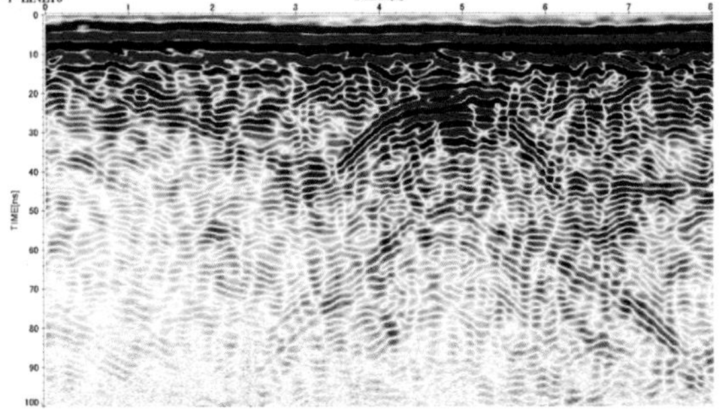

Figure 2: Reflection profiles. A visualization example applied on data correction method, this red line shows the underground structure (N-LINE32). The usual reflection profile on spot C, we can find two high amplitude curved lines (Y-LINE10).

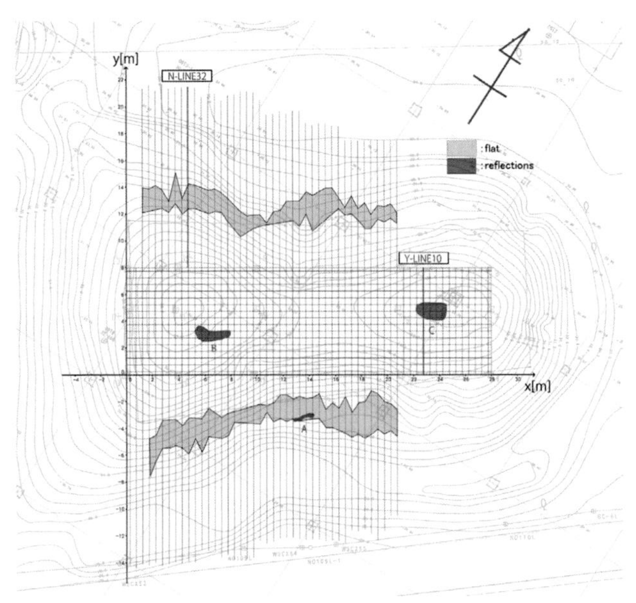

Figure 3: Results with archaeological interpretations.

Figure 2 (Y-LINE 10) shows the reflection profile on Spot C (see Figure 3). We can see two high amplitude curved lines. This reflection shows that the main buried area such as a stone coffin may be lying here. If the coffin is empty, the height of this coffin is calculated as more than 3m. This size is far bigger than other Kofuns' coffins in this period. The authors interpret that the chamber is not empty.

Bibliography

Conyers, L. B. (2015) Ground-penetrating radar data analysis for more complete archaeology interpretations. *Archaeologia Polona,* **53**: 272-275.

Conyers, L. B. and Goodman, D. (1997) *Ground-Penetrating Radar: An Introduction for Archaeologists.* Alta Mira Press, Latham, Maryland.

Higure, F. (2014) *Excavation of archaeological site in Inzai city.* Chiba: Inzai city Board of Education.

Kita, H. (2011) *Dousaku Ichigoufun (second). Excavation report 295th series.* Chiba: Inba Cultural Property Center.

Ui, Y. (2007) *Dousaku Kofun-gun. Excavation report 249th series.* Chiba: Inba Cultural Property Center.

Acknowledgements

We would like to thank Mr. Takeshi Nemoto and Mr. Keiichi Suzuki from Inzai City Board of Education for their kindly help. This project was supported by Inzai City Hall, Chiba, Japan.

An Achaemenid site in south-east Iran. A magnetic survey at Afraz (Bam-Baravat fault), Kerman

Kourosh Mohammadkhani[1] and Raha Resaleh[2]

[1]Department of Archaeology, Shahid Beheshti University, Tehran, Iran; [2]Free University of Marvdasht, Iran

K_mohammadkhani@sbu.ac.ir

Introduction

After an earthquake in January 2003 partly destroyed the historical city of Bam (Kerman province, SE-Iran), Heritage Organizations increased efforts to promote new archaeological research in the area around Bam citadel. An archaeological survey was carried out by Shahryar Adle, an Iranian archaeologist, in the desert areas around Bam. During this survey, two sites of Achaemenid date (ca. 550-330 BCE), a considerable number of Qanats as well as numerous archaeological sites dating from historic periods were documented (Adle 2005). Shahram Zare and Mohammad Taghi Ataei (Bam Research Foundation) expanded the area of the archaeological survey towards the southern Bam-Baravat fault (Ataei and Zare 2016, Zare 2007). In 2012, a magnetic survey was carried out at Afraz, the southernmost part of the 12km long Bam-Baravat fault (Fig. 1). The surveyed area (UTM coordinates: 637277.00 m E, 3209980.00 m N) is found to have been covered with evidence of earlier settlements. A particularly dense concentration of ceramics was documented at the site of Afraz, where in addition to ceramic finds; walls were identified with their heights preserved up to 70cm above ground.

Results of the Magnetic Prospection

The magnetic survey was carried out using a G858 CESIUM gradiometer (Geometrics) over a surface of 52,500m^2, divided into squares of 50 x 50m. In this survey the traverse interval was 1m. On the magnetic map, there are some linear anomalies which, in part, matched with wall structures noticed

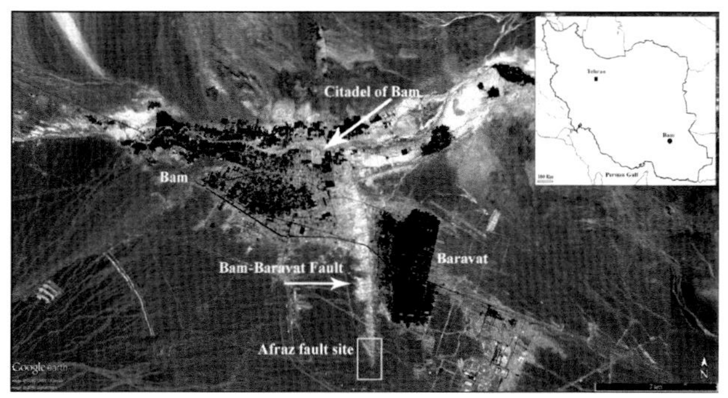

Figure1: Map of Bam region in Iran, showing the location of Afraz fault site (map © Google earth 2017).

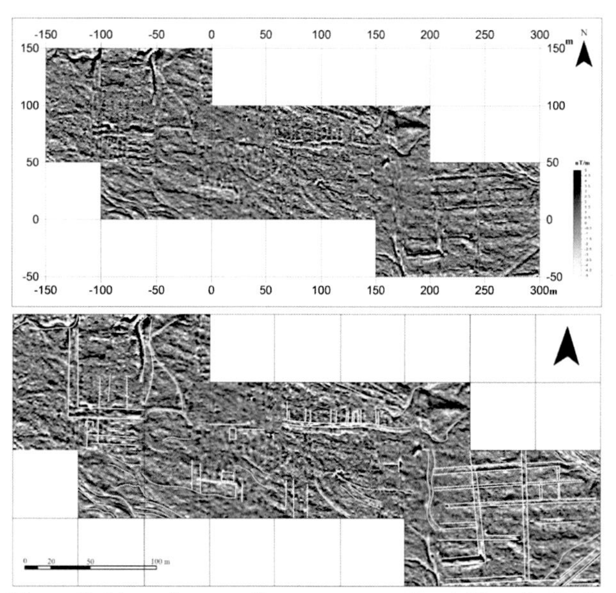

Figure 2: Map of magnetic surveys results at Afraz fault site and interpretation of anomalies.

during the earlier archaeological survey (Fig. 2). One such structure of considerable size (58 x 65m) was identified in the southwestern part of the surveyed area and certainly corresponded with an important building structure, as the walls were nearly 7m thick. We could document anomalies along the western and southern walls, and the eastern wall appeared to have been of a more irregular shape. The internal division of the building structure is not clear, but there are some weak linear anomalies indicating that this space was once divided into several sections. At an area in the south, 22 x 38 m, we identified five rooms. There were noticeable linear anomalies in the centre of the magnetic map. In the northern part of the surveyed area, another structure of considerable size (100 x 30m) with rectangular rooms was identified. It was not possible to identify the northern boundary of this structure; perhaps it was destroyed by a water creek in the north.

The largest structure (115 x 100m) was identified in the eastern part of the surveyed area. Divided into 7 sections running in the north-south direction and 5 sections running east-west, one part of a wall section of this structure is still visible on the surface. According to the magnetic map, the structure once had 15 rooms. To the west, there are 9 rectangular rooms, with about 30m long and about 9 to 12m wide. At least 4 spaces in the north-western part of the structure have different dimensions (23 x 10m, 10 x 11m, 11 x 12m and 27 x 11m). The thickness of the walls of this structure ranges between 1.70m to 4m. In the south-eastern part of this structure, the divisions of interior spaces are visible. There is only one linear anomaly of a diagonal fort with northeast-southwest orientation in this section. The different curved anomalies are probably due to water cuts and are not related to archaeological remains.

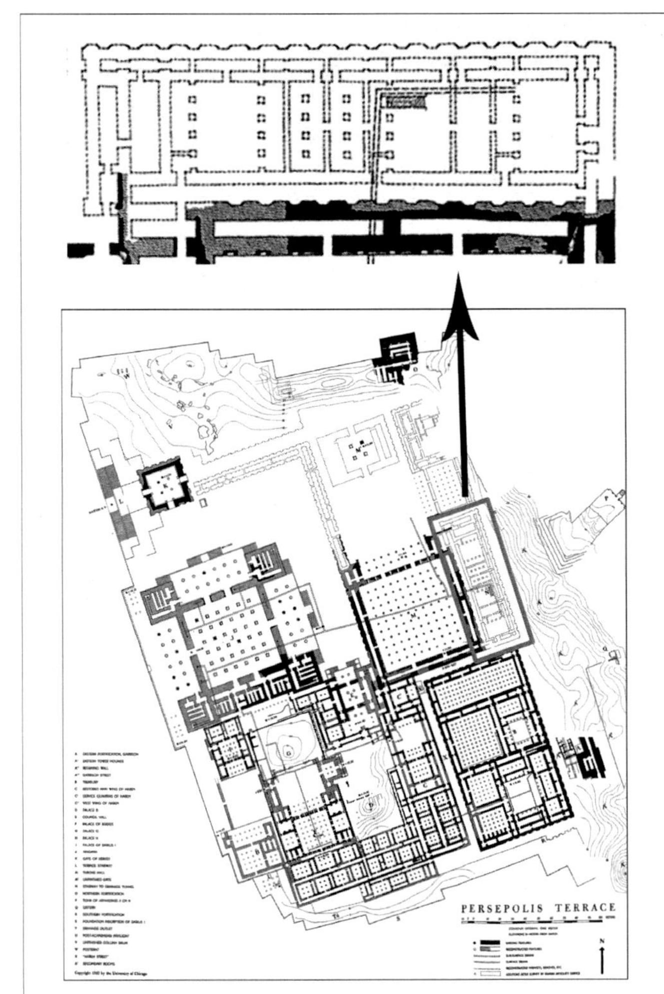

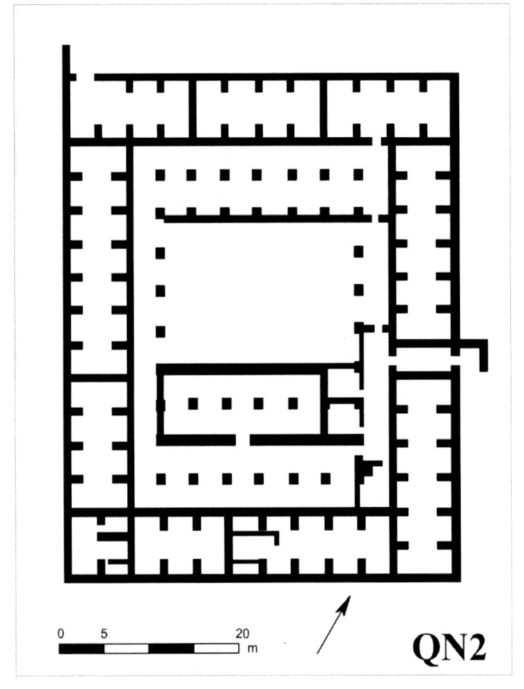

Figure 3: Left: Map of Persepolis Terrace and location of Garrison (Schmidt 1953: fig. 23). Right: Plan of Building QN2 at Dahaneh-e Gholaman (Mohammadkhani 2014: fig. 3-7).

Conclusion

Structures in the end of Afraz faultare spread over 6 hectares. There are water carrying creeks in the north and the south of the structures. These creeks may have been responsible for the loss of some of the structures. The northern structure is similar to a building located East of the Hall of the Hundred Columns at the Achaemenid site of Persepolis (Fig. 3), which has been identified as a garrison or place of gathering for soldiers. The function of two large structures in the east and west cannot be verified yet we argue that these served as administrative buildings, not as residences for a living community. A parallel of such a public structure with multiple divisions has been identified in building number 2 at the Achaemenid site of Dahaneh-e Gholaman in Sistan, south-east Iran (Fig. 3), where Achaemenid pottery helped to secure chronology. This is the first time an Achaemenid site has been identified in the Bam region. The site of Afraz and another possible Achaemenid site identified about 2km north of Afraz (*Ghāleh Khargoushān*) (Zare 2007) will help us to better understand the Achaemenid period in the region of Bam.

Bibliography

Adle, C. (2005) Qanats of Bam: An archaeological perspective irrigation system in Bam, its birth and evolution from the prehistoric period up to modern times, in *Qanats of Bam a Multidisciplinary Approach*, UNESCO Tehran Office, 33-85.

Ataei, M. T. and Zare, S. (2016) Bam in the First Half of the First Millennium B.C.: Archaeological Evidence for the State Formation in Eastern Iran, in *Bāstānpazhūhi*, Persian Journal of Iranian Studies (Archaeology), **9**(18-19): 76-92.

Mohammadkhani, K. (2014*) Étude de l'urbanisme des villes achéménides : reconnaissances de surface et prospection géophysique à Dahaneh-e Gholaman (Sistan, Iran),* Thèse de doctorat «Longues, Histoire et Civilisations des Mondes Anciens», Université Lumière Lyon.

Schmidt, E. F. (1953) *Persepolis I. Structures, Reliefs, Inscriptions*, Oriental Institute Publications 68. Chicago: The University of Chicago Press, 297.

Zare, S. (2007) *Archaeological survey at Bam fault*, Master Thesis, Tarbiat Modaress of Tehran University (In Persian), unpublished.

Archaeological Seismic survey: a Case study from Millmount, Drogheda, Ireland

Igor Murin(1) and Conor Brady(2)

(1)Faculty of Natural Sciences, Comenius University, Bratislava, Slovakia; (2)Institute of Technology, Ireland

murini@fns.uniba.sk

Introduction

A seismic method, *seismic transmission tomography* (STT), has been used as a part of a multi-method archeogeophysical prospection survey to noninvasively explore the inner structure of Millmount monument (LH024-041009-). The Millmount mound stands 16m high, is 62.5m in diameter at its base and has a flat summit 27m in diameter. Conventionally, the site which dominates the panorama of the town of Drogheda is classified as a 12th century AD earthwork motte-and-bailey castle but there are legends that this mound was originally a prehistoric burial place (Fig. 1 a).

This seismic tomography survey (STT) was conducted as a complementary survey to the previous geophysical surveys of electrical resistivity tomography (ERT) and ground penetrating radar (GPR).

Application of seismic methods is rare in archaeology, but when combined with ERT it allows a deeper investigation than other standard archaeo-geophysical methods (Domenido *et al.* 2005, Leucci *et al.* 2007) in quite a simple and efficient way (Vafidis *et al.* 1995). The STT method is convenient for investigation of mound-shaped sites as it allows placing geophones in a circle at constant elevations and applying a simple transmission of seismic waves through the surveyed object to create horizontal slice images (Forte and Pipan 2008, Polymenakos and Tweeton 2015).

Data acquisition and processing

The Geometrics Geode seismograph system was applied for the Millmount survey. Geophones with a natural frequency of 14Hz were used. We measured (STT) at four planes-spreads (L30m, L32m, L35m and L38m), which were designed in horizontal circles around the mound. The spreads were positioned at four different elevations, each with a different diameter and different distances between geophones (Fig. 1.b). The measuring points were set out and surveyed using a Trimble R8 GPS system using the Irish National Grid reference.

The processing of the seismic data was done using Sandmeier ReflexW software. The simultaneous iterative reconstruction technique (SIRT) was applied to create the final models. First, travel-times were picked from first break arrivals in recorded data files. After configuring the geometry of the survey spreads, the seismic data inversion process was iterated several times to obtain an estimated velocity model for each of the four horizontal sections-spreads (Fig. 2).

Results

Figure 2 shows the STT results of the four spreads (L30m, L32m, L35m, L38m). They indicate the presence of four types of velocity fills (F1, F2, F3, F4):

F1: represents velocity fill: 350 – 500 m/s, which can be interpreted as topsoil. It occurs in all four planes. It rings the perimeter of the planes, close to the surface.

F2: represents velocity fill: 500 - 750 m/s, which can be interpreted as soil with occasional gravel & rocks. This fill appears at planes L38m and L35m, at L38m it comprises the entire plane, at 35m most of its composition.

F3: represents velocity fill: 750 – 1000 m/s and can be interpreted as clay, compact clay. This fill

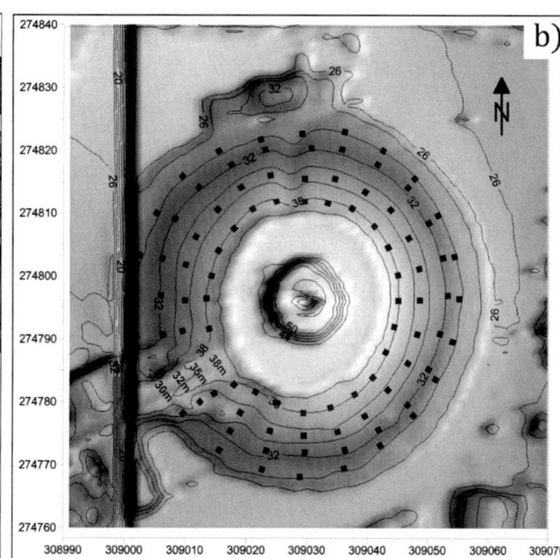

Figure 1: a) Location of the survey, Millmount site, Drogheda (Google Earth image, 2017); b) Lidar image of the site with plotted positions of measuring points (four planes-spreads).

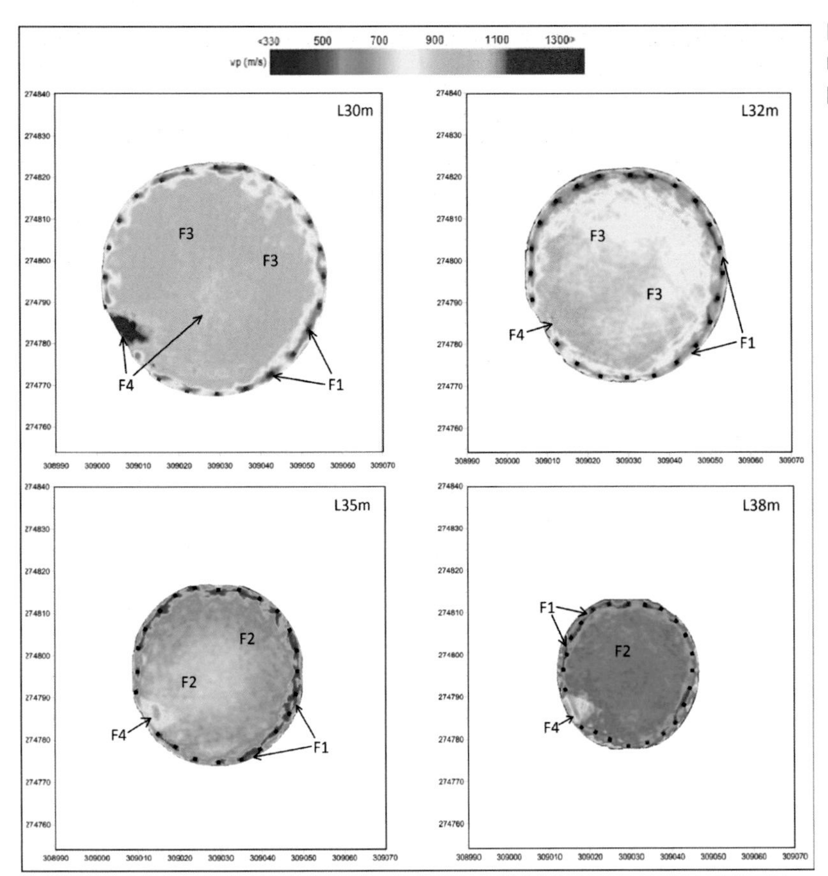

Figure 2: The generated final velocity models of the four horizontal seismic planes-spreads.

comprises most of the composition in planes L30m and L32m, is slightly detectable in central part of the plane L35m.

F4: represents velocity fill: over 1000>, can be interpreted as stone structures, walls or bedrock. There are two anomalous zones of this fill. One zone is located at the SW side in all four planes and is caused by the stone-built base of the entry ramp and stair to the mound summit. The second anomalous zone is visible at the lowest plane L30m and is located at the centre. It is a weak, not strongly defined anomaly.

The previous earth resistivity tomography (ERT) survey supports the results from the seismic tomography and shows a similar composition based on resistivity values.

Conclusion

It seems that the surveyed monument Millmount was originally a clay mound with a height ca. 4 - 6m. Later, this was extended up to 16m and in more recent times a stone fortification was added on the top. At the lowest plane-spread (L30m in Fig. 2), there is a weak anomaly with higher values close to the centre. It probably has a natural origin and could be interpreted as bedrock.

The results of the seismic tomography survey succeeded in identifying and imaging the inner structure of the surveyed object. The previous ERT survey confirms the interpretation from the seismic survey and proves the ability of these two methods to investigate the deeper parts of elevated features as Millmount monument.

Bibliography

De Domenico, D., Giannino, F., Leucci; G. and Bottari, C. (2005) Integrated geophysical surveys at the archaeological site of Tindari (Sicily, Italy). *Journal of Archaeological Science* **33**: 961-970

Forte, E. and Pipan, M. (2008) Integrated seismic tomography and ground-penetrating radar (GPR) for the high-resolution study of burial mounds (tumuli). *Journal of Archaeological Science* **35: 2614–2623.**

Leucci, G., Greco, F., De Giorgi L. and Mauceri, R. (2007) Three-dimensional image of seismic refraction tomography and electrical resistivity tomography survey in the castle of Occhiola (Sicily, Italy). *Journal of Archaeological Science* **34**: 233-242

Polymenakos, L. and Tweeton, D. (2015) Reevaluating a seismic traveltime tomography survey at Kastas tumulus (Amphipolis, Greece). *Journal of Arch.Science* **4**: 434-446

Vafidis, A., Tsokas, G. N., Loukoyiannakis, M.Z., Vasiliadis, K., Papazachos, C. B. and Vargemezis, G. (1995) Feasibility Study on the use of seismic methods in detecting monumental tombs buried in tumuli. *Archaeological Prospection*, **2**:119-128

Motorized archaeological geophysical prospection for large infrastructure projects – recent examples from Norway

Erich Nau[(1)], Lars Gustavsen[(1)], Monica Kristiansen[(1)], Manuel Gabler[(1)], Knut Paasche[(1)], Alois Hinterleitner[(2)] and Immo Trinks[(2)]

[(1)]Norwegian Institute for Cultural Heritage Research, Oslo, Norway; [(2)]Ludwig Boltzmann Institute for Archaeological Prospection & Virtual Archaeology, Vienna (LBI ArchPro), Austria

Erich.nau@niku.no

Introduction

While rescue archaeological investigations and potential subsequent archaeological excavations in relation to large infrastructure development projects are time- and cost-consuming, they provide valuable archaeological source material that otherwise certainly would be neither known nor available. During the past 20 years, exploration archaeology in Norway on previously untouched arable land has traditionally been carried out using topsoil stripping across large areas. Usually, parallel trenches of ca. 2 m width and a distance between 8-12 m from one to another are used to search for potentially present buried archaeological features, and to initially map these. This traditional approach implies that a hardly satisfying percentage of merely 10–15% of the entire areas concerned are actually investigated archaeologically.

Since 2015, the *Norwegian Institute for Cultural Heritage Research* (NIKU) has been using motorized geophysical prospection at a number of large infrastructure development projects in order to supplement and partially replace the traditional archaeological registration approach. The previous and ongoing research collaboration between NIKU and the Vienna based Ludwig Boltzmann Institute for Archaeological Prospection and Virtual Archaeology (LBI ArchPro) has led to the gathering of essential experience concerning both the organization and realization of large-scale archaeological prospection projects, as well as the efficient processing and handling of the huge amount of data involved and their subsequent archaeological interpretation.

Archaeological structures have shown a rather complicated detectability in Norwegian geological and sedimentological settings, which often are complicated by the presence of a mixture of unsorted glacial moraines, shallow bedrock, and marine or coastal deposits. Therefore, a good understanding of the environmental settings has shown to be crucial to obtain reliable results. Additionally, present buried prehistoric structures in Norwegian soils are often rather small, weakly expressed and thus difficult to detect (minor postholes, pits, ditches), which complicates the situation and limits the chances to detect associated features in the data, or to confidentially interpret the data images. So far, high-resolution GPR prospection has shown the greatest potential to successfully detect and map buried archaeological remains in the investigated study areas. Admittedly, together with the majority of earlier gained prospection experiences, these stem from only a small part of the country, primarily the region of Vestfold County.

Methods

NIKU is utilizing a motorized 16-channel MALÅ Imaging Radar Array with a crossline channel spacing of 10.5 cm and an inline sampling distance of approximately 5 cm at operational speeds of 7km/h and 40Hz sampling frequency. The system allows for an average daily coverage of ca. 3–4 ha. Data processing is carried out with the LBI ArchPro's in-house developed software package ApRadar, and subsequent archaeological data interpretation is carried within ArcGIS using custom built visualization and interpretation toolboxes and geodatabases.

As the outcome of the archaeological investigations have direct impact and consequences for both follow-up archaeological excavations, or the design and definition of development areas, or changes thereto, the underlying investigation and prospection results need to be highly reliable. Therefore, a strategy for further investigations of areas with somewhat unclear prospection results was developed. In the case of a positive geophysical archaeological prospection response in form of clearly identifiable archaeological structures visible in the data (often in the form of remains of burial mounds or distinct settlement patterns) no further measures are necessary. If potentially relevant features are detected but cannot be clearly identified and interpreted (often individual pits that could have different causes) small, targeted test trenches are

Figure 1: NIKUs motorized MALÅ MIRA system applied at Marstein, Møre and Romsdal county, Norway.

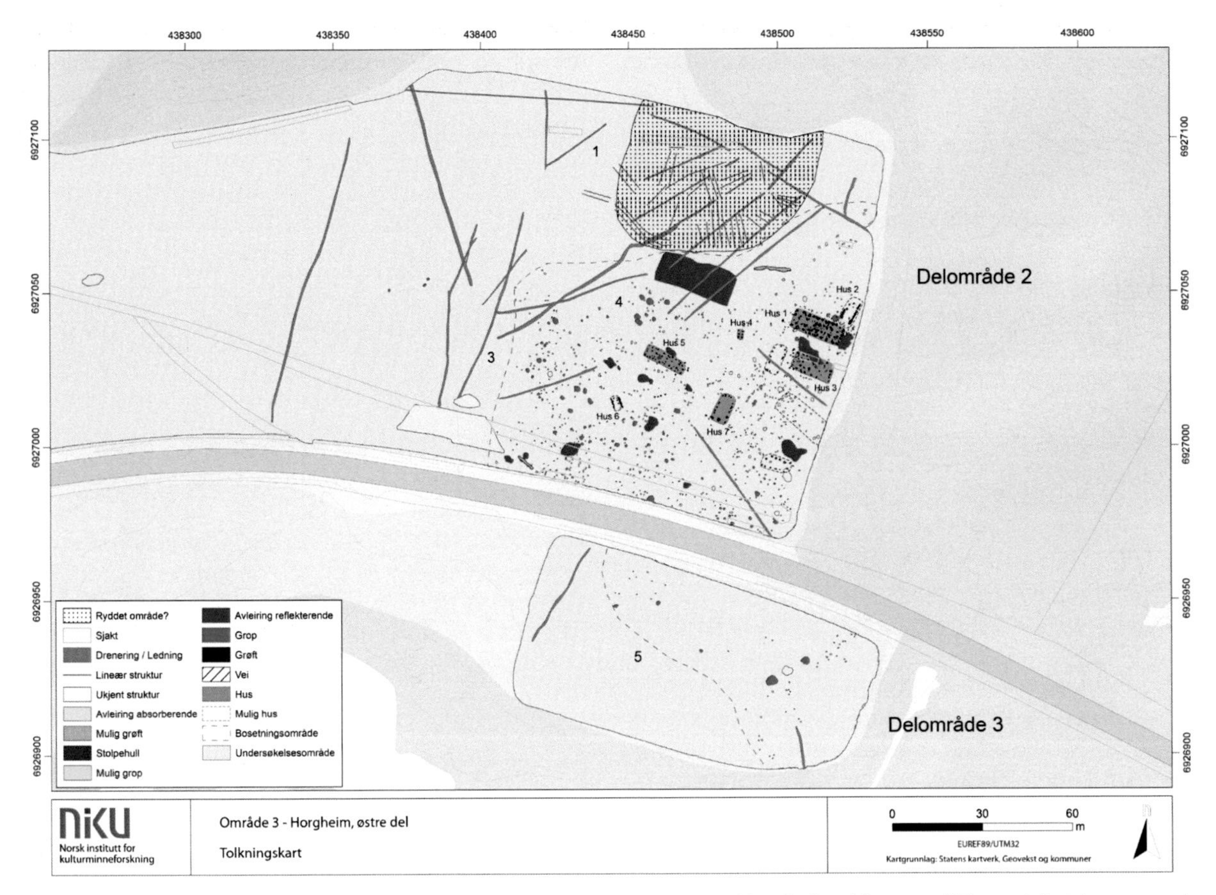

Figure 2: Archaeological interpretation of a detected settlement area at Horgheim, Møre and Romsdal. A large number of single pits and postholes as well as a number of clearly identified three-aisled post building could be detected and mapped.

used in order to clarify the archaeological relevance of the detected structures. Areas that do not show any archaeological or potentially archaeological features are most difficult to deal with, as the absence might be due to a limited detectability (principle of negative evidence). These areas might be subjected to larger test-trenching in order to reliably determine the presence of absence of archaeological remains.

Large-scale Infrastructure Development Projects

NIKU is currently involved in several large-scale railroad and road development projects taking place throughout Norway. We will present geophysical archaeological prospection results from a new double-track intercity railroad development project underway in Vestfold County, as well as from road construction projects in Rogaland and Møre and Romsdal counties.

The planned development area of the intercity railroad track in Vestfold, which stretches over 24 km and includes over 100 ha of arable land, will be completely investigated using high-resolution GPR surveys. To this day (February 15th 2017) about 55 ha have been surveyed using GPR, while the remaining areas are scheduled to be covered over the course of winter and spring 2017. The project area is situated in the vicinity of previous large-scale prospection projects conducted in the framework of the LBI ArchPros Case Study Vestfold, and the available experience led to the decision with the awarding authority to apply GPR prospection as the main investigation and registration method within this project. Results from the first survey campaign in autumn 2016 resulted in a minor number of actual archaeological findings, which were directly used to refine the planned course of the railroad track in order to avoid costly large-scale archaeological excavations.

GPR prospection in the framework of the highway construction in Rogaland and Møre and Romsdal were carried out as part of a collaboration project conducted together with the Norwegian Public Road Administration. The goal of this project is to evaluate the potential of large-scale GPR prospection for archaeological surveys in other parts of Norway and under different geological and sedimentological settings. The pilot project in Rogaland is situated in a coastal area and its geological background comprises exclusively unsorted moraine material. First results show that the prevailing geology is very much complicating the detectability of archaeological features. Prospection data analysis in combination with corresponding test trenching is ongoing.

In contrast to the rather poor results from Rogaland, the first large-scale prospection datasets generated in Møre and Romsdal County show a large number of clearly identifiable archaeological structures. The survey areas located in the valley of Romsdal are situated on top of sandy, fluvial deposits. A wide range of archaeological features ranging from individual postholes, entire building structures, larger settlement areas, and mound cemeteries could be detected.

Results

The ongoing transition towards a standardized application of geophysical archaeological prospection methods in Norwegian exploration archaeology is facilitated by the earlier standardized use of topsoil stripping as a method for archaeological registrations consequently applied in development projects. High-resolution motorized GPR prospection offers a great potential for time and cost-efficient non-invasive surveys and is therefore well appreciated amongst contractors. Frequently, archaeologists used to the traditional way of archaeological investigations are somewhat more sceptical to the geophysical prospection approach. However, with an increasing number of reliable archaeological results to show for the method, it is gaining increasing acceptance and demand is increasing.

First analysis shows that cost savings of the order of up to 70% can be achieved by the use of geophysical prospection methods compared to systematic topsoil stripping. However, potentially necessary further test trenching is not yet included in this calculation since its extent heavily depends on the outcome of the geophysical prospection.

The actual organization and realization of large-scale geophysical prospection in the framework of large infrastructure projects, usually within a very limited timeframe, has proven to be challenging. A large number of different parties are involved in the process, and a good communication and reporting strategy is essential: local heritage authorities are responsible for fulfilling the obligation for archaeological assessment and investigation of any area affected (§9, Norwegian cultural heritage law). Local or national museums are responsible for follow-up archaeological excavations, and thereby the clearing of the areas. National heritage authorities are responsible for general review and approval of all measures. Furthermore, the interest and priorities of the contracting authorities, independent consultants, and last but not least, a large number of individual land owners need to be considered.

Additionally, a growing interest in further use of the acquired geophysical data, aside of archaeology, was registered. Information about mapped utilities, drainage systems, depth to solid bedrock and old river and creek systems is increasingly requested and can be of direct use in the planning process of infrastructure development projects.

Sussing out the super-henge: a multi method survey at Durrington Walls

Wolfgang Neubauer[1,3], Vincent Gaffney[2], Klaus Löcker[1,4], Mario Wallner[1], Eamonn Baldwin[5], Henry Chapman[5], Tanja Trausmuth[1], Jakob Kainz[6], Petra Schneidhofer[1], Matthias Kucera[1], Georg Zotti[1], Lisa Aldrian[1] and Hannes Schiel[1]

[1]Ludwig Boltzmann Institute for Archaeological Prospection and Virtual Archaeology, Vienna, Austria; [2]University of Bradford, School of Archaeological Sciences, Faculty of Life Sciences, Bradford, UK; [3]University of Vienna, Vienna Institute for Archaeological Science, Vienna, Austria; [4]Zentralanstalt für Meteorologie und Geodynamik, Vienna, Austria; [5]University of Birmingham, Classics, Ancient History and Archaeology, Birmingham, UK; [6]Landesverband Westfalen-Lippe, Archäologie für Westfalen, Münster, Germany

Mario.Wallner@archpro.lbg.ac.at

Introduction

Since 2010 the Stonehenge Hidden Landscapes Project (SHLP) has undertaken extensive archaeological prospection across much of the landscape surrounding Stonehenge. The sheer scale, resolution and complexity of the data produced are unprecedented. The results range from discoveries of new prehistoric monuments to the very detailed mapping of extensive multi-period field-systems and modern complexes such as Royal Air Force Stonehenge. Inevitably, there is particular interest in the landscape context of Stonehenge itself from the 3rd millennium BC, our knowledge of which has greatly increased as a result of the project. In this interpretative context, the significance of the Durrington Walls 'super-henge', located c. 3 km to the north-east of Stonehenge, cannot be overrated.

The roughly circular henge enclosure consists of an internal ditch up to 5.5m deep and 18m wide, and an external chalk rubble bank surviving up to 1.5m high and up to c. 32m wide, with an overall diameter of c. 480 metres. It encloses a number of other structures, including two timber circles excavated by Wainwright on the east side of the enclosure (Wainwright and Longworth 1971, 204-34).

The 'Stonehenge Riverside Project' investigations have significantly changed our understanding of the monument. It is now clear that a settlement existed prior to the henge construction, dated to c. 2525-2440 BC, and it has been suggested that it was inhabited by up to 4000 people (Parker Pearson 2012, 109-111), although this extrapolation of the excavated south-east entrance area data is conjectural and direct evidence for wider occupation is limited.

The SHLP multi method survey at Durrington Walls

In September 2013, the LBI ArchPro conducted a magnetometer survey at Durrington Walls using a motorised 8-channel Förster fluxgate magnetometer array with a line spacing of 25cm, mounted on a non-magnetic cart. The survey resulted in a highly detailed visualisation of the entire monument, showing the exact location of the bank and ditch construction, several internal enclosures, old excavation trenches and pit-like features.

Subsequent Electromagnetic Induction (EMI) surveys were carried out over the full area of the enclosure by the University of St. Andrews, using a CMD Explorer (GF Instruments) with horizontal coil orientation and a GNSS receiver for real-time positioning. Line spacing was set at a maximum of 4m with measurement intervals at 0.3 sec (approximately 0.5m along lines). Three coil spacings were simultaneously recorded for both conductivity and in-phase values. In horizontal coil orientation the approximate depth of survey for the three coils are 2, 4, and 6m.

In addition a high resolution EMI survey was carried out by Ghent University using a motorised multi-receiver EMI survey with a DualEM-21S sensor (De Smedt *et al.* 2014) covering the complete monument in 2013.

Electrical Resistivity Tomography (ERT) was used to obtain geo-electric cross-sections through the site using a Lund Terrameter SAS4000. For these surveys an electrode spacing of between 1 and 5m was tested with protocols consisting of a modified Wenner-Dipole array. Data was interpreted for 0.5m bins with a maximum penetration of 25m depth. For integration with the surface EMI data the cross-section resistivity values were inverted.

An initial GPR survey with a motorised SPIDAR system from Sensors&Software was conducted in 2013 and 2014, using an array of six pairs of 500MHz antennae with a line spacing of 25cm, towed by an ATV Quad. The first interpretation of these GPR

Figure 1: The LBI ArchPro 16-channel MALÅ Imaging Radar Array system surveying at Durrington Walls.

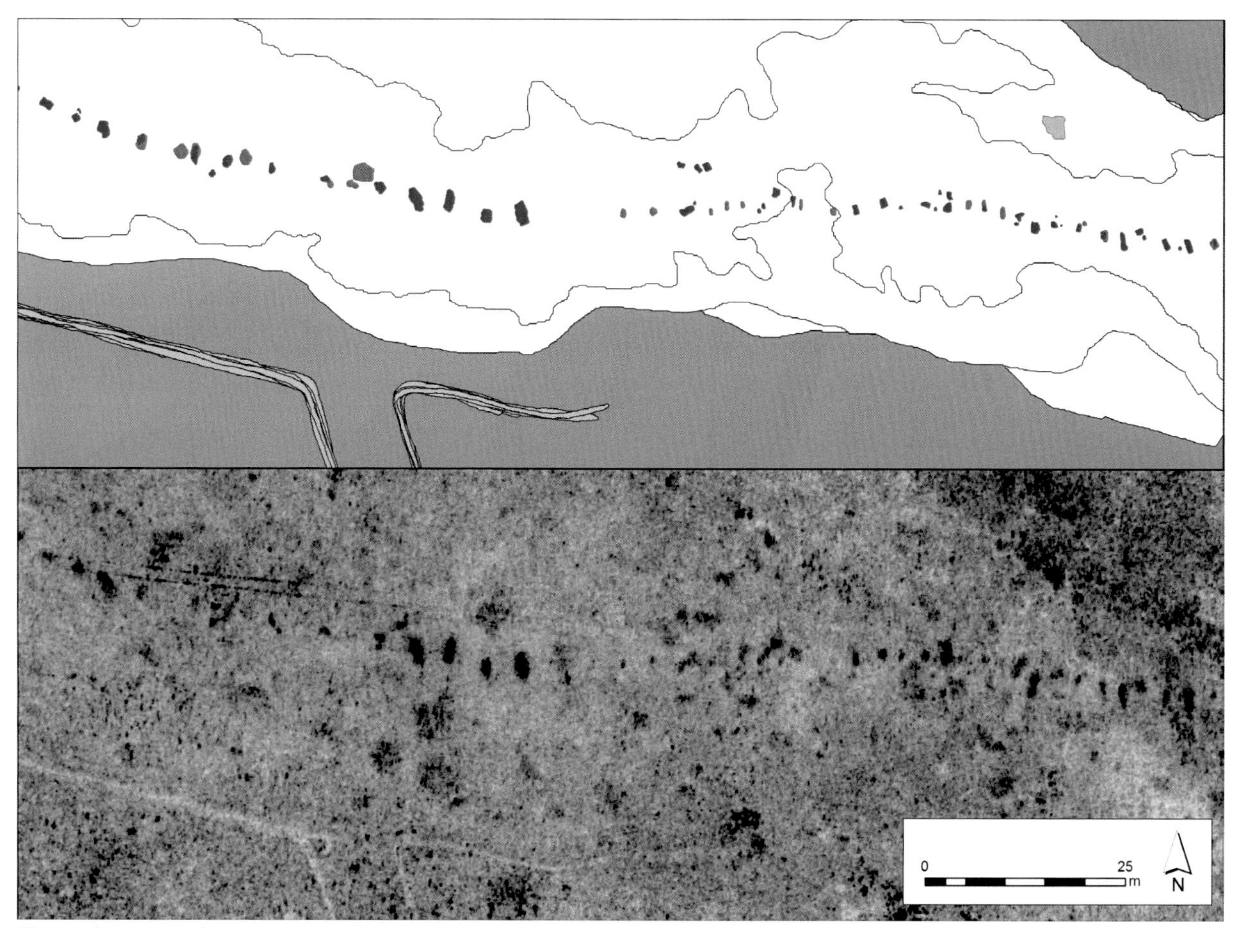

Figure 2: Detail of the archaeological interpretation (above) and a depth slice of the GPR data from Durrington Walls (below: depth 1.0–1.5m).

data outlined the detailed structure of the henge monument and a previously unknown series of aligned anomalies underneath the remnant bank on the enclosure's south side. Up to 90 large features, variable in nature and alignment, were identified. The GPR responses suggested that the largest of these features are up to 4.5m long and c.1.5m wide, composed of a strongly reflecting material such as a large boulder or the heterogeneous backfill of dug features. The features seem to be buried beneath 0.7-1.0m of bank and topsoil material.

Following these observations, another GPR survey was undertaken re-surveying the initial area using a motorised 16-channel 400MHz MALÅ Imaging Radar system (Trinks *et al.* 2010) and a spatial resolution of 8x10cm. The survey area was extended to the north and east in order to cover the complete monument and surrounding areas in 2015 and 2016.

Interpretation and Modelling of the Geophysical Data

An extensive archaeological interpretation of the geophysical datasets made for the discovery of a new construction phase and use of the Durrington Walls henge monument. Up to 300 aligned anomalies can be interpreted beneath or just outside of the surviving bank, most likely encircling the entire monument on all sides. The integrated analysis of all datasets allowed detailed assumptions about the location, superposition and characteristics of material of the newly detected features in contrast to surrounding soil and bedrock. Within all datasets an alignment of rather intensive, high-contrast and clearly confined anomalies of comparable size (approx. 3x1m) and shape could be observed. To gain more insight into the superposition of these anomalies by the remnant bank and into their volumetric extent, a subset of the 2015/2016 GPR survey data was selected for 3D visualisation in Voxler 4 (Golden Software). This subset covered four of the most discernible anomalies and was volume-rendered for further visual analysis. The subset was also visualised based on selected isosurface values, which were edited and converted into point clouds using Geomagic. The resulting volumes indicate a significant material contrast between the observed anomalies and the surrounding material within a well-defined and clearly representable shape. Visual analysis furthermore suggested the heterogeneous character of the embedded material based on the presence of continuously strong reflections inside the entire body of the anomaly. Such reflections are typically caused by a rapid change in electric properties of different subsurface elements and thus often indicate heterogeneous and/or loosely packed materials. Nevertheless a distinction between fragmented large boulders such as sarsen stones, which might have suffered from erosion leading to the partial disintegration of the former solid rock, and heterogeneous backfill, necessitated a targeted excavation, which took place in summer 2016.

Results

At Durrington Walls the combined results of multiple archaeological surveys have added an entire new phase of construction and use to a monument that has been studied intensively for many years. In doing so, the results exemplify the validity of core premises of the project team that multiple survey methods should always be utilized in order to extract the most information from such work, and that the greatest benefits of carrying out such research accrue at a landscape scale of enquiry (Gaffney C. *et al.* 2013; Gaffney V. *et al.* 2013, Löcker *et al.* 2013). This does not dismiss the need for excavation, refining the results of survey, and recovering the evidence needed for chronological, environmental and social interpretation.

Bibliography

De Smedt, P., van Meirvenne, M., Saey, T., Baldwin, E., Gaffney, C., and Gaffney, V. (2014) Unveiling the prehistoric landscape at Stonehenge through multi-receiver EMI. *Journal of Archaeological Science* **50**: 16-23.

Gaffney, C., Gaffney, V., Neubauer, W., Baldwin, E. and Löcker, K. (2013) An integrated geophysical approach to Stonehenge. In W. Neubauer, I. Trinks, R. B. Salisbury and C. Einwögerer (eds) *Archaeological Prospection. Proceedings of the 10th International Conference on Archaeological Prospection Vienna 2013.* Wien: Verlag der Österreichischen Akademie der Wissenschaften, 230.

Gaffney, V., Gaffney, C., Garwood, P., Neubauer, W., Chapman, H., Löcker, K. and Baldwin, E. (2013) Stonehenge Hidden Landscapes Project: Geophysical investigation and landscape mapping of the Stonehenge World Heritage Site. In W. Neubauer, I. Trinks, R. B. Salisbury and C. Einwögerer (eds) *Archaeological Prospection. Proceedings of the 10th International Conference on Archaeological Prospection Vienna 2013.* Wien: Verlag der Österreichischen Akademie der Wissenschaften, 19–23.

Löcker, K., Baldwin, E., Neubauer, W., Gaffney, V., Gaffney, C. and Hinterleitner, A. (2013) The Stonehenge Hidden Landscape Project - Data acquisition, processing, interpretation. In W. Neubauer, I. Trinks, R. B. Salisbury and C. Einwögerer (eds) *Archaeological Prospection. Proceedings of the 10th International Conference on Archaeological Prospection Vienna 2013.* Wien: Verlag der Österreichischen Akademie der Wissenschaften, 107–109.

Parker Pearson, M. (2012) *Stonehenge: exploring the greatest Stone Age mystery.* London: Simon & Schuster.

Trinks, I., Johansson, B., Gustafsson, J., Emilsson, J., Friborg, J., Gustafsson, C. and Nissen, J. (2010) Efficient, Large-scale Archaeological Prospection using a True Threedimensional Ground-penetrating Radar Array System. *Archaeological Prospection* **17**(3): 175-186.

Wainwright, G. J. and Longworth, I. H. (1971) *Durrington Walls: excavations 1966-1968.* London: Society of Antiquaries Research Report 29.

Urban prospections in The Netherlands, successes and failures

Joep Orbons[1, 2]

[1]ArcheoPro, Eijsden, The Netherlands; [2]Saxion University, Deventer, The Netherlands

j.orbons@archeopro.nl

Introduction

Many of the best geophysical surveys are obtained at open field-sites over large areas with the clearest results. These sites have a very boring archaeological taphonomy, very often these are sites that have a thin cover of sand or clay and are more-or-less intact in the subsoil. If the geology and the soil is also unchanging over the study area, you have found the ideal survey area for geophysics with nearly guaranteed good results.

In an urban area, the situation is totally different. Urban zones contain a multilayered sandwich of archaeological structures from many different time periods, one cutting into the other, due to the continuation of habitation. The medieval layers are very often buried rather deep below post medieval layers. The geomorphology of growing medieval cities means that all soil types – sand, clay and peat – can be found. The geophysical contrast of the base-soils is mostly dominant over the archaeological geophysical contrast. So, urban areas are on all aspects a very unfavorable location to do archaeological geophysical prospection.

Urban Geophysics as a subject is discussed during symposia and university lectures, but is generally considered a complicated matter both practically in the field (pun intended) as well as in the interpretation.

Still, geophysics can be helpful in urban areas. In Dutch archaeology, many geophysical surveys have been carried out in urban areas over the last couple of decades. The success rate of these surveys is not as high as in open rural mono-layered areas but still the results of these geophysical measurements were useful in a large number of prospection surveys. The presentation gives nine examples of prospection surveys in urban areas over the last 5 years.

Results of Urban Prospections

All geophysical surveys were part of a combined prospections survey. As many prospection methods as possible were used, and combined to try to obtain as much information about the underground archaeology as possible. The prospection methods that were used are: desktop study, geophysics, augering, surveying and trial trenching. Not all methods were used in each case, dependent on the specific circumstances.

The selection of the nine studied areas produces a representative sample with three giving good results, two acceptable results and four surveys that yielded no usable results. The success rate of this selection of urban geophysical surveys is, therefore, about 50%.

To understand why half of the surveys produce unusable results, it is necessary to analyze in detail the actual on site situation.

The three successful studies were:

- Fort Willem, Maastricht: Finding a buried entrance gallery under about one metre of building debris. The area has (metal) playground equipment and trees. The gallery could partly be accessed from below so we carried out a combined geophysical (EM and ERT) and 3D laser survey. It was located successfully.
- Vreeland Castle: A medieval castle in an open grass field that is part of a small medieval city. The site has been extensively used over several centuries. The area is covered in a layer of building debris of 70 cm. A desktop study, high resolution LiDAR, resistivity, magnetometry, EM and augerings were used. The moat and walls were located successfully.
- Sluice in Volendam: The centre of Volendam is situated on a large dyke with a road on top and houses on both sides. The aim was to locate an expected late medieval sluice inside the dyke. A combination of a desktop study, EM and GPR was used to locate the location of the former sluice.

Very poor results

- Huveneersheuvel: This sand dune in a peaty/ clay area had a medieval church on top that was demolished in 1825. A desktop study, EM, magnetometry, resistivity, ERT and augerings did not produce clear results, only vague indications of the walls. Trial trenches proved walls in several stages, at different depth and at varied stage of preservation. Furthermore, the varied geology complicated the interpretation.
- Castle Groenewoude: A small castle in a wide moat with two separate open yards with moats. The moats are filled with household waste and the area is landscaped. A desktop study and EM, resistivity and augerings were carried out. The filled moat and landscaping ruined detailed surveys but the EM31 produced a good visualization of the moat and castle zone.

Figure 1: Vreeland, result of magnetic survey.

No success

- Havezathe Altena: A small castle converted into a farmyard in a wet area. Later roads were built over the site. A desktop study, resistivity, magnetometry, EM and augerings were carried out. No good results, the signals of the medieval castle and moats are probably too small in relation to the later landscape-activities and the soil differences.
- Fortifications Maastricht: A large area with post medieval (Vaubun) fortifications with wide walls, deep and wide moats and huge embankments. The location at the time of the survey was an industrial site with roads, buildings, cables, pipes, concrete floors with steel reinforcements. A desktop study, EM and GPR survey was carried out. Only positive results were gained in areas without industrial buildings. Everywhere else the results were hopeless.
- Kasteel Plettenburg: Finding a medieval castle underneath a later Napoleonic fort. Using a desktop study resistivity, EM and augering methods. The results are very poor.
- City fortifications Harderwijk: The (post) medieval city fortifications are covered with roads and buildings and harbour equipment. A desktop study, EM and auger surveys were carried out. No usable results were obtained. Some parts were excavated afterwards and produced huge medieval walls that were completely missed by the prospection.

Conclusions

From these nine cases some conclusions can be drawn.

- A good desktop study is needed to understand the survey location, make a good selection of prospection methods and state the (archaeological) questions.
- Combined prospection surveys are essential. Not only geophysics, but also augerings, surveyors work, LiDAR, trenches, etc. are most important to understand the location and the measurements.
- Understanding what happened to the site in the last two centuries (taphonomy) is essential because of the intensive use of the locations in the last 200 years.
- Layers of building debris can complicate the surveys. The state of homogeneity of the debris layer decides the outcome. An

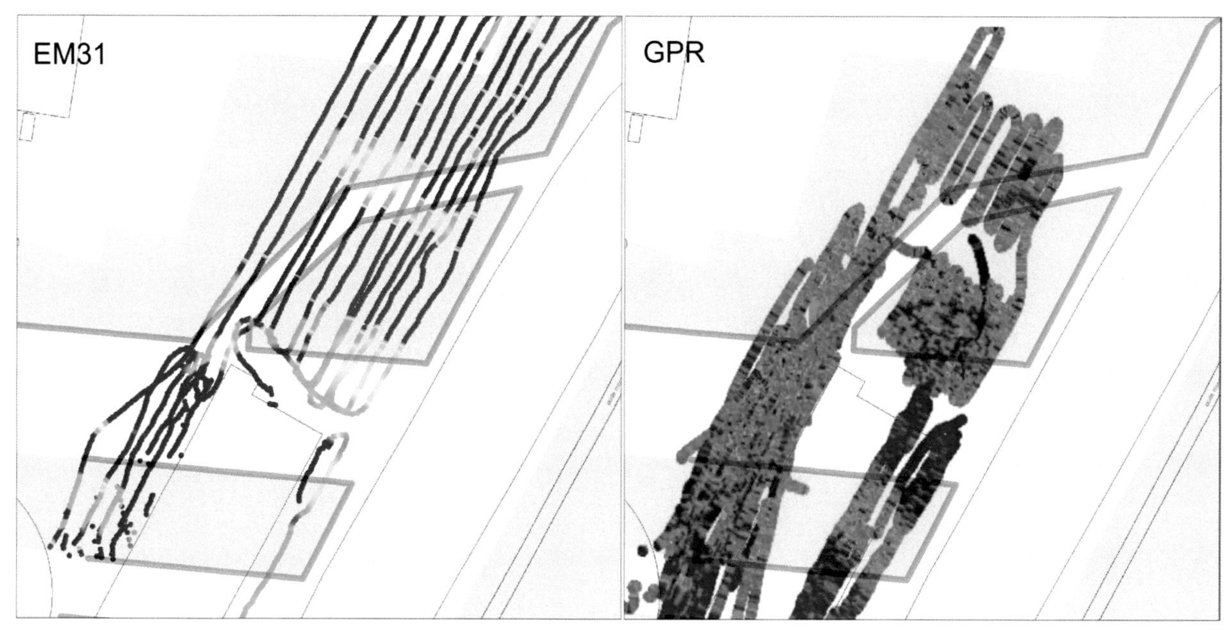

Figure 2: Fortifications Maastricht. EM31 and GPR results. Expected structures are transparent.

inhomogeneous layer will ruin the end result, while a more or less homogeneous layer with not too much metal and contamination is not a major problem.

- Geological and soil transitions can ruin the results but can sometimes be filtered.
- There is always a 50/50 change of success or failure.

In preparing a geophysical prospection in an urban environment, it is important to honestly give an estimation of 50/50 results. It is very difficult to predict the outcome of the surveys as the complexity of the situation cannot be foreseen in advance.

Ultra shallow marine geophysical prospection in the prehistoric site of Lambayanna, Greece

Nikos Papadopoulos[(1)], Julien Beck[(2)], Kleanthis Simyrdanis[(1)], Gianluca Cantoro[(1)], Nasos Argyriou[(1)], Nikos Nikas[(1)], Tuna Kalayci[(1)] and Despoina Koutsoumpa[(3)]

[(1)]Laboratory of Geophysical-Satellite Remote Sensing, IMS-FORTH, Rethymno, Greece; [(2)] Département des sciences de l'Antiquité, University of Geneva, Geneva, Switzerland; [(3)]Ephorate of Underwater Antiquities, Athens, Greece

nikos@ims.forth.gr

Introduction

The Early Bronze Age submerged settlement of Lambayanna (Fig. 1.a) is located in the eastern Peloponnese, about 170km southwest of Athens (Greece). The initial underwater survey during 2015 indicated the expansion of the settlement in a marine area of at least 12,000 square meters with a possible in-land continuation. This mapping revealed scattered walls and stones all over the bay as well as fortification walls in the opposite direction of the coastline in a water depth ranging between 1m to 3m. At least three U-shape large foundation walls (18x10m) aligned along the direction of the fortification wall were verified and more than 6,000 artefacts were recovered from the sea bottom. These preliminary results revealed parallel stone traces in an almost constant distance of 3.5m indicating a potential structured and built urban environment. In order to test and verify the hypothesis of the existence of a potential structured and built environment a comprehensive high resolution geophysical survey involving marine 3D electrical resistivity tomography (ERT), magnetic gradiometry mapping, GNSS bathymetry and terrestrial ground penetrating radar (GPR) was completed (Fig. 1.b, c, d).

Methodology

The bathymetry of the bay was mapped by moving a rover GNSS unit (in constant connection with the base for differential correction) along almost parallel west-east and south-north tracks. For the deeper parts of the bay (>1.6m) a total station and a prism attached to a 3-meter pole was moved along west-east tracks. A total area of about 1.5 hectares was finally mapped with an average spatial resolution of 0.6m. The digital bathymetry model ranges between 0.12 meters above mean sea level (amsl) at the eastern part of the bay and 2.03 meters below mean sea level (bmsl) towards the western part with an average depth of 1.0 meter (Fig. 2.a)

A total marine area of 0.97 hectares was scanned with a multisensor system. Eight sensors were placed every 0.25m on a plastic frame, which was mounted on a custom-made floating apparatus that also carried the GPS and the electronic box during the offshore geomagnetic survey. The design and construction of the apparatus made possible the vertical adjustment of the frame which facilitated the placement of the magnetic sensors in different heights with respect to the sea bottom. The magnetic

Figure 1: a) Location of Lambayanna in Peloponesse, Greece and geophysical coverage with magnetic (b), ERT (c) and GPR (d) techniques.

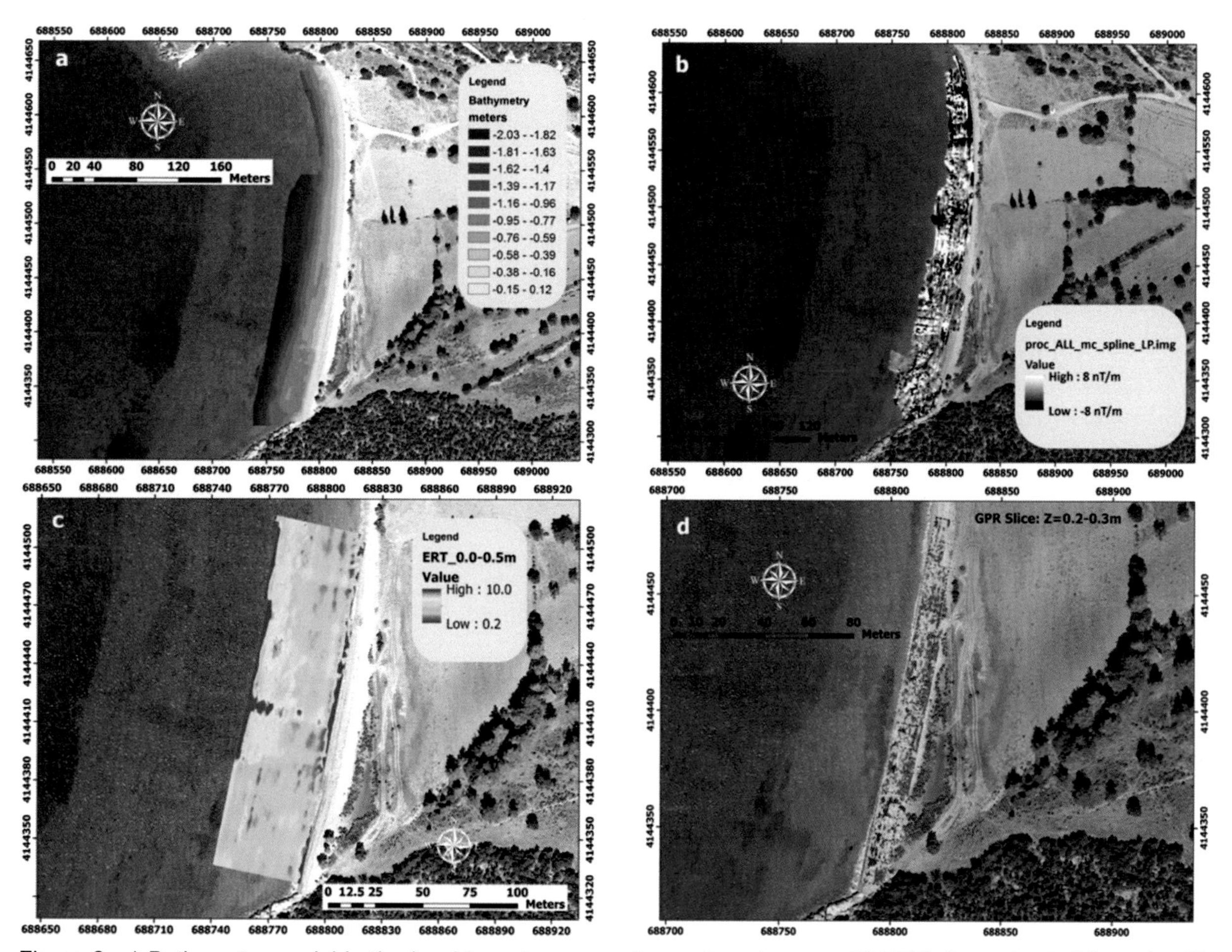

Figure 2: a) Bathymetry model in the bay.b) marine magnetic gradiometry map. C) ERT slice extracted from the 3D inversion model. D) GPR depth slice.

survey was concentrated up to the point where the seawater thickness was below 1.5m. This specific field strategy in combination to the placement of the sensors close to the sea bottom should make feasible the mapping of potential archaeological features in depth of about 1-1.5m below the sea bottom. Specific de-stripping algorithms were used to remove stripes caused by overlapping transects or potential drifting of a specific sensor and low pass filters smoothed out the low frequency anomalies.

ERT survey covered a marine section of 0.86 hectares overlapping more than 75% of the respective magnetic area. The field strategy involved the collection of the tomographic data along 184 parallel 2D lines placed 1m apart, where the basic inter electrode spacing along each traverse was set to 1m. The submerged resistivity mode with a pole-dipole protocol in forward and reverse mode was chosen as the most appropriate to maximize the resolving capabilities of the ERT in identifying cultural material (Simyrdanis et al. 2013). The cumulative length of all the lines was 8.65km, rendering this specific marine archaeologic ERT survey the largest known in Europe up to date. The data were processed with 2D and 3D inversion algorithms (Papadopoulos et al. 2011) incorporating the thickness and the resistivity in the processing procedure in order to constrain the final 2D and 3D resistivity models.

The GPR survey with the 250 MHz antenna was concentrated only at the southern and central part of the coast covering an area of 0.2 hectares with a spatial resolution of 0.5 by 0.025m. The signal of the individual GPR transects was enhanced with standard processing routines (trace reposition, time zero correction, dewow filter background subtraction). Finally the vertical sections were synthesized to compile slices in different depths assuming a velocity for the electromagnetic waves of 0.1m/nsec.

Results

The magnetic gradiometer map showed intense magnetic anomalies related to scattered metal fragments mainly in the southern and northern sections of the bay. At the same time the signal in the central and northern part was influenced by natural processes and the material deposited in the sea by the flow of the seasonal stream. Despite these disturbances in the signal, the magnetic

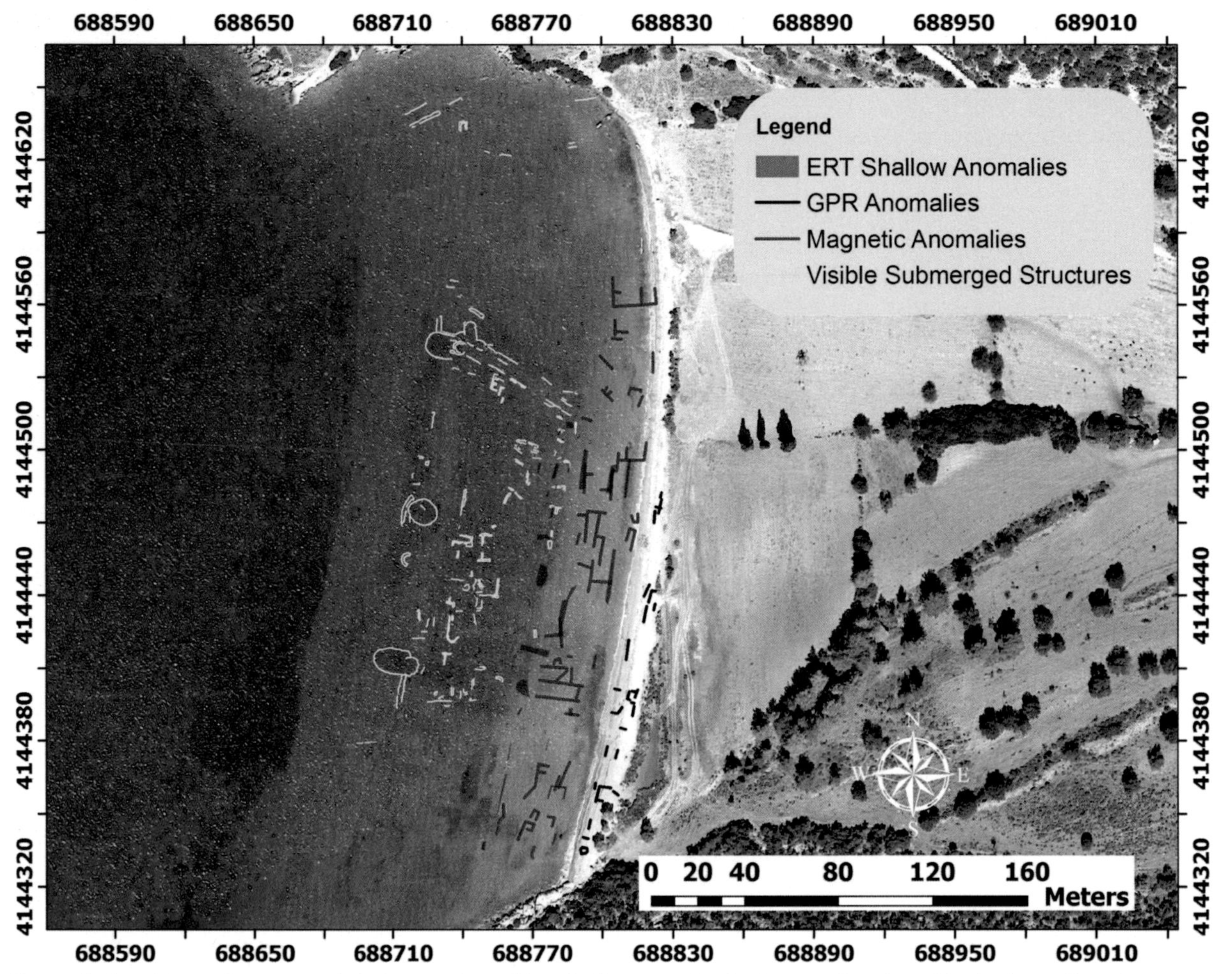

Figure 3: Combined diagrammatic interpretation of the shallow geophysical anomalies (<1m) in the bay of Lambayanna.

map outlined a number of linear, rectilinear and rectangular anomalies having similar orientation that can be attributed to cultural remains buried below the sandy material of the sea bottom (Fig. 2.b).

The combined interpretation of the individual ERT lines resulted in the compilation of a simplified conceptual geological model of the stratigraphy below the sea level, showing that any remains of archaeological importance lie at most 1m below the sea bottom within a layer of fine sediments. The ERT depth slices up to 1m below the sea floor outlined mainly linear segments related to walls or roads as well as compact resistive regions related either to collapsed remains or gathered piles of rocky material (Fig. 2.c).

Despite the saline nature and the increased conductivity of the background matrix on the coast GPR was also quite successful in identifying strong linear reflectors showing potential buried anthropogenic structures, thus starting to complete the image for the landscape on the land (Fig. 2.d).

Conclusions

In general, the marine geophysical campaign of 2016 at Lambayanna provided promising results in the sections that were investigated. Independently to modern or natural interventions to the site, the high resolution geophysical approach was able to indicate a number of candidate targets thus completing to a certain degree the picture for the submerged built environment of the archaeological site. At the same the results of this comprehensive geophysical survey gave strong indications to support the hypothesis of a submerged structured and built urban environment in Lambayanna (Fig. 3).

Bibliography

Papadopoulos, N .G., Tsourlos, P., Papazachos, C., Tsokas, G. N., Sarris, A. and Kim, J-H. (2011) An Algorithm for the Fast 3-D Resistivity Inversion of Surface Electrical Resistivity Data: Application on Imaging Buried Antiquities. *Geophysical Prospection*, **59**: 557-575.

Simyrdanis, K., Papadopoulos, N., Kim, J. H., Tsourlos, P., and Moffat, I. (2015). Archaeological Investigations in the Shallow Seawater Environment with Electrical Resistivity Tomography. *Journal of Near Surface Geophysics, Integrated geophysical Investigations for Archaeology*, **13**: 601 – 611.

Recent trends in shallow marine archaeological prospection in the eastern Mediterranean

Nikos Papadopoulos[(1)], Kleanthis Simyrdanis[(1)] and Gianluca Cantoro[(1)]

[(1)]Laboratory of Geophysical-Satellite Remote Sensing and Archaeoenvironment, IMS-FORTH, Rethymno, Greece

nikos@ims.forth.gr

Introduction

Geoinformation technologies have been extensively used for the non-destructive mapping of onshore buried antiquities, thus contributing in the management and promotion of the cultural heritage (Sarris 2015). However, these imaging technologies have so far made a minimal contribution towards the understanding of the past dynamics in littoral and shallow off-shore environments.

The mapping of ancient shipwrecks or the extraction of valuable information for the stratigraphy of the seafloor have been mainly conducted though acoustic marine geophysical methods (Papatheodorou *et al.* 2014). Coastal areas with water thickness less than 20m have been magnetically mapped with relatively low resolution to identify port installations (Cocchi *et al.* 2012). On the other hand, similar magnetic surveys in shallower coastal sites are, to date, relatively rare (Wunderlich *et al.* 2015). The efficiency of Electrical Resistivity Tomography (ERT) for the reconstruction of submerged cultural material in ultra-shallow marine environment has been verified through numerical simulations (Simyrdanis *et al.* 2015) and successful field applications (Simyrdanis *et al.* 2016).

In more recent years, the important value of photogrammetry in the reconstruction of sea-floors and underwater artefacts (mainly shipwrecks) has been repeatedly proven (Westaway *et al.* 2001). Whilst the number of scientific papers on deep water photogrammetry is consistent, still limited are the applications of photography for shallow underwater photogrammetric reconstructions.

Case Studies

Elounta, Ancient Olous (Crete; Fig. 1) exhibited a continuous habitation spanning from Minoan to Venetian period and was a flourishing harbour town with a sanctuary, a necropolis and its own coinage since its earlier phases. An integrated geoinformatic

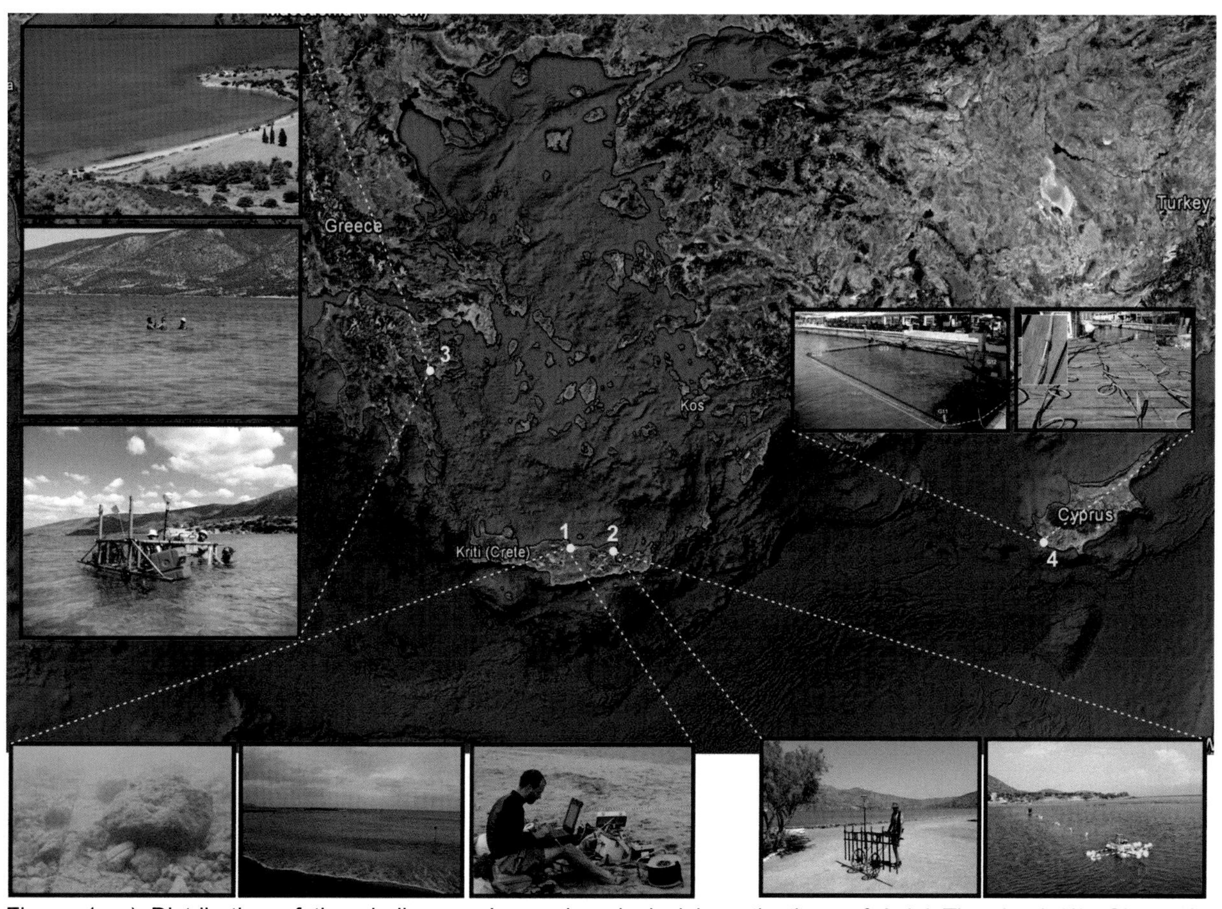

Figure 1: a) Distribution of the shallow marine archaeological investigations of Agioi Theodoroi (1), Olous (2), Lambayanna (3) and Paphos (4) where the integrated geoinformatics were pioneeringly employed.

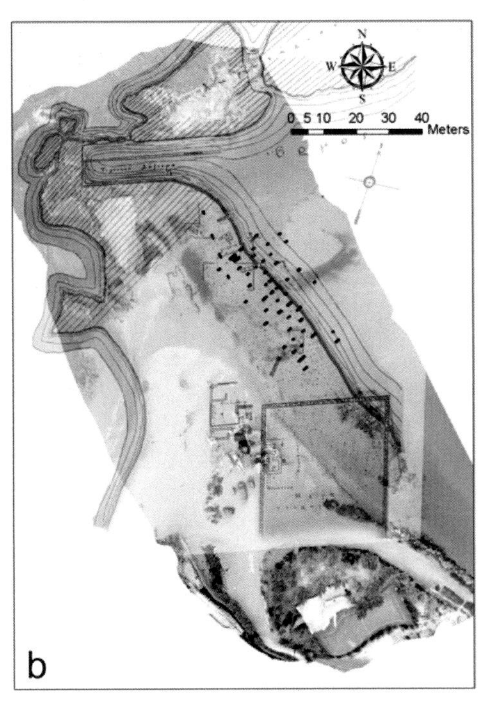

Figure 2: a) Multisensor magnetic gradiometry results from the coastal front of Elounta, ancient Olous. b) Rectification of the old excavation plan with bathymetry (black lines) from Agioi Theodoroi over satellite image of the site along with the interpretation from the ERT survey in the site.

approach was undertaken to map and classify submerged and coastal remains and generate the first comprehensive site plan with the visible and non-visible remains of Olous. Particular landmarks of the site are the large Venetian salt pans that have already been schematically illustrated in early Venetian maps. Later cartographic maps compiled by Spratt (1865) show the spatial distribution of visible architectural relics at that time. New low altitude aerial images taken with a drone were photogrammetrically processed to produce the digital elevation and bathymetry models of Olous. The visible underwater and on-shore archaeological structures (walls, roads, buildings) were mapped with an RTK-GPS system compiling the first archaeological map of the site. High resolution 3D ERT survey was focussed on a marine area of about 0.7 hectares. Vertical sections and horizontal slices extracted from the 3D inversion model outline submerged public buildings with inner compartments, possible walls, and roads within a depth of no more than one meter below the sea bottom. Parts of the coast were also scanned with magnetic gradiometry to map the continuation of submerged relics towards the land (Fig. 2a).

The Minoan coastal site of Agioi Theodoroi (Crete) (Fig. 1) was subject to systematic excavations during

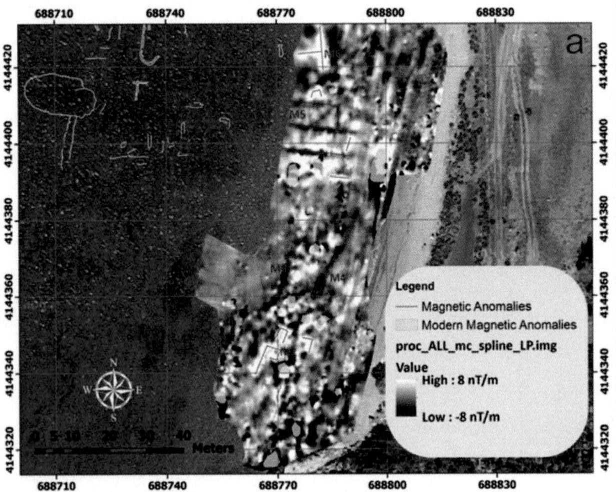

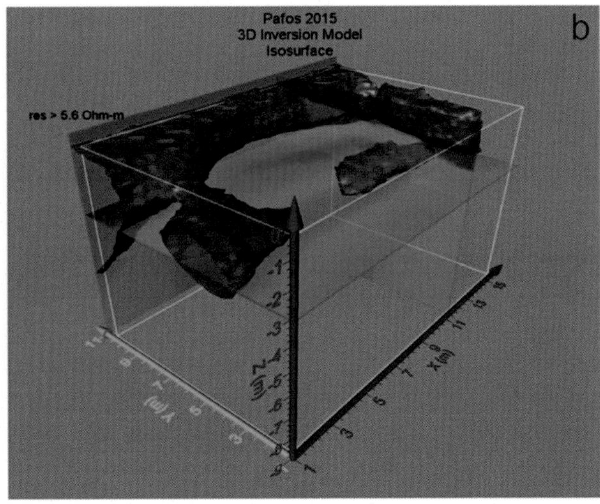

Figure 3: a) Magnetic gradiometry results and diagrammatic interpretation resulting from the survey in Lambayanna. b) 3D isosurface-inversion model representing the submerged walls in the harbour of Pafos.

the early 20th century revealing seaside buildings and wall constructions continuing towards the sea. Recent archaeological surveys included the mapping and photo capturing of the submerged structural relics. A marine ERT survey composed of individual 2D lines was used to map the continuation of the visible on-shore relics below the sea bottom. The rectification of older topographic and excavation plans of the site on satellite imagery completed the picture of the structured environment with both the excavated in-shore relics and those mapped though the geophysical, GPS and underwater survey (Fig. 2b).

The Early Bronze Age coastal settlement of Lambayanna (Peloponnese; Fig. 1) exhibits submerged walls, buildings, and traces of a thicker surrounding wall at a depth of 1-3m. An integrated exploration composed of RTK-GPS and total station, 3D ERT and magnetic gradiometry was used to survey the shallow marine area (<1.5m) covering an area of about 1.4 hectares. The subsequent data analysis revealed a wealth of hidden structures supporting the hypothesis of a submerged built urban environment in the site (Fig. 3a).

A relatively small area ($15x10m^2$) in the harbour of Paphos (Cyprus; Fig. 1) was subject to a very detailed 3D submerged ERT survey composed of orthogonal 2D lines. The 3D resistivity inversion model showed that the archaeological layer was extended up to 2m below the sea bottom with architectural relics (linear features such as walls) with a relatively good preservation level (Fig. 3b).

Conclusions

The possibility of applying well-known methods to un- (or under-) explored archaeological shallow-water contexts was very important from the methodological point of view and for the obtained results. Although the adaptation of commonly used tools to depths below 2 meters presented some challenges and required the creation or customization of equipment, the final results support such efforts and provide useful information for the understanding of complex and pluri-stratified archaeological sites encountered in the Eastern Mediterranean. Ultimately, the results of this work can be regarded as a first step toward the development of an effective interdisciplinary research model that could be applied to similar archaeological surveys in coastal or shallow-water environments.

Bibliography

Cocchi, L., P., Carmisciano, S. C., Caratori Tontini, F., Taramaschi,L. and Cipriani, S. (2012) Marine Archaeogeophysical Prospection of Roman Salapia Settlement (Puglia, Italy): Detecting Ancient Harbour Remains. *Archaeological Prospection* **19** (2): 89–101.

Papatheodorou, G., Geragam M., Christodoulou, D., Iatrou, M., Fakiris, E., Heath, S. and Baika, K. (2014) A Marine Geoarchaeological Survey, Cape Sounion, Greece: Preliminary Results. Mediterranean Archaeology and Archaeometry 14 (1): 357–71.

Sarris, A. (ed.) (2015) *Best Practices of Geoinformatic Technologies for the Mapping of Archaeolandscapes.* Oxford: Archaeopress.

Spratt, T. A. B. (1865) *Travels and Researches in Crete.* London, J. van Voorst.

Simyrdanis, K., Papadopoulos, N. and Cantoro, G. (2016) Shallow off-shore archaeological prospection with 3-D electrical resistivity tomography: The case of Olous (modern Elounda), Greece. *Remote Sensing* **8**: 897

Simyrdanis, K., Papadopoulos, N., Kim, J. H., Tsourlos, P. and Moffat, I. (2015) Archaeological Investigations in the Shallow Seawater Environment with Electrical Resistivity Tomography. *Journal of Near Surface Geophysics, Integrated geophysical Investigations for Archaeology* **13**: 601 – 611.

Westaway, R. M., Lane, S. N. and Hicks, D. M. (2001) Remote Sensing of Clear-Water, Shallow, Gravel-Bed Rivers Using Digital Photogrammetry. *Photogrammetric Engineering and Remote Sensing* **67** (11): 1271-1281.

Wunderlich, T., Wilken, D., Riedel, P.-B., Karle, M. and Messal, S. (2015) Marine magnetic and seismic measurements to find the harbour of the early medieval Slavic emporium Groß Strömkendorf/Reric, Germany .In T. Herbich and I. Zych (Eds.) *Proceedings of the 11th International Conference on Archaeological Prospection, 15.-19.9.2015, Warsaw (2015). Archaeologia Polona* **53**: 543-546.

Dynamic 3D electrical resistivity tomography for shallow off-shore archaeological prospection

Nikos Papadopoulos[(1)] and Kleanthis Simyrdanis[(1)]

[(1)]Laboratory of Geophysical-Satellite Remote Sensing and Archaeoenvironment, IMS-FORTH, Rethymno, Greece

nikos@ims.forth.gr

Introduction

During the last 30 years, electrical resistivity methods have undergone major improvements in terms of specialized instrumentation, modified field survey strategies and automated processing algorithms (Loke *et al.* 2013). Especially in archaeological prospection, 3D electrical resistivity tomography (ERT) has gained increased attention for imaging archaeological structures (Papadopoulos *et al.* 2011). Resistivity surveys have also been successfully employed in off-shore areas covered either with brackish or saline water, posing new challenges in terms of data acquisition (floating or submerged/static or moving modes) and data processing (Loke and Lane 2004). On the contrary, the use of resistivity tomography is rather uncommon in shallow maritime archaeology (Passaro 2010; Simyrdanis *et al.* 2015).

Underwater Survey Modes

A water covered area can be surveyed by placing the electrodes either on the water surface (floating) or underwater attached on the bottom (submerged). Each of these modes can be combined with a fixed cable position (static) or a moving array (dynamic) along predefined survey lines. The final choice of a specific survey mode is strongly related to the thickness of the water layer, the anticipated dimensions and burial depth of the archaeological structures, the bathymetry variation, the desired vertical and horizontal resolution and the time constraints.

Static configurations involve anchoring both ends of a floating or a submerged survey line thus rendering it rather time consuming. On the other hand, dynamic mode (float or submerged) is characterized by increased area coverage since it involves the towing of a cable behind a boat, the integration of the resistivity meter with a GPS and the employment of electrode configurations suitable for multichannel resistivity instrumentations (Fig.1). The floating arrangement avoids underwater obstacles but its resolving capability is degraded by increasing the water layer thickness.

Numerical Simulations

A 3D marine resistivity survey composed of 16 lines oriented along the X-direction and being 1m apart was considered. In every line 21 electrodes were placed with a basic electrode distance of 1m. The subsurface resistivity model was composed of eight layers, each one being 0.5m thick. The background resistivity was 1 Ohm-m (e.g. sandy saline layer) and a resistive (5 Ohm-m) rectangular prism 1m thick defining a submerged archaeological structure that could be placed in various depths (Fig. 2).

The floating survey mode was simulated by assigning a low resistivity value (0.2 Ohm-m) to the superficial layers thus assuming sea water layer with various thickness (Fig. 2.a1). A similar set-up was followed for the simulation of the submerged mode. However, in this case the thickness of the water column was incorporated in the model as the difference between the free water surface (0m) and a constant bathymetry (Fig. 2.b1).

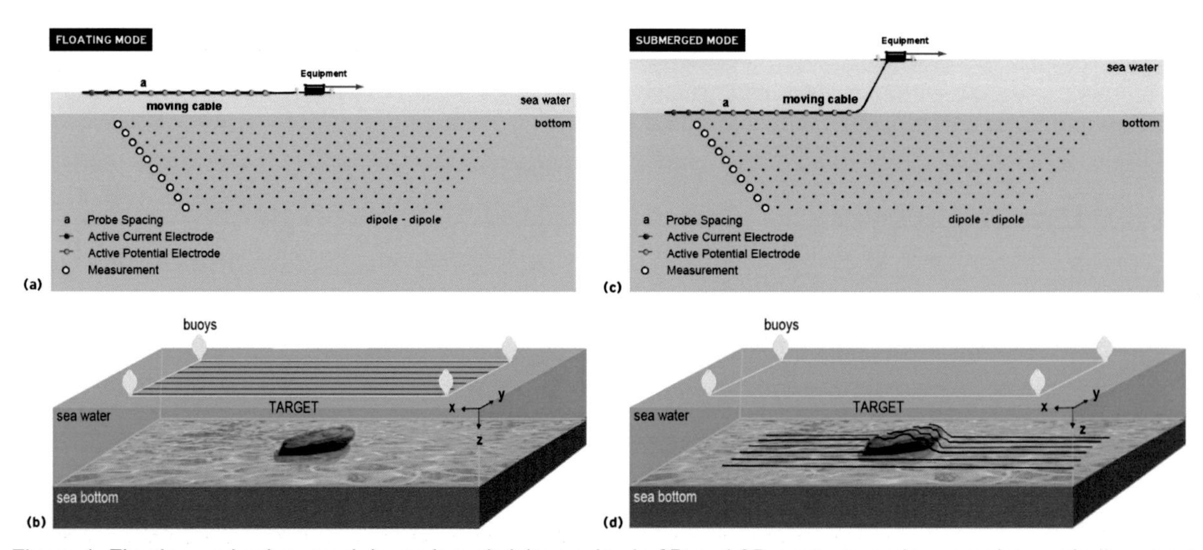

Figure 1: Floating and submerged dynamic resistivity modes in 2D and 3D context used to complete a shallow marine survey with dipole-dipole array.

The forward calculations were performed with RES3DMODx64 that utilizes a finite element solver to retrieve the response of the resistivity model using a dipole-dipole configuration with a maximum distance of 10m (Nsep=1 to 10) between the current and potential dipoles. The synthetic apparent resistivity data were corrupted with 3% Gaussian noise and inverted with RES3DIVNx64 employing a smoothness constrain iterative algorithm. The 3D inversion was also constrained with the a-priori knowledge of the thickness and the resistivity value of the water layer.

Regarding the floating mode, when the water layer is 0.5m thick the final inversion model clearly reconstructs the outline of the structure and its burial depth, but it fails to represent the small internal walls of the building (Fig. 2.a2). More tested models where the water thickness exceeds 1m (i.e. the smallest electrode spacing) showed that the resolving capability of the resistivity image is severely degraded eliminating any possibility of a valid archaeological interpretation. Additional tests with floating simulated surveys showed that the resistivity model is distorted in cases of over- or under estimating the water thickness and resistivity by more than 50%.

Attaching the electrodes on the sea bed increases dramatically the spatial resolution and response of the resistive feature regardless the thickness of the water column (Fig. 2.b2). At the same time the vertical resolving power of submerged arrays is less than the smallest electrode spacing used in the survey. On the other hand, submerged modes are more sensitive to erroneous incorporation of a priori knowledge since over- or under estimation of more than 20% for sea water resistivity and thickness can completely corrupt the inversion model.

Experimental Case Study

An experimental floating dynamic survey was completed in a small part of the submerged archaeological site of Olous (Crete), where the average thickness of the sea water layer was less than 0.3m. The off-shore grid (40m by 20m) was laid out above a known submerged wall in an effort to validate the efficiency of the approach. A custom made plastic boat carrying the resistivity meter, a toughpad and an external battery was pulled along 21 parallel transects 1m apart. The dipole-dipole array with 1m electrode spacing and Nsep=10 was used to collect the tomographic data. The first depth slice (0.00-0.25m) extracted from the 3D

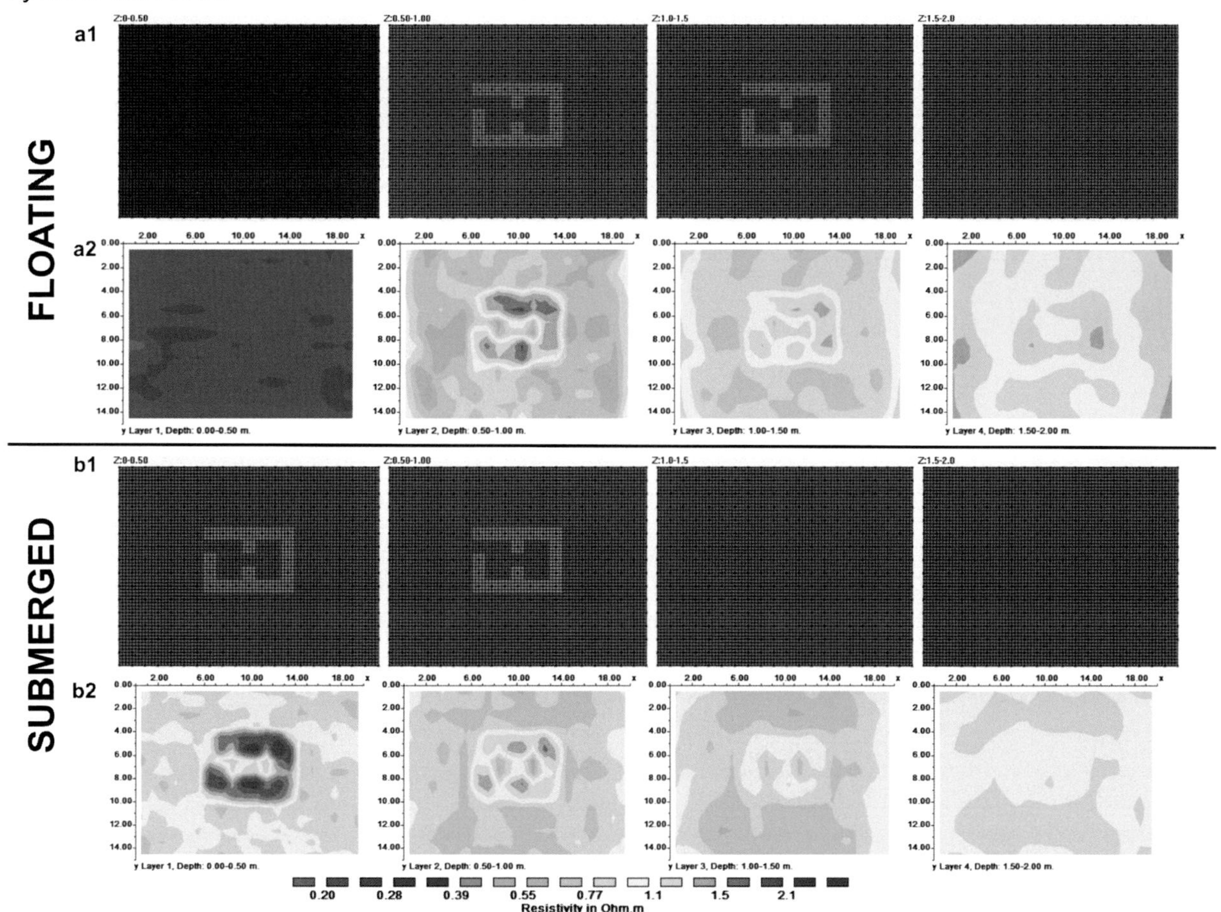

Figure 2: Original resistivity model (a1) and inversion model (a2) assuming a floating survey mode. Original resistivity model (b1) and inversion model (b2) assuming a floating survey mode. In the submerged model the depth slices refer below the sea bottom.

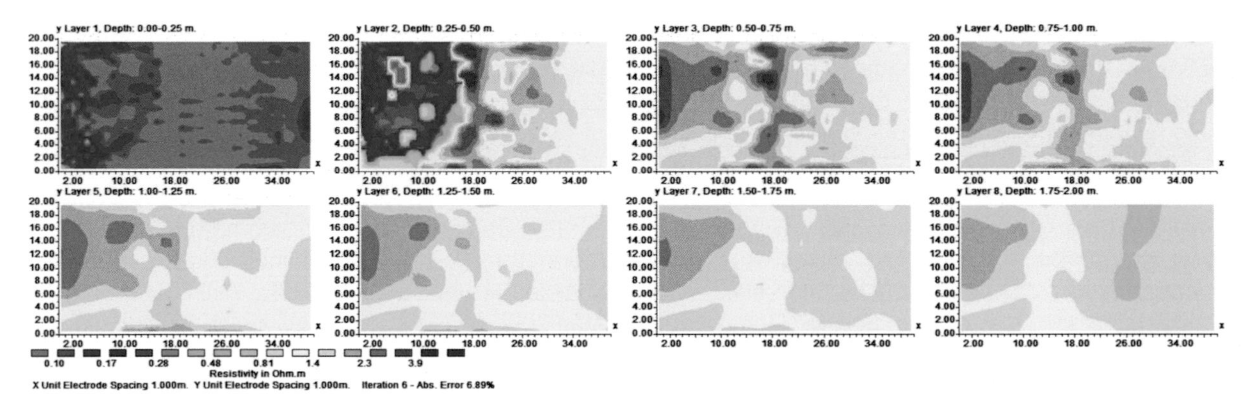

Figure 3: 3D inversion model from the archaeological site of Olous. The depth slices indicate depth below the free water surface.

model corresponds to the water layer which also continues to the left part of the second slice. The layer 0.25-0.50m outlines a long linear resistive feature that corresponds to the visible wall which is crossed with at least three new vertical segments that were not visible with visual inspection. The archaeological structure seems to reach the depth of 1m below the free water surface.

Conclusions

The simulation of floating or submerged 3D ERT surveys through synthetic modelling documented the benefits of such an approach in reconstructing structured cultural material in shallow off-shore environments. In such surveys, it is important to accurately account the effect of the water layer thickness and resistivity value within the inversion procedure. The synthetic modelling results were verified with an actual field case study enhancing the conclusion that 3D marine ERT is capable of providing a robust method to reconstruct submerged archaeological structures related to the ancient built environment.

Bibliography

Loke, M.H., Chambers, J.E., Rucker, D.F. Kuras, O., Wilkinson, P.B. (2013) Recent developments in the direct-current geoelectrical imaging method. *Journal of Applied Geophysics* **95**: 135-156.

Loke M.H., Lane Jr J.W. (2004) Inversion of data from electrical resistivity imaging surveys in water-covered areas. *Exploration Geophysics* **35**: 266–271.

Passaro, S. (2010) Marine electrical resistivity tomography for shipwreck detection in very shallow water: a case study from Agropoli (Salerno, southern Italy). *Journal of Archaeological Science* **37**: 1989–1998.

Papadopoulos, N.G, Tsourlos, P., Papazachos, C., Tsokas, G.N., Sarris, A., and Kim, J-H. (2011). An Algorithm for the Fast 3-D Resistivity Inversion of Surface Electrical Resistivity Data: Application on Imaging Buried Antiquities. *Geophysical Prospection* **59**: 557-575.

Simyrdanis, K., Papadopoulos, N., Kim, J.-H., Tsourlos, P., Moffat, I. (2015) Archaeological investigations in the shallow seawater environment with electrical resistivity tomography. *Near Surface Geophysics* **13**: 601-611.

Designing workflows in the Paphos Agora project: first results of an integrated methodological approach

Ewdoksia Papuci-Władyka[1], Tomasz Kalicki[2], Wojciech Ostrowski[3], Martina Seifert[4], Łukasz Miszk[1], Weronika Winiarska[1], Nikola Babucic[4] and Michael Antonakis[4]

[1]Department of Classical Archaeology, Jagiellonian University Krakow, Poland; [2]Department of Geomorphology, University of Kielce, Poland; [3] Department of Photogrammetry, Warsaw University of Technology, Poland; [4]Department of Culture and Arts, Hamburg University, Germany

martina_seifert@uni-hamburg.de

Abstract

A main issue of modern classical archaeology is how to design an appropriate workflow to investigate ancient city remains. In times of political disturbances, numerous archaeological sites are unavailable or face the risk of destruction. Even in stable countries, full-scale excavations are sometimes difficult to perform. However, the recent rapid development of natural science based methods used for archaeological purposes has significantly increased the capabilities of research.

An Integrated approach has become the most important methodological approach, particularly in the area of a widely understood landscape archaeology using a broad range of available methods for data collection ('total archaeology' cf. Campana *et al.* 2012; Pearce 2011, 86; Donati 2015). The concept of looking at the history of mankind not only from the perspective of historical events, but also by identifying the changes over many generations including also the geographical development, derives from the Annales School from the 1960s (Braudel 1949; Bintliff 1991, 2008, 2011; Pluciennik 2011). The broad research perspective (long, medium and short duration) not only improved the understanding of the processes occurring in the studied communities (Bintliff 2010), but also forces a multidisciplinary research methodology.

The project presented takes place at the Archaeological Park of Nea Paphos in Cyprus, which covers most of the ancient city area (ca. 75ha) and was included in the UNESCO World Heritage List in 2010 (Papuci-Władyka and Dobosz 2016). The paper under discussion focuses on the first results modelling the development of the cityscape of the Hellenistic-Roman settlement by 'total archaeology'. Krakow University's intensive experience at the site for many years at the Paphos Agora Project allows the establishment of an extensive multidisciplinary research on a larger scale (*longue durée*) than in any other urban site in the eastern Mediterranean. Considering political commitment and economic feasibility, it is based on non-invasive methods using the full rank of geophysical prospection, the analysis of the architectural and infrastructural history of the site as well as remote sensing research (satellite, aerial and UAV images) verified by subsequent archaeological research (Kalicki *et al.* 2015; Antonakis *et al.* 2016; Miszk *et al.* 2016).

Magnetic survey will be carried out as a decisive method on the entire park area, with further magnetic susceptibility measurements for specific material allocation. GPR measurements will be accomplished on a larger scale, where the topographical conditions allow, for supplementary three-dimensional reconstruction. Additional resistivity measurements should be performed depending on specific research questions (e.g. detection of the NW-Harbor).

By conducting remote sensing methods, performing classical field survey and evaluating old maps and aerial photographs, an interference surface map as well as high-resolution elevation maps and orthofotos will be realized, which will form the basis for the classification and standardized validation of all geophysical results, ultimately leading to the interpretation and reconstitution of the whole site.

Initial investigations from the years 2015 and 2016 yield a positive outlook and enable structured planning for the coming years (Antonakis *et al.* 2016).

Figure 1: Location of research areas investigated by the Paphos Agora project.

The team of researchers aims to implement a holistic view, simultaneously recognizing the changes occurring in Nea Paphos from three research perspectives – environmental changes affecting the functioning of the town, changes taking place in spatial organization which are the emanation of the social changes, and the individual events, traces of which can be found in the archeological remains. Advantages and problems in this digital field of classical archaeology should be discussed during the conference.

Bibliography

Antonakis, M., Babucic, N. and Seifert, M. (2016) Testing Methods: preliminary results of the geophysical campaign at Paphos 2015, In: E. Papuci-Władyka and A. Dobosz (eds.) *In the Heart of the Ancient City*. Instytut Archeologii Uniwersytetu Jagiellonskiego: Krakow, 60-62.

Bintliff, J. (1991) The contribution of an Annaliste/ structural history approach to archaeology. In: J. Bintliff (ed.) *The Annales School and Archaeology.* New York: New York University Press, 1–34.

Bintliff, J. (2008) History and continental approaches. In: R. A. Bentley, H. D. G. Maschner, and C. Chippindale (eds), *Handbook of Archaeological Theories*. Lanham: AltaMira Press, 147–164.

Bintliff, J. (2010) The Annales, events, and the fate of the cities. In: D. J. Bolender (ed.) *Eventful Archaeologies: New Approaches to Social Transformation in the Archaeological Record.* Albany N.Y: The State University of New York Press, 117–131.

Braudel, F. (1949) *La Mediterranée et le monde mediterranéen à époque de Philippe II*. Librairie Armand Colin: Paris.

Campana, S., Sardini, M., Bianchi, G., Fichera, G.A., and Lai, L. (2012) 3D Recording and total archaeology: from landscapes to historical buildings, *Journal of Heritage in the Digital Era,* **1**(3): 443–460.

Donati, J. C. (2015) Cities and Satellites: Discovering Ancient Urban Landscapes through Remote Sensing Applications. In: A. Sarris (ed.) *Best Practices of Geoinformatic technologies for the mapping of archaeolandscape*, Oxford: Archaeopress, 127– 136.

Kalicki, T., Krupa, J. and Chwałek, S. (2015) Geoarchaeological studies in Paphos, *Studies in Ancient Art and Civilization*, **19**: 233 – 254.

Papuci-Władyka, E. and Dobosz, A. (eds) (2016) *In the heart of the ancient city. Five years of Krakow archaeologists' research at the Paphos Agora on Cyprus (2011-2015), International Symposium and Exhibition of Photographs by Robert Słaboński, 21-22 January 2016, JU Institute of Archaeology,* Kraków: Instytut Archeologii Uniwersytetu Jagiellońskiego 2016.

Pearce, M. (2011) Have rumors of the 'Death of Theory' been exaggerated? In J. Bintliff and M. Pearce (eds) *The death of archaeological theory?* Oxford: Oxbow, 80–89.

Pluciennik, M. (2011) Theory, Fasion, Culture. In J. Bintliff and M. Pearce (eds), *The death of archaeological theory?* Oxford: Oxbow, 31–47.

Miszk, Ł., Ostrowski, W. and Hanus K. (2016) Close range and UAV-based photogrammetry in the Paphos Agora Project (PAP) research and documentation process. In E. Papuci-Władyka and A. Dobosz (eds) *In the Heart of the Ancient City. Instytut Archeologii Uniwersytetu Jagiellonskiego*: Krakow: Instytut Archeologii Uniwersytetu Jagiellońskiego, 22 – 26.

A new semi-automated interpretation of concave and convex features in digital archaeogeophysical datasets

R Pašteka[(1)], S HrončEk[(2)], M Felcan[(3)], P Milo[(4)], D Wilken[(5)] and R Putiška[(1)]

[(1)]Comenius University, Bratislava, Slovakia; [(2)] Proxima R&D, Svätý Jur, Slovakia; [(3)]Slovak Academy of Sciences, Nitra, Slovakia; [(4)]Masaryk University Brno, Czech Republic; [(5)]Christian-Albrechts University, Kiel, Germany

pasteka@fns.uniba.sk

Introduction

During the processing and interpretation of large volumes of geophysical and remote-sensing data in archaeological prospection, there is still a great need to find semi-automated methods for the interpretation of important features, which are useful for the final archaeological evaluation of researched sites. Various methods for separation of anomalies and edge detection (based on processing of geophysical potential fields and geomorphological surfaces) have been published and presented. In this contribution, we would like to present results obtained by means of a new approach (Dirstein *et al.* 2013; Pašteka *et al.* 2015) of geophysical and/ or geomorphological dataset analysis and compare it with other a-priori information and results of standard interpretation methods, which are used in archaeological prospection.

Method

Presented here is a new approach of digital dataset analysis which respects theories of differential geometry, but rather than direct application of differential geometry formulas it uses digital geometry and numerical methods. A key aspect of the method is the calculation of the so called Dupin's indicatrices (a special recognition tool from differential geometry; Dupin 1813) for every discrete point of a digital surface. This tool enables the calculation of all normal curvatures for the whole surface. Principal curvatures and gradient-related curvatures are then used to extract features from the analysed surface. The main detected and extracted features are convex and concave shapes of the analysed dataset along with their morphometric, geometric and statistical properties. Recognition of such features can be very helpful in the subsequent archaeological interpretation. A great advantage of this approach is the fact that these features can also be recognized in cases with strong trends and interference from neighbouring anomalies/ influences in the interpreted fields/surfaces. Thanks to these properties, small-amplitude manifestations of objects can also be recognized in the interpreted data-sets and extracted as isolated objects.

All detected features are stored in a database for further interpretation processes. Multiple geometric and morphometric attributes are calculated for every object, enabling selection of objects based on combination of different criteria or spatial relationships. Detected objects are automatically extracted and analysed on several "levels-of-detail", providing so called "multi-scale" information (Dirstein *et al.* 2013). Such objects are then archived in a form suitable for GIS-platforms (ESRI shape files) and further interpretation is based on selections using the obtained attributes. Such an approach can be a great advantage in the processing of large volume datasets obtained from geophysical and remote-sensing exploration methods in archaeogeophysics.

Results

The presented method can be used on any digital dataset typically used in archaeological prospection (usually saved in a regular grid). Here we present results from the analysis of a LiDAR dataset acquired across the area of the early Iron Age hill-fort Smolenice-Molpír in SW Slovakia (Stegmann-Rajtár, 2005) (Fig. 1). Proposed convex/concave shape analysis of this dataset (Fig. 1.a, with a selected part in 1b) revealed a set of detected features (selected convex shapes in Fig.1.c). Here a strong detection ability of the presented method can be seen – detected objects I, II, III are hardly visible in the shaded relief of the LiDAR dataset (Fig. 1.b). These objects were then checked by the review of historical photos (Fig 1.d) and verified by in-situ site inspection (Figure 1.e) – these could be interpreted as deformed barrows from the Iron Age settlement (this part of the researched hill-fort is positioned on a relatively steep slope, so hill-side deformation processes are very intensive).

For the verification of one of these detected anomalies, a hand-held high-definition magnetometer survey (Cs-vapour sensor) has been applied (in the frame of the blue rectangle in Figure 1.b,c).

Results show that the major object I does not show a real manifestation of some magnetometric structure (it maybe a naturally formed small elevation), but in the case of the adjacent object Ib, a part of a ring-shaped anomaly was detected (Figure 2, left hand part). This acquired anomaly has an identical position as the object Ib from this new presented morphometrical approach and it can be linked to the internal structure of a barrow (probably partly destroyed on its left side).

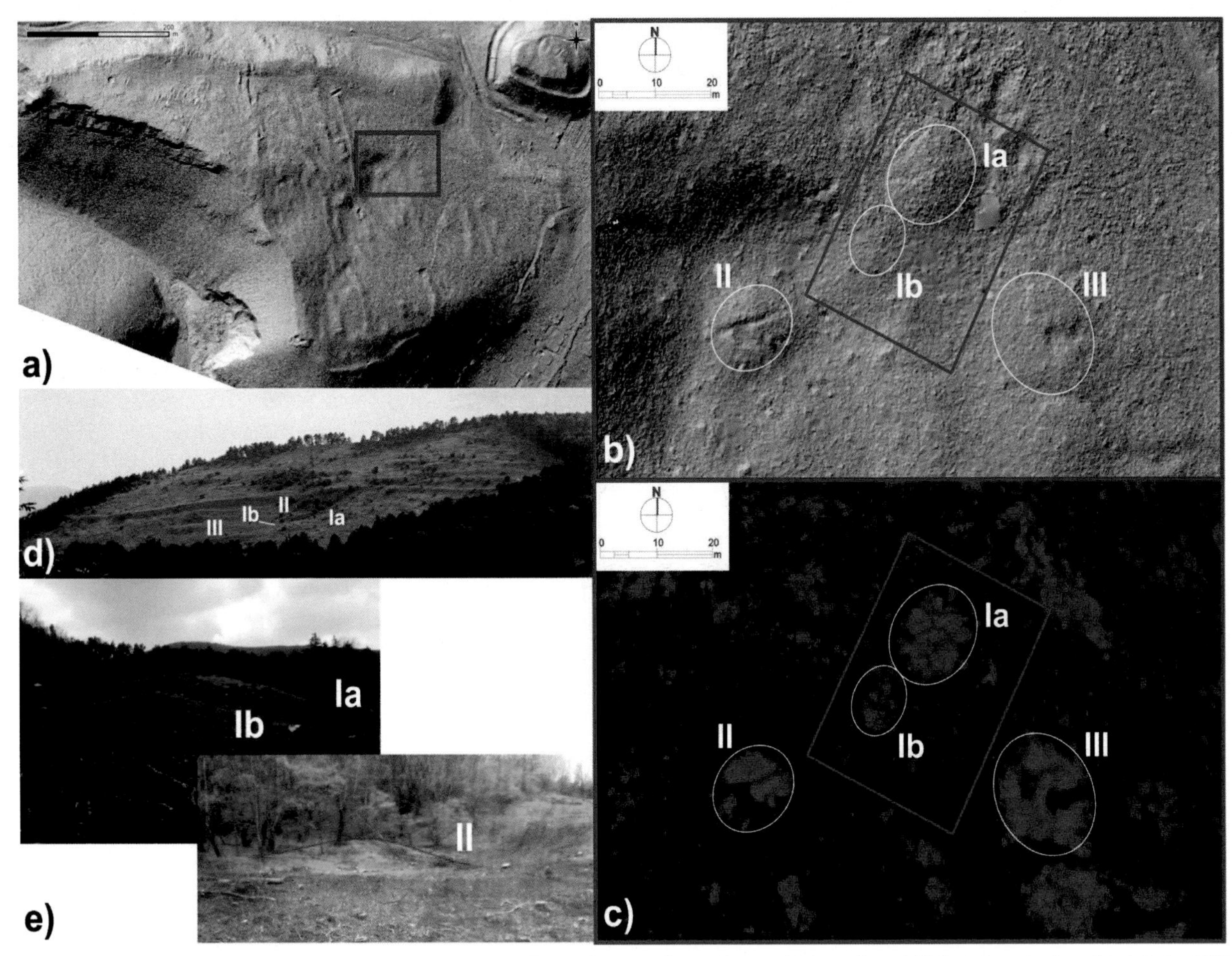

Figure 1: Results from the presented method; a) LiDAR dataset from the early Iron Age hill-fort Smolenice-Molpír in SW Slovakia; b) selected detail from the LiDAR dataset (shaded relief with lighting from SW); c) results from the presented morphometric analysis – automatically selected local convex features; d) historical photo of the hill-fort; e) photos from in-situ site inspection.

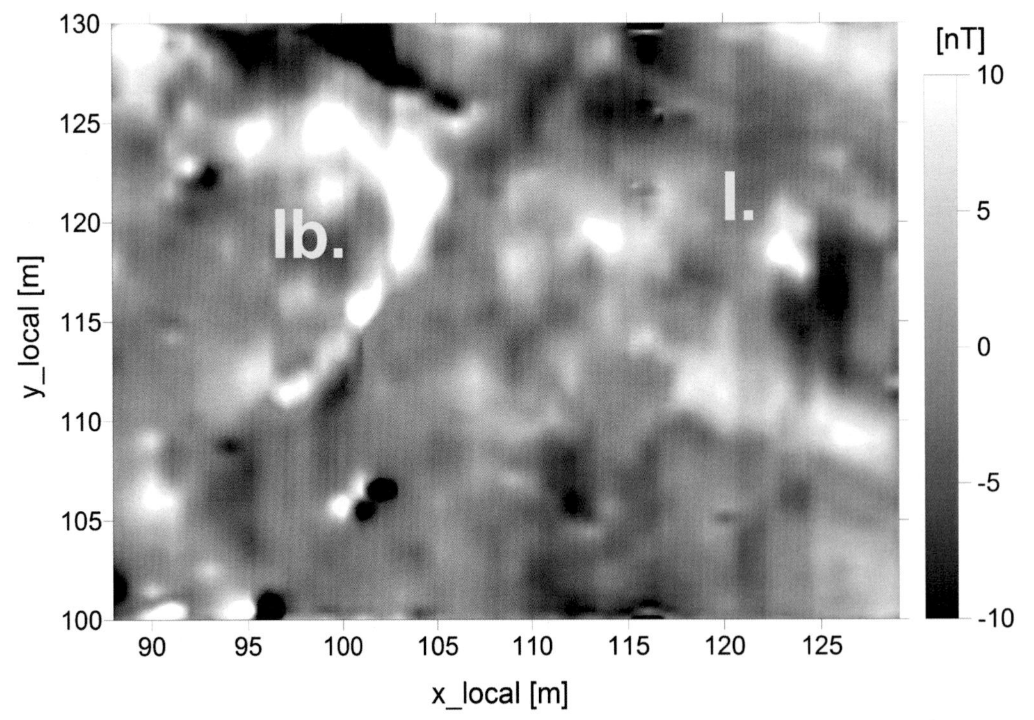

Figure 2: Map of anomalous magnetic field (result from a high-definition magnetometer Cs-vapour sensor) – area of the blue colour rectangle in Fig.1.

Conclusions

These results show the great potential of a new concave/convex feature analysis method, based on the evaluation of Dupin's indicatrices. Comparison with the standard methods of LiDAR data analysis and edge detectors in potential fields in archaeogeophysical interpretation shows it has greater sensitivity to the recognition of small amplitude parts of the interpreted anomalous fields/surfaces. Our presented example from the site of Molpír in SW Slovakia is in our opinion a good demonstration of extracting anomalous objects in LiDAR datasets.

A very important fact is that the presented algorithm is fully automated and objective and outputs are vector objects (not raster format, like in standard edge detectors). This is a great advantage in the processing of large datasets of geophysical and remote-sensing exploration methods in archaeogeophysics.

Bibliography

Dirstein, J., Ihring, P. and Hroncek, S. (2013) GeoProxima – a new approach for digital data analysis. *ASEG conference extended abstracts*, 5.

Dupin, Ch. (1813) *Déveleppements de géometrie.* MmeVeCourcier, Imprimeur-Librairepour les Mathématiques, quai des Augustins, Paris, No 57. 373.

Pašteka, R., Hornček, S., Ihring, P., Bošanský, M., Putiška, R. (2015) Concave and Convex Features Analysis of Bouguer Gravity Field with Following Qualitative Interpretation. *Extended abstracts, 77th EAGE Conference & Exhibition 2015*, EAGE, DB Houten, P2-07, 5.

Stegmann-Rajtár, S. (2005) Smolenice-Molpír. In: *Reallexikon der Germanischen Altertumskunde.* Band 29. Berlin: Walter de Gruyter, 146-156.

Integrated geophysical, archaeological and geological surveys for the characterization of Tusculum archaeological site (Rome, Italy)

Salvatore Piro[(1)], Elisa Iacobelli[(2)], Enrico Papale[(1)] and Valeria Beolchini[(3)]

[(1)]Istituto per le tecnologie Applicate ai Beni Culturali, CNR, Monterotondo Sc, Rome, Italy; [(2)]Università degli studi di Roma "La Sapienza", Rome, Italy; [(3)] Escuela Española de Historia y Arqueología en Roma, Rome, Italy

salvatore.piro@itabc.cnr.it

Introduction

Tusculum is one of the largest Roman cities in the Alban Hills, located in the Latium region of Italy, about 25km south-east of Rome. This settlement was inhabited from pre-Roman times until the middle ages.

The aim of the project is to characterise Tusculum site from archaeological and geoarchaeological points of view. With these aims, multimethodological geophysical surveys were carried out during the summer and winter of 2016 to locate the archaeological remains in a portion of the studied area and to reconstruct the orientation of an underground aqueduct system.

GPR and gradiometric surveys were carried out to investigate the archaeological remains, while ERT surveys were employed to find evidence of undiscovered parts of an underground aqueduct system. These results, together with geological studies and historic evidences, were used to understand how water was supplied to such a large, populated settlement.

Tusculum Archaeological Site

Tusculum is one of the largest Roman cities in the Alban Hills, located in the Latium region of Italy, about 25 km south-east of Rome. Its ruins are located on Tuscolo hill, whose summit is 670 metres above sea level, more specifically on the northern edge of the outer crater ring of the Alban volcano. This settlement was inhabited since pre-Roman times until the middle ages. During Roman times, it was important from strategic and military points of view, but with the consolidation and extension of Rome's domination over Europe and the entire Mediterranean basin, it became, above all, the setting for patrician villas. In the archaeological and cultural park, there are nowadays the remains of a forum, a theatre and Tiberius' Villa, as well as stretches of Roman roads, remains of other patrician villas, a necropolis, numerous funerary monuments and a medieval tower that once belonged to the Counts of Tusculum, who were the lords of this area in the Middle Ages.

An important consideration for this study is also the presence of several cisterns for water storage together with a complex system of drainage tunnels (cuniculi) which were built to supply water to Tusculum (Capulli 2008). Some of these tunnels have already been mapped, while others remain unexplored due to collapses.

Geologically the area is part of the Tuscolano – Artemisio composite lithosome which includes the Tuscolo succession which is made by scoria, welded scoria, clastogenic lavas and lava units (Giordano *et al.* 2006).

Geophysical Surveys

The geophysical surveys were carried out in two stages. In the first stage surveys were performed to detect buried structures in the lower part of the archaeological site. Surveys were carried out using the Ground Penetrating Radar and gradiometric methods. The selected area was divided in three sectors: Area 1 with dimension 40x100 m; Area 2 with dimension 31x50 m; Area 3 with dimension 15x19 m. For the measurements, a SIR3000 GPR system (GSSI) equipped with a 400 MHz (GSSI) bistatic antenna with constant offset was employed. The horizontal spacing between parallel profiles at the site was 0.50 m. In the investigated areas, a total of 175 adjacent profiles across the site were collected alternatively in forward and reverse directions, employing the GSSI cart system equipped with odometer. All radar reflections within the 80 ns (twt) time window were recorded in the field as 16-bit data and 512 samples per radar scan.

Part of the area overlapping the surveyed GPR area has been investigated employing the gradiometric method. This was divided in 7 squares of 20x20 m and 5 squares of 10x10 m where the vertical gradient of the magnetic field was measured using a fluxgate gradiometer FM256 (GEOSCAN Research, UK)

Figure 1: GPR time slices at the estimated depth of 0.70m.

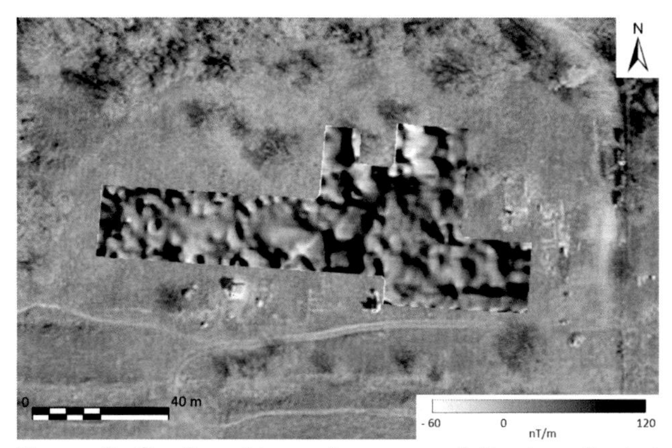

Figure 2: Gradiometric contour map of the investigated area. Range -60, +120 nT/m.

along parallel profiles with a horizontal spacing of 1 m and with a sampling interval of 0.5m.

The second stage, carried out during the winter time, involved the use of the electrical resistivity tomography (ERT) method to detect deep unexplored drainage tunnels in the upper part of the archaeological site. The measurements have been acquired in two areas where the presence of these drainage tunnels has been hypothesised. A total of 5 ERT profiles have been acquired, using an IRIS SYSCAL Junior Switch-72; 48 electrodes with a spacing of 1 m for all the profiles was used, employing the Wenner - Schlumberger and Dipole-Dipole array configurations.

Data Processing and Results

All the GPR data were processed with GPR-SLICE v7.0 Ground Penetrating Radar Imaging Software (Goodman 2016). The basic radargram signal processing steps included: (i) post processing pulse regaining; (ii) DC drift removal; (iii) data resampling; (iv) band pass filtering; (v) background filter and (vi) migration. With the aim of obtaining a planimetric vision of all possible anomalous bodies, the time-slice representation technique was applied using all processed profiles (Goodman and Piro 2013). Time slices for a depth of about 0.70 m are represented in Fig. 1. Some of the anomalies detected by the GPR method matched the results of the archaeological excavations that took place in the central part of the survey area. Following the excavations, some buildings came to light. The walls of these buildings can be correlated to the anomalies detected in a sector of the GPR time slices.

The magnetic data was processed with GEOPLOT 3.0 software (GEOSCAN Research). After de-spiking, filtering and rearranging processes, the data were assembled in a contour of the vertical gradient of the total magnetic field (Fig. 2). This contour map is characterized by the presence of surface noise which does not allow a proper interpretation of the anomalies.

The magnetic data were processed using the 2D cross-correlation technique to enhance the S/N ratio and to better define the spatial location and orientation of the possible targets (Piro *et al.* 2009). This method is a measure of the similarity between the raw data and calculated synthetic anomalies.

The processing for the ERT profiles are in progress to match the location of anomalies in the ground with the representation of the explored tunnels inside the ground. Figure 3 shows an example of the obtained tomography after the inversion. This information, regarding the presence of the drainage tunnels in the investigated area can enable us to determine the extension, depth, thickness and electrical characteristics of subsurface features including geological units.

Bibliography

Capulli, R. (2008). I cuniculi del Monte Tuscolo (Roma-Lazio). *Atti VI convegno nazionale di speleologia in cavità artificiali; Opera Ipogea,* **1/2**.

Giordano, G., De Benedetti, A., Diana, A., Diano, G., Gaudioso, F., Marasco, F., Miceli, M., Mollo, S. and Funiciello, R. (2006) The Colli Albani mafic caldera (Roma, Italy): Stratigraphy, structure and petrology. *Journal of Volcanology and Geothermal Research,* **155:** 49–80.

Goodman, D. and Piro, S. (2013) *GPR Remote sensing in Archaeology, Springer: Berlin.*

Piro, S. and Gabrielli, R. (2009) Multimethodological Approach to investigate Chamber Tombs in the Sabine Necropolis at Colle del Forno (CNR, Rome, Italy). Archaeological Prospection, **16: 111-124**.

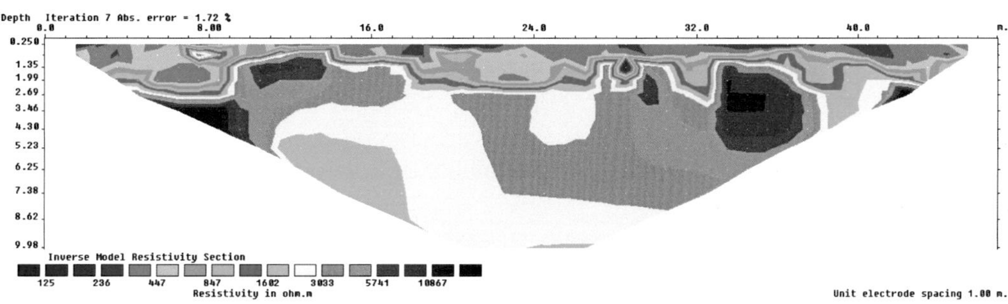

Figure 3: Electric resistivity tomography obtained after the inversion showing an anomaly (on the right) produced by one of the drainage tunnels.

Integrated geophysical and archaeological surveys to study the archaeological site of Cerveteri (Rome, Italy)

Salvatore Piro[(1)], Enrico Papale[(1)], Daniela Zamuner[(1)] and Vincenzo Bellelli[(2)]

[(1)]Istituto per le tecnologie Applicate ai Beni Culturali, CNR, Monterotondo Sc, Rome, Italy; [(2)]Istituto di Studi sul Mediterraneo Antico, CNR, Monterotondo Sc., Rome, Italy

salvatore.piro@itabc.cnr.it

Introduction

The archaeological research in the southern Etruria territory (between Tuscany and Latium regions, Italy) has made a lot of progress in the last ten years. Cerveteri is one of the main important town of this territory, located between Veio in the south, and Tarquinia in the north. The amount of knowledge related to the ancient town and its environment has become remarkable and complex, and one of the objectives of the present project is the reconstruction of the urban plan of the Cerveteri etruscan town in relation to its territory. The National Research Council (CNR), through two of its institutes (ISMA and ITABC), in collaboration with the Soprintendenza Archeologica planned an urban archaeology research project on the based on the direct archaeological explorations and landscape geophysical surveys.

Cerveteri Archaeological Site

The selected area, part of the present project, is located on the north-west urban plateau, outside the actual modern town of Cerveteri, between the Manganello creek and Vignali plateau, where archaeological evidence for occupation is known from the Iron Age. This area is characterised by the presence of a sanctuary (Etruscan sanctuary of Manganello) which represents the main development hub and urban organization hub of the eastern part of the urban plateau, very close to the main part of the ancient acropolis of Cerveteri. The sacred area is delimited by "*temenos*" structures, which separate the sacred building from the service buildings. Between this sacred area and the eastern part of the ancient town there is a portion of the plateau which can be considered as a hinge between them. The surface archaeological surveys have indicated that this area has been used from the Iron Age until the Roman Imperial era, and is characterised by the presence of external fortified walls and hydraulic structures. To enhance the knowledge of this area a first stage of integrated geophysical surveys was made during 2014, employing ground penetrating radar (GPR), gradiometric and electrical resistivity tomographies (ERT).

Geophysical Surveys

The geophysical surveys were conducted to detect buried structures as remains of buildings and portion of drainage tunnels. Surveys were carried out using the ground penetrating radar, the differential magnetometric method and electrical resistivity tomographies ERT. For GPR surveys the area was divided in two sectors: Area 1 of 34x17 m and Area 2 of 95x54 m. Measurements were collected, along parallel profiles, employing the SIR3000 (GSSI) system, equipped with a 400 MHz (GSSI) bistatic antenna with constant offset. The horizontal spacing between parallel profiles at the site was 0.50 m. In the investigated areas, a total of 260 adjacent profiles across the site were collected in alternating forward and reverse directions, employing the GSSI cart system equipped with odometer. All radar reflections within the 75 ns (twt) time window were recorded in the field as 16-bit data and 512 samples per radar scan.

Part of the area, overlapping those surveyed with GPR, has been investigated employing the differential magnetic method. In particular, the area was divided in three 30x30 m squares and three 10x10 squares where the vertical gradient of the magnetic field was measured using a fluxgate gradiometer FM256 (GEOSCAN Research, UK) along parallel profiles with a horizontal spacing of 1 m and with a sampling interval of 0.5.

The second stage, carried out at a later date, involved the use of the electrical resistivity tomography (ERT) method to detect deep unexplored drainage tunnels in the upper part of the archaeological site. The measurements have been acquired in an area where the presence of these drainage tunnels has been hypothesised. A total of 21 parallel profiles, with a length of 48 m, have been acquired, using an IRIS SYSCAL Junior Switch-Pro72; 48 electrodes with a spacing of 1 m for all the profiles was used, employing the Wenner - Schlumberger and Dipole-Dipole array configurations.

Data Processing and Results

All the GPR profiles were processed with GPR-SLICE v7.0 Ground Penetrating Radar Imaging Software (Goodman 2013). The basic radargram signal processing steps included: (i) post processing pulse regaining; (ii) DC drift removal; (iii) data resampling; (iv) band pass filtering; (v) migration and (vi) background filter. With the aim of obtaining a planimetric vision of all possible anomalous bodies, the time-slice representation technique was applied using all processed profiles (Goodman and Piro 2013). Time slices for a depth of about 0.80 m are represented in Figure 1.

Figure 1: Cerveteri – Manganello. GPR time slices for a depth of about 0.80m.

The magnetic data was processed with GEOPLOT 3.0 software (GEOSCAN Research). After de-spiking, filtering and rearranging, the data were assembled in a contour of the vertical gradient of the total magnetic field Fig. 2. The contour map representing the results of the magnetic method is characterized by the presence of many dipole anomalies that can be related to the presence of a portion of walls.

The magnetic data is currently being processed using the bi-dimensional cross-correlation technique in order to enhance the S/N ratio and to better define the spatial location and orientation of the possible targets (Piro *et al.* 1998). This method is a measure of the similarity between the raw data and calculated synthetic anomalies.

Figure 2: Cerveteri - Manganello Magnetic map of the investigated area. Range -40, +40 nT/m.

The processing for the ERT profiles are in progress to match the location of anomalies in the ground with the representation of the explored tunnels inside the ground. Figure 3 shows an example of the obtained tomography after the inversion. This information, regarding the presence of the drainage tunnels in the investigated area can enable us to determine the extension, depth, thickness and electrical characteristics of subsurface features including geological units.

To enhance the understanding of the obtained anomalies, the data sets have been processed with different integrated approaches which consists in combining and fusing the results from all the employed methods. For the processing, we have used the following techniques: graphical integration (overlay and RGB colour composite), discrete data analysis (binary data analysis and cluster analysis) and continuous data analysis (data sum, product, and PCA). The first type of integration yields only new images and not new data that can be further analysed; the discrete solutions give new datasets which emphasize strong anomalies but tend to ignore more subtle ones. Continuous integration allows both subtle and robust anomalies to be expressed simultaneously producing a high information content especially in the case of the sum and the principal component analysis.

This integrated processing and results, will be presented and discussed and it allowed us to produce an interpretation of the investigated area.

Bibliography

Goodman, D. and Piro, S. (2013) *GPR Remote sensing in Archaeology*, Springer: Berlin.

Loke, M. H. and Barker, R. D. (1996) Rapid least-squares inversion of apparent resistivity pseudosections using a quasi-Newton method. *Geophysical Prospection* **44**: 131–152.

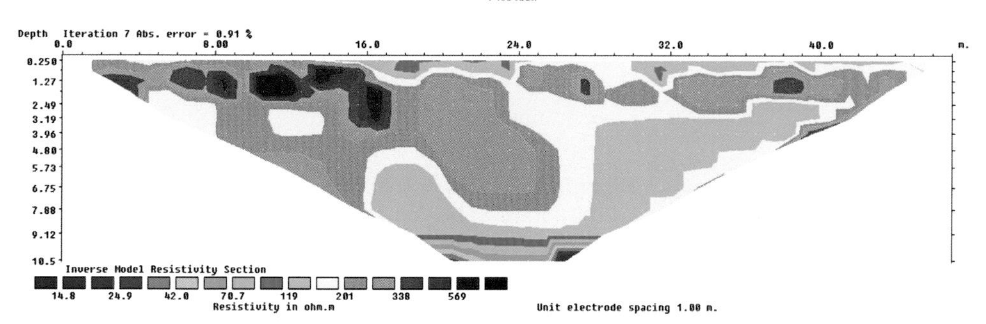

Figure 3: Cerveteri – Manganello. Electric resistivity tomography obtained after the inversion showing an anomaly (on the left) produced by one of the drainage tunnels.

The challenges of reconstructing the archaeological landscape around the castle in Gołuchów, Poland

Michał Pisz[1] and Inga Głuszek[2]

[1]University of Warsaw, Warsaw, Poland; [2]Nicolaus Copernicus University, Toruń, Poland; National Museum in Poznań, Poznań, Poland

michal.pisz@uw.edu.pl; ingag@umk.pl

Introduction

In 2016 the Polish Ministry of Culture and National Heritage decided to co-fund a non-destructive archaeological survey project in Gołuchów, led by the National Museum in Poznań. Gołuchów, also called the Pearl of the Calisian Land, is famous for its Renaissance castle located in a picturesque park on the bank of Trzemna river (Fig. 1). The castle itself has been rebuilt and expanded a few times in the past and has been an object of architectural studies. However, not much was known about its historical surrounding, except several objects which have been preserved until this day.

The Non-destructive Survey in a Park

A parkland might be considered a very convenient setting for a non-destructive survey as it does not impede the measurements on the well-maintained surface where they are taken. While there is no need to deal with overgrown vegetation, uneven ground or unexpected obstacles, a number of other challenging features occur. For example, many areas may be inaccessible due to the presence of flowerbeds, shrubberies or copses (Parkyn 2009, 230-231). The edges of the latter cause interference to the GPS operation, thus it is sometimes necessary to switch between gridded and RTK-positioned survey in some parts.

Another big challenge arises during the data interpretation stage. The old parks have been rebuilt many times in the past, therefore the earthworks could have left some marks that can be confusing when interpreting the data. Furthermore, a lot of modern infrastructure may distort the measurements, especially those performed with the magnetic method (Parkyn 2009, 231).

Figure 1: The castle in Gołuchów – left: present day drone picture, right: drawing from 1838 with the depiction of a watermill on the side of the river.

Notwithstanding all of the challenges mentioned above, a successful non-destructive geophysical survey could result in interesting and meaningful findings from the data analysis.

Reconstructing the Non-existent Objects and Landscape

The main object of interest in Gołuchów is the castle's surroundings. The castle itself is considered to have been built at the end of 16th century. Its present form is a compilation of three construction phases (Kąsinowska 2011). Presumably, since the castle's erection, a surrounding settlement was located nearby. Historical sources indicate that in the later period, between the 18th and 19th century, there was a settlement located in the neighbourhood. However, with the exception a few buildings that survived to the present day, there are no traces of either the nearby Dybul settlement, or any buildings – including a watermill – located right next to the castle. The river Trzemna itself was also regulated in the past. It probably first occured when countess Izabella Działyńska, who purchased the residence, decided to re-design the castle and landscape the park. The geophysical survey was supposed to be the first attempt to shed some new light on the surrounding of the Gołuchów castle.

Multi-method Survey and the GIS Database

The first of the field tasks was to create the orthophotomap of the whole park, with the focus on all the accessible areas. The orthophotomap was made with DJI Phantom 4 UAV and processed with AgiSoft Professional software, producing an output image with the resolution greater than 5 cm per pixel.

The survey was carried out using two geophysical methods: magnetic and earth resistance. The survey in Gołuchów could be considered a Level II survey (according to Gaffney and Gater 2003, 88-90). However, since it was the very first survey of this area, the field researchers decided to focus, to the extent possible, on the full scope of the area rather than strive for obtaining high resolution imagery.

Magnetic measurements have been performed, with the use of Sensys MXPDA 5-channel fluxgate gradiometer with the Satlab SL600 RTK GPS. In some cases, while the problems with the fixed GPS signal occurred, some fields were measured within the grids. The total area of the magnetic survey, divided into 11 different fields, exceeded 6,5 ha with 0,5 m spacing between profiles and 5 cm sampling rate (Fig. 2).

The last field task was the complementary earth resistance survey. Seven fields were chosen to be surveyed with this method, covering in total over 3,5 ha, with a 1 x 1 m resolution. The Wenner (a=0.5m) array was solely used for the majority of the surveyed areas with a single exception where, besides the Wenner, the area had been measured also with the twin-probe array with a 1m separation distance between the mobile electrodes.

The results of the geophysical measurements have been processed and visualised in accordance with the EAC guidelines (Schmidt *et al.* 2015: 100-116). Magnetometry data was treated with MAGNETO Arch, a software dedicated for Sensys magnetometer data treatment and visualised in Surfer in various dynamic range of anomalies (from +/-2 to +/-20 nT) and different gray and colour scales. The earth resistance data was processed and visualised with Geoplot 4 software. For each surveyed region, many variants of the visualisation have been exported to GIS (including raw, despiked, high pass filtered data in various colour scales and ranges).

The final results of the prospection have proven that combining at least two complementary methods in this kind of research is essential. On one hand, the magnetic survey has revealed the most interesting zones of possible historical human activity. On the other hand, it has also shown where landscape transformations may have occurred – the results of

Figure 2: The results of the magnetic survey on the satellite image of the park.

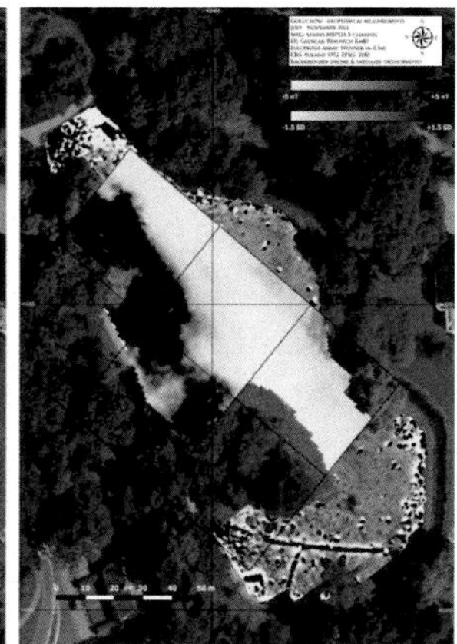

Figure 3: One of the survey areas – comparison of magnetic and earth resistance survey.

the magnetic measurements seem to indicate the old extents of the Trzemna river (Fig. 3).

In some areas, the zones of intensified small dipolar and positive anomalies occurrence turned out to be regular resistance anomalies (Fig. 2). High resistant, regularly-shaped anomalies could be interpreted as the presumable relics of the architecture (including the watermill mentioned above), while the low resistance anomalies, coinciding with the zones of small dipoles concentration, could be interpreted as the possible garbage-pits.

A Tool for Heritage Management

The GIS database has been created in the open-source QGIS system. It contains the results of the non-destructive survey (various maps); shapefile layers with the analysis and interpretations of the geophysical data; georeferenced archival maps; and raster layers, such as satellite imagery, DEM, DTM or archival aerial pictures. The database was the integral part of the report from the survey, presented to the officers of the local Heritage Management service. It is hoped that the database could help in the effective management of archaeological resources in the further research in Gołuchów.

Bibliography

Gaffney, C. and Gater, J. (2003) *Revealing the buried past. Geophysics for Archeologists.* Stroud: Tempus publishing. (reprint: 2010)

Kąsinowska, R. (2011) *Gołuchów. Rezydencja magnacka w świetle źródeł.* Gołuchów: Ośrodek Kultury Leśnej.

Parkyn, A. (2009) A Survey in the Park: methodological and practical problems associated with geophysical investigation in a late Victorian municipal park. *ArcheoSciences* **33**: 229-231.

Schmidt, A., Linford, P., Linford, N., David, A., Gaffney, C., Sarris, A. and Fassbinder, J. (2015) *EAC Guidelines for the Use of Geophysics in Archaeology: Questions to Ask and Points to Consider.* Namur: EAC, AISBL.

The Wenner array not as black as it is painted - surveying shallow architectural remains with the Wenner array. A case study of surveys in Szydłów, Poland, and Tibiscum, Romania

Michał Pisz[(1)] and Tomasz Olszacki[(2)]

[(1)]University of Warsaw, Warsaw, Poland; [(2)] Trecento, Łódź, Poland

michal.pisz@uw.edu.pl; tolszacki@o2.pl

Introduction

The Wenner electrode configuration is one of the first arrays used in archaeological prospecting with the earth resistance method. It is a Type A electrode array, which means that the potential electrodes are placed between the current electrodes (Schmidt 2013, 40). It is considered to be the simplest and most easily understandable electrode configuration (Gaffney and Gater 2003, 29). According to studies on this method, it can produce multiple peaks over a single feature, thus the response from the measurements could be muddled and ambiguous in some cases.

In 2016, an earth resistance survey with the Wenner array has been carried out in Szydłów (Poland), and at the *Tibiscum* Roman fort, near Caransebeş in Romania. Both surveys have been carried out with the a=0.75m separation distance between electrodes, 1m separation distance between the profiles and a 1m sampling interval. The measurements were performed with Geoscan Research RM85 Meter. Applying such parameters resulted in clear and unequivocal measurements – something what is rather supposed to be atypical for the Wenner array.

Theoretical Model

The phenomenon of multiple peaks typically occurs when a highly resistant object is buried shallow beneath the ground surface. This relation is precisely described by the theoretical resistivity response model, presented by Schmidt (2013, 57-61). According to the model, a shallow-buried, insulating object produces a high peak anomaly over an object's centre. In addition, two "shoulders" on the sides of the main peak are apparent. That feature of the Wenner array is considered to be characteristic for this kind of electrode configuration which could lead to misinterpreted results.

Tibiscum Roman Fort Survey

In April 2016, a vicus area of the Roman fort Tibiscum was surveyed with four geophysical methods. Besides earth resistance, performed with Geoscan Research RM85 Meter, also magnetic

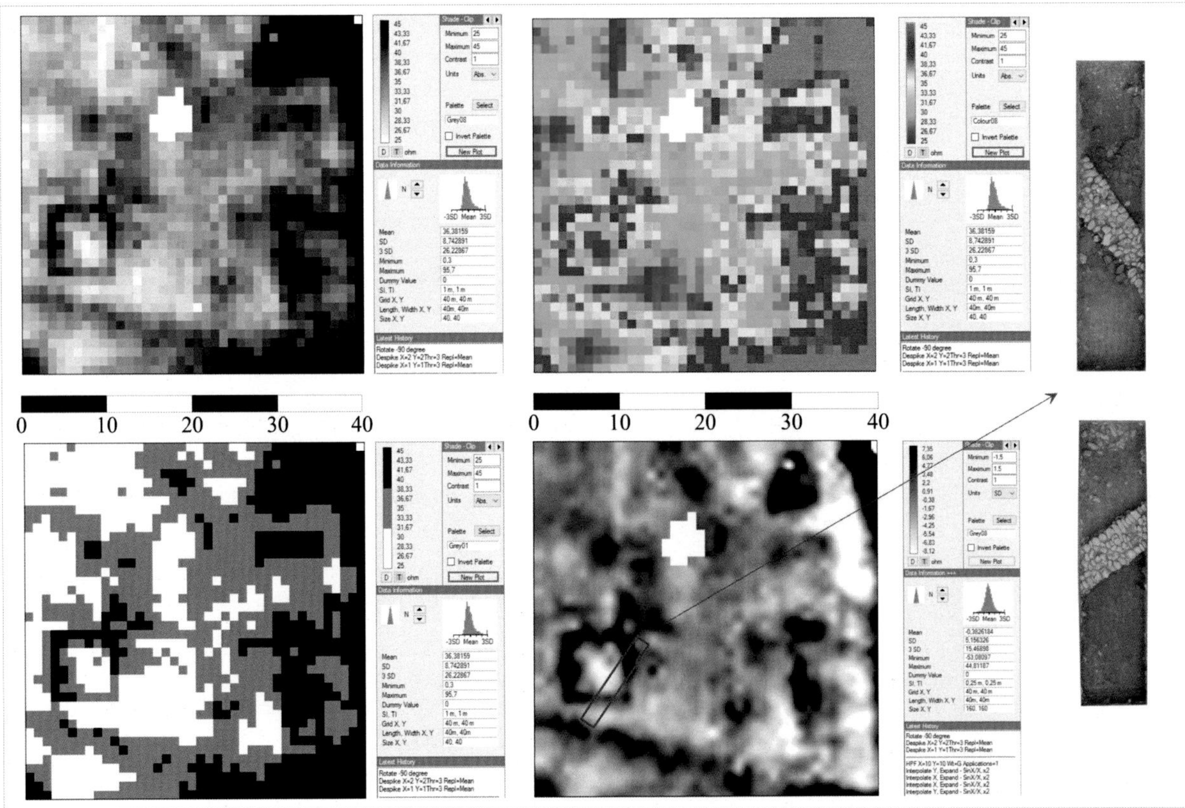

Figure 1: Earth resistance measurements results from Romania (Tibiscum) presented in various colour scales. Top left, top right and bottom left are the raw data with Despike filter applied, bottom right with High Pass filter and double interpolation. In the right part of the image the orthophoto of the excavated structure is presented.

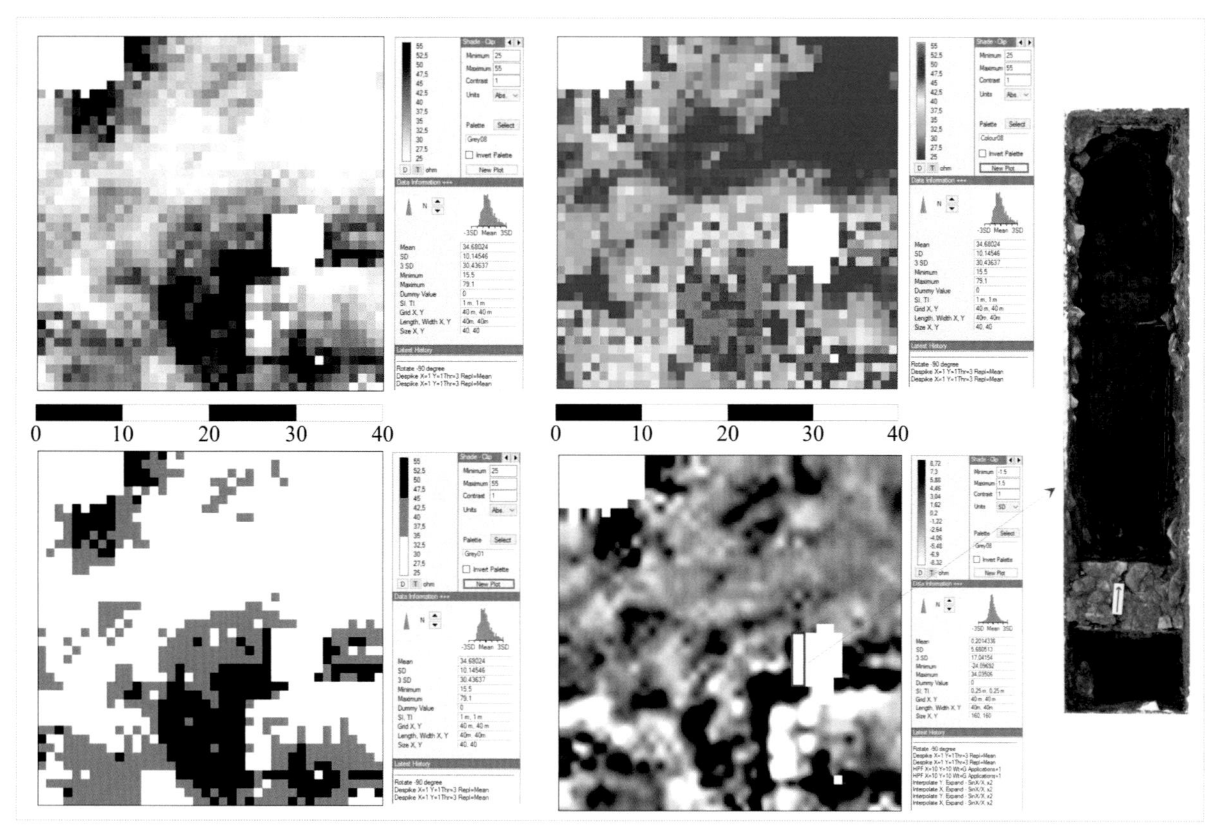

Figure 2: Earth resistance measurements results from Poland (Szydłów) presented in various colour scales. Top left, top right and bottom left are the raw data with Despike filter applied, bottom right with High Pass filter and double interpolation. In the right part of the image the orthophoto of the excavated structure is presented.

measurements with Geometrics G-858 caesium vapour magnetometer, GPR measurements with GSSI SIR3000 with 270 MHz antenna and ERT measurements with GEO-TOM instrument (double-dipole array) were carried out. One clear and regular anomaly become a subject of further investigations. In August 2016, a test trench over the object was excavated. The anomaly was caused by the shallow-buried wall made of cobble, with the ceiling approximately 20 cm from the ground surface and the floor approximately 90 cm below the ground surface.

The most interesting fact about this object and the results of geophysical prospection was that the earth resistance measurements with a=0.75m Wenner array did not produce any noticeable multiple peak responses. The image of the structure was quite sharp and clear (Fig. 1).

Szydłów – 19th Century Stone Walls

In December 2016, a survey of the Jewish District in Szydłów, Poland was carried out. Its main aim was to conduct a preliminary field analysis for the archaeological excavations scheduled to take place in 2017. Again, the Wenner array with a 0.75m electrodes separation and a 1 x 1 m sampling resolution distance was chosen. During the survey a test trench was excavated so it was easy to identify those anomalies which had particular structures. The excavated object was a wall made of limestone. Again, the anomalies registered during this survey turned out to be quite clear and sharp (Fig. 2).

Unexpected Phenomenon

In both cases, the shallow buried remains did not produce any ambiguous or unclear imagery. The captured anomalies seemed to be sharp and, apparently, no significant multiple peak effect was registered.

What could have caused the surprisingly good results obtained with the Wenner array? One of the possible explanations is that the final good result might have been only an effect of a lucky coincidence and 1 x 1 m resolution of the sampling.

Another theory is that it could depend on the thickness of the object itself. However, in those two cases, two different depths of the wall's floor occurs.

Advantages and Disadvantages of Wenner Array

The Wenner array is considered to be a little bit unwieldy and thus slow to be used for archaeological purposes (Gaffney and Gater 2010, 29). To increase speed and efficiency a twin-probe configuration is usually employed for area evaluation surveys (Schmidt et al. 2015, 69). Notwithstanding, it also possess significant advantages over other arrays

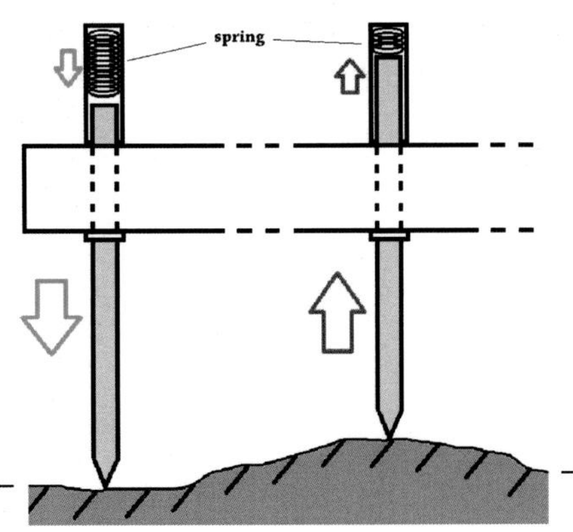

Figure 3: RM85 meter mounted of a self-made frame with a 230 cm beam (Wenner array, a=0,75) – left, and the concept of the adaptive-suspension electrodes – right.

which could make it a noteworthy choice for use in the field in some cases.

The biggest advantage of the Wenner array is that when the distance between electrodes is a = 0.5 or 0.75 m, then a single person would suffice to operate a measurement device (RM85 meter or similar device). This makes it a very easy and handy solution, especially if there is no more qualified personnel available on the ground. The other important feature of Wenner array is the comfort it provides when taking measurements – there is no cable, no remote electrodes, thus no need to move them, which could be a complicated procedure (Gaffney and Gater 2003, 32-34) or keep the 30 x a separation distance. That should be particularly useful when surveying on a small sites and in complex settings, such as yards of the strongholds. Another noticeable difference is that there is no need for grid balancing – something that is often applied in twin-probe survey results (Schmidt 2013, 135-140).

The Wenner array has certain disadvantages as well. The most commonly described issue in literature is the problem with multiple peaks. However, as the examples above have shown, it is attainable to obtain quite clear and sharp image of anomalies, even those caused by the strongly insulating, shallow-buried structures. Some might find the size of the beam also be slightly unhandy (for a=0.5m the beam is almost 1.5m, when for a=0.75 it reaches almost 2.3m). It is not that onerous as long as the field area remains flat where all electrodes could be easily inserted into the ground. Otherwise it could be sometimes problematic to find a spot to insert all four aligned electrodes into the ground. However, that could be tackled with a an adaptive electrode suspension system (Fig. 3).

Considering all of the above, it is concluded that the Wenner array can in certain cases be a valid and robust alternative to a twin-probe array.

Bibliography

Gaffney, C. and Gater, J. (2010) *Revealing the buried past. Geophysics for Archaeologists.* Stroud: The History Press.

Schmidt, A. (2013) *Earth Resistance for Archaeologists.* Lanham AltaMira Press.

Schmidt, A., Linford, P., Linford, N., David, A., Gaffney, C., Sarris, A. and Fassbinder, J. (2015) *EAC Guidelines for the Use of Geophysics in Archaeology: Questions to Ask and Points to Consider.* Namur: EAC, AISBL.

Live-streaming for the real-time monitoring of geophysical surveys

F Pope-Carter[1], C Harris[1], G Attwood[1] and T Eyre[1]

[1]Magnitude Surveys, Bradford, United Kingdom

f.pope-carter@magnitudesurveys.co.uk

For commercial archaeological geophysics surveys in the United Kingdom, the development-led nature of the project typically demands a rapid and cost-effective completion of the survey. Site-wide magnetometer surveys will often feedback into the creation of trenching plans, which can add pressure for the quick turnaround of positionally accurate high quality collected data. Any unidentified errors in data collection or compromises to the survey environment can cause unnecessary and expensive delays.

Traditional dataloggers must be downloaded at the end of the survey day or after survey completion to a computer for further processing. The requirement for the field team to download the day's work from the datalogger to the computer, and send back to the office-based team for processing, visualisation and interpretation, introduces an efficiency gap in the turnaround time for field survey feedback of results to clients. In recent years, the development of GNSS-positioned cart-based systems has improved the speed at which surveys can be undertaken but the use of dataloggers to store the collected measurements still limits the turnaround of the results to the client.

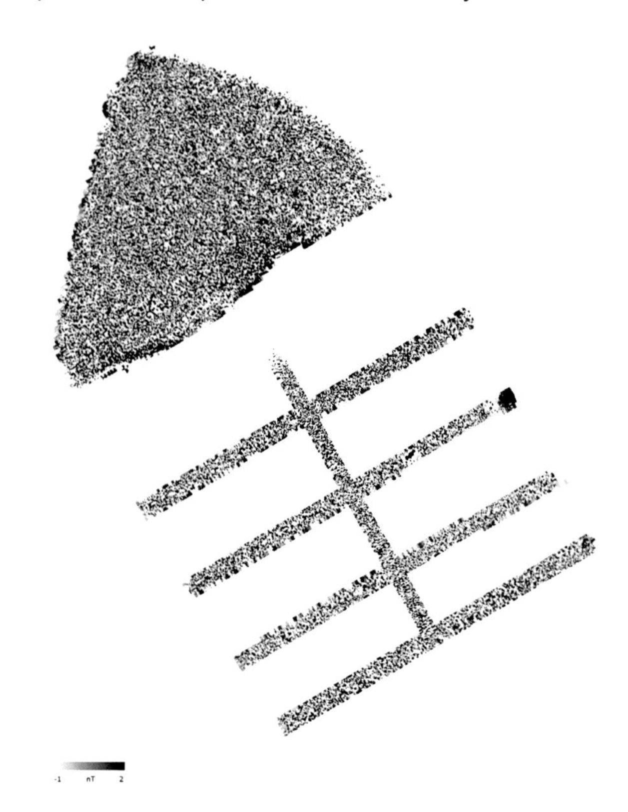

Figure 1: Modified survey strategy following the detection of adverse site conditions.

Magnitude Surveys has developed a methodology for the near real-time processing and visualisation of magnetometer data as they are being collected. This methodology combines the latest digital fluxgate gradiometers, ultrafast data links from site to the office, and dedicated monitoring and processing servers. This allows Magnitude Surveys to perform real-time quality assurance on data during data collection while facilitating a rapid turnaround of results to clients.

The latest Digital Three Axis Fluxgate Gradiometers from Bartington Instruments operate at 200Hz and collect nine readings simultaneously. This is two orders of magnitude greater than a traditional fluxgate cart system. While a traditional system may collect 10MB of raw data per day, the Magnitude Surveys cart utilising three-axis sensors will collect this much data per sensor traverse. To put this into perspective, downloading one day's raw data from the Magnitude Surveys cart would take approximately 30 hours over a traditional RS232 serial connection.

In addition, the Bartington Grad13 Digital Three Axis Fluxgate Gradiometers deviate from historical fluxgates such as the Bartington 1000L and Geoscan Research FM series in that they do not have a built in calibration procedure and always output raw data values, which are calibrated during the pre-processing stage. The advantage of this system is that data suffering from a poor calibration procedure do not have to be re-collected, merely re-calibrated against a new set of calibration coefficients. It does however mean that quality assurance checks cannot be undertaken on the data until the pre-processing steps have been undertaken.

It is possible to store the calculated calibration coefficients within the processing software and evolve these values as a survey continues, to correct for sensor variations induced by movement or temperature variation. This requires a considerable amount of data processing power, which would not be possible on a small battery powered computer running in the field. To meet this requirement, dedicated servers were set up to constantly monitor for incoming data and perform on-the-fly calibration evolution and filtering in near real-time. The quantity of data to be interrogated and backed up necessitates these servers be located within our office facilities, with a fast link between these sites and the field teams. Enterprise grade, dual network, 4G connections in the field coupled with long range outdoor WiFi antennae and redundant fibre optic

Figure 2: Quad towed Digital Three Axis Fluxgate Gradiometers

links into our office allow for fast, high bandwidth communication between the office and field teams.

A dedicated project officer working from the office can continually monitor data being streamed in from the field teams. This allows for rapid quality assurance assessments before field teams leave the site or survey area. Survey and coverage issues can be identified immediately and corrected without the need for revisits or deployments.

Potential survey or site issues can be identified and mitigated easily in the field, saving time and money. For example, a portion of a larger field was identified as being covered by a uniform spread of high-contrast ferrous material, impeding on the detection of archaeological features. Because the data were being monitored in the office in real-time, the client could be notified of the situation and a mediation strategy could be immediately put into place.

Processed data can be analysed while the survey is being undertaken, allowing for preliminary findings to be reported back to clients immediately. This allows clients to plan and map potential mediation strategies and trial trenches before the survey is complete.

The methodologies outlined above have developed over the past two years as Magnitude Surveys has come to grips with the complications associated with collecting large datasets in remote environments. Further developments including automated quality assurance warnings to field operatives and real time streaming to client desktops are planned on the next development cycle.

Urban archaeology in Affile (Rome-Italy): preliminary results of the ground penetrating radar survey

Valeria Poscetti[(1)] and Davide Morandi[(2)]

[(1)]Vienna Institute for Archaeological Science - Vienna University, Vienna, Austria; [(2)]IDS GeoRadar s.r.l., Pisa, Italy

valeria.poscetti@univie.ac.at

Introduction

Affile is a small town situated about 70km east of Rome (Italy), in a hilly area in between the *Monti Affilani* and *Monti Ernici* mountain systems. The village, probably developed on a pre-Roman *oppidum* (Mari 2005), was, in Roman times, a *municipium* (*Afilae*) with a quite large territory including a number of *villae rusticae* and the well-known Trajan's villa (Fiore and Mari 2000, Piro *et al.* 2003), whose remains are located 6km south-east of Affile. There are only a few visible remains of the Roman town *Afilae*, like the structures of a large cistern and a number of *opus mixtum* walls integrated in medieval and modern buildings. Former researches about Affile mainly focused on the well-preserved medieval part of the village (Travaini 1990), while the Roman town is still largely unknown.

A small project has been recently initiated (February 2017) with the goal of collecting more information about the ancient settlement; procedures include non-invasive prospections in the urban area. The project is supported by the Affile Municipality and conducted in cooperation with the University of Vienna and IDS GeoRadar.

The paper presents the preliminary results of the high resolution Ground Penetrating Radar (GPR) survey recently conducted in the Affile's main square, which is supposed to be the place of the ancient *forum* (Mari 2005).

GPR Survey and Preliminary Results

For the GPR survey in the urban area of Affile the new multi-channel Stream C from IDS GeoRadar has been tested (Fig. 1). The new compact array solution includes 34 antennae with a frequency of 600 MHz and is conceived for real-time 3D mapping of underground features with high accuracy, due to the very dense sample interval of 4 x 2 cm (respectively, cross line and in line spacing).

The main square of Affile, a paved area of about 1700 m^2, has been efficiently investigated in about 1 hour by pushing the motor-assisted multi-channel GPR system with integrated positioning system along parallel profiles. In this first step of the research, we have focused our attention on the central and south-eastern parts of the square (Fig. 2). In these areas, we have presumed the probable presence of buried archaeological structures, based on a few visible remains of Roman walls and bases incorporated in the medieval church situated on this side of the square. In this part of the square we also conducted some tests by using a multi-frequency system from IDS (RIS MF Hi-Mod), equipped with 600 MHz and 200 MHz, with the goal to achieve a better interpretation of the subsurface.

The good penetration of the GPR signal radiated through the Stream C system (600 MHz) has allowed the subsurface investigation down to about 2 m. From the GPR 3D data block a sequence of 5 cm depth slices has been generated and archaeo`structures (Roman buildings? Later structures?). However, it is interesting to note the north-west/south-east orientation of most of the observed anomalies, which is similar to that of the old houses in the northern and eastern part of the square, which were probably built on ancient foundations. This situation seems to offer a starting point for the understanding of the topography of the ancient town, which is largely unknown.

Near the medieval church two linear features (Fig. 3.B) and a strongly reflective rectangular feature about 2 x 1 m (Fig. 3.C) have been detected. Both, the linear features and the strong rectangular anomaly have different orientations compared to the GPR anomalies described above. This suggests the hypothesis that we are probably dealing with a multiphase complex.

Figure 1: Data acquisition in Affile by means of the multi-channel Stream C from IDS GeoRadar.

Figure 2: Investigated area in the main square of Affile. Background satellite imagery © Google Earth.

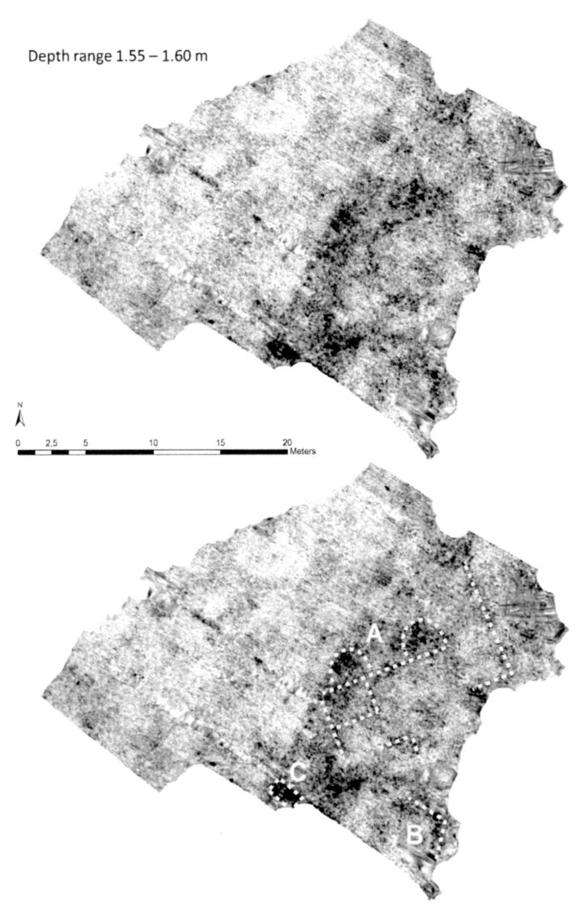

Figure 3: GPR depth slice in the depth range 1.55-1.60 m. In the image below, the detected features have been outlined (dashed lines). GPR data image: Courtesy of Alois Hinterleitner.

First Conclusions and Future Directions

The GPR survey in Affile has demonstrated the efficiency of the new multi-channel Stream C from IDS GeoRadar for urban archaeology. In this specific case, we have a quite difficult archaeological situation because of an almost total lack of information about the ancient town and the relatively small size of the investigable areas. Despite the non-optimal conditions, the first archaeological geophysical survey in Affile allowed the rapid acquisition of a relatively large amount of information useful for further investigations, like excavations in selected areas and archaeological mapping.

In the future, further GPR surveys are planned in small agricultural areas within the village, with the goal of achieving the first GIS-based archaeological mapping of the ancient town. We expect that the information that will be collected in these areas will contribute to the further understanding of the archaeology of this small village, giving at the same time input to larger scientific projects, aimed at the studying the archaeological landscape in the region through the integrated and systematic use of non-invasive methods and new technologies.

Bibliography

Fiore, M. G. and Mari Z. (2000) Villa di Traiano ad Arcinazzo Romano: Risultati della prima campagna di scavo. Ostia Antica(Roma): Publidea.

Mari, Z. (2005). La topografia degli Equi della Valle dell'Aniene, in A. M. Dolciotti and C. Scardazza (eds.) *L'ombelico d'Italia, Popolazioni preromane dell'Italia Centrale. Atti del Convegno, IV Giornata per l'Archeologia, Roma, Complesso monumentale di S. Michele a Ripa, 17 maggio.* Rome: Gangemit Editore spa, 117-146.

Piro, S., Goodman, D. and Nishimura, Y. (2003). The study and characterization of Emperor Traiano's Villa (Altopiani di Arcinazzo, Roma) using high-resolution integrated geophysical surveys. *Archaeological Prospection* **10**(1): 1-25.

Travaini L. (1990). *Due castelli medioevali. Affile e Arcinazzo Romano.* Roma: Team Stampa.

The late-Roman site of Santa Margarida d'Empúries. Combining geophysical methods to characterize a settlement and its landscape

Roger Sala[(1)], Helena Ortiz-Quintana[(1)], Ekhine Garcia-Garcia[(2)], Pere Castanyer[(3)], Marta Santos[(3)] and Joaquim Tremoleda[(3)]

[(1)]SOT Archaeological Prospection, Barcelona, Catalonia; [(2)]Affiliation, Town, Country; [(3)]Museu d'Arqueologia de Catalunya-Empúries, L'Escala, Catalonia

roger_sala_bar@yahoo.es

The archaeological site of Empúries is an interesting case of the historical evolution of a settlement from the early Iron Age to medieval times. From a small Greek commercial site of VIth century BC (Palaiapolis), it evolved in the V-IVth centuries to a bigger site (Neapolis) that acted as a node for Greek commerce with the eastern Iberian Peninsula tribes. In the context of the second Punic War, Emporion was occupied by the Roman army, becoming a bigger Roman city called Emporiae, with an extent of 30Ha. Although the abandonment of the Roman city during the IIIrd century AD is clearly shown by archaeological record, the continuity of smaller settlements in the late Roman period is testified by historic fonts and archaeological evidence.

The TerAmAr project, leaded by investigators of MAC Empúries and ICAC, aimed to establish a sequence of the Empúries region landscape evolution. The project included a core drilling campaign that revealed important changes in the landscape and coastal line and floodable areas around Empúries throughout antiquity.

As part of this multidisciplinary study, a geophysical survey was conducted in two phases around the medieval church of Santa Margarida, placed 500m from the Roman city limits. Excavations on the church remains carried out in 1994 revealed previous building phases, which projected its chronology through the late Roman period. As a consequence, the Santa Margarida site was considered an interesting research area to study how the changes in landscape affected the settlements from late antiquity to the early medieval period.

A first phase consisted in a GPR survey of Field 1, aiming to describe the building remains around Santa Margarida church. A second phase carried out in 2016 aimed to enlarge the context of the settlement revealed by GPR data, using a new magnetic survey and three ERT profiles, as well as a new GPR survey in field 2 (Figure 1).

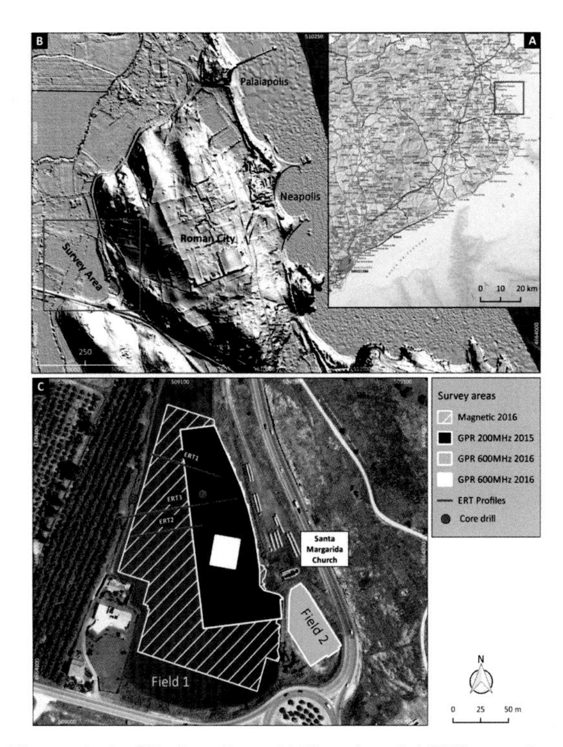

Figure 1: A. Site location at L'Escala, at 140Km northeast from Barcelona. B. Hillshade plot of digital terrain model of Empúries archaeological sites. C. Explored areas and methods in Santa Margarida site.

Methodology

The main survey area (field 1) is a flat cultivation field with a sandy topsoil, at 2.8m ASL. The GPR survey of 2015 was completed using an IDS streamX module (200MHz). A second GPR survey covered field 2 near Santa Margarida and a specific building (group 10) imaged in 2015, used a new custom IDS, 600MHz, 5 antenna array.

The magnetic survey in 2016 used a Bartington G-601 fluxgate gradiometer to cover 22,258m^2, aiming to bring new qualitative information on the building area and exploring the western limits of the site.

Three ERT profiles were produced to obtain information on the geological base of the settlement and to get possible clues to understand its relation with possible ancient shore lines. Two ERT profiles (ERT1 and ERT2) covered 62.5m using a double dipole array with an electrode spacing of 0.5m. ERT3 was used to generate a deeper section, using 2.5m spacing in a 107.5m profile.

Results

The GPR survey of 2015 covered 8,900m^2. The resultant dataset showed a poor penetration (1.5m), interpreted as an influence by high soil moisture and salt contents in the soil.

As is shown in Figure 2, the GPR time-slice sequence revealed a well-defined group of buildings

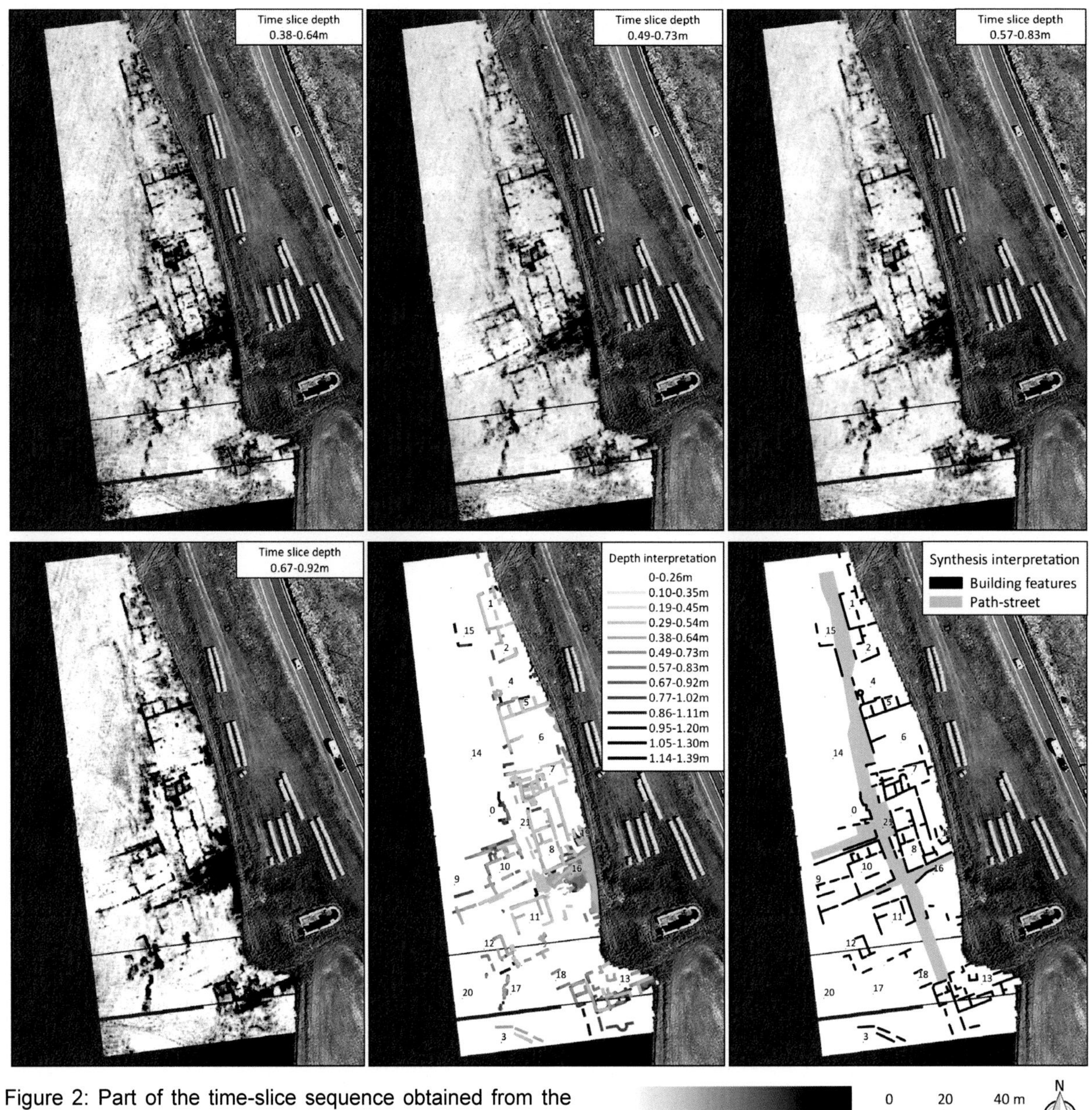

Figure 2: Part of the time-slice sequence obtained from the GPR survey of 2015. At bottom centre, depth interpretation diagram. At bottom right, synthetic interpretation diagram.

aligned to a possible street that runs parallel to the slope of Empúries Roman city hill, from SSE to NNW. The depth interpretation diagram shows how most of detected features lay below 0.3m depth, except for western structures (group 9) which are detected from 0.8m depth. This was interpreted as evidence that the settlement was originally placed in a descending slope to the west.

The interpretation of the building remains allowed the recognition of a planned construction, or at least a progressive construction sequence that maintained the geometrical axis marked by the central street or path.

The magnetic survey of 2016 (Figure 3B) covered 22,258m², including the previously GPR surveyed area and extending to the western limits of field 1.

The resulting magnetic map described new features in the building area, revealing thermo-altered areas, possible ceramic soils or a group of subtle features that indicate a possible extension of the site westwards (groups 22, 23 and 24).

An extensive, slightly magnetic anomaly was described in the north-west limit, called group 31. It consists of a fringe of 95x12 m interpreted as a product of magnetic sediment deposits. Taking into account the core drillings, one of the possible interpretations of this feature is the fillings of a buried channel or a flooded area, but other explanations cannot be discarded.

The core-drillings collected in field 1 revealed a succession of high organic content and sandy layers which were interpreted as a sequence of

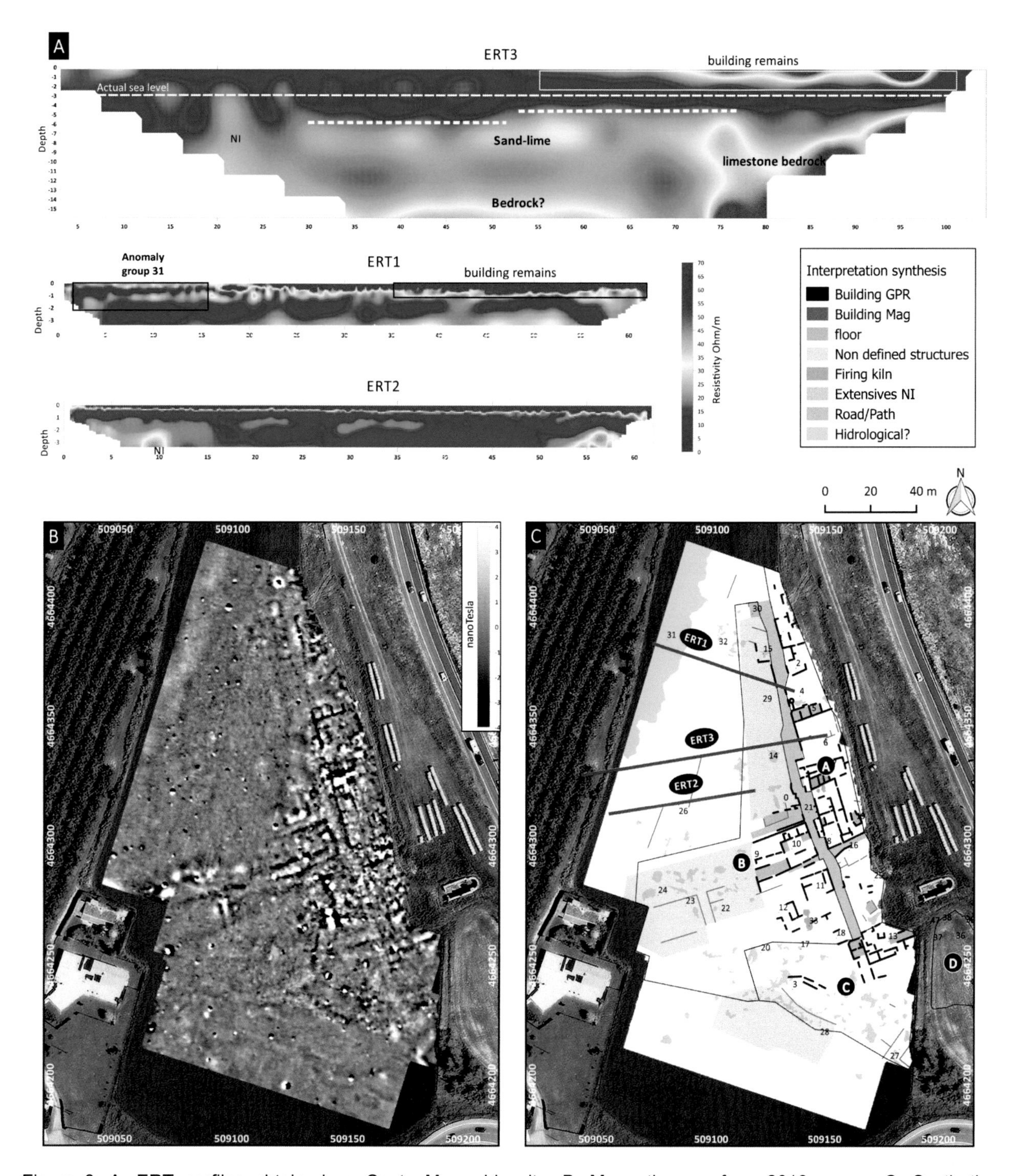

Figure 3: A. ERT profiles obtained on Santa Margarida site. B. Magnetic map from 2016 survey. C: Synthetic interpretation diagram.

marshy and coastal landscape periods, but none of them were located within the anomaly group 31.

The three ERT profiles were used to study the geological background of the settlement, also aiming to bring information on the sparse evidence for an ancient nearby shore area (Figure 3.a). The results from ERT3 confirmed a descending trend of the basement, showing a resistive lime-stone bedrock base at the east of the profile, covered by a descending layer of possible sands at 4-7 m depths. The shallower data show a resistive body near surface, interpreted as building remains, standing over a conductive layer, interpreted as clays and limes.

The profile ERT1 was placed in order to cross the anomaly group 31. The part of the profile crossing group 31 shows a fringe of mid values in at 1.3m depth, over a deeper layer of conductive values.

The 2016 survey was completed with two GPR extensive surveys in Field 2 and in a specific area from the 2015 survey results. The data obtained in Field 2 showed a lower quality than the previous campaign, mainly because of the recent ploughing of the field, leaving a rugged surface. The results revealed a new building group, maintaining a similar orientation to the ones located in Field 1.

Conclusion

The two phases of the geophysical survey brought a first characterization of the site of Santa Margarida, solving the delimitation of the site and the description of building remains.

A synthetic interpretation diagram has been produced using the GPR and magnetic interpretation maps, including both information about building structures and other anomalies related with human activity, such as possible kilns or working areas.

The synthetic diagram also proposes a division of the known site in 4 zones: A, B, C and D (Figure 3.c). Although these zones don't necessarily give a definitive archaeological division of the site, they help to understand the geometry of the settlement and further architectural analysis.

The possible connection of the west of Field 1 with water volumes suggested by drillings seems to be consistent with the anomaly group 31 described in the magnetic survey. In the same region, the ERT1 profile shows a flat, resistive body at 1.3m depth that could be interpreted as a layer of sands, but this interpretation only would be confirmed by new drillings on the same area.

Bibliography

Annan, A. P. (2009) Electromagnetic Principles of Ground Penetrating Radar. In H. M. Jol (Ed.), *Ground Penetrating Radar Theory and Applications*. Amsterdam: Elsevier, 1–40

Aspinall, A., Gaffney, C. and Schmidt, A. (2008) Magnetometry For Archaeologists. Plymouth: Alta Mira Press.

Julià, R., Montaner, J., Castanyer, P., Santos, M., Tremoleda, J. and Ferrer, A. (2015) Évolution paléoenvironnementale de la limite sud du site archéologique d'Empúries à partir de la séquence holocène de Horta Vella (l'Escala). In F. Olmer and R. Roure (eds.) *Les Gaulois au fil de l'eau, (actes 37e coll. de l'AFEAF, Montpellier 2013)* 1. Communications, Ausonius Éditions, Mémoires 39, 35–52.

Nolla, J. M., Tremoleda, J., Sagrera, J., Vivó, D., Amich, N., Buscató, Ll., Pons, Ll., Aquilué, X., Castanyer, P., Santos, M., Cobos, A., Nolla, J. M. and Tremoleda, J. (eds.) (2015) *Empúries a l'antiguitat tardana*, Monografies Emporitanes 15, 2 vols.

Acknowledgements

The authors want to thank the participants in the survey fieldwork, Angels Pujol, Jordi Gibert, and Eduard Ble. This work would not have been possible without the help of the team of GS Ingenieria Roger Juanola, Anna Cano and Joan Claveria for their work in the ERT data collection and professor Ramon Julià for his support in the core data interpretation.

Exploring the urban fabric of ancient Haliartos, Boetia (Greece) through remote sensing techniques

Apostolos Sarris[(1)], Tuna Kalayci[(1)], Manolis Papadakis[(1)], Nikos Nikas[(1)], Matjaž Mori[(2)], Emeri Farinetti[(3)], Božidar Slapšak[(2)] and John Bintliff[(3)]

[(1)]GeoSat ReSeArch Laboratory, Foundation for Research and Technology (FORTH), Rethymno, Crete, Greece; [(2)] Ljubljana University, Slovenia; [(3)] University of Leiden, the Netherlands

asaris@ret.forthnet.gr

Introduction and Past Research

The geophysical research campaign constitutes a continuation of the survey that was initiated in 2000 by the University of Ljubljana and Leiden University teams (Ancient Cities of Boeotia Project by John Bintliff and Božidar Slapšak) aiming towards the study of the intra-site planning and landuse of the ancient cities in the region of Boeotia. The 2016 geophysical season at Haliartos was jointly funded by a grant from the McDonald Institute of Cambridge University to Professor John Bintliff and Professor Anthony Snodgrass, and from the ERC project Empire of 2000 Cities co-ordinated by Professor Luuk de Ligt.

A preliminary archaeological survey that was carried out in 1980s under the auspices of the Boeotia Project indicated that the Bronze Age settlement was confined to the top of the hill. Subsequently, the hill was occupied as an acropolis in the later Classical period. The dense distribution of the Archaic, Classical and Early Hellenistic Period (c. 600BC to 200BC) sherds is spread all over the region, both on the acropolis and the smooth slopes around the acropolis, filling almost all space (~30 ha) within the walls that ran around the settlement (LC-Lower city) of that period. In 171 BC the city was destroyed by the Romans, as were several other cities of the Boeotian League.

The remote sensing investigations in Haliartos were initiated in 2004 by the University of Ljubljana through blimp photography by Rafko Urankar and Jure Krajšek, followed in 2006 by aerial flights by Darja Grosman. A number of architectural remains were identified in all sections of the city, and an initial plan of the city was possible – especially for the central lower city section.

The 2009 geophysical survey employed caesium magnetometers and covered a total area of 3.2ha in the SE section of the city (Mori 2009). In 2016, geophysical investigations focused to the south of the acropolis, in the plateau where the lower city extends. A manifold approach (Sarris 2012) was applied in order to address specific questions regarding the city planning. A multi-sensor magnetic survey was employed for the wide coverage of the site. The Ground Penetrating Radar (GPR) was used to explore sections that seem to be affected by heavy erosion, as there were a number of bedrock outcrops rising to the surface. A handheld magnetic survey coupled with soil resistance measurements was carried out in the central north section of the lower city, where dense vegetation did not allow the use of the multi-sensor magnetic survey.

Results of the Geophysical Approaches

According to the results of the geophysical survey, the city plan of the lower city seems to follow an NE-SW orientation forming a large number of insulae separated by NE-SW and NW-SE roads (Fig.1). The plan of the city seems to be oriented in a way that the NE-SW streets are directed towards the acropolis. More than 60 insulae are suggested by the geophysical and aerial images (Fig. 2). The average spatial extent of the insulae (based on those that are almost fully described) is around 1000-1500 square meters and even if there is a trend of larger deviations to the north, some of the smaller sized insulae are defined as such due to some other internal divisions of them (probably of a later period).

To the north, the insulae seem to expand gradually upon the inclines of the hill of the acropolis. Two more observations can be made. First, as Grosman (2007, 2009) suggested from the photointerpretation of the aerial images, the grid of the city exhibits regularity but not an orthogonality; very few insulae are of orthogonal shape and most of them have the shape of an inclined parallelogram. Second, dissolution of the strict arrangements of the insulae is witnessed towards the east, suggestive of the possible location of the agora in that particular section of the city. This can be also justified due to the smoother terrain and the easier communication through the city provided via the E-W roads. If so, the agora was located towards the main gate to the east, namely the entrance of the road leading from Tanagra and Thebes.

The city blocks are defined by the cross sections of about 13 NE-SW (~4m wide) and 7 SE-NW (~5-6m wide) roads. It is obvious that the SE-NW roads are the main corridors of communication through the city, whereas the narrower NE-SW roads are directed mainly towards the acropolis and the upper city. A couple of features seem to diverge from the strict layout of the road system. First, a curvilinear road runs in the middle of the survey zone, intruding insulae D3, D4, D5, D6 and C11. Second, another curvilinear anomaly can be traced to continue further to the north passing through the eastern sections of

Figure 1: Integration of the magnetic data (both 2009 and 2016 survey seasons) overlaid on the WorldView2 satellite image of the site.

insulae F3, E5, D8 and C12 and crossing insulae B12, A11 and A10. Taking into account that the feature goes through the suggested urban blocks, it may be suggested that it also belongs to a later period than the original urban layout.

If we take into account the finds of the surface survey that indicate some Late Roman material exists at the eastern sections of the acropolis and the sloping terraces below it, we may propose that a Late Antique Roman settlement was established on the eastern fringes of the original settlement, after a long interval due to the Senate's ban on resettlement of the city after its destruction in 171 BC. This can then be linked to a full re-walling of the Acropolis at that time. If so, the curvilinear features may represent new communication corridors connecting the reoccupied sections of the lower city. The difference in their magnetic signature may also

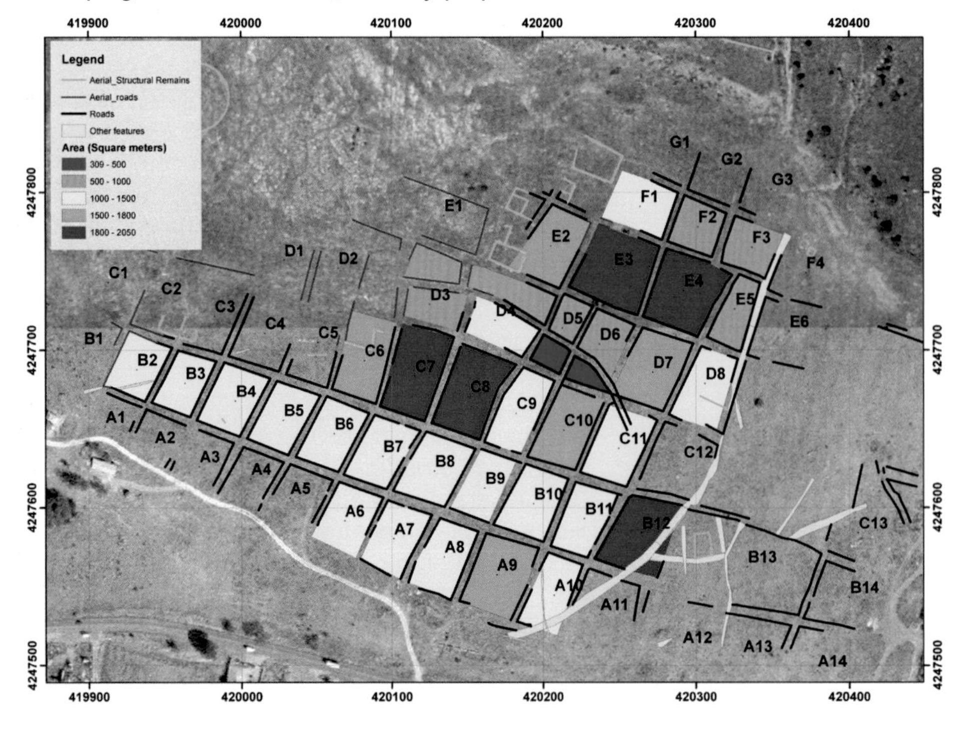

Figure 2: The suggested insulae of ancient Haliartos, based on the interpretation of the geophysical and aerial data. The classification of the insulae is based on their areal extent.

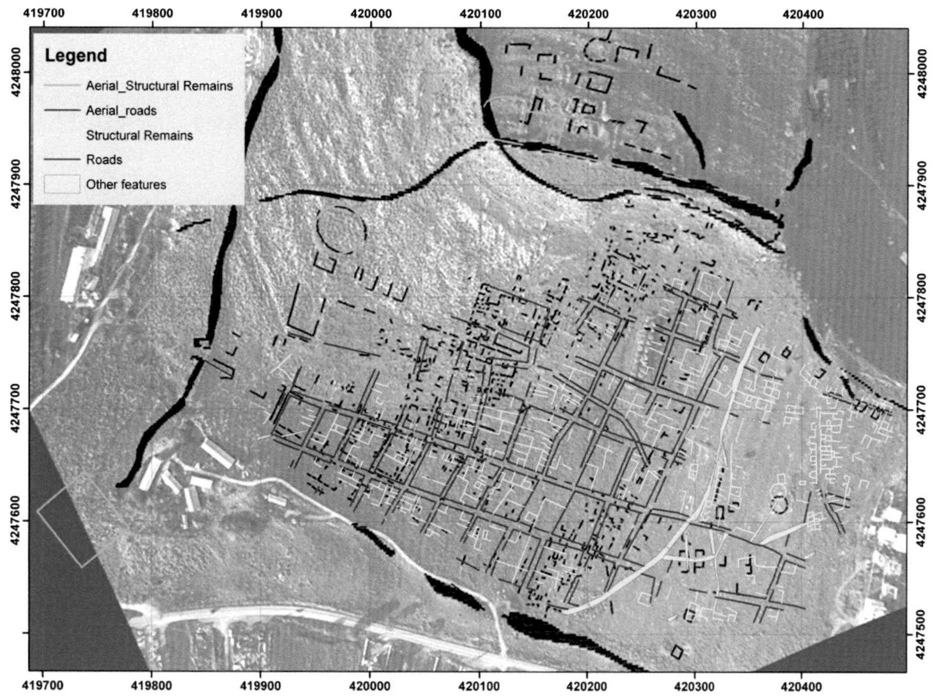

Figure 3: Interpretation of the architectural relics suggested by both geophysical results and aerial images.

be indicative of a different construction material, with the bifurcating feature to the east consisting of a looser material (compacted earth?) than the one to the central north section of the settlement. This is also supported by the high resistance signature of the latter. The curvilinear features, cutting roads at various angles and crossing through various building blocks indicate an evolution of the original urban block layout, which was already adapted to the local topology as it does not follow a strict orthogonal Hippodamian plan.

Most of the blocks contain a high density of buildings, the foundations of which are aligned in the same direction as the city blocks (Fig. 3). For some of the blocks there is further information in terms of internal partitioning and possibly empty spaces. The central west section of the site seems most regularly populated with structures distributed within the layout of the blocks of the city. Even if it is hard to outline in detail the internal compartments of these structures, there are obvious deviations regarding their size, possibly representing the existence of residential quarters, stoas and gardens or yards.

As shown especially from the GPR survey, the strict regular pattern of the road and block grid is getting dissolved at the eastern fringes of the city, exhibiting a relatively empty area (100x80m) surrounded by a dense distribution of structures. The eastern boundary of the open area, which is actually the only potential location for the agora of the city, is defined by a N-S road which seems to bend to the south taking a NW-SE orientation. At both sides of the road there is evidence of the existence of small (5x5m) structures. If indeed the area to the west of the road designates the location of the agora, it may be possible that the series of the small structures to the west of the road may define a stoa with small commercial units. To the SE and NE however the city grid is recapturing its dominant directionality which may continue for more than 100m eastwards where large surface blocks may pinpoint the location of the East fortifications of the city. The different orientation of the assumed agora and the accompanying high density of buildings to the east can suggest an organic development of the city and a differentiation between the residential and the more commercial sections of the city.

Bibliography

Grosman, D. (2007) Aerial Reconnaissance in Boeotia. Preliminary Report on the 2006 Test Season. *Unpublished Technical Report.* Ljubljana.

Grosman, D. (2009) Aerial Prospection in Boeotia. Report on the 2009 Field Season. *Unpublished Technical Report.* Ljubljana.

Mori, M. (2009) Preliminary Report on Magnetometry survey at Haliartos, *Unpublished Technical Report.* Ljubljana.

Sarris, A. (2012) Multi+ or Manifold Geophysical Prospection? in G. Earl, T. Sly, A. Chrysanthi, P. Murrieta-Flores, C. Papadopoulos, I. Romanowska, and D. Wheatley (eds.) *Archaeology in the Digital Era. Papers from the 40th Annual Conference of Computer Applications and Quantitative Methods in Archaeology (CAA).* Amsterdam: Amsterdam University Press, 761-770.

Revealing the structural details of the minoan settlement of Sissi, eastern Crete, through geophysical investigations

Apostolos Sarris(1), Meropi Manataki(1), Sylviane Déderix(2) and Jan Driessen(2)

(1)GeoSat ReSeArch Laboratory, Foundation for Research and Technology (FORTH), Rethymno, Crete, Greece; (2) Université Catholique de Louvain, CEMA INCAL, Louvain, Belgium

asaris@ret.forthnet.gr; mmanataki@gmail.com, sylviane.dederix@uclouvain.be, jan.driessen@uclouvain.be

Scope of the Survey

The Sissi Archaeological Project was initiated in 2007 by the Université Catholique de Louvain (Prof. Jan Driessen) under the auspices of the Belgian School of Athens (EBSA). The scope of the 2015 geophysical survey was to locate and map architectural remains on the south-eastern terrace of the hill, where a potential court-centred building (Building F) had been identified in 2011 (Area C) (Fig. 1). In 2016, investigations were focused in Area A, i.e. to the north of the large compound on the summit of the hill (Building CD), and in Area B, i.e. between Building BC and Area C. The ground penetrating radar (Noggin Plus GPR), the electrical resistance (GeoScan Research RM85) mapping and the magnetic gradiometry (Bartington G601) techniques were used in a complementary way.

Results of the Geophysical Approaches

Area A was surveyed using the 250MHz antenna. It was scanned in two directions, i.e. along both the x-axis and the y-axis. Data were processed individually for each direction, and also combining the datasets of both directions. The high sampling of the measurements (0.25mx0.025m) resulted in a detailed outlining of the architectural remains that seem to extend down to about 1-1.2m below the surface.

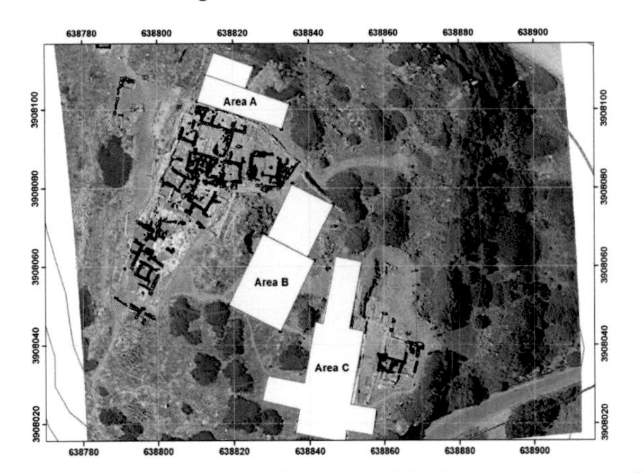

Figure 1: The geophysical survey grids that were surveyed at Sissi in 2015 (Area C) and in 2016 (Area A and Area B) fieldwork campaigns.

A series of small rooms extend to the north of the massive façade of Building CD (Fig. 2). The orientation of these rooms is similar to that of the compartments excavated in Building CD (see Rooms 4.5, 4.8, 4.9 and 4.10 in Fig. 2). Room A1 is ~2x2m large. To the north, A2 and A4 have a long axis of ~4.5m, but both seem to be subdivided by inner walls. Still further north, GPR signals indicate the existence of a much larger room (A3), ~8x5.30 in size. There is evidence for more compartments (A5, A6 and A7) with internal divisions between A3 and 4.8/4.5. A thick wall to the east of A6 and A7 appears to separate these rooms from another unit comprised of A8, A9 and A10. Room A9 seems to be the largest of them, measuring ~2.8x3.8m. To the east of A9, the elongated feature A11 probably suggests a segment of a road or path.

Area B was also surveyed using the 250MHz antenna and scanned in two directions (along the x-axis and the y-axis) with a sampling of 0.25mx0.025m. Structures B1, B2, B3, B4 and B5 are the most clearly defined (Fig. 3). Their walls are oriented N-S and E-W. Rooms B1 and B2 seem to belong to the same complex, ca. 5m by 2.5m. Further to the north, rooms B3 and B4 measure ~2.3x2.5m. B5, which can be traced for a distance of ca. 7m, consists of two or even three rooms, all ca. 2m wide. There is also evidence for two additional small rooms (B8 and B9) in the northern part of Area B.

In Area C geophysical survey was carried out with magnetic, soil resistance and GPR (250MHz and 500MHz antennas) methods. To the north, information coming mainly from GPR and partially from magnetometry indicates the existence of a number of compartments with an orientation roughly similar to the rooms excavated to the SE (Fig. 4). The most evident feature is the two-room structure C1, ca. 5x3.5m. More linear anomalies (C2) suggest the continuation of the habitation quarters to the north and to the west.

There is clear evidence of multiple walls delimiting several compartments in the vicinity of C3 and C5. However, the most striking feature identified by means of geophysics is a large area, ca. 30m long and maximum 14m wide, devoid of anomalies or reflectors – with the exception of the small anomaly at C7. This large area corresponds to a court, the north and eastern edges of which were identified during the 2011 excavation campaign (Driessen *et al.*, 2012:142-151). The uniform signature of the court is clear in GPR and soil resistance data, both of which show an absence of internal built features. The outline of the central court is clearly defined

Figure 2: Results of the GPR survey in Area A. Top left: X-Y direction survey. Rectified depth slice of 80cm. Top right: Y direction survey. Rectified depth slice of 80cm. Bottom: Diagrammatic interpretation of the GPR reflectors.

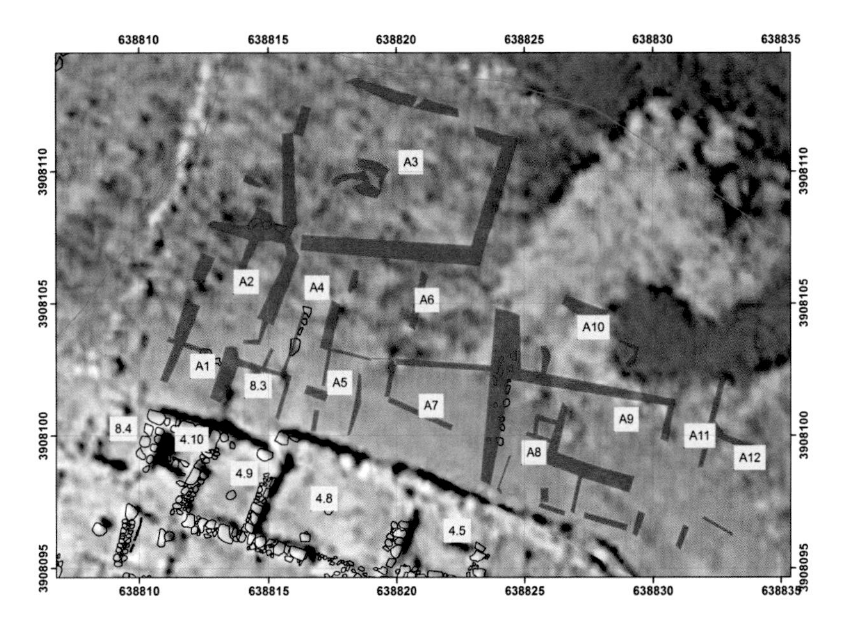

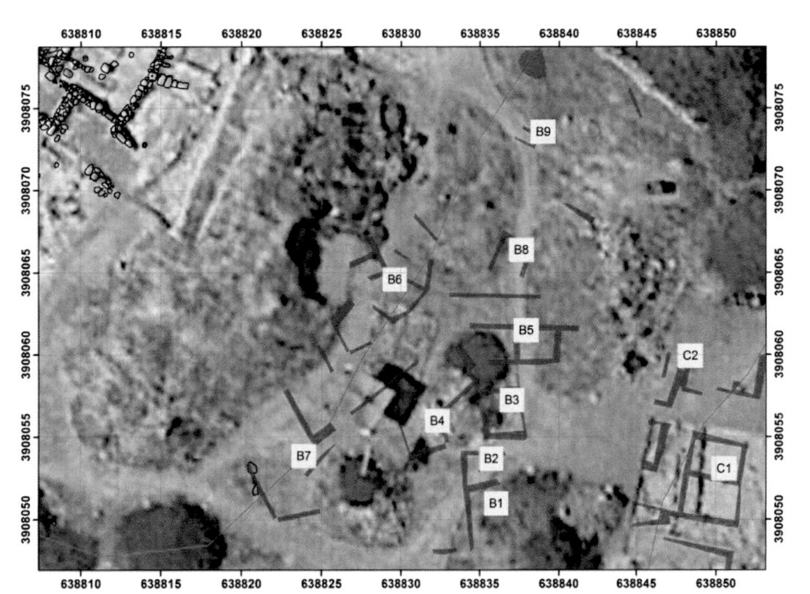

Figure 3: Results of the GPR survey in Area B. Top left: X direction survey. Rectified depth slice of 50cm. Top right: X direction survey. Rectified depth slice of 80cm. Bottom: Diagrammatic interpretation of the GPR reflectors.

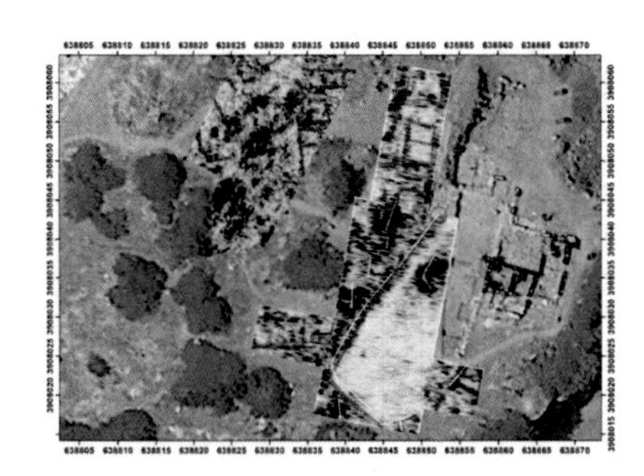

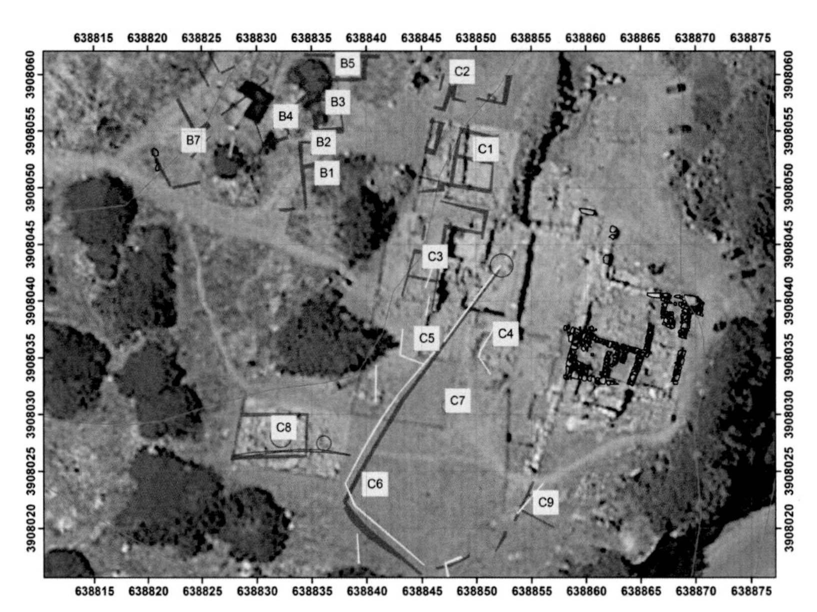

Figure 4: Results of the GPR survey in Area C. Top left: 500MHz antenna survey along the X direction. Rectified depth slice of 60cm. Top right: 250MHz antenna survey along the X direction. Rectified depth slice of 60cm. Bottom: Diagrammatic interpretation of the GPR reflectors.

to the west and to the south by linear segments that meet at a right angle to the SW. A similar linear segment is also evident to the south-east of the court, where Room C9 was also partly outlined. The latter segment probably extends north up to the façade of the rooms excavated to the east of the court, which however follow a clearly distinct orientation.

Finally, GPR and magnetic data allowed the identification of an independent rectangular building (C8) in the south-western part of Area C. The magnetic data also indicate a couple of extreme magnetic features (thermal targets suggesting residues of burning) in the same region.

Final Remarks

Independently of the past and recent anthropogenic interventions to the site and the increased levels of noise, the high resolution manifold geophysical approach was able to indicate a number of candidate targets. In this respect we have to emphasize a) the extent of the structural remains to the north of building CD (Area A); b) the sparse disturbed architecture in Area B; and c) the irregularly shaped court surrounded by built features in Area C.

In terms of the evaluation of the geophysical methods, GPR techniques proved the most effective in delineating the horizontal and vertical extent of the architectural remains. The application of very high resolution sampling (25-50cm x 2.5cm) was crucial to mapping the buildings. Magnetic measurements (50x12.25cm) were able to identify thermal targets and partly confirm the results of the GPR technique. Soil resistance measurements (Areas A and C) were the least successful, independent of the sampling resolution (1x1m or 50x50cm), most probably due to the very dry conditions of the subsoil.

Bibliography

Driessen, J., Schoep, I., Anastasiadou, M., Carpentier, F., Crevecoeur, I., Déderix, S., Devolder, M., Gaignerot-Driessen, F., Jusseret, S., Langohr, C., Letesson, Q., Liard, F., Schmitt, A., Tsoraki, C. and Veropoulidou, R. (2012) *Excavations at Sissi III. Preliminary Report on the 2011 Campaign* (Aegis 6), Louvain-la-Neuve.

Acknowledgements

Geophysical exploration at Sissi was made possible thanks to the Andante Travels Archaeology Award 2016.

What you see is what you get? Complimentary multi-scale prospection in an extant upland landscape, Yorkshire Dales National Park, UK

Mary K Saunders[1]

[1]University of Bradford, Bradford, UK

m.k.saunders@bradford.ac.uk

In the UK, the use of geophysical survey in the context of upland landscape archaeology remains unusual. Cost, difficulties in access, terrain, methodological challenges and the lack of developer funded projects, combine to create a paucity of work in these kinds of areas. In places where a high number of extant features are present, it is also easy to believe that *everything* is visible and that surface recording methods, particularly walkover survey, high resolution LiDAR or photogrammetry can provide a complete picture of the archaeology present.

Within the uplands of the Yorkshire Dales National Park, UK, the use of geophysical survey is uncommon, in line with the pattern observed in other upland parts of England in Jordan's 2009 regional overview (Jordan 2009, 80). Within the Yorkshire Dales, archaeological investigation in advance of commercial developments, such as water pipelines or proposed roads, has almost always taken the form of walkover survey (Saunders 2017, 28-31).

In many respects, the dominance of this technique is unsurprising given the volume of extant archaeology present in the region and which was recorded in some detail through aerial photographic transcription by the Yorkshire Dales Mapping Project (YDMP) in the late 1980s (Horne and MacLeod 1995). There does, however, appear to be a latent assumption that because so many archaeological features remain visible, that there is less of a need to undertake subsurface investigation.

This project involved the use of multi-method, multi-scale prospection over approximately 9 ha of what has traditionally been interpreted as a well preserved prehistoric field-system near Grassington, Yorkshire Dales National Park, UK (Fig. 1). The importance of this area was first recognised by a group of antiquarians during the late 19th century (Harker 1892, Speight 1895) and was most well studied during the early and mid 20th century by Arthur Raistrick (Raistrick and Chapman 1929; Raistrick 1937). Despite being so close to several large universities, academic interest in the area has never really been forthcoming. The survey area lies at a height of 330m O.D. and is situated on a limestone terrace high above the River Wharfe. Currently used as rough grazing, the area was improved with the addition of lime following 19th century enclosure. Although vegetation and ground conditions were reasonable, the survey area is littered with extant archaeology, predominantly in the form of stone features of various types (Fig. 2).

Transcriptions from the YDMP indicated that the survey area contained part of one of the many large-scale field-systems which characterise the archaeology of the Yorkshire Dales and this new research began with the assumption that a single field- system was present. Initially, the project utilised broad-scale magnetic susceptibility and a non-standard approach to electromagnetic induction (EMI) survey, in an effort to identify zonality and function evidence within this field-system, however, walkover survey immediately showed that the YDMP transcription vastly underestimated the number of extant features present. Following high resolution magnetic survey, the completion of a detailed GPS based walkover survey, together with the interrogation of 25cm LiDAR and landscape scale photogrammetry, the sheer complexity and more interestingly, the likely temporal depth, present within the survey area became evident.

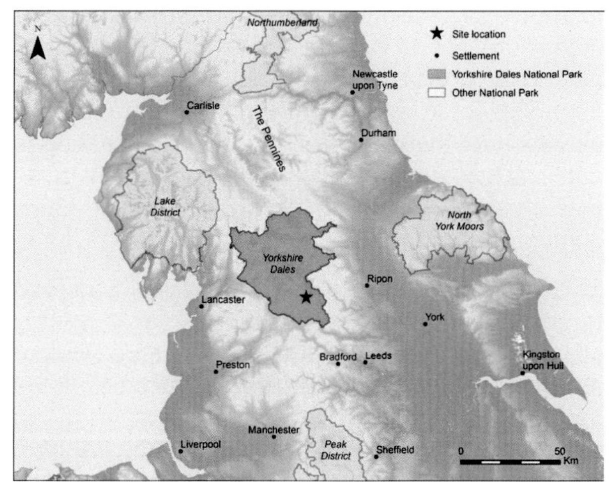

Figure 1: Location of the Yorkshire Dales National Park on the high ground of the Pennines.

Figure 2: Daniel Watkeys undertaking EMI survey amongst the complex extant archaeology.

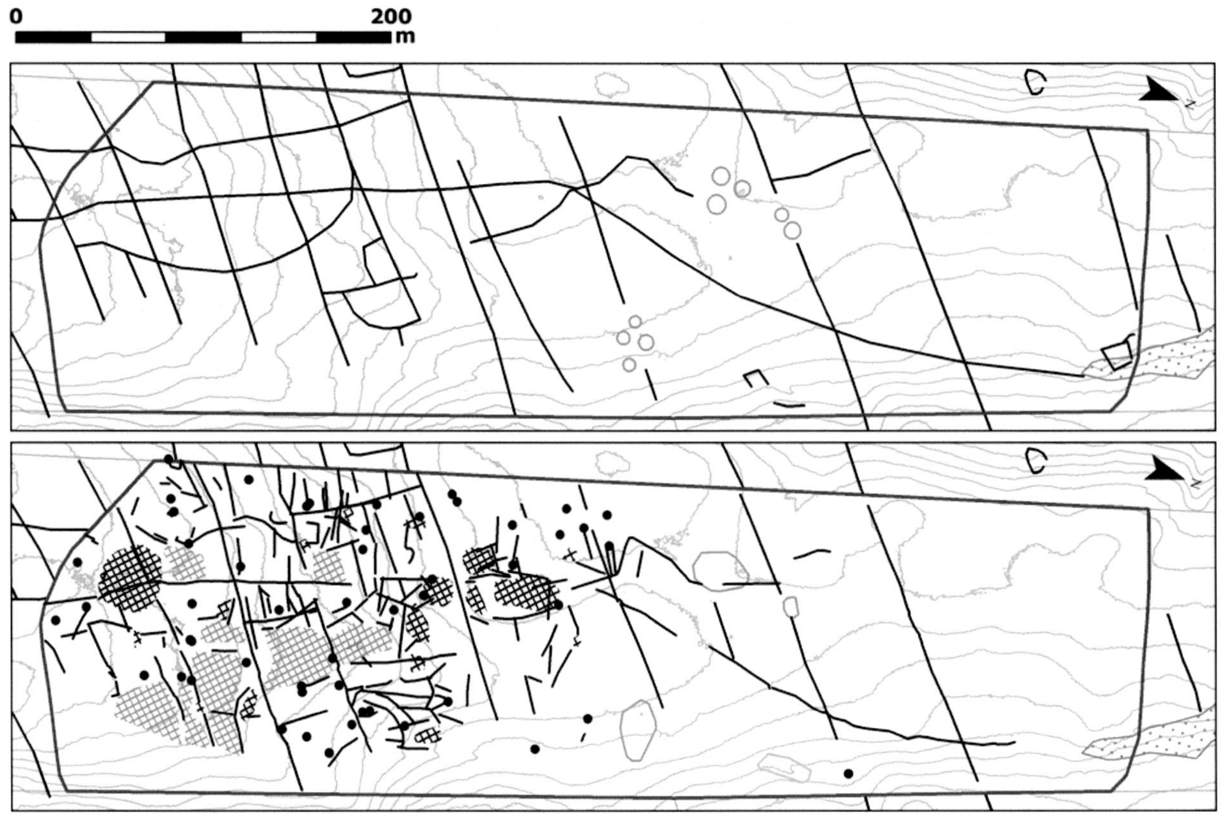

Figure 3: Comparison of the original YDMP transcription (top) and results of the walkover survey (bottom).

A comparison of the YDMP transcription and full walkover results are shown in Figure 3. The picture becomes even more involved when the results from all datasets are combined.

Although LiDAR and photogrammetry provided information beyond the limits of the survey area, they did not pick out the level of detail afforded through thorough walkover survey, while conversely, this intensive on-the-ground investigation was not useful in the identification of landscape scale features easily seen in the former datasets. In particular, the supposed field-system was readily apparent in the LiDAR and photogrammetry, however the walkover survey and magnetic survey indicated the presence of many layers of features and anomalies and the continued remodelling and re-use of the survey area. This then fed back into the interpretation of the broad-scale geophysical survey results – when taken in the context of the higher resolution data, the interpretation of what appeared initially to be evidence of zones of distinct activity within a singular 'field-system' became much less clear cut.

The survey area does not contain part of a single field-system, instead there is a whole series of systems, several elements of which appear to have been constructed many hundreds of years apart. Amongst these agrarian features, other types of site are also present, most notably funerary monuments and some suggestion of settlement. Because so many different datasets were collected or utilised, it was also possible to begin to pick apart the archaeological phasing of the survey area, for example, physical relationships between features were present in some places and these could be extrapolated elsewhere, while geophysical anomalies could only relate to features constructed at a certain point within the history of the survey area because of their relationship to extant remains. Although it proved impossible to use geophysical survey as a means of identifying zonality within any single phase of field-system, the geophysical responses, principally from the magnetic survey, gave an added dimension to the interpretation of the survey area, in particular by providing evidence for domestic activity or intensive manuring. This upland landscape was a continually evolving area, potentially with its roots in the Neolithic and has been subject to reuse, remodelling, clearance and alteration for thousands of years. It is fascinating to note just how much additional information this multi-method approach adds to the original oblique aerial photographic transcription provided by the YDMP.

In this environment, although physically difficult to undertake, geophysical surveys, at a variety of scales, provided vital pieces in a much bigger puzzle. The surveys were always intended to be part of a wider archaeological interpretation of the survey area and to provide a springboard from which theories surrounding its development, usage and demise could be constructed. Ultimately, the aim of this work was to develop a new hypothesis for the evolution of prehistoric society in the Yorkshire Dales.

The datasets used in this project proved entirely complementary to each other and crucially, in terms of future upland research, there was no single technique that provided all the answers. Although this project relied heavily on walkover data, the same rich archaeological tapestry would not have been revealed without combining this with both multi-scale geophysical survey and remote sensing data. In an upland environment such as this, in which extant archaeology is prevalent, the use of such a multi-method approach is clearly warranted. Here, the additional levels of returned information justified the associated time and cost implications, but in landscapes where a single blanket methodology will never be appropriate, the challenge for upland archaeologists, particularly in the commercial sector, is finding a way in which this combined approach can be made adaptable, workable and acceptable to clients.

In our wonderful uplands, what you see is most definitely not always what you get.

Bibliography

Harker, B. J. (1892) Discovery of Pre-historic remains at Grassington, in Craven, Yorkshire. *The Antiquary* **26**: 147 – 9.

Horne, P. and MacLeod, D. (1995) *Yorkshire Dales Mapping Project: A report for the National Mapping Project (PDF version 2011).* RCHME (unpublished).

Jordon, D. (2009) How effective is geophysical survey? A regional review. *Archaeological Prospection* **16**(2): 77 – 90.

Raistrick, A. and Chapman, S. E. (1929) The lynchet groups of Upper Wharfedale, Yorkshire. *Antiquity* **3**(10): 165 – 181.

Raistrick, A. (1937) Prehistoric cultivations at Grassington, West Yorkshire. *Yorkshire Archaeological Journal* **33**(2): 16-74.

Saunders, M. K. (2017) Walking through time: A window onto the prehistory of the Yorkshire Dales through multi-method, non-standard survey approaches, PhD Thesis, School of Archaeological Sciences, University of Bradford (unpublished).

Speight, E. E. (1895) Upper Wharfedale Exploration Committee. First annual report (1893). Communicated January 16th 1894 (with an additional note dated March 1895). *Proceedings of the Yorkshire Geological and Polytechnique Society* **1891-1894 XII**: 374 – 384.

A king and his paradise? A major Achaemenid garden palace in the Southern Caucasus

M Scheiblecker[2,3], J W E Fassbinder[1,2], F Becker[1], A Asăndulesei[4], M Gruber[3] and K Kaniuth[3]

[1]Bavarian State Dept. of Monuments and Sites, Archaeological Prospection, Munich, Germany; [2] Ludwig Maximilians University, Dept. Earth and Environmental Sciences, Section Geophysics, Munich, Germany; [3]Ludwig Maximilians University, Institute of Near Eastern Archaeology, Munich, Germany; [4]Alexandru Ioan Cuza University Research Department – Field Sciene, Iaşi, Romania

scheiblecker@geophysik.uni-muenchen.de

The Achaemenid Empire (539-331 BC) had an immense impact on cultures from the Ancient Near East. For more than 200 years, the Achaemenids controlled the entire Near East from Egypt and Anatolia to the Pamir and Indus. However, although its influence and importance on the ancient cultures is without question, it has regularly been referred to as "elusive" or "archaeologically invisible" (Wiesehöfer 2001, Sancisi-Weerdenburg 1990). Outside its centres Persepolis, Pasargadae, and Susa with its monumental stone buildings, the Achaemenid architecture is difficult to grasp archaeologically.

Archaeological excavations from Dr. Florian Knauss (Staatliche Antikensammlung und Glyptothek, Munich) and Prof. Dr. Ilyas Babaev (Azerbaijan Academy of Sciences) in the Southern Caucasus at Karačamirli (Azerbaijan, Fig. 1) started in 2006, 35 years after the discovery of an Achaemenid column base (Knauss 2007). These investigations conclusively proved that Persian artisans were present even in the remote parts of the empire (Knauss et al. 2013). A huge residence with palace and garden site was discovered, comparable to Pasargadae in size and shape (Figs. 2 and 3, Benech et al. 2012). Since 2013, an extensive survey by Dr. Kai Kaniuth (Ludwig Maximilians University, Munich) has been up these ground breaking studies, both archaeological and geophysical.

According to the ceramic finds, the site Karačamirli was occupied during the whole Achaemenid period, with earlier components from the Iron Age period. It is the location of a major Achaemenid palatial structure, which has been under excavation since 2006 by Dr. Florian Knauss and Prof. Dr. Ilyas Babaev (Knauss *et al.* 2013). The site covers an area of roughly 4 square kilometres, of which more than 1.5 square kilometres were covered by geophysical surveys (Fig. 2).

Hence the aim of our survey was to further investigate the surrounding landscape through archaeological and geophysical means in order to clarify the integration of this building complex within the micro-region, both structurally and chronologically, and to examine further possibilities for a continued examination of the site as a whole. The primary prospection areas from 2013 were measured by a Scintrex caesium magnetometer, and are located in the north and adjacent to the Palace area. Further prospecting areas were selected simply according to the accessibility and the suitability for a high resolution magnetometer prospection. All areas were scanned with a sampling interval of 10 x 50 cm. In order to speed up the measurements in 2015 and 2016, survey areas were extended and measured by two 4 probe Förster magnetometers with a sample interval of 10 x 50 cm. For a better and fuller understanding of the magnetometer

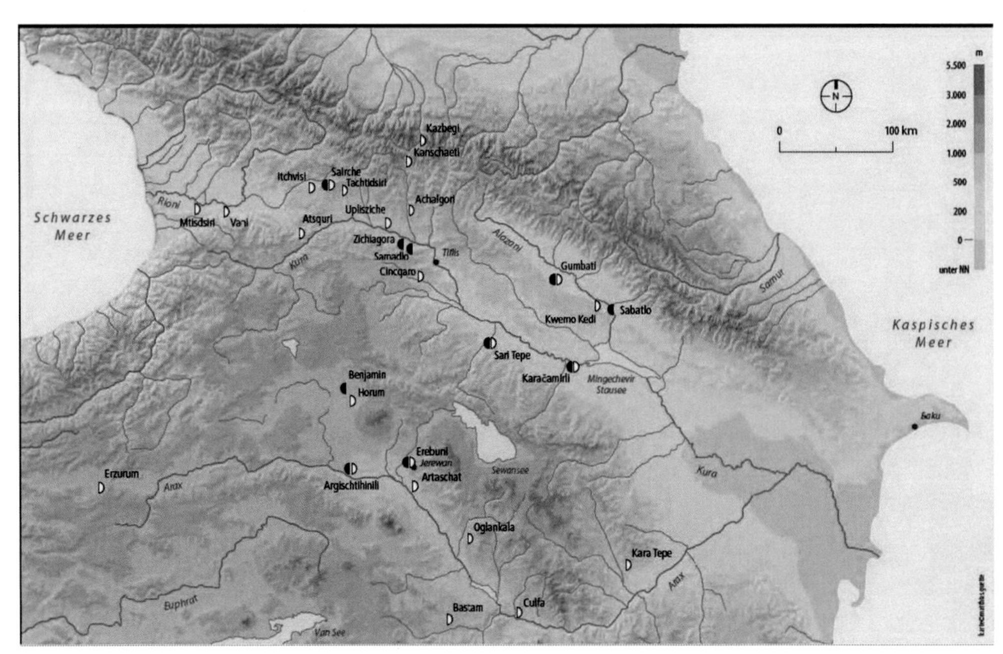

Figure 1: Achaemenid sites in the Caucasus region (Knauss *et al.* 2013, 3, Fig. 2).

Figure 2: Karačamirli. Satellite overview of the total survey area overlaid by the grey shade images of the magnetometer survey areas. Note: grey scale patchwork is due to application of different instruments with different sensitivities/dynamics (caesium and fluxgate magnetometer).

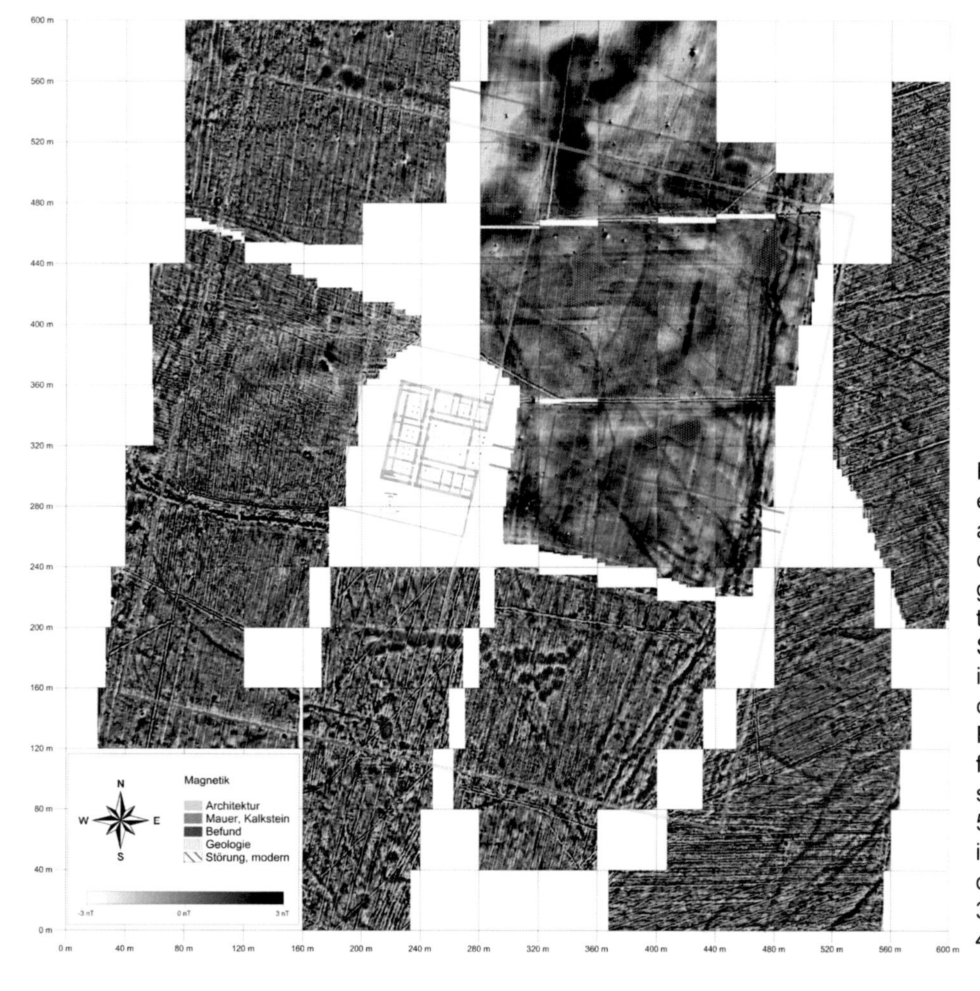

Figure 3: Karačamirli. The excavated garden palace and the inner enclosure of the Achaemenid garden site. Caesium total field magnetometer Scintrex, SMG-4 special in duo-sensor variometer configuration and Ferex Förster, four sensor fluxgate magnetometer, sampling density 25 x 50 cm resp. 10 x 50 cm, interpolated to 25 x 25 cm, dynamics +/- 12 resp. 3 nT in 256 grey scales, 40 m grid.

Figure 4: Karačamirli. Excavation trench across the surrounding wall of the garden palace.

measurements survey, we complemented the survey with some trench excavations and by additional mineral magnetic and susceptibility measurements of typical soil layers (Fig. 4).

Due to increasing and extensive cultivation of the land, the archaeological finds are considerably affected and threatened. The resulting magnetometer images thus at first sight dominated by ploughing furrows and small but regular irrigation canals.

Nevertheless, a detailed sophisticated analysis of the results reveals for the first time the complete layout of the site. The inner core of the palace site was enclosed by a mudbrick wall of exactly 420 x 420 m. The foundation of the wall consists of gravels (10-20 cm in diameter) which appear in the magnetogram by their strong magnetic remnant magnetization and high magnetic susceptibility. Furthermore we detected the very last traces of an alley by a slight negative compaction of the soil running from the gate in the east (excavated, Babaev et al. 2006) symmetrically to the west. The western gate however is extremely destroyed by modern ploughing and can be only hypothesized.

Further features in the surroundings of the Palace are extremely rare. Several pits and remains of pit houses or storage cellars could belong very probably to the ancient building lots which were only temporary in use. In addition, the data reveals numerous natural riverbeds and ancient streamlets, and also ancient irrigation canals which belong very probably to the Achaemenid garden site, but could also have been in use for longer periods. Inside the garden we found a further wall, which divides the garden area from north to south, and numerous old, and some more recent, irrigation canals. Outside of the garden enclosure, in the north and in the north-east, we discovered more, previously unknown, monumental buildings. These buildings are oriented symmetrically with respect to the ground plan of the Archaemenid garden and palace structures, indicating that these buildings formed a significant part of this site.

Bibliography

Babaev, I., Gagošidse, I. and Knauss, F. (2006) Ein Perserbau in Azerbajdžan. Ausgrabung auf dem Ideal Tepe bei Karačamirli 2006. Erster Vorbericht. *Archäologische Mitteilungen aus Iran und Turan* **38**: 291-330.

Benech, C., Bouchalat, R. and Gondet, S. (2012) Organisation et aménagement de l'espace Pasargade. *ARTA* **2012**(3): 1-37.

Knauss, F. (2007) Ein Perserbau auf dem Ideal Tepe bei Karacamirli (Aserbaidschan). *ARTA* **2007**(2): 1-51.

Knauss, F., Gagošidse, I and Babaev, I. (2013) Karačamirli: Ein persisches Paradies. *ARTA* **2013**(4): 1-28.

Sancisi-Weerdenburg, H. (1990) The quest for an elusive Empire. In Sancisi-Weerdenburg Kuhrt (eds.) *Centre and Periphery*, Achaemenid History 4, Leiden: Netherlands Institute for the Near East, 263-274.

Wiesehöfer, J. (2001) *Ancient Persia.* London, New York: I. B. Tauris Publishers.

Acknowledgements

The authors appreciate funds given from the Ludwig Maximilians University Munich, Deutsche Forschungsgemeinschaft and Fritz-Thyssen-Stiftung.

Large-scale high-resolution magnetic prospection of the KGAs Rechnitz, Austria

Hannes Schiel[(1)], Wolfgang Neubauer[(1,3,4)], Klaus Löcker[(1,2)], Ralf Totschnig[(2)], Mario Wallner[(1)], Tanja Trausmuth[(1)], Matthias Kucera[(1)], Immo Trinks[(1)], Alois Hinterleitner[(1,2)], Alexandra Vonkilch[(1)] and Martin Fera[(1,3)]

[(1)]Ludwig Boltzmann Institute for Archaeological Prospection and Virtual Archaeology,Vienna, Austria; [(2)]Zentralanstalt für Meteorologie und Geodynamik, Vienna, Austria; [(3)]Vienna Institute for Archaeological Sciences, Vienna, Austria; [(4)] University of Vienna, Vienna, Austria

hannes.schiel@archpro.lbg.ac.at

Introduction

The small town of Rechnitz is located close to the Austrian-Hungarian border in south-east Austria's federal state of Burgenland. During the systematic analysis of vertical aerial photographs taken by the Austrian Airforce by the Austrian Arial Photography Archive some other characteristic shapes were detected and confirmed during reconnaissance flights. Two circular enclosures, known as "*Kreisgrabenanlagen*" (KGA) were located within 500m distance. In 2014 the field that covers the western of these two KGAs was surveyed with a motorised multichannel magnetometer system by the Ludwig Boltzmann Institute for Archaeological Prospection and Virtual Archaeology (LBI ArchPro) in cooperation with the Zentralanstalt für Meteorologie und Geodynamik (ZAMG). Around the circular enclosure the prospection data showed an area with traces of a presumably multi-phased settlement from the early Neolithic, as well as workshop areas dating to the Iron Age. In this first survey campaign it was not possible to either prospect the entire KGA nor to cover the entire settlement area or to define its limits. Therefore, in 2016 the LBI ArchPro started a large-scale geophysical prospection survey using magnetometry and high-resolution ground penetrating radar (GPR) measurements and defined an initial area of interest covering one square kilometre, of which 52.5 ha were surveyed over the course of six days in late October 2016.

Method

The archaeological prospection surveys were conducted with two different magnetometer systems and two different GPR systems, using sensors of different manufacturers. More than 95% of the surveyed areas were mapped by magnetic prospection. A motorised 8-channel Fluxgate-type magnetometer array mounted on a custom built, non-magnetic cart equipped with eight Förster FEREX CON650 gradiometer probes, a 10-channel EasternAtlas AD converter and a ruggedized field computer for data acquisition using in-house developed data logging and navigation software was used in combination with RTK-GNSS for exact data positioning. The system is towed by an ATV Quad bike. This survey setup permits for a sampling resolution of 25 cm cross-line and 10 cm in-line direction at speeds of upto a maximum 20 km/h. For appropriate data positioning at this speed the RTK-GNSS was used with a 5 Hz NMEA string output rate and a PPS fed as time-stamp marker into the magnetic data. Data is recorded in XML format and visualized as coverage map or simple greyscale image in real-time on a screen in front of the operator.

The second motorized magnetometer system used at Rechnitz is based on eight Scintrex CS3 optically pumped total field Cesium magnetometers connected via two custom built four-channel AD converters from PicoEnvirotec. This system was likewise mounted on a non-magnetic cart pulled by an ATV Quad bike, with data positioning and acquisition identical to the above described Förster/ Fluxgate setup.

In addition to the magnetic prospection, GPR surveys were conducted at selected areas, using a 16-channel 400 MHz MALÅ Imaging Radar Array (MIRA). It consists of 17 radar antennas (9 transmitters, 8 receivers) arranged in two rows and shifted by half the antenna width. Every receiver records the reflections of their two neighbouring transmitter antennas, resulting in a resolution of four by eight centimetres at a speed of 13 km/h using a four-fold trace stacking factor. The Mira system was operated using the same RTK GNNS for positioning as the magnetometers.

The second GPR system used at the site of Rechnitz was a manually pushed, RTK-GNSS positioned, three-channel SmartCart from Sensors and Software equipped with three pulseEKKO PRO 500 MHz antenna-pairs at 25 cm crossline spacing. While data acquisition took place on a Sensors and Software Digital Video Logger, an additional tablet computer was used with the LBI ArchPro navigation solution LoggerVis in order to permit efficient area mapping.

All data collected at Rechnitz was processed and visualised with the software ApMag and ApRadar a collaborative in-house development of the LBI ArchPro and its partner ZAMG.

Results

The most prominent structures in the collected magnetic data are unarguably the two double ditch KGAs dating to the Middle Neolithic period 4850/4800 – 4600/4500 BC (Neubauer 2012;

Figure 1: Fluxgate magnetometry data from the Rechnitz survey. Dynamic range: -4nT (white) to +6nT (black). (LBI ArchPro, Hannes Schiel).

Figure 2: Archaeological interpretation of the magnetic prospection data. (LBI ArchPro, Hannes Schiel)

2010). The two KGAs consist of two concentric, circular, normally v-shaped and initially 4 to 6m deep ditches with diameters of about 82 and 96 meters. They show concentric internal wooden palisades or rings of timber uprights. Access to the interior was only possible by narrow entrances, interruptions in the ditches and the palisade rings. The monuments show at four radial opposed entrances, more or less regularly arranged formal routes of admission into the central space.

The majority of buildings and settlement pits visible in the data can be connected to a large settlement made up of longhouses which seems to predate the two KGAs. These timber buildings are typical Early Neolithic longhouses related to the Linearband Culture (LBK) (Sauer 2006). A part of the settlement is surrounded by a very prominent double ditched earthwork that seems to be respected by most of the structures. It might be interpreted as a fortification, probably dating to the final phase of LBK which did see many conflicts as observed at several other contemporary sites. The superposition of longhouses will make it possible to discern an initial phasing of the site to be verified by targeted future excavations.

Besides the three large circular structures and the extensive settlement, whose complete extent is still unknown, two more interesting structures

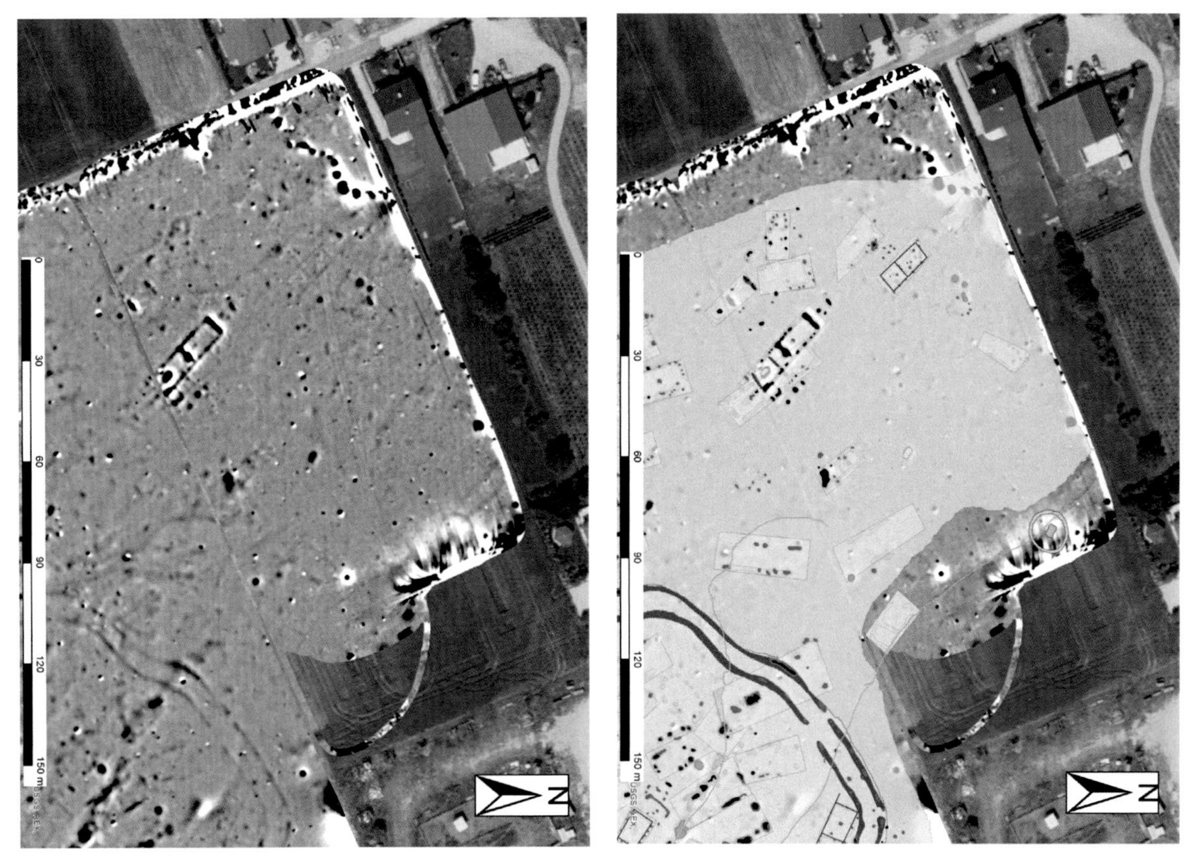

Figure 3: Detail showing the burnt longhouse to the east of the western circular earthwork. For figure legend see Figure 1. (LBI ArchPro, Hannes Schiel)

are prominent in the magnetic data. The first is a longhouse north of the western KGA obviously destroyed by fire. The longhouse, which must have been rebuilt after conflagration, offers a very detailed picture of this type of building due to its thermoremanent magnetisation, and even provides a clue where the devastating fire started. The second observed structure is a burial mound enclosed by a circular ditch on the same field south east of the longhouse, most likely dating to the Iron Age. The mound will be further investigated by GPR measurements and will be excavated in 2017.

The area of Rechnitz is also well known for the mass production of the famous *Ferrum Noricum* in the second and first century BC at the end of the Iron Age. Therefore it is not surprising to find furnaces, slag tips and workshop areas all over the site. Also, the Romans left their traces in form of remains of a single building. Furthermore there are also some medieval field structures recognisable in the data.

Conclusion

Within the prospected area of 52.5 ha numerous structures dating to different periods shed new light on the spatiotemporal development of the area of Rechnitz. Large-scale, high-resolution archaeological prospection helps in recognising extensive structures while permitting the imaging of details, such as superimposed houses or small workshop huts. By conducting such extensive surveys with efficient motorised systems the research as well as local community gains means to better understand the development of their region through times, fostering their interest in protecting endangered historical sites from destruction.

Bibliography

Manoschek, W. (ed.) (2009) *Der Fall Rechnitz. Das Massaker an Juden im März 1945.* Vienna: Braumüller.

Neubauer, W. (2012) Kreisgrabenanlagen – Middle Neolithic Ritual Enclosures in Austria 4800-4500 BC. In A. Gibson (ed.) *Enclosing the neolithic. Recent studies in Britain and Europe British.* BAR International Series 2440.

Neubauer, W. (2010) Archäologische Auswertung der systematischen Prospektion. In P. Melichar and W. Neubauer (eds) *Mittelneolithische Kreisgrabenanlagen in Niederösterreich. Geophysikalisch-archäologische Prospektion – ein interdisziplinäres Forschungsprojekt*, Mitteilungen der prähistorischen Kommission der österreichischen Akademie der Wissenschaften **71**: 56-135.

Sauer, F. (ed.) (2006) *Fundstelle Rannersdorf. Die archäologische Grabung auf der Terasse der S1. Wien.* Bad Vöslau : Bundesdenkmalamt, Wien. 10.

When the time is right: the impact of weather variations on the contrast in earth resistance data

Armin Schmidt[(1)], Robert Fry[(2)], Andrew Parkyn[(3)], James Bonsall[(4)] and Chris Gaffney[(5)]

[(1)]GeodataWIZ, Remagen, Germany; [(2)]University of Reading, UK; [(3)]AECOM, Texas, USA; [(4)]IT Sligo, Ireland; [(5)]University of Bradford, UK

A.Schmidt@GeodataWIZ.com

Introduction

The successful detection of archaeological features through electrical methods of survey requires a contrast in electrical conductivity between these features and their surrounding soil matrix. Several studies have attempted to identify the 'right time' to do earth resistance surveys, but with limited success. Practioners need a simple method using widely available precipitation data for the forecasting of good electrical contrast in archaeological electrical surveys. By applying a simplified soil and moisture model to seasonal survey series in different years and at different sites we show how an index formed from the last 15 days of precipitation and its monthly average can be used as a predictor of contrast in many archaeological situations.

Background

Although the relationship between soil moisture and electrical conductivity is complicated, the subject has been widely studied and data for many soils are now available in the literature. Far more difficult to evaluate is the influence of weather parameters (e.g. rain, wind, temperature) on soil moisture. In particular, the *actual* evapotranspiration and downward percolation are very difficult to estimate; a general model that describes all relationships in detail is, therefore, very difficult to formulate. Instead we developed a simplified model that can predict broad trends as a useful alternative.

In this model, the soil electrical conductivity is represented by a factorial expression:

$$\sigma_A = \sigma_m \, a \, S \, \phi$$

where σ_A is the apparent conductivity of the measured soil, σ_m is the conductivity of the interstitial moisture, a is the admittance ratio that describes any non-linear relationship between moisture and conductivity (particularly for clay soils; sand by contrast would have a=100%), S is the saturation (i.e. the percentage filling of pores with moisture) and ϕ the porosity (i.e. the relative amount of pore-space in the soil). For a site with a ditch that gradually silted from the surrounding soil (resulting in a different pore structure) the conductivity response simplifies to $\sigma/\sigma_o = (S\,\phi) / (S_o\,\phi_o)$, where the subscript 0 indicates measurements at a reference position. This confirms that even on a site where the saturation of a ditch and its surrounding soil are the same (e.g. fully saturated after prolonged rain), the conductivity response does not disappear due to the difference in porosity. To describe the contrast seen in an earth resistance survey the symmetrical conductivity contrast is introduced as $(\sigma-\sigma_o)/(\sigma+\sigma_o)$, where the apparent conductivities are derived from earth resistance measurements. This quantity varies symmetrically between +1 and -1 (i.e. positive and negative contrasts are described in the same way) and it is the negative of the similarly formed symmetrical resistivity contrast. This symmetrical contrast is used throughout this study for the comparison of different sites.

The net amount of moisture in the soil, and hence its conductivity, depends on the amount of precipitation that enters the soil, on the evapotranspiration of moisture through bare soil and through crops, and on downward drainage. The simplified model is mainly concerned with *changes* of the net moisture and therefore evapotranspiration and downward drainage can be assumed to occur at a nearly constant rate, while the rainfall is taken from a local weather record. It is often mentioned by practitioners that the best time for detecting a ditch in earth resistance data is after a dry period that follows after considerable rain. This can be explained by the higher net amount of moisture that a ditch retains compared to the surrounding soil even after a similar saturation during rain (e.g. nearly 100%), due to the frequently larger pores. In addition, the silted bottom fill of some ditches impedes downward drainage of the trapped moisture, unlike the usually better draining surrounding soil. To quantify these considerations a precipitation ratio $p(d)$ was introduced that is formed as the ratio of the average rainfall in the last d days prior to a survey date and the average rainfall over the whole month prior to the survey. If this ratio is smaller than one it indicates that the last d days were drier than the period before. This measure can then be correlated with the symmetrical resistivity contrast calculated for a sequence of measurements to test its suitability as a predictor of the resistivity contrast: if the correlation is positive it means that a negative resistivity contrast (e.g. a wet ditch) is correlated with a period of dry weather after some rain.

To evaluate the suitability of the precipitation ratio $p(d)$ for the prediction of electrical contrast it was calculated for different numbers of days d prior to each survey event for nine case studies in which the variation of earth resistance data was monitored on a monthly basis over the course of a whole year.

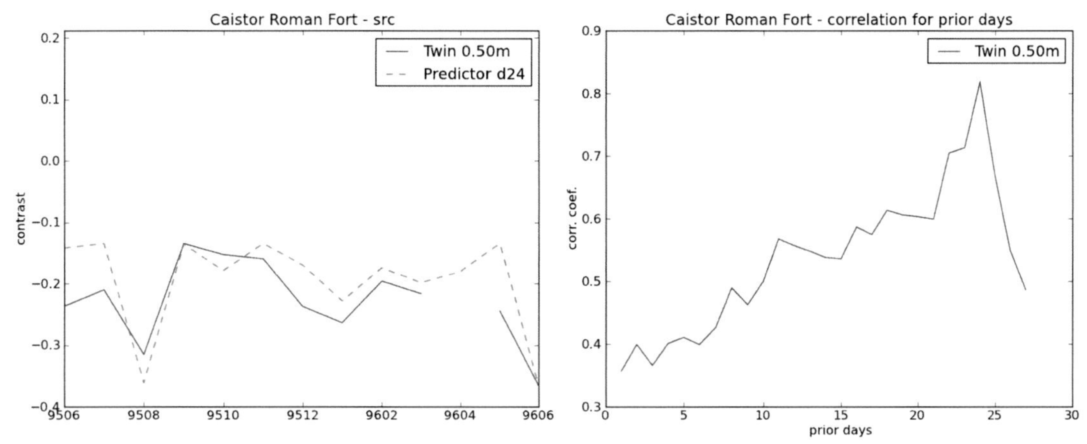

Figure 1: Monthly change of electrical contrast over the ditch at Caistor Roman Fort.

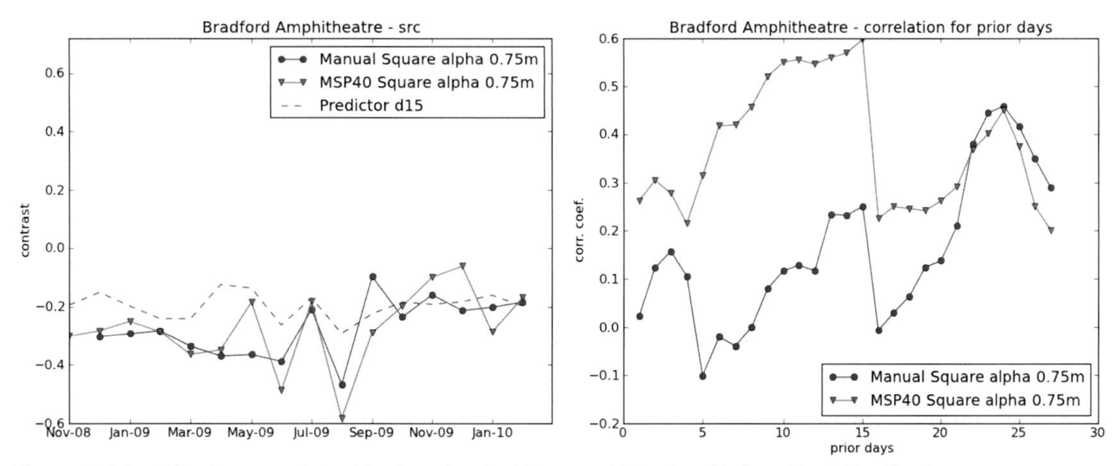

Figure 2: Monthly change of electrical contrast at the amphitheatre, University of Bradford.

Figure 1 shows the results for a large Roman ditch at Caistor, Roman Town, Norwich, UK. A twin-probe pseudosection was measured across the ditch and the symmetrical resistivity contrast was calculated from the 0.5 m twin-probe data. Figure 1a shows the variation of this symmetrical resistivity response (solid line) and a predictor (broken line, scaled to fit the display). Figure 1b graphs the Pearson correlation coefficient between the resistivity contrast and the precipitation ratio *p(d)* for different numbers of days *d* and shows clearly that the correlation is strongest, if the precipitation ratio is taken for day 24, which was used as the predictor in Figure 1a. In other words, if there was less average rainfall in the 24 days prior to a survey than in the week preceding these 24 days then the ditch would have a pronounced lower resistivity compared to the background; and the opposite was true for wet periods.

Similar results are shown in Figure 2 for the modern amphitheatre in the grounds of the University of Bradford, UK. Measurements were undertaken with square arrays of 0.75 m electrode separation, both mounted on a frame and manually inserted and as the wheels of a MSP40 cart by Geoscan Research. The contrast was evaluated from a small area of low apparent resistivity, probably a former waste pit. Figure 2a shows that the precipitation ratio at day 15 (broken line) is a good predictor for these electrical contrasts, even though the two square array measurements are not identical and the predictor shows a slight shift in the spring of 2009. The correlation (Figure 2b) has its absolute maximum at day 15 (as plotted in Figure 2a) and a second maximum at day 24.

Conclusions

Based on the results from eight of the seasonal measurements it was found that the newly introduced precipitation ratio, taken approximately on day 15, is a good predictor for the strength of a resistivity contrast in earth resistance surveys. Calculated from daily rain data it allows practitioners to predict the strength of resistivity responses from buried ditches. However, there will be sites on which such a generalised predictor is not suitable. For example, in the ninth site of this study the correlation between the resistivity contrast and the precipitation ratio was found to be negative, which was attributed to the extremely well draining site conditions, both for the ditch and its surrounding matrix.

SQUID-based magnetic geoprospection: a base technology of multimodal approaches in applied geophysics

M Schneider[1,2], S Linzen[2], M Schiffler[2], S Dunkel[2], R Stolz[2] and D Baumgarten[1]

[1]Institute of Biomedical Engineering and Informatics, Technische Universität Ilmenau, Germany; [2] Leibniz-Institute of Photonic Technology, Jena, Germany; [3]Supracon AG, Jena, Germany

m.schneider@tu-ilmenau.de, michael.schneider@leibniz-ipht.de

Introduction

In the last few years, geomagnetic prospection measurements have oftentimes proven their advantages during large-scale investigations of ground monuments and other archaeological structures which are buried in the subsoil. The passive and non-invasive documentation of extended structures and their mutual as well as inner relations is very important. Especially for the characterization of the contemporary environmental situation and the embedding of structures into the local and regional countryside, an efficient and optimized prospection is very useful.

In this context, geomagnetic investigations can identify structural dependencies between topographical, geological and archaeological formations. But as a method which is dealing with physical potential fields, magnetic geo-prospection includes the disadvantage of ambiguity relating to the causative source distributions. One way to restrict such equivocality problems are two-dimensional measurements of more than one independent component of the tensor which mathematically describes the local Earth's magnetic field gradients. Another possibility is the parallel or simultaneous application of multiple geophysical methods with different operating effects. The latter approach is called *multimodal* and enables *joined* or *constrained* inversion as well as interpretation methods for the acquired data. Such approaches are used to eliminate the mentioned ambiguities. Furthermore, the acquisition of more consistent information enhances the quality of statements about the investigated situation.

After a brief introduction into the applied SQUID-based prospection system, the topic of this contribution firstly covers the advantages of multi-component acquisition of magnetic gradients and the associated benefits of *real gradient* acquisition in comparison to so-called *pseudo-gradients* calculated from total field sensors or vector magnetometers. Beginning with the joint investigation of topographical and magnetic data sets which are simultaneously acquired with the mentioned system, the idea of multimodal prospection is deduced from a methodical discussion. After the mentioned symbiosis, the next logical step towards a fast and efficient prospection is the employment of electrical and electromagnetic methods. These systematic supplements could operate with a minimum of soil contact; therefore, their application is very fast.

Additionally, our previously introduced and innovative inversion algorithms can generate plausible 3D reconstructions of the subsoil situation with the help of our advanced, SQUID-based prospection system. These methods are based on a well-founded theoretical background that contains analytical signature descriptions, geometrical source body parametrizations and numerical optimization algorithms. The discussion is closed by an overview about the supports and improvements of our established inversion algorithms using the mentioned methodical combinations.

Multimodal Methods in Geophysical Context

The basic modality of this contribution is a SQUID-based geomagnetic prospection system which consists of various high-sensitivity magnetic magnetometers and gradiometers (Linzen *et al.* 2007; Zakosarenko *et al.* 1996). Additionally, a differential global positioning system (dGPS) and an inertial unit are integrated into this system. Hence, this sensor configuration enables a fast and precise simultaneous acquisition of different magnetic gradient components together with the localization and orientation of the entire measurement platform. Therefore, effective 3D magnetic anomaly mapping is realized which is very beneficial for archaeological studies.

Figure 1 provides a brief overview of the mentioned system components. The core of this system are SQUID sensors which have a very high sensitivity of below 100 fT/(m*sqrt(Hz)) under realistic field conditions. This sensor type operates very fast with a sampling rate of 1 kHz. Therefore, the data acquisition speed is not limited by the sensors but rather by the operation crew in the all-terrain vehicle which usually tows the measurement system. Aspects of data quality and system stability under various prospection conditions led to an empirically determined mean measurement speed of 20-40 km/h. This acquisition speed results in an average area-efficiency of 10 ha per day. The effectiveness of the system is enhanced by the simultaneous acquisition of at least two different gradient tensor components. Particularly, the used configuration reduces the requirements for a priori information about the estimated situation. Such valuations were

Figure 1: Overview of the SQUID based measurement system of IPHT Jena and Supracon AG which is normally pulled by an all-terrain vehicle. A combination of high-sensitive magnetic sensors, precise localization and orientation reading ensures effective, and high-resolution geo-magnetic prospection measurements for archaeological investigations with a mapping efficiency of up to 100.000 sqm per day.

usually needed before direction sensitive sensors could be successfully applied. Therefore, the risk of missing structures caused by a limited sensor perspective is eliminated with the help of the current sensor configuration.

By means of the mentioned dGPS, the system offers the possibility of a simultaneous topographical mapping of the investigated area in addition to the magnetic maps that are usually called magnetograms. Such an additional local digital terrain model (DTM) supports data interpretation by identifying small topographical relations. The lateral sensitivity of the system relating to the mobile base station is in the range of centimeters whereas the vertical accuracy reaches sub-decimeter values. The dGPS support clarifies structures that could be missed on the measurement sites otherwise. Figure 2 illustrates the comparison of a DTM with the

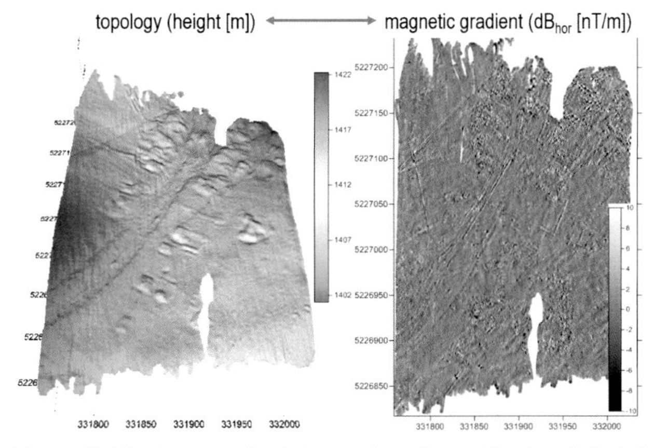

Figure 2: Illustration of a data combination of the local digital terrain model and the associated magnetic map. Both data sets are simultaneously acquired with the presented geoprospection system. The resulting maps are intrinsically georeferenced and could therefore be embedded in all usual GIS systems.

associated magnetogram. Both data sets result from simultaneous measurement from 2009 in Mongolia.

Resulting from the underlying physics, magnetic gradiometry accentuates the signatures of structure edges parallel to the sensor's viewing direction. Therefore, the vertical gradient of the vertical magnetic field component (dBz/dz) would be the most favored sensor arrangement. This configuration enables an associated imaging of the entire top view footprint of an investigated structure. In our case, the geometrical reproduction is realized in a comparable quality by combining both lateral components of the vertical magnetic gradient (dBx/dz and dBy/dz).

As previously mentioned, electromagnetic and geoelectric measurements could enhance the benefits of multimodal geophysical prospection. Such methods insert or induce electrical currents in the subsoil and thus can disclose connections inside or between different subsoil structures. Thereby, extensive electromagnetic surveys preferentially show the lateral connections. Furthermore, line-based geoelectrical investigations illustrate the vertical relations of structural units of the sub-soil.

The combination of these methods with the magnetic prospection permits the reconstruction of structure boundaries as well as inner and outer relations. Hence, the multimodal characterization of the subsoil situation becomes more realistic than a monomodal one.

Inversion of Geophysical Results in Archaeological Context

Magnetograms usually contain several types of signatures, like point-shaped, linear or extended anomalies. This diversity is required to be characterized and classified to evaluate the relevance regarding to the respective archaeological problem. Subsequently, the source structures of the selected anomalies are requested to be characterized or reconstructed. In this context, our innovative approaches for point-shaped or linear signature were presented in Schneider *et al.* (2013, 2014), and new methods for the handling of extended structures were discussed in Schiffler *et al.* (2015, 2016). Each algorithm is based on an associated source model. Such models contain all essential source properties, an adaptable parametrization as well as well-founded assumptions of the magnetic situation. The discussed methods enable the forward calculation of the magnetic field distribution out of the modelled sources. Subsequently, an optimization procedure adapts the synthetic configuration to the acquired data sets.

Finally, Figure 3 presents a chart to illustrate the multimodal support of inversion algorithms

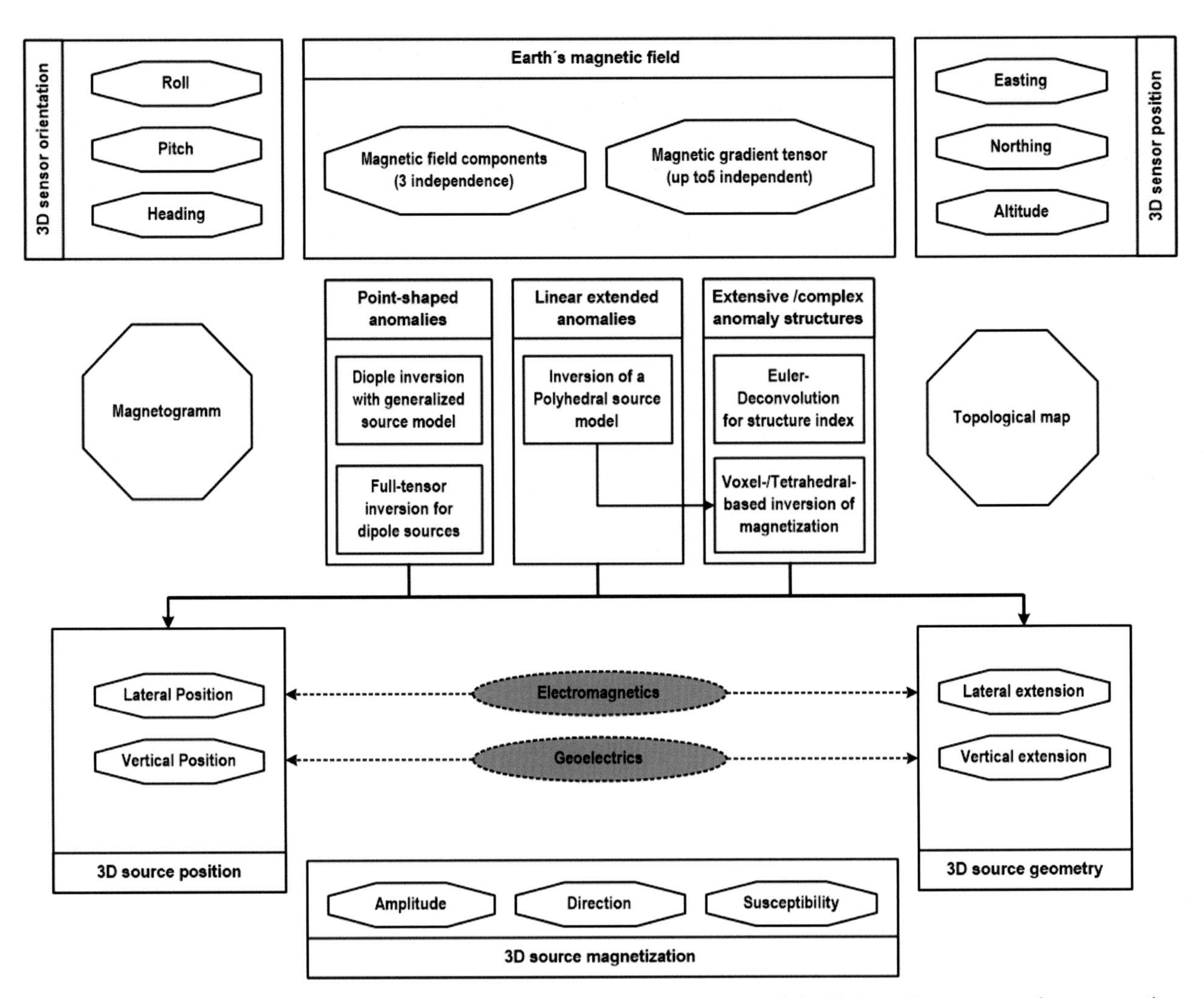

Figure 3: Overview about the possibilities of inversion resulting from the SQUID-based geomagnetic prospection system. The top layer represents the main parts of the acquires data sets which consists of magnetic field data, position and orientation. The central layer illustrates the various inversion algorithms which could be applied to the mentioned data sets. Finally, the bottom layer illustrates the achievable inversion results which are classified in statements about the position, geometry and magnetization of an investigated source structure.

by providing realistic structural boundaries or material properties that are usable as initial values or variation constraints during the optimization processes. The more independent information about the investigated situation are known in preparation of an inversion process the more detailed the analytical description can be formulated and the more realistic are the results of the recalculations.

Conclusion

The advantages of SQUID-based geomagnetic prospection as a base technology for multimodal geophysical prospection approaches are discussed in relation to different field studies. The beneficial combination of topographical and magnetic maps can be impressively illustrated at the city district of the former capital of the Mongolian Empire. Additional multimodal achievements will be presented by means of different case studies in Franconia as parts of the DFG priority program 1630.

Bibliography

Linzen, S., Chwala, A., Schultze, V., Schulz, M., Schüler, T., Stolz, R., Bondarenko N. and Meyer, H.-G. (2007) A LTS-SQUID system for archaeological prospection and its practical test in Peru. *IEEE Transactions on Applied Superconductivity* **17**(2): 750-755.

Schiffler, M., Queitsch, M., Stolz, R., Goepel, A., Malz, A., Meyer, H.-G. and Kukowski, N. (2015) *First inversion results of airborne full tensor magnetic gradiometry measurements in Thuringia.* Hannover, Deutschland: 75. Jahrestagung der Deutschen Geophysikalischen Gesellschaft.

Schiffler, M., Schneider, M., Queitsch, M., Stolz, R., Meyer H.-G. and Kukowski, N. (2016) *Euler Deconvolution using Full Tensor Magnetic Gradiometry Data.* Münster, Deutschland: 76. Jahrestagung der Deutschen Geophysikalischen Gesellschaft.

Schneider, M., Stolz, R., Linzen, S., Schiffler, M., Chwala, A., Schulz, M., Dunkel, S. and Meyer, H.-G. (2013) Inversion of geo-magnetic full-tensor gradiometer data. *Journal of Applied Geophysics* **92**: 57-67.

Schneider, M., Linzen, S., Schiffler, M., Pohl, E., Ahrens, B., Dunkel, S., Stolz, R., Bemmann, J., Meyer, H.-G. and Baumgarten, D. (2014) Inversion of Geo-Magnetic SQUID Gradiometer Prospection Data Using Polyhedral Model Interpretation of Elongated Anomalies. *IEEE Transactions on Magnetics* **50**(11): 6000704.

Zakosarenko, V., Warzemann, L., Schambach, S., Bluethner, K., Berthel, K. H., Kirsch, G., Weber, P. and Stolz, R. (1996) Integrated LTS gradiometer SQUID systems for unshielded measurements in a disturbed environment. *Superconductor Science and Technology* **9**: A112–A115.

Acknowledgments

This work was supported by the German Federal Ministry of Education and Research (BMBF) as part of the Innoprofile Transfer project 'MAMUD' (03IPT605X) and German Research Foundation (DFG, ET 20/8-1 and LI 2724/1-1).

Investigating the influence of seasonal changes on high-resolution GPR data: the Borre Monitoring Project

Petra Schneidhofer[1], Christer Tonning[2], Vibeke Lia[2], Brynhildur Baldersdottir[2], Julie Karina Øhre Askjem[2], Lars Gustavsen[3], Erich Nau[3], Monica Kristiansen[3], Immo Trinks[1], Terje Gansum[2], Knut Paasche[3] and Wolfgang Neubauer[1]

[1]Ludwig Boltzmann Institute for Archaeological Prospection and Virtual Archaeology, Vienna, Austria; [2]Kulturarv Vestfold fylkeskommune, Tønsberg, Norway; [3]The Norwegian Institute for Cultural Heritage Research, Oslo, Norway

petra.schneidhofer@archpro.lbg.ac.at

Introduction

The Viking Age royal burial site of Borre, located in Vestfold county– Norway – some 80 km south of the capital Oslo, hosts nine monumental burial mounds and is the largest assemblage of this type in Scandinavia (Myhre 1992; Myhre 2015)(Fig. 1). In total, at least 50 burials are known from the Borre area dating the site to the late Nordic Iron Age (ca. 600-900 CE). Together with the famous Oseberg (Brøgger *et al.* 1917; Gansum 1995; Nordeide 2011) and Gokstad ship burials (Nicolaysen 1882; Bill *et al. 2013)*, Borre represents a key site for understanding the social elite of Viking Age Scandinavia.

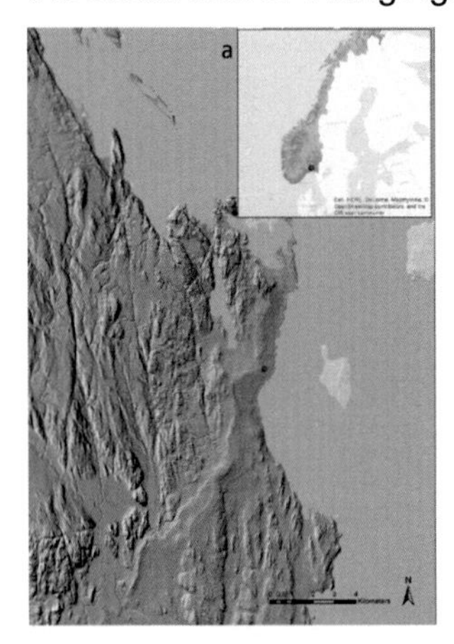

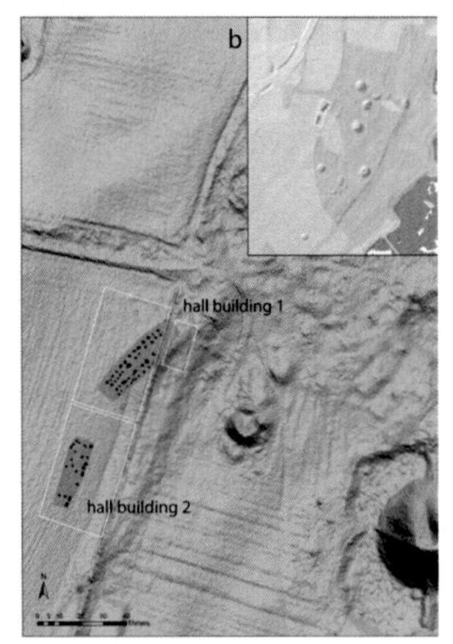

Figure 1: a) Location of Borre in Vestfold County, Norway. b) The three test sites in- and outside of Borre National Park. c) The Borre National Park contains the largest assemblage of monumental burial mounds in Northern Europe.

In 1932, the core area of the Borre site was turned into a National Park and made publicly accessible. Much of the surrounding area of Borre National Park is today agriculturally used. In 2007, test surveys using magnetometry and ground penetrating radar conducted by the archaeological prospection unit of the Swedish Central National Heritage Board (Trinks *et al.* 2007) on behalf of archaeologists from Vestfold County discovered the remains of two large hall buildings in a field just outside of the National Park. This promising discovery, amongst other successful archaeological prospection pilot surveys conducted in Vestfold County (Trinks *et al.* 2010), led to the formation of a research collaboration between Vestfold fylkeskommune (Vfk), the Norwegian Institute for Cultural Heritage Research (NIKU) and the Ludwig Boltzmann Institute for Archaeological Prospection and Virtual Archaeology (LBI ArchPro) with the aim of applying non-invasive geophysical prospection techniques on a broader basis in Norwegian heritage management, rescue archaeology and archaeological research (http://lbi-archpro.org/cs/vestfold/publications.html.).

One of the focal points of this endeavour concentrated on the arable land surrounding Borre National Park. Several small and large-scale GPR surveys have been conducted since 2007. In 2008, GPR test surveys covering 1.5 ha and using a stepped frequency device from 3D Radar were conducted by the system manufacturer in collaboration with the authors. In March 2013, a large-scale GPR survey on solid snow cover was realized with a 500 MHz six-channel Sensors & Software SPIDAR system towed by a snowmobile. This survey resulted in the discovery of another large hall building dating presumably to the Viking Age. Following these promising results, an additional GPR survey was performed in September 2015 after considerable rainfall using a 400 MHz 16-channel MALÅ Imaging Radar Array with a high spatial sampling of 4x8 cm. In stark contrast to the previous surveys, the results did not show any discernible archaeological features. This finding was alarming, in particular regarding the large-scale intended implementation of GPR in heritage management. Therefore, a project was formulated to study the influence of seasonal effects on high-resolution GPR data.

The Borre Monitoring Project

Preliminary investigations using freely available local precipitation data from *eKlima* (Meteorologisk institutt, eklima.met.no/) quickly showed a correlation between increased precipitation rates and limited visibility of the archaeological features. In order to investigate this phenomenon in more detail, the Borre Monitoring Project was initiated in 2016 by Vfk, NIKU and LBI ArchPro. The aim of this project is to establish

Figure 2: Test pit 3 displaying a massive posthole belonging to the remains of one of the Iron Age hall buildings. In order to measure the contrast between surrounding beach deposits and the archaeological material, four soil moisture sensors were installed.

the most suitable conditions for the recording of high-resolution GPR data with respect to the prevailing environmental settings (soils, sediments, geology) and seasonal changes (precipitation, temperature) at Borre. The approach includes systematically monthly repeated GPR measurements of test areas over the duration of 14 months. Simultaneously, ground conditions are monitored for diagnostic parameters, such as soil water content, electrical conductivity and temperature.

Methods

Two test areas were defined immediately outside Borre National Park in a field that has been used for agriculture. They cover 30x50 m each, and include the remains of two of the Viking Age hall buildings discovered in 2007. A third, smaller test area is situated inside Borre National Park, measuring 10x20 m.The test areas are surveyed once every month using a handheld 500 MHz single channel antenna system (Noggin) from Sensors & Software with a spatial sampling of 25x2.5 cm (4-fold trace stacking).

In order to collect measurements of physical properties relevant for the use of GPR surveys, three test pits were dug. Two of the pits focus on small areas that apparently had been free of any substantial archaeological structures, and which were meant to provide the natural background for the Borre area. A third test pit targets the actual contrast between archaeology and natural background (Fig. 2). Water content, electrical conductivity and ground temperature are measured in-situ using soil moisture sensors and real-time data loggers (Decagon). Additionally, a local weather station was installed at the test sites, including precipitation and solar radiation monitoring. During winter, snow heights as well as the depth of frozen ground are measured at regular intervals. Data acquisition will be finished at the end of September 2017, after an operational period of 14 months. Upon removal of the soil moisture sensors, samples for soil and sedimentological analyses, including chemical and mineralogical composition and grain size distribution, will be collected.

Preliminary Results

So far, a total of seven monitoring GPR surveys have been undertaken at all three test sites at monthly intervals. Preliminary results show a significant increase in visibility of the anomalies associated with the Iron Age hall buildings in the September to January data sets, when compared to data sets acquired in June, July and August. This is demonstrated in particular by the postholes belonging to hall building 2, which had only been weakly visible in the original data from 2007 – which have become much more discernible in autumn and winter. Factors responsible for this phenomenon are still being analysed, but are potentially connected to a slight increase in precipitation and subsequently soil moisture. Additionally, frozen ground provides a smoother surface on the otherwise more uneven ground.

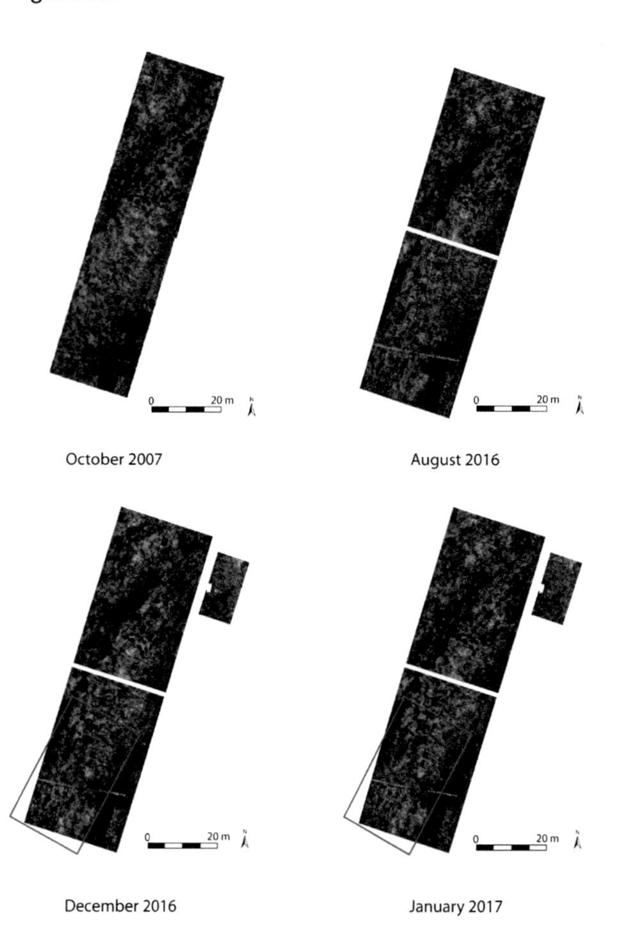

Figure 3: Comparison between GPR data collected in a) October 2007, b) August 2016 c) December 2016 (no snow cover) and d) January 2017 (no snow cover but frozen ground).

In the coming months, systematic collection of geophysical as well as of soil and sedimentological data will continue. Based on these data sets, we will demonstrate how seasonally changing ground conditions can influence the outcome of GPR surveys, and why a thorough understanding of the processes involved is significant for the successful application of non-invasive GPR techniques in efficient reliable heritage management.

Bibliography

Bill, J., Nau, E., Trinks, I., Tonning, C., Gustavsen, L., Paasche, K. and Seren, S. (2013) Contextualising a Monumental Burial - The Gokstad Revitalised Project. In W. Neubauer, I. Trinks, R. B. Salisbury and C. Einwögerer (eds) *Archaeological Prospection. Proceedings of the 10th International Conference on Archaeological Prospection. Wien, Austria, 29.05.-02.06 2013*. Vienna: Verlag der Österreichischen Wissenschaften,134–36.

Brøgger, A. W., Falk, H. J. and Schetelig, H. (1917) *Osebergfundet*. Bind 1-5. Krisiana: Norske Stat, Universitetes Oldsaksamlin.

Gansum, T. (1995) *Jernaldergravskikk I Slagendalen: Oseberghaugen Og Storhaugene I Vestfold - Lokale Eller Regionale Symboler? En Landskapsarkeologisk Undersoekelse*. Oslo: Universitet i Oslo.

Myhre, B. (1992) The Royal Cemetery at Borre, Vestfold: A Norwegian Centre in the European Periphery. In M. Carver (ed.) *The Age of Sutton Hoo. The Seventh Century in North-Western Europe*. Woolbridge: Boydell Press, 301–313.

Myhre, B. (2015) *Før Viken Ble Viken. Borregravfeltet Som Religiøs Og Politisk Arena*. Edited by Kjersti Løkken. Norske Oldfunn XXXI. Tønsberg: Vestfold fylkeskommune.

Nicolaysen, N. (1882) *Langskibet Fra Gokstad Ved Sandefjord*. Kristiana: Cammermeyer.

Nordeid, S. W. (2011) *The Oseberg Ship Burial in Norway: Introduction*. Acta Archaeologica 82(1): 7–7.

Trinks, I., Karlsson, P., Hinterleitner, A., Lund, K., and Larsson, L.-I. (2007) *Preliminary Results of the Archaeological Prospection Survey at Borre, October 2007*. Survey report. RiksantikvarieämbetetStockholm.

Trinks, I., Gansum, T. and Hinterleitner, A. (2010) Mapping iron-age graves in Norway using magnetic and GPR prospection. *Antiquity Project Gallery*, **84** (326): Online publication only, available at http://www.antiquity.ac.uk/projgall/trinks326/.

A ghostly harbour? How delusive gradiometric data can be and how seismic waveform inversion might help

Michaela Schwardt[1], Daniel Köhn[1], Tina Wunderlich[1], Dennis Wilken[1], Wolfgang Rabbel[1], Thomas Schmidts[2] and Martin Seeliger[3]

[1]Christian-Albrechts-Universität zu Kiel, Kiel, Germany; [2]Römisch-Germanisches Zentralmuseum, Mainz, Germany; [3]Universität zu Köln, Köln, Germany

mmerz@geophysik.uni-kiel.de

Introduction

For many years, refraction and surface wave seismics have been used for imaging the subsurface. Although seismic methods have been used to address geoarchaeological problems, they are generally not widespread in archaeological prospecting on land. This is mainly due to the great effort needed concerning measurements and analysis.

Different approaches have been introduced to image the near surface using seismic methods in the past. Their resolution, however, is limited so that small scale features cannot be reproduced, although the data itself is sensitive to small structural changes. To use this information and, therefore, significantly increase the resolution of seismic measurements, the process of full waveform inversion (FWI) can be used.

An example of the application of the FWI in archaeological prospection is given by a dataset recorded in Enez, Turkey over the course of the Priority Program *Harbours from the Roman Period to the Middle Ages* (von Carnap-Bornheim and Kalmring, 2011). The town of Enez can be found in the former settlement area of the Thracian harbour city Ainos. It is located at the mouth of the river Hebros, which forms the border between Greece and Turkey. The city of Ainos was important during the Roman and Byzantine times as it formed an economic hub between the inland and the sea. Although numerous archaeological investigations were conducted during the 1980s, it still is unclear how the area has been used as a harbour.

From 2012 to 2014, on- and offshore geomagnetic measurements were carried out to find potential harbour structures. Investigations east of the city of Enez, adjacent to the lagoon Taşlık Gölü, show distinct, linear anomalies, being orthogonal to the shore (Fig. 1) and were at first interpreted as harbour structures. However, no constructional features could be verified by multiple corings (Fig. 1) making the origin of the magnetic anomaly an interesting riddle for geophysics. Therefore, this area was subject to further investigations using different geophysical methods (GPR, EMI, ERT, seismics; Fig. 1). The common methods yielding a depth resolution (GPR,ERT) suffer from high conductivity associated with lagoon sediments. Therefore, seismic methods have been used. Especially, a 2D full waveform approach for SH-wave data has been carried out, hopefully allowing the imaging of the first few meters of subsoil and resolving the origin of the magnetic signal.

Methodology

The approach of FWI aims at the estimation of elastic parameters by using the whole seismic record. By minimizing the misfit between measured

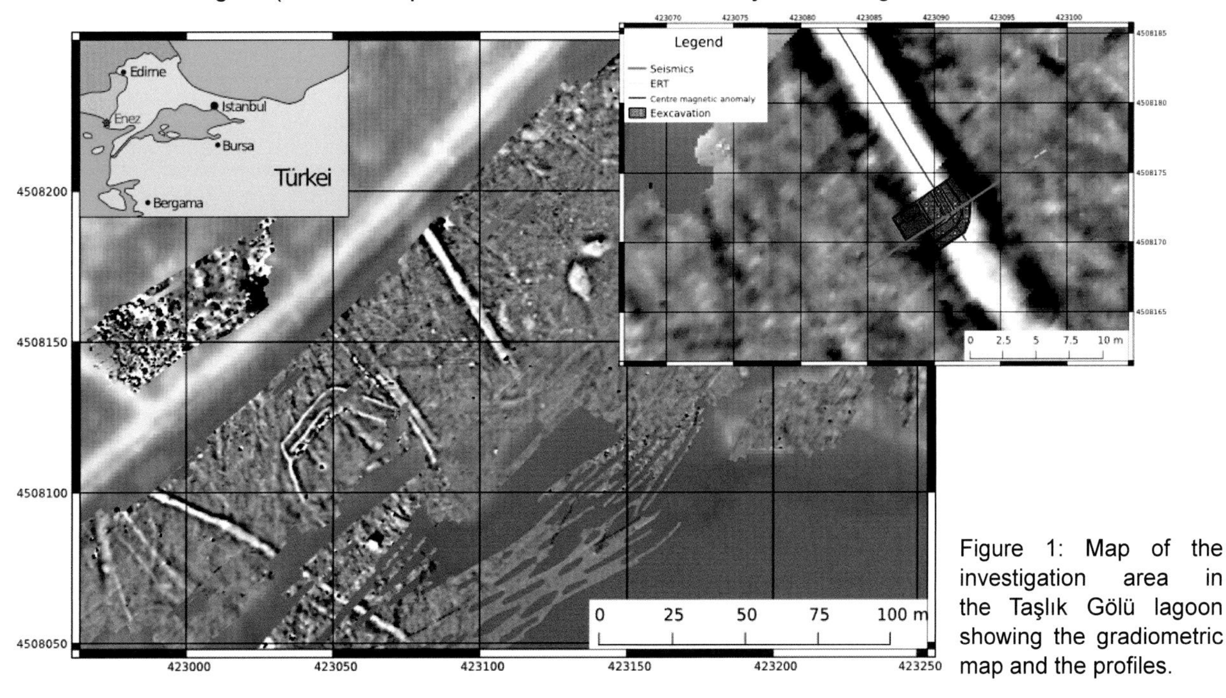

Figure 1: Map of the investigation area in the Taşlık Gölü lagoon showing the gradiometric map and the profiles.

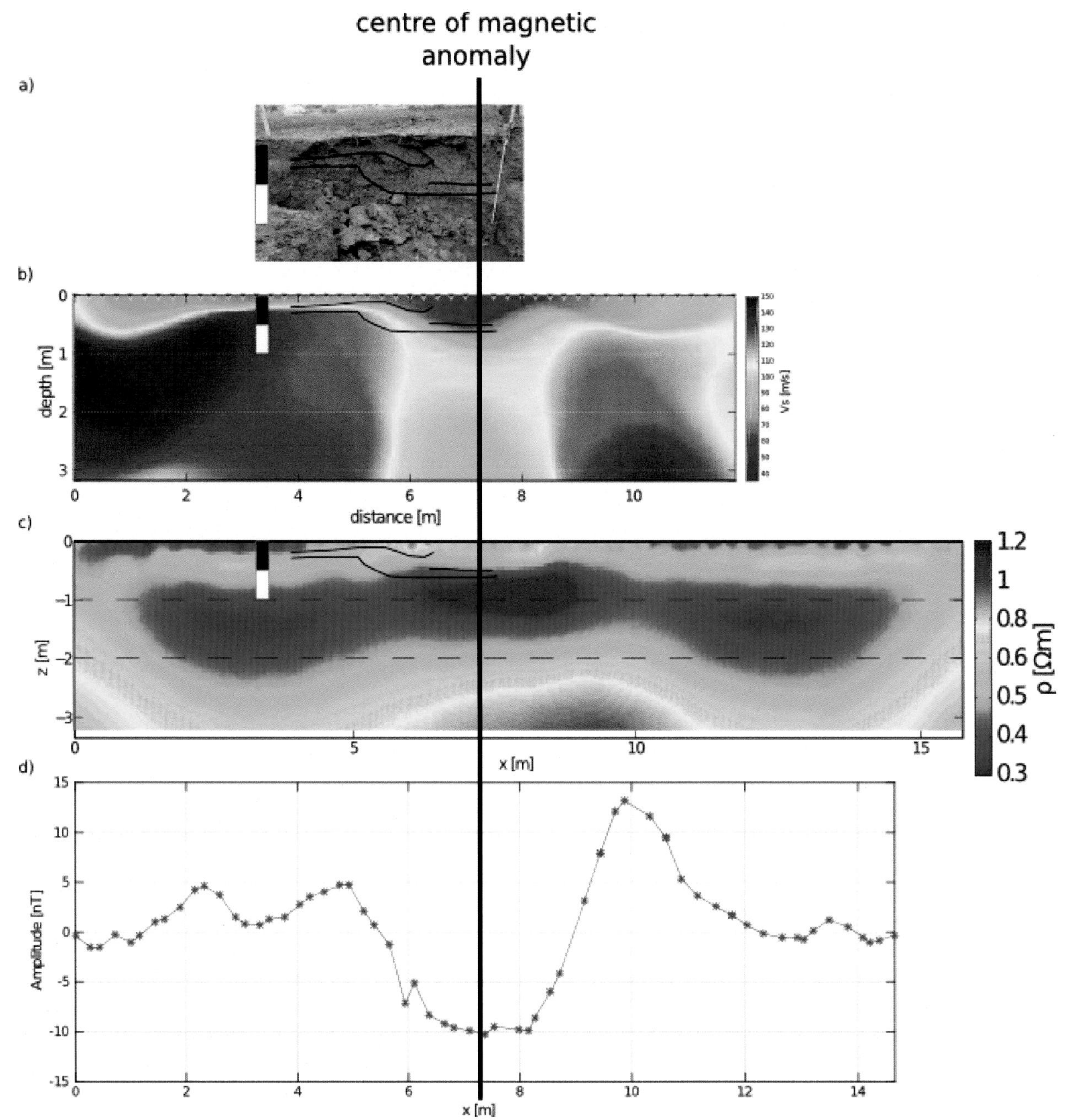

Figure 2: Comparison of the geophysical results and the excavation. The centre of the magnetic anomaly is indicated by a vertical black line. a) Part of the excavation. Black lines indicate changes in the subsurface. b) Results of the FWI. Blue colours show low velocities; red colours show high velocities; black lines show changes in the subsurface derived from the excavation. c) Resulting profile of the ERT measurement. Black lines show changes in the subsurface derived from the excavation. d) Magnetic anomaly along a profile parallel to the seismic and electric profiles.

and modelled data, FWI tries to find a model that best explains the whole measured dataset, i.e. phases, travel times and amplitudes (Köhn 2012). Furthermore, viscoelastic wave propagation effects like attenuation and dispersion are considered. Besides the model parametrization the initial model has a major impact on the quality of the inversion result (Köhn 2012). To further estimate the depth to which the results can be trusted as well as the resolution checkerboard tests are applied (Dokter *et al.* in review).

It has been shown by previous studies, that the FWI is able to resolve small scale structures near the surface with an adequate resolution. The use of the whole recorded wavefield increases the resolution significantly and allows the imaging of features smaller than the seismic wavelength (Köhn 2012).

Results

Several different initial models have been tested for the FWI profile. In the area of the magnetic anomaly, a zone of low seismic velocities (60-70 m/s) being about 2.5 m wide and 0.7 m deep is observed in

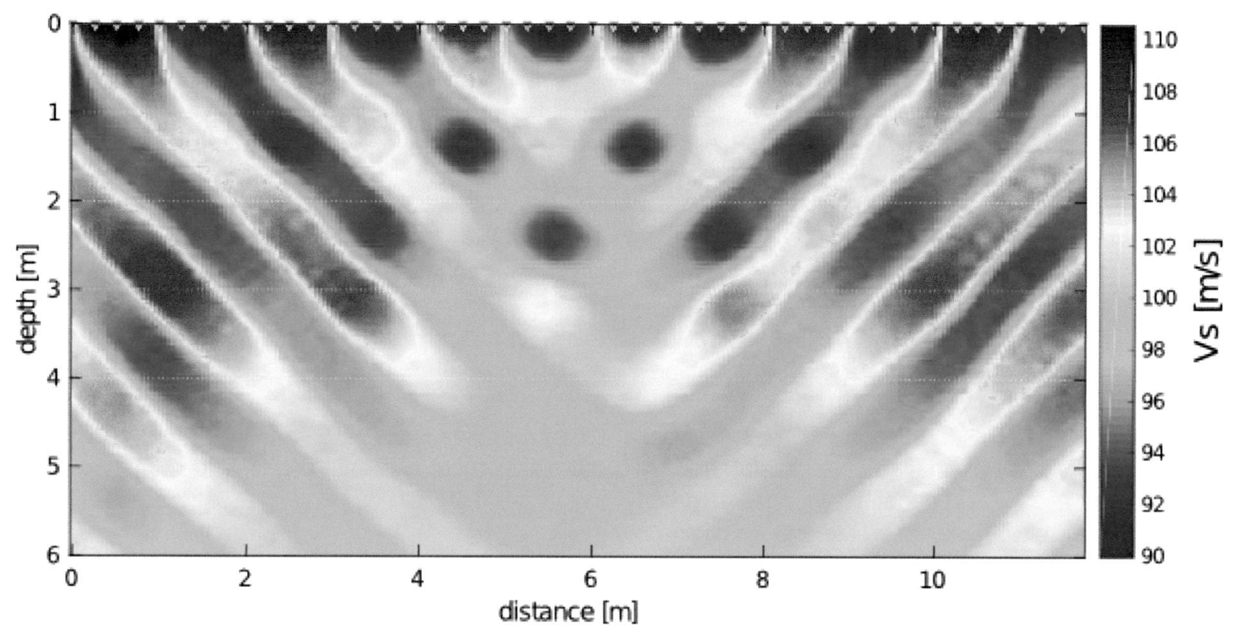

Figure 3: Resulting pattern after the checkerboard test for alternating high and low velocity structures of 1m x 1m in size.

the majority of resulting models (Fig. 2). Below, higher velocities occur. These results have been verified by drillings and an excavation in the area of the anomaly, which show a dipping structure, being about 3 m wide and 0.8 m deep (Fig. 2).

The applied checkerboard tests resolve structures of 1 m – 2 m in size to a depth of about 4 m– 6 m. Structures of 0.5 m in size can be resolved to a depth of about 3 m (Fig. 3).

Conclusion

Although the setting is quite difficult for most depth resolving geophysical methods, the achieved resolution is adequate to permit an interpretation. Particularly the result of the FWI coincides well with the excavation results. It further provides the most probable interpretation of all applied geophysical investigations in the presented case study.

The achieved good resolution together with the low computational costs as well as the promising results justify the higher effort necessary for this method. Therefore, the 2D SH-FWI seems to be a promising supplemental depth resolving tool in archaeological prospecting especially in areas of high electric conductivity.

Bibliography

von Carnap-Bornheim, C. and Kalmring, S. (2011) *DFG-Schwerpunktprogramm 1630 Häfen von der Römischen Kaiserzeit bis zum Mittelalter. Zur Archäologie und Geschichte regionaler und überregionaler Verkehrssysteme*. Schleswig: Annual Report Centre for Baltic and Scandinavian Archaeology: 28-31.

Doktor, E., Köhn, D., Wilken, D., De Nil, D. and Rabbel, W. (In Review) Full-waveform inversion of SH- and Love-wave data in near surface prospecting. *Geophysical Prospecting*.

Köhn, D. (2011) *Time Domain 2D Elastic Full Waveform Tomography*. Kiel, PhD-Thesis, Kiel University.

Köhn, D., De Nil, D., Kurzmann, A., Przebindowska, A. and Bohlen, T. (2012) On the influence of model parametrization in elastic full waveform tomography. *Geophysical Journal International* **191**(1): 325-345.

Köhn, D., Kurzmann, A., De Nil, A. and Groos, L. (2014) DENISE – User manual. Available online at http://www.geophysik.uni-kiel.de/~dkoehn/software.htm

Acknowledgements

This work was funded by the German Research Foundation (DFG) within the frame of the Priority Program 1630 „Harbours from the Roman Period to the Middle Ages“ (von Carnap-Bornheim and Kalmring 2011). The FWI inversions were performed on the NEC-HPC Linux Cluster at Kiel University. The 2D SH-FWI code DENISE-SH is available at *https://github.com/daniel-koehn*.

Testing boundaries: integrated prospection from site to lanscape in western Sicily

Christopher Sevara[1], Michael Doneus[1,2], Erich Draganits[2,3], Rosa Cusumano[6], Cipriano Frazzetta[1], Barbara Palermo[6], Filippo Pisciotta[6], Rosamaria Stallone[6], Ralf Totschnig[4], Sebastiano Tusa[5] and Antonina Valenti[6]

[1]University of Vienna, Department of Prehistoric and Historical Archaeology, Vienna, Austria; [2]Ludwig Boltzmann Institute for Archaeological Prospection and Virtual Archaeology, Vienna, Austria; [3] University of Vienna, Department of Geodynamics and Sedimentology, Vienna, Austria; [4]Zentranstalt für Meteorologie und Geodynamik, Vienna, Austria; [5]Soprintendenza del Mare, Regione Siciliana, Palermo, Italy; [6]Associazione PAM, Partanna, Italy

christopher.sevara@univie.ac.at

Introduction

The Prospecting Boundaries project (http://mazaro.univie.ac.at/) seeks to explore the diversity of land use in the region along the Mazaro river corridor in western Sicily, through the use and improvement of integrated archaeological prospection techniques. Centred on the Mazaro river, the project encompasses an area of roughly 70km^2, stretching inland from the river mouth at modern-day Mazara del Vallo (Fig. 1). Of particular interest is the exploration of this area as a potential zone of interaction and a boundary area between shifting coastal, interior, indigenous and colonial interests from the Bronze Age to the end of the First Punic War in 241 BC. However, the project is concerned with human activity during all periods in the region, working backward from the present to document and deconstruct modern and historical land use.

One key research area within the project boundaries is the site of Guletta, an area of dense multi-period activity situated on the plain above the Mazaro gorge, and the terraces below that descend down to the river (Fig. 1). Prior to 2004, the area around Guletta had, from an archaeological standpoint, been known chiefly for its Roman-era settlement component and the presence of Bronze Age and Paleochristian rock-cut tombs along the cliff face overlooking the river (Calafato *et al.* 2001, 39; Tusa 1999). In 2004, a large multi-ditched structure was identified at Guletta during an aerial photography campaign carried out by the University of Vienna Aerial Archive. Thus, the complex archaeological and environmental nature of the area around Guletta makes it a suitable spot to investigate both methodological and interpretive questions related to our project goals.

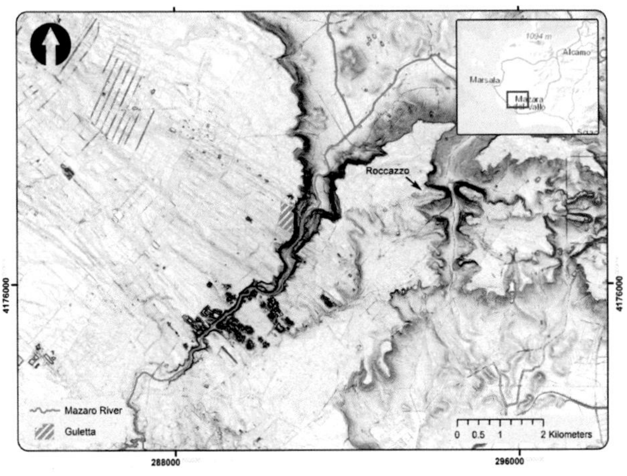

Figure 1: Project area and location of Guletta and Roccazzo.

Integrated Prospection at Guletta

Geophysical prospection at Guletta was carried out in the spring of 2016 and included the collection of roughly 9 ha of magnetic and 2.5 ha of ground penetrating radar (GPR) data. The environment is stony and difficult to prospect using motorised equipment; therefore all prospection was done using handheld systems. Results from corresponding in-field survey and subsequent artefact analysis indicate the possibility of human occupation from at least the 8th to 6th centuries BC. The ditches themselves also appear in ALS data and orthoimagery, extending to the edge of the plain where they are clearly visible as rock-cut structures. GPR data indicate that the remains of the ditch structures could reach a depth of at least 1.5 m below the modern ground surface. The enclosure appears to have a large south-facing opening, and the interior of the innermost ditch is filled with a series of rectangular and subrectangular structures, a number of which are interpreted as domestic structures based on their form and corresponding surface artefacts (Fig. 2). In addition to the double ditch system, a third and more ephemeral structure seems to enclose the wider area around the ditches. Corresponding mortuary activities are indicated by numerous rock-cut tombs, now plundered, along terraces below the settlement and clustered at both riverside edges of the outer ditch.

A completely unexpected result of the geophysical prospection was the indication of a large rectilinear structure to the north-east of the central ditched enclosure. Morphologically, the interpretation of the magnetic and radar data suggests an early Greek period rural habitation structure (Fig. 2). Numerous finds recovered from surface survey in the area are consistent with domestic activity from that period. However, extensive postdepositional disturbance in the area means that the likelihood of substantial remaining subsurface material is low.

Figure 2: Interpretation of geophysical prospection results at Guletta. Extremely high parallel magnetic anomalies indicate areas where materials from former vineyards remain in the soil.

Prospecting Landscapes: The Wider Context of Western Sicily

Geophysical prospection at Guletta has been conducted in conjunction with a wider interpretation of the landscape around the Mazaro river corridor, based largely on the use of airborne laser scanning (ALS), historical maps, and historic and modern aerial imagery coupled with in-field archaeological and geological survey and evaluation of existing site information. For example, at the multiperiod site of Roccazzo, some 3 km to the east-northeast of Guletta (see Fig. 1), there is evidence of activity from the Upper Palaeolithic, Copper, Late Bronze, and Archaic/Classical Greek periods. Initial spatial analyses suggest intervisibility between both sites, as well as with sites such as Mokarta, a key Late Bronze Age settlement in inland western Sicily. Relict traces of transport activity such as paths and grooves, probably from animal-drawn carts, are partially visible in ALS data and imagery as worn surfaces cut into exposed bedrock, and the nature of these as well as exploration of their possible use in different periods is helping us to build a picture of past movement between zones of activity in the landscape. In such a densely utilised landscape, however, subsequent activity often masks or altogether obliterates prior traces of land use. Accounting for this bias is sometimes difficult, and one way in which we are attempting to do so is through the use of historic digital elevation models (hDEMs), allowing us to track over 60 years of landscape development through accurate terrain reconstruction from historic aerial photographs (Sevara 2013, 2016; Verhoeven *et al.* 2013) (Fig. 3).

Conclusions

Through the use of integrated archaeological prospection techniques, this project seeks to contribute to the understanding of land use along the Mazaro river corridor at multiple spatial and temporal scales. In addition to deepening our understanding of structures already visible via other means, geophysical prospection and surface survey at Guletta have yielded information about previously unknown archaeological features, including structures and finds indicative of settlement activity from multiple periods. Coupled with existing archaeological data, ALS and historic orthoimage-based data interpretation, this information is helping us to understand how Guletta may have acted as a central node in the landscape. In turn, new applications of historic and modern remote sensing datasets are also helping us to

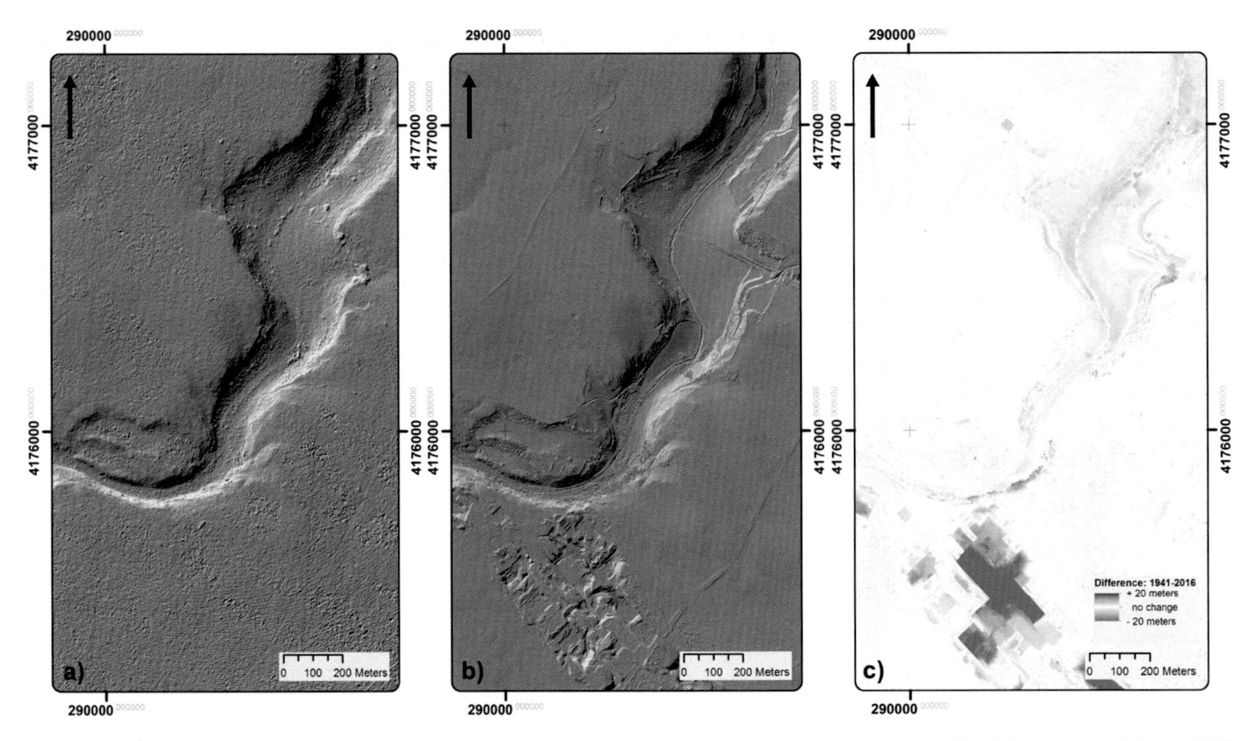

Figure 3: a) Hillshade of hDEM calculated from imagery collected in 1941. b) Hillshade of DEM generated from ALS data collected in 2016. c) Difference DEM showing physical change in terrain between 1941 and 2016. Areas along the river, which contain significant archaeological sites, have been heavily affected by modern land use.

broaden and deepen the application of integrated prospection approaches in complex Mediterranean contexts.

References

Calafato, B., Tusa, S., and Mammina, G. (2001) *Uomo e ambiente nella storia di Mazara del Vallo - Indagine topografica nell'agro mazarese.* Palermo: Grafistampa Palermo.

Sevara, C. (2013) Top Secret Topographies: Recovering Two and Three-Dimensional Archaeological Information from Historic Reconnaissance Datasets Using Image-Based Modelling Techniques. *International Journal of Heritage in the Digital Era*, **2**(3): 395–418.

Sevara, C. (2016) Capturing the Past for the Future: an Evaluation of the Effect of Geometric Scan Deformities on the Performance of Aerial Archival Media in Image-based Modelling Environments. *Archaeological Prospection,* **23**: 325-334.

Tusa, S. (1999) *La sicilia nella preistoria* (2nd ed.). Palermo: Sellerio editore.

Verhoeven, G., Sevara, C., Karel, W., Ressl, C., Doneus, M. and Briese, C. (2013) Undistorting the past: New techniques for orthorectification of archaeological aerial frame imagery. In C. Corsi, B. Slapšak and F. Vermeulen (eds.), *Good practice in archaeological diagnostics.* Cham: Springer, 31–67.

Acknowledgements

The Prospecting Boundaries project is funded by a grant from the Austrian Science Fund (Project ID: P-28410) and is carried out in cooperation with the Soprintendenza per I Beni Culturali ed Ambientali di Trapani. We would like to thank our project partners at the Associazione PAM (Prima Archeologia del Mediterraneo), who conducted surface survey at Guletta as well as subsequent artefact analysis. Thanks also to Sirri Seren & Erol Bayirli of Archeo Prospections (Vienna) for conducting the geophysical prospection data collection. Sheba Schilk contributed to field survey and remote sensing data interpretation & analysis. The Ludwig Boltzmann Institute for Archaeological Prospection and Virtual Archaeology (LBI ArchPro) has also provided support for this research.

Potential of multi-frequency electromagnetic induction in volcanic soils for archaeological prospection

François-Xavier Simon[(1)], Alain Tabbagh[(2)], Bertrand Douystessier[(3)], Mathias Pareil-Peyrou[(4)], Alfredo Mayoral[(5)] and Philippe Labazuy[(4)]

[(1)]Inrap, Paris, France; [(2)]UPMC – Metis Laboratory, Paris, France; [(3)]Maison des Sciences de l'Homme de Clermont-Ferrand, Plateforme Intelespace, Clermont-Ferrand, France; [(4)]Magma et Volcan Laboratory, OPGC, Clermont-Ferrand, France ; [(5)] Université Clermont Auvergne, CNRS, GEOLAB, F-63000 CLERMONT-FERRAND, France

frx.simon@gmail.com

Volcanic soils cover only 1% of the emerged lands, but present a strong interest for human settlement: the high fertility of these soils induced by their physical, chemical and mineralogical properties make them very attractive for agriculture. This results in a high density of population and a high concentration of archaeological sites. However, this environment is challenging for geophysical exploration. Magnetic survey is considerably affected by the random orientation of the remnant magnetization which, despite some exceptions (Hesse *et al.* 1997, Deberge and Dabas 2009, Herles and Fassbinder 2015,), blurs the clear characterization of archaeological remains and/or geological background. GPR and resistivity are more adapted to this context, but harder to implement in the field. The EMI technique seems to be a good alternative, nevertheless it still needs to be better evaluated (Pareilh-Peyroux 2016, Guillemoteau *et al.* 2016).

Context

The site of Corent (Puy-de-Dôme, France) is an archaeological site with important Neolithic, Late Bronze Age and Gallo-Roman settlements, but mainly known to be a large *oppidum*, probably the main place of the *Arverni*. This long-term settlement is located on the top of a basaltic plateau resulting from the relief inversion of an ancient lava flow 3 million years old, deposited at the top of 200 m of Oligocene sedimentary rocks. Lowlands around the plateau mainly support soils developed in alluvium or in volcano-sedimentary colluvium. This heterogeneous context makes it complex for any geophysical investigation, and requires attention to the choice of the most adapted techniques.

Methodology

In order to overcome all of these difficulties we suggest using a multi-frequency EMI instrument which presents advantages of a multi-component characterization of remains and soils (Simon *et al.* 2015). This instrument is used for archaeological prospection and soil/sediments survey on the top of the plateau and on the alluvial soils nearby the plateau. Measurements were done with the EMP400 from GSSI. It is a multi-frequency device working at low frequencies and with a small coil offset (1.21 meter) both in HCP (VMD) or VCP (HMD). The user may select up to three working frequencies. The 15, 8 and 5 kHz frequencies were used in order avoid noise existing at lower frequencies. Associated with a short coil spacing, it also avoids electrical polarization effects (clearly met for larger coil spacing or higher frequency on the in-phase measurement). Data are processed through our own python processing code implemented in Qgis. This processing allows determining (i) the electrical conductivity and the magnetic susceptibility at the three frequencies, (ii) the conductivity by the difference between 15 Khz and 5 kHz and (iii) the determination of magnetic viscosity (i.e. the imaginary part of the complex magnetic susceptibility). Data are then plotted with Qgis and Wumap. For sediment mapping, acquisition was done every 5 meters with GPS positioning, and for archaeology every one meter along profile with a continuous measurement and a grid positioning.

EM drift was not processed as it looked small compared to the noise and because acquisition duration for one grid and/or field plot took around one hour. However, we observed a strong drift for longer time of acquisition with this instrument. In addition, we met strong offset effects from one grid to another which would justify a robust calibration procedure (Thiesson *et al.* 2014). Finally, it was necessary to test several offset trials for each survey in order to fit the median and interquartile values of the conductivity and of the magnetic susceptibility at the different frequencies.

Test site and results

Two cases studies are presented, allowing an estimation of multi-frequency EMI efficiency on two different soil contexts and for different archaeological remains. The first one is a small basin on the basaltic plateau where EMP400 was used to determine the extent of the sedimentary infilling, mainly clayey sediments. The second one is a Gallo-Roman villa on the bottom of the plateau, built with basaltic stone in a high clay content deposit context resulting from sedimentary processes.

On the first site (Fig. 1), EMI allows a good characterization of both electrical conductivity and magnetic susceptibility for both orientation (HCP and VCP). This clear determination results from very high values of the geophysical properties. EMP400 measurements well delineate the clay deposit extent and basin shape. Also, the high value of magnetic susceptibility here is clearly correlated

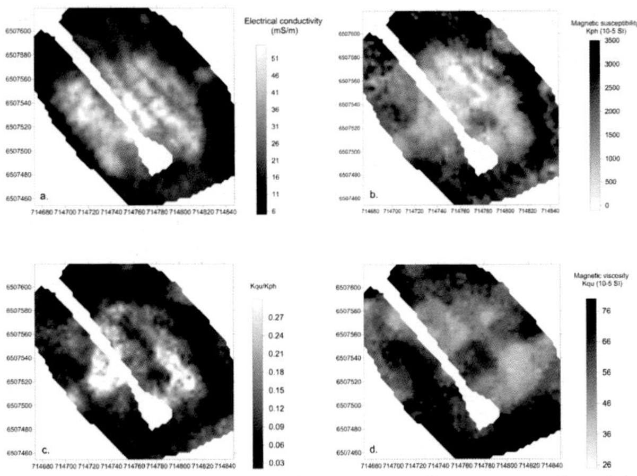

Figure 1: Interpretation of EMI multi-frequency signal in HCP geometry. a: electrical conductivity (15-5 kHz), b: magnetic susceptibility (15kHz), c: Kqu/Kph (15 kHz), d: magnetic viscosity (15 kHz)

with a high value of magnetic viscosity (it represents 6 to 7% of the in-phase magnetic susceptibility).

On the second site, an EMI device is useful as the map of magnetic susceptibility seems to be the best option for the determination of buried walls (Figure 2). Nevertheless, electrical conductivity in VCP is not so effective to evidence archaeological remains but gives information on the soils clay content (this is probably due to the greater volume). Finally, the magnetic viscosity map does not show high contrasts similar to those observed with in-phase magnetic susceptibility. In this site, magnetic viscosity is not strong enough to be precisely determined in the highly conductive soil context.

Bibliography

Deberge Y. and Dabas M. (2009) An area of the gallic oppidum of Gondole (Le Cendre, Puy-de-Dôme, France) revealed by magnetic survey: extrapolation from excavation data. *Archéosciences, revue d'archéométrie*, suppl. **33**: 287-288.

Guillemoteau J., Tronicke J. and Simon F.X. (2016) 1D sequential inversion of portable multi-configuration electromagnetic induction data. *Near Surface Geophysics*, **14**(5): 423-432.

Herles M. and Fassbinder J. W. E. (2015) Magnetic prospecting on basaltic geology: the lower city of Erebuni (Armenia), *Archaeologia Polona*, **53**: 292-296

Hesse A., Barba L., Link K. and Ortiz A. (1997) A magnetic and electrical study of archaeological structures at Loma Alta, Michoacan, Mexico. *Arcaheological Prospection*, **4**(2): 53-67.

Pareilh-Peyrou M. (2016) Optimisation des méthodes à induction électromagnétique pour l'ingénierie des sols, *Thèse Université Blaise Pascal*, Clermont-Ferrand

Simon F.-X., Tabbagh A., Thiesson J. and Sarris A. (2015) Mapping of quadrature magnetic susceptibility/magnetic viscosity of soils by using multi-frequency EMI, *Journal of Applied Geophysics, **120**: 36-47.*

Thiesson J., Kessouri P., Schamper C. and Tabbagh A. (2014) Calibration of frequency-domain electromagnetic devices used in near-surface surveying, *Near Surface Geophysics*, **12**(4): 481-491.

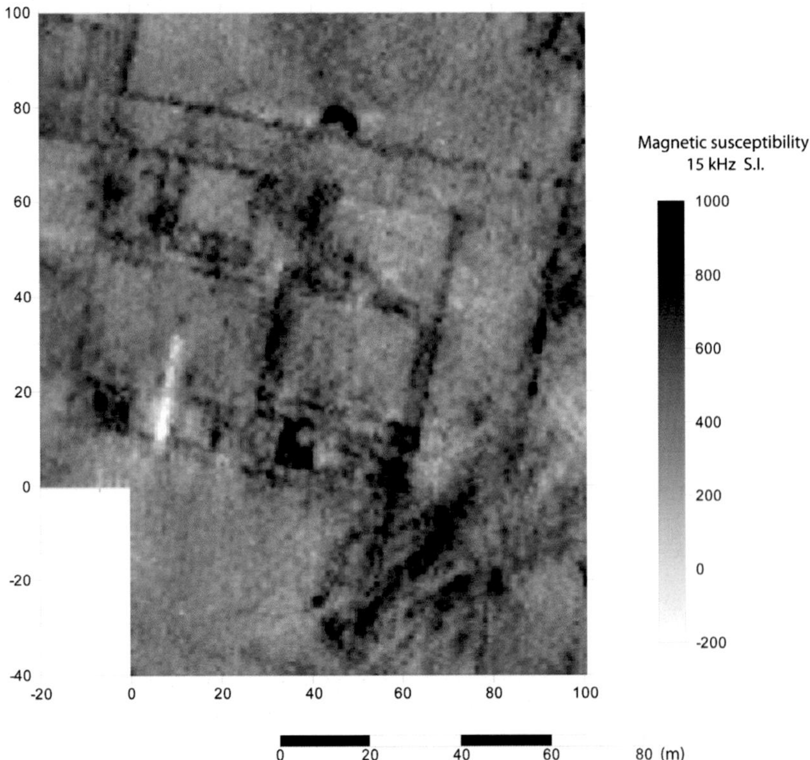

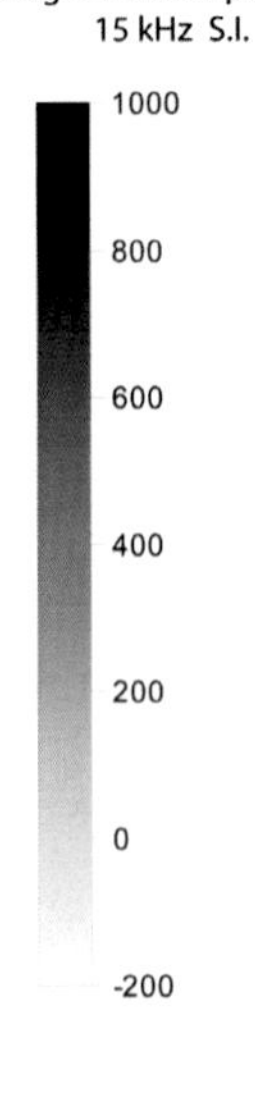

Figure 2: Magnetic susceptibility (15 kHz – VCP) Gallo-roman villa of Lieu-Dieu

3D electrical resistivity imaging in shallow marine environment: case study at the harbor "KATO pafos", cyprus

Kleanthis Simyrdanis[(1)], Nikos Papadopoulos[(1)] and Gianluca Cantoro[(1)]

[(1)]Laboratory of Geophysical-Satellite Remote Sensing, IMS-FORTH, Rethymno, Greece

ksimirda@ims.forth.gr

Introduction

Electrical resistivity tomography (ERT) is a well-established method used for onshore archaeological prospection (Papadopoulos *et al.* 2011). Recently, there is a tendency for incorporating this technique within offshore geophysical surveys since it can be effective in imaging faults or weak zones beneath the water layer (Kwon *et al.* 2005) or characterizing waterbed sediments (Orlando 2013). Data acquisition can be carried out by using either submerged (Kim *et al.* 2002) or floating electrodes in conjunction with continuous resistivity profiling.

For archaeological purposes, especially in shallow marine environments (water depth less than 2m) little has been accomplished. Passaro (2010) managed to outline the shape of a military shipwreck as low resistivity values from a towed floating mobile configuration. A comprehensive feasibility study on the efficiency of ERT in reconstructing submerged archaeological relics in shallow seawater environment, based on extensive numerical models, verified by subsequent field experimentation is given by Simyrdanis *et al.* (2015).

The purpose of this work is to investigate any possible underwater archaeological relics dating from Hellenistic to Roman period. The study was conducted at the port of 'Kato Pafos' at Cyprus (Fig. 1.a) where the geoelectrical method was applied in an area of 15x10m^2 (Fig. 1.b).

Methodology

The corners of the survey area (G10, G11, G12, G13) were stacked using RTK-GPS enabling high positioning accuracy (Fig. 2.a, b). Multiple ERT parallel lines were developed in two vertical directions (x, y) in order to increase the density of the acquired information. Parallel to the the 'x' direction 21 lines (0.5m apart with 31 electrodes per line) and parallel to the 'y' direction 16 lines (1.0m apart with 21 electrodes per line) were installed. The static submerged mode was employed for the survey, where the cable was sunk and laid on the bottom of the seabed during data collection. The submerged cable was kept fixed with two anchors at both ends during the measurement to avoid unnecessary movement due to the waves (Figure 2d).

The bathymetry model of the area was measured manually with an elongated plastic calibrated stick on every probe (sensor) position of the cable (Fig. 2.c). The conductivity value of the water was measured with a handheld equipment in various positions and an average value was then calculated. A 10-channel resistivity meter, an external battery and a laptop were used to inject the current below the sea bottom through the sensors based on a predefined electrode protocol and measure the resulting potential.

The average resistivity value of the sea-water (0.19 Ohm-m) and the column depth information (above each electrode location), which was extracted from the respective bathymetry model, were also incorporated into the inversion model to account for the effect of the highly conductive saline water layer.

The 'pole dipole' (PD Forward, PD Reverse) and 'gradient' (GRD) arrays were used with #425 and

Figure 1: (a) Map and (b) photo of the survey area in 'Kato Pafos' port, at South West Cyprus.

Figure 2: (a) Survey area, (b) defining the survey boarders with RTK-GPS method, (c) measuring the column depth and (d) cable used for the survey where the distance between the metal cylinder stakes was set to 0.5m.

#721 measurements for each line, accordingly. A commercial 3D inversion software was used for data processing. All survey lines were analysed individually and at the end they were joined to produce a 3D resistivity model. For the 3D model 651 electrodes in total were incorporated and a robust inversion method (Loke *et al.* 2003) was applied for the deconvolution of the apparent resistivity data.

Results

The sea-water column depth and the inverted resistivity values varied between 0.3 to 2.3m and 0.15 to 50 ohm-m, accordingly. The calculated RMS % error was less than 8%. Depending the array used (pole-dipole with/or gradient), the penetration depth analysis varied from 4.5 to 6.5m below water surface level.

According to the inversion results three different areas of interest can be identified, indicated as 'R1', 'R2' and 'R3' (Fig. 3.a). The approximate depth of the targets is estimated at 0.57 - 1m below seabed. Target 'R1a' is parallel to the harbor wall with direction SW to NE. The average width of the target is less than 0.5m. Another smaller target attached to the aforementioned target (indicated as 'R1b') is located vertical to the main axis of 'R1a'. Target 'R3' is located on the NE side of the area, vertical to the main axis of 'R1a' with average estimated width approximately 1m. A rectangular structure ('R3b') is attached to 'R3'. On the SW side of the survey area the target 'R4' is located. In Figure 3.b, a 3D representation of the same results using an isosurface rendering mode is shown where the areas with high resistivity values (>5.6 ohm-m) are depicted with red color.

Conclusions

The geophysical prospecting in marine environments to map archaeological remains is considered a major challenge arising from the difficulties related to the dynamic environment where the data are collected. The geophysical survey using electrical resistivity tomography in the port 'Kato Paphos' was able to

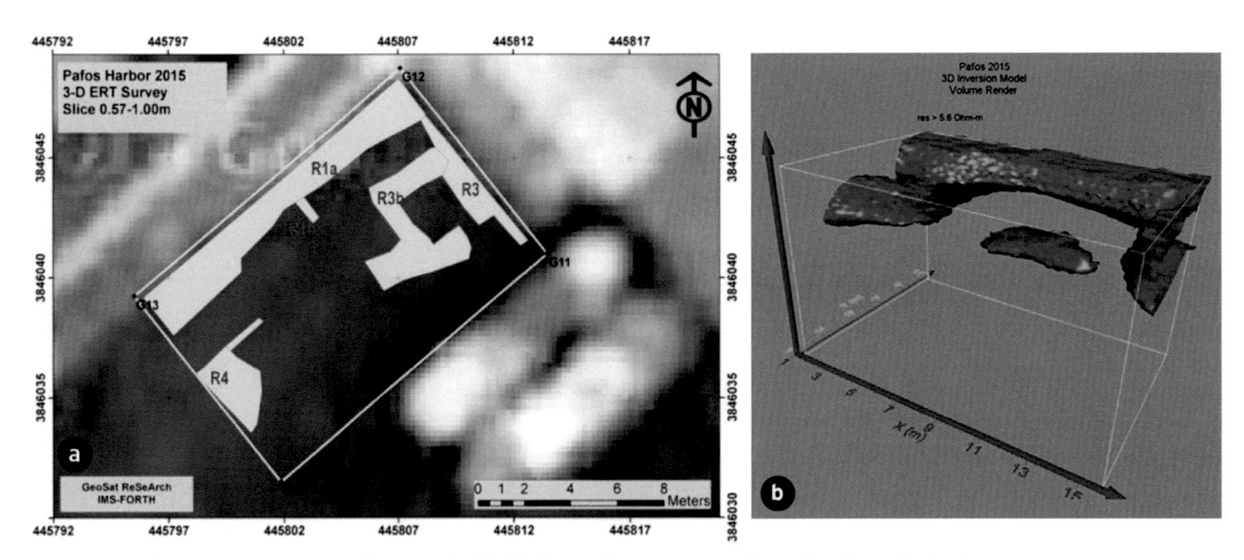

Figure 3: (a) Schematic location of "targets", (b) 3D isosurface presentation of archaeological targets.

provide solid evidence on possible architectural structures related to buildings and structures that are buried at a depth of no more than 2 meters under the seabed. Within this layer the geophysical survey was able to map potential building walls, which are in relatively good conservation level.

Bibliography

Kim, J.-H., Yi, M.-J., Song, Y., Cho, S.-J., Chung, S.-H. and Kim K.-S. (2002) *DC Resistivity Survey to Image Faults Beneath a Riverbed*, SAGEEP, Las Vegas, 13IDA10.

Kwon, H. S., Kim, J.-H., Ahn, H. Y., Yoon, J. S., Kim, K.-S., Jung, C. K., Lee, S. B. and Uchida, T. (2005) Delineation of a fault zone beneath a riverbed by an electrical resistivity survey using a floating streamer cable, *Exploration Geophysics* **35**: 50-58.

Loke, M. H., Acworth, I. and Dahlin, T. (2003) A comparison of smooth and blocky inversion methods in 2D electrical imaging surveys. *Exploration Geophysics* **34**: 182-187.

Orlando, L. (2013) Some considerations on electrical resistivity imaging for characterization of waterbed sediments, *Journal of Applied Geophysics* **95**: 77-89.

Papadopoulos, N. G., Tsourlos, P., Papazachos, C., Tsokas, G. N., Sarris, A. and Kim, J.-H. (2011) An Algorithm for the Fast 3-D Resistivity Inversion of Surface Electrical Resistivity Data: Application on Imaging Buried Antiquities. *Geophysical Prospection* **59**: 557-575.

Passaro, S. (2010) Marine electrical resistivity tomography for shipwreck detection in very shallow water: a case study from Agropoli (Salerno, southern Italy). *Journal of Archaeological Science* **37**: 1989–1998.

Simyrdanis, K., Papadopoulos, N., Kim, J.-H., Tsourlos, P. and Moffat, I. (2015) Archaeological Investigations in the Shallow Seawater Environment with Electrical Resistivity Tomography. *Journal of Near Surface Geophysics, Integrated geophysical Investigations for Archaeology* **13**: 601 – 611.

Skills and protocols for archaeological interpretation in a multispectral geophysical survey world

Lewis Somers[1, 2]

[1]Archaeophysics LLC; [2]Geoscan Research USA

somers@mcn.org

Context Based Interpretation of Geophysical Features

We follow Philip Barker's reflection that "the whole of our landscape, rural and urban, is a vast historical document - - - Every archaeological site is itself a document. It can be read by a skilled excavator".

Today our document is filled with graphics, images from multispectral geophysical surveys and our goal is to perform a feature by feature archaeological / cultural reading from the geophysical survey record. This is both guided by and constrained by the feature's context: cultural, soils, stratigraphy, instrument characteristics etc. It is a messy business and will always be imperfect but there are approaches that can be made.

One is to parse the archaeological meaning of each geophysical phrase, that is to say, understand the archaeological meaning or implication of the feature's geophysical data within the feature's context: Cultural, Soils, Stratigraphy, Materials, Survey Instrument etc. We briefly review two case studies: First, the floor of an individual Spanish mission room block and Second, a cluster of similar room blocks with their immediate surround.

Spanish Mission Room Block

This is essentially a static protocol where we overlay a graphic of one measured geophysical property upon another and derive a plausible archaeological interpretation/ understanding based on: (1) our knowledge of the cultural and material context of the site and (2) the relative spatial distribution of these properties with respect to each other and within each other.

Spanish Mission Room Cluster

This is a more interactive protocol made possible with the presence of radar survey. As with the static protocol we overlay a graphic of one measured geophysical property upon another and

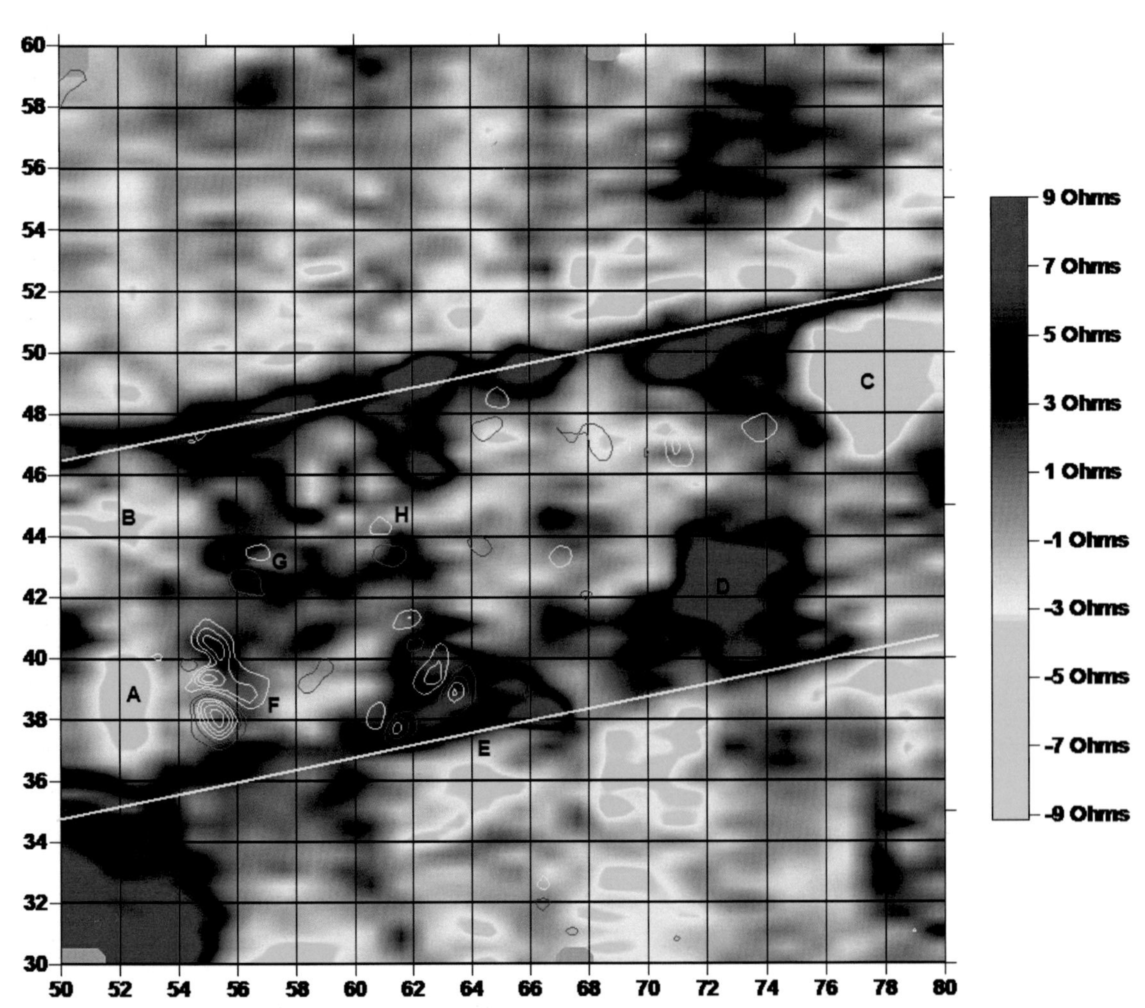

Figure 1: Combined Magnetic and Resistance Map Dormitory.

Figure 2: Animal Fat Rendering and Candle Making, mission San Antonio de Padua

derive a plausible archaeological interpretation/ understanding based on the cultural and material context of the site as well as the relative spatial distribution of these properties. Various overlay combinations are possible, one of particular interest overlays earth resistance data and/or magnetic data on a radar depth-slice animation.

It is possible to start at the surface and "excavate" one radar depth-slice a time, discovering the radar related features while simultaneously viewing their earth resistance, magnetic, magnetic susceptibility surrounds. We are able to "excavate" in geophysical space without harm to the archaeological resource. This protocol is used to solicit attendee critique and suggestions.

Finally, we encourage training in archaeological feature formation from the geophysical properties point of view. Archaeologist are intimately concerned with soils and most geophysical instruments are electro or magnetic the introduction of an electromagnetic soils triangle would be useful.

Figure 3: Lavenderia and East end of Dormitory wall, Mission Santa Inés.

Mission Candle and Soap Manufacture?

Mission Santa Inés

It is reasonable to expect the earthen floors of the neophyte dormitory to contain a good geophysical and geochemical record of activity within. The high quality resistance and magnetic data from the Mission survey presents an opportunity to read this record and to develop a plausible cultural interpretation.

This cultural interpretation exploits the over lapping resistance and magnetic survey data of the Dormitory. Within this area it is possible to recognize three significant features: 1) a 4 x 4-m, very low resistance feature in the northeast corner of the southern dormitory, "C"; 2) a 3 x 4-m, very high resistance feature along the interior of the south wall, "D"; and 3) a triangular cluster of high resistance features 4 - 5-m west of "D", also along the south wall, of the southern dormitory, "E".

In addition, Feature E contains a number of individual, highly localized, magnetic features co-located and within the resistance feature, the red and yellow contour lines. The magnetic signature is complex. They are possibly an in situ fired earthen feature (a cluster of bricks, oven, etc.) with small iron objects.

One interpretation revolves about soap and candle manufacture. Specifically the large multi-component magnetic feature (individual iron objects in the midst of a triangular high resistance feature "E") located on the interior of the south wall is consistent with an industrial-level tallow rendering feature. That is to say, consistent with a large permanent installation for heating/rendering gallons of animal fat. A few meters along this wall toward the Lavenderia is a correspondingly large high resistance area ("C") devoid of any significant magnetic objects. High resistance suggests the absence of soil moisture and mobile ions which can be understood as soils saturated with oil, e.g., spilled tallow. Across the room, along the interior of the north wall, the low resistance feature can be understood as a floor saturated with mobile ions, e.g., soil moisture plus hydroxides, salts etc. Note: electrical resistance in soils is primarily governed by the presence and concentration of water/ moisture soluble compounds, the more mobile ions the lower the resistance and the fewer mobile ions the higher the resistance to the flow of electrical currents.

Given these possible ingredients for soap and candle manufacture, one might find this interpretation plausible. It is further supported by proximity to the lavenderia and known economic activities at Inés.

The photograph of a tallow rendering and candle making facility is taken from Mission San Antonio de Padua. The other photograph is the Inés lavenderia with the standing stone wall at the eastern end of the dormitory under the tree.

The status, role and acceptance of geophysical methods in Norwegian archaeology

Arne Anderson Stamnes[1]

[1]Institute of Archaeology and Cultural History, The NTNU University Museum, The Norwegian University of Science and Technology, Trondheim, Norway

arne.stamnes@ntnu.no

The usage of geophysical methods within Norwegian Archaeology has increased within the last decades, and especially within the last seven years. This relates to an increased research emphasis on geophysical methods by the Norwegian Institute of Cultural Heritage Research (NIKU) in collaboration with the Ludwig Boltzmann Institute for Archaeological Prospection and Virtual Archaeology in Vienna (LBI ArchPro) and Vestfold County Council, as well as at the NTNU University Museum in Trondheim, Norway. While the number of geophysical surveys increase, of which many contribute with new and important cultural historical information, there is still a reluctance in the acceptance for use of geophysical methods by the Cultural Heritage Management (CHM) Sector. This presented work was initiated to study the status, role and potential of geophysical methods in Norwegian Archaeology, as well as how Norwegian Archaeology has been impacted by geophysical methods. An important aspect of the work was examining how geophysical methods have been discussed or mentioned in public documents such as cultural heritage management strategies, budgeting guidelines for archaeological registrations and excavations, and similar documents issued by various institutions related to Norwegian archaeology.

The Norwegian Cultural Heritage Management system is not privatized, but is a public system with a series of institutions having legal responsibilities such as the Directorate for Cultural Heritage Management, regional archaeological museums and county authorities. All archaeological monuments and objects older than 1537 have a legal protection, no matter if they are still undiscovered in the ground as buried structures or are visible on the surface. The Directorate for Cultural Heritage has overarching responsibility for the Norwegian Cultural Heritage Management on behalf of the Government and the Ministry of Climate and Environment. When a planning permission is filed, this reaches the County Authorities, who employ County Archaeologists whose responsibility it is to judge whether the planned development is a threat to known or hitherto unknown archaeology or not, as well as to decide the cultural-historical importance of the site. It is up to the County Archaeologist to choose what method(s) he or she finds best applicable to locate and delimit any archaeology on a case by case basis, weighted against the size of the planned development, its cost and importance to society. Certain regulations and guidelines do exist. It is usually the Directorate that issues such guidelines for CHM. The quality of the archaeological registration should locate and delimit the archaeological site and monuments for entry in the national monument registry called "Askeladden" and area regulations, as well as provide information sufficient for proper budgeting and planning of a potential archaeological excavation. If there is a conflict between an archaeological site and the planned development, the developer can apply for an exemption of the Norwegian Cultural Heritage Act to be allowed to remove the site or monument following an archaeological excavation for which the developer in most cases have cover the expenses of (developer pays principle).

In relation to the status, role and acceptance for the use of geophysical methods, it is interesting to identify and investigate the CHMs own description of geophysical methods, as well as their impression of the main usage of such methods. Geophysical methods are often described as non-destructive or non-intrusive methods, but also often described as new technologies, new methods for registration and documentation, and as high-technological and advanced methods of investigation. The description

TYPE	NUMBER	ROLES AND FUNCTIONS
INVESTIGATION	51	registration, documentation, investigation, excavation, mapping etc.
MANAGEMENT	45	integration in CHM/handling of planning permissions, efficiency/cost-effectiveness/speed, tool for prognosis etc.
KNOWLEDGE	14	knowledge, interpretation and analysis
PRESERVATION	9	non-intrusiveness/non-destructiveness, reduce physical impact, limit conflict

Figure 1: The total number of geophysical surveys in Norwegian archaeology including 2015. Source: The AGIN database compiled by NIKU and the NTNU University Museum

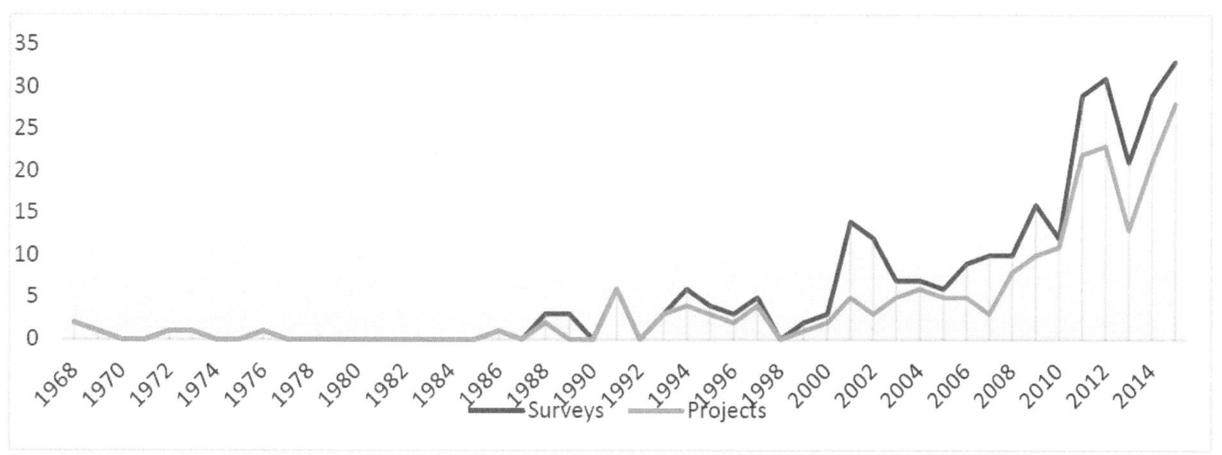

Figure 2: The most emphasised roles and functions of geophysical methods by actors in Norwegian Cultural Heritage Management.

as "new" methods is clearly a misunderstanding, as the amount of known surveys over time shows that this, at least for Norway, is not the case. What is new, is the increased research emphasis by the beforementioned institutions NIKU (in collaboration with the LBI ArchPro) and the NTNU University Museum in Trondheim. By denoting geophysical methods as "new" methods, it is obvious that there is a lack in the general knowledge and understanding of geophysical methods within the CHM system in general, which in turn also leads to a reluctance in actually utilizing geophysical methods as part of the daily cultural heritage management.

The list in figure two demonstrates that even if the most used description of geophysical methods is as non-destructive and non-intrusive methods, this function is one of the least described roles by actors within the CHM in Norway. The actors within the Norwegian CHM system rather envision geophysical methods as methods for documentation, registration, investigation etc., as well as a noted emphasis of their potential for increased effectivity. This becomes even clearer when investigating the perceived role and purpose each actor envision that geophysical methods can provide. In order from the most mentioned role and function to the least, the types of roles and functions envisioned is as detailed in Figure 2.

There is also often an emphasis on the potential geophysical methods have in replacing traditional archaeological registration methods such as mechanical test trenching, which is one of the preferred methods of archaeological registration methods. This emphasis of replacing rather than seeing geophysical methods as a supplement and management aid is an element that hinders an increased integration of geophysical methods in Norwegian Cultural Heritage Management. The attitude of judging the effectiveness of geophysical methods purely on its effectiveness to identify archaeological features without additional investigation, but not recognising its potential for increased knowledge in a non-intrusive manner early in the planning stages, hinders proper integration.

The slow implementation and integration of geophysical methods in Norwegian Archaeology can be explained by several factors. These are mainly an inherent scepticism and an increased focus on costs – where geophysical methods are seen as an extra cost compared with traditional registration methods of which the archaeologists have greater experience. There are major challenges in properly implementing and integrating geophysical methods in the day- to day CHM. This relates especially to issues such as costs versus knowledge potential, dissemination and increased training to government officials and actors related to the daily CHM, and the will to see the value of the non-intrusive potential of geophysical methods as an integrated part of the CHM as a longer process than purely the detection of sites and monuments. The consequences of this is a political rather than a knowledge based management, slow implementation of conventions and strategies, less focus on the value of the cultural history as well as their values for creating identity and knowledge, in addition to increased damage and loss of cultural historically important sites and monuments (Stamnes 2016).

Bibliography

Stamnes, A. A. (2016) *The Application of Geophysical Methods in Norwegian Archaeology: A study of the status, role and potential of geophysical methods in Norwegian archaeological research and cultural heritage management.* PhD. Trondheim, Norway: Norwegian University of Science and Technology.

Integrating GPR and excavation at Roman Aeclanum (Avellino, Italy)

Guglielmo Strapazzon[(1)], Ben Russell[(2)] and Girolamo F De Simone[(3)]

[(1)] Independent researcher, Italy; [(2)]University of Edinburgh, Edinburgh, United Kingdom; [(3)] Accademia di Belle Arti di Napoli, Naples, Italy

guglielmostrapazzon@gmail.com

This contribution provides a comprehensive report of the ongoing investigations through GPR and excavations at the Roman city of Aeclanum, in the southern Italian Apennines. These activities provide new insights into the usefulness of GPR for the reconstruction not only of the urban grid, but also of the phases of spoliation and natural erosion that the city suffered after its abandonment.

Introduction

Ancient cities are often characterized by several phases of unbroken inhabitation which result in a continuous transformation of buildings and – on a larger scale – of the urban fabric itself. Ultimately, the superimposition of these events generates a deeply stratified archaeological deposit. After the abandonment of the cities, other factors modify these archaeological deposits; among these, the spoliation of masonry, damage caused by modern agricultural activities, and processes of natural erosion connected to the geological conformation of the areas on which the cities lie.

The information about how and where these factors influenced buried structures of an ancient city can be obtained by means of geophysical techniques. Among these, thanks to its ability to provide a high-resolution 3D model of the subsoil, GPR (Ground Penetrating Radar) can be reckoned as the most useful. In fact, although it is mainly used to produce maps of built structures, it also shows proxies which can be interpreted as the result of loss, damage, and erosion. Thus, GPR can be used to evaluate which areas within the boundaries of an ancient city will benefit from further investigation. This type of spatial information is of paramount importance not only for understanding post-depositional activity but also planning for long-term rescue and safeguarding practices.

Context

The Roman city of Aeclanum is a clear example of site where the interaction of all these factors created a complex archaeological deposit.

Aeclanum lies beyond the shores of the Bay of Naples in inner Campania, and more precisely in the province of Avellino, in the lower Italian Apennines. According to the ancient sources, a first city-like settlement was established in the 3rd century BC under the Samnites. In 89 BC the city was sacked by Sulla and turned into a colony under Hadrian in AD 120, and lastly developed into an important Christian bishopric that survived until the destruction of the city, ordered by the Emperor of Byzantium Constans II in AD 662.

Although the site is by no means small (at least 18 hectares) and encircled by still-visible city walls, only a few buildings have been brought to light in the past centuries, noticeably part of a *macellum*, an early Christian church, as well as the Roman baths and a small portion of a probable theatre. Vast areas between these buildings, however, remain unexplored, and it is difficult to reconstruct the original layout of the city. This picture is further complicated by the unusual setting of the site, which is spread across a series ridges and the steep hillsides between them. The reason for this curious location is the presence of the important Via Appia, which runs along the top of the highest ridge on which the city is located. The city seems to have developed along the length of this ridge and the hill tops branching off it, and then spilled down the slopes to the west over time. To deal with this extreme topography, the city was probably built on terraces, but the exact arrangement of these is unclear. Later geomorphological changes, including localized landslides, have severely affected the topography of the site. All of this makes the area an interesting test case for the potential of GPR for reconstructing both the city's built environment and its original topography.

Methods

For the first season at Aeclanum, the project aimed to carry out an extensive GPR survey and to combine this with thorough documentation of the preserved standing remains in order to elaborate a working hypothesis of the original city grid and its development throughout time.

A high resolution map of the archaeological deposits has been obtained by collecting parallel GPR profiles 25cm apart with 200MHz and 600MHz antennas, while for the standing remains, SfM (Structure from Motion) techniques were employed in order to acquire a three-dimensional view of these complex structures. Archaeological features exposed by new excavations have also been documented using SfM, which was later used to test the GPR survey results.

The combination of these two techniques into a GIS produced a complete picture of both standing structures and the subsoil in the 1.6 hectares surveyed so far.

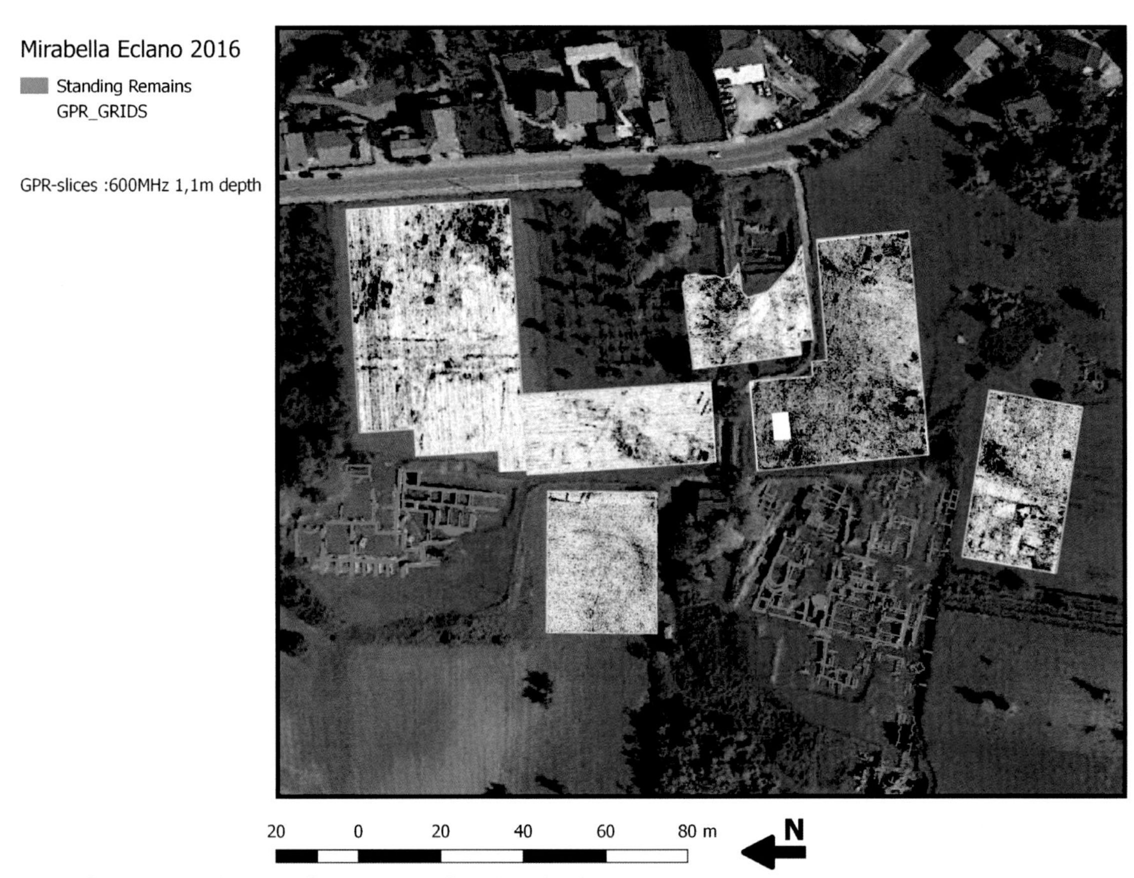

Figure 1: GPR results from the first season of work at Aeclanum.

Figure 2: Interpretation of the GPR survey.

Results

The results of the GPR survey are promising and will be presented in detail. The survey shows many clear features through which a tentative plan of at least the eastern half of Aeclanum can be reconstructed, while other features probably testify to compromised archaeological deposits, destroyed by human or natural events.

At the northern end of the surveyed area, to the east of the excavated baths, a major road can be identified. Between this road and the baths several open spaces, probably surrounded by porticoes and associated with the baths and the building to the south are visible. This suggests that the baths were a much larger complex than previously thought, and that the building to the south was a monumental (cultic?) building. Both structures opened on to this substantial road, 5m wide, running north-south through the city. It is tempting to identify that as the Via Appia, since portions of it have been exposed in the past, near the city, with a similar north-south orientation.

The southern end of the surveyed area shows a large area covered with paving only in its north-western corner, while the rest had probably been robbed. This open area was flanked by the excavated *macellum* to the east and a road to the south. Beyond this road lies a vast (public?) building, whose walls are clearly visible for at least 2m in depth. The open space looks very much like a forum, while the other building might be interpreted as baths or a temple.

At the centre of the site, in an area compromised by agricultural and robbing activities, the GPR shows a curious curved building that was also partially excavated during the first season. The curved features visible in the GPR match the curve of the structure recently brought to light and seem to indicate the remains of a (heavily spoliated) theatre.

Alongside these built structures, the high resolution of the GPR survey enabled the identification of key areas within the site where robbing and agricultural activities have profoundly damaged the archaeological deposits. Two main areas of agricultural damage are visible on the main ridge (highlighted in yellow on Fig. 2), the largest one of which has effectively levelled the top of the ridge, removing all built structures. The area closest to the modern road has also been the most heavily spoliated. This is clearest in the area of the theatre, where almost no walls survive, and is confirmed by excavation; this entire complex was spoliated down to ground level. On the western side of the main ridge, in the area shaded in green on Figure 2, very few structures could be identified. This is a sharply sloping hillside, the exact topography of which probably results from a more recent landslide. It seems likely that a terrace wall of some sort originally divided the eastern ridge of the site from this lower area and that the collapse of this terrace has obscured all of the structures in this area. Some sort of terrace wall certainly existed to the west of the excavated baths (to the north) and the residential area (to the south).

In sum, GPR has been used at Aeclanum not simply to identify the locations of ancient buried structures but also to explore the impact of post-depositional factors of the preservation of Roman-era buildings. Agriculture and later spoliation have heavily shaped the archaeology of the site, but natural events also seem to have played a role. Further experimentation with this technique on the lower areas of the site will be able to confirm whether natural events have totally obscured the urban plan in this area, or whether pockets of the city remain visible to GPR in this area.

Prospection at the Medamud (Egypt) site: building archaeological meaning from the geophysical *in situ* measurements

Julien Thiesson[(1)], Felix Relats Montserrat[(2)], Christelle Sanchez[(3)], Roger Guérin[(1)] and Fayçal Réjiba[(1)]

[(1)]UMR 7619 – Milieux environnementaux, transferts et interactions dans les hydrosystèmes et les sols, UPMC, Paris, France; [(2)]UMR 8167 – Orient et Méditerranée, composante Mondes-pharaoniques, Paris, France; [(3)]UMR 6249 Chrono-environnement, APRAGE (Approches Pluridisciplinaires de Recherche Archéologique du Grand-Est), Besançon, France

julien.thiesson@upmc.fr

Introduction

Medamud is a well-known upper Egypt temple excavated between 1924 and 1939 dedicated to the god Montu. Bisson de la Roque and Robichon, the French excavators of the site, have published a rich documentation but they only have taken into account the temple's remains, and the surrounding area (the "kom") was mainly ignored (Bisson de la Roque 1946). A recent project aimed to reconsider the site. In 2015, the project team has performed a ceramic and geophysical survey in order to increase our knowledge of the kom. The numerous firing wastes found all over the area prove that Medamud was one of the biggest production centres in Upper Egypt (Relats Montserrat *et al.* 2017). In the frame of the project, the geophysical survey was intended to bring elements on three main questions:

- Is the link between the temple and the Nile river evidenced in the soils?
- Are the remains found in the former excavations detected by the method(s) used?
- Does the unstudied area present remains of interest in their relationship to the temple?

This paper will focus on the third point. We will describe the devices used and especially how the maps obtained were interpreted. After this step, the geophysical anomalies were clustered by statistical methods. The cluster maps were then combined according to the *a priori* expected geophysical signature of some remains. Finally, we propose maps of features of potential archaeological interest.

Material and Method

A 6090 m² area was surveyed by the electromagnetic induction method, and over 5500 m² by the magnetic method. The measurements were done with a CMD MiniExplorer from GF Instruments in VCP (vertical co-planar coils) configuration (frequency 30000 Hz, three Tx-Rx distance: 0.32, 0.71 and 1.18 m) for the electrical conductivity and the magnetic susceptibility. The magnetic field was recorded with a G858 magnetometer from Geometrics (two sensors in vertical gradient mode). Each map was acquired in continuous mode walking along profiles separated with a 0.5 to 1 m space. The maps are given in apparent magnetic susceptibility and electrical conductivity for the data gathered with the CMD.

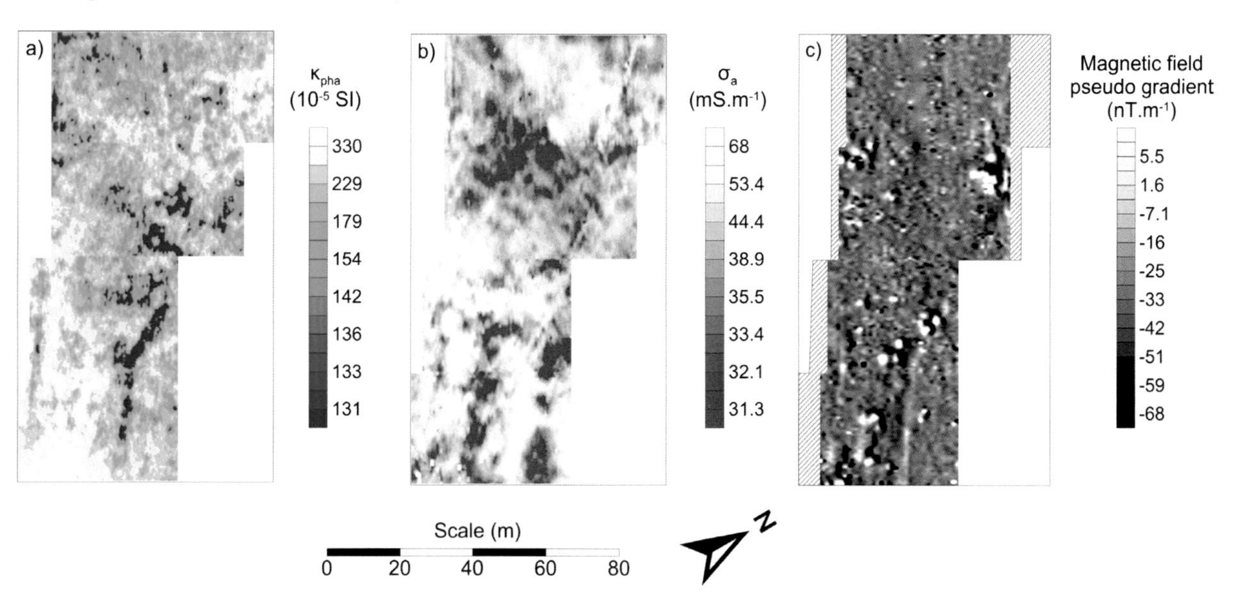

Figure 1: Results obtained over the surveyed area: a) magnetic susceptibility (CMD MiniExplorer Tx-Rx=0.71 m); b) electrical conductivity (CMD MiniExplorer Tx-Rx=0.71 m); c) magnetic field pseudo vertical gradient (G858 magnetometer).

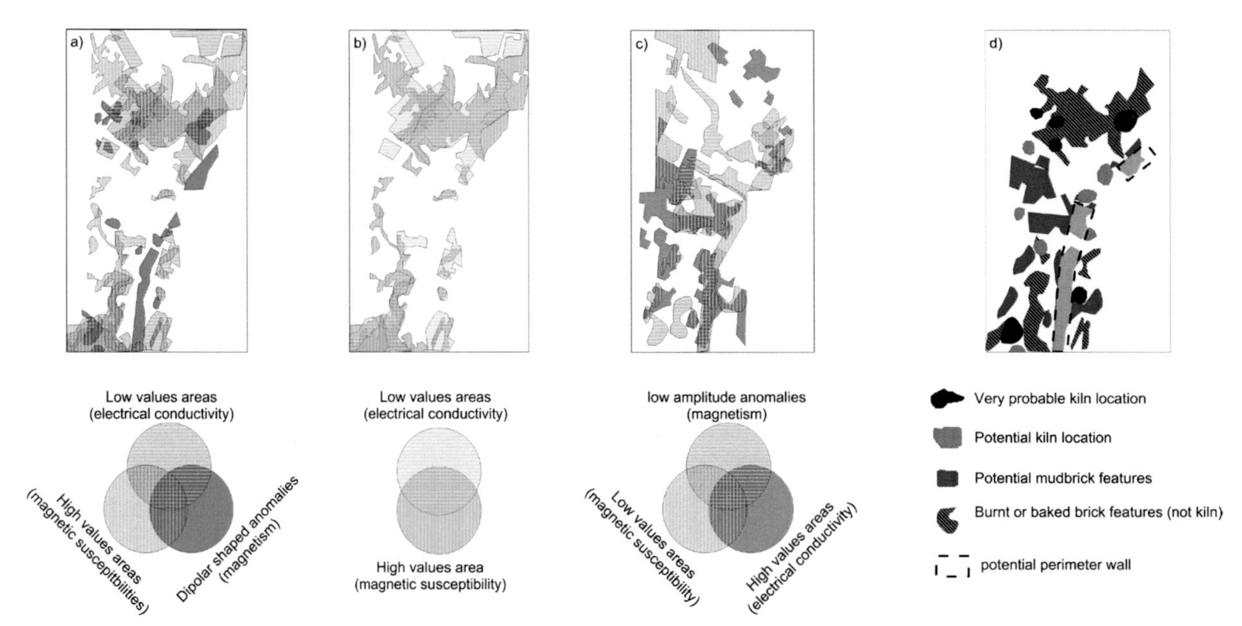

Figure 2: Qualitative maps: a) kiln features; b) baked brick features; c) mudbrick features; d) archaeological synthetic map.

Geophysical Results

Figure 1 shows the results over the whole area. First it appears to be very rich in remains. The magnetic susceptibility shows some large low values areas in the middle of the survey which are continued towards the SE direction in an elongated shape. This organisation is less visible on the conductivity map showing clearly only the elongated feature as a more conductive body. Finally, on the magnetic map, the elongated features also appear and with a polarity corresponding to a lower magnetization in agreement with the low magnetic susceptibility anomaly on the CMD map.

On the northern part of the apparent electrical conductivity map, there is a wide area of low value with rectangular shapes. It seems to correspond to some very faint magnetic features. Dipolar shaped anomalies of high magnitude are shown by Figure 1c. Nonetheless, they do not resolve into clear patterns.

Extracting some Archaeological Meaning

We have established an *a priori* classification of the signature of the archaeological remains which can be found over the site. It forms on a table where the expected remains type were associated with a cluster of values of the geophysical parameters. This scheme was adapted according to the results obtained on the area with the known remains (essentially in term of detectability). For the type of anomalies which appear to have good potentialities to be detected (pottery workshops, mudbricks and baked brick features) we produced the maps shown as Figure 2a, 2b and 2c. The qualitative scale was adapted for each prospection using basic statistics (weak, respectively strong, means inferior, respectively superior, to the first, respectively third, quartile, for example). The qualitative maps for each type of features were obtained by overlapping all the qualitative maps. Finally, we obtained a map (Figure 2d) with anomalies classified by expected type of archaeological features.

Conclusion and Perspective

The prospection over the site of Medamoud was made to answer several archaeological questions. Over a previously unstudied area, the geophysical prospection evidenced a lot of archaeological remains. We used the combined results of electromagnetic and magnetic survey to realize some qualitative archaeological maps. These maps were combined with the archaeological clues gathered other the entire site (ceramic gathering, small excavations) and permits the identification of a new artisanal district specialised on ceramic production, and an unknown surrounding wall. The results highlight the potential of the area and will be used to plan future excavations and surveys.

Bibliography

Bisson de la Roque, F. (1946) Les fouilles de l'institut français à Médamoud. *RdE* **5**: 25-44.

Relats Montserrat, F., Thiesson, J., Sanchez, Chr., Réjiba, F. and Guérin, R. (Accepted manuscript) Une première campagne de prospection à Médamoud : méthodologie et résultats préliminaires (Mission IFAO/Paris-Sorbonne/ Labex Resmed de Médamoud), *BIFAO 116, 2017*, IFAO, Cairo, (accepted article).

Acknowledgements

This research was founded by the MEDAMOUD (SU-14-R-SCPC-16-2-IA) project of the Convergences proposal from Sorbonne Universités.

Results of the GPR survey of former Roman churches in Slovakia

J Tirpak, M Bielich, M Martinak and D Bešina

jtirpak@ukf.sk; mariobielich@seznam.cz

The aim of the article is to present the examples of geophysical and archaeological prospection on deconsecrated religious buildings in the south of Slovakia. This topic is presently being researched by Dr J Tirpak (2012, 2016).

Based on the study of written sources and the field survey we identified the locations of former medieval churches. Initial landscape surveys were conducted through topographic, aerial and satellite image analysis. Subsequently, we have selected the locations for georadar measurement survey (GPR measurement), with the aim of identifying the presence of foundations of dispensed church.

As an example, we have chosen three measurements differing by methodology of archaeological prospection (Fig.1:A). The first locality (Ondrejovce) was identified on the basis of written sources. The second (Levice – Bratka) was examined by archaeological research in the years 1958 – 1963 (Habovsiak 1963, 407-458). Unfortunately, the research wasn't precisely measured from a geological point of view, and it wasn't known exactly where the religious building was situated. The last locality (Trnovec nad Vahom, part Velky Jatov) was discovered during field survey. Besides medieval ceramics, fragments of human bones were also collected. The site was the remnant of cemetery situated around the church.

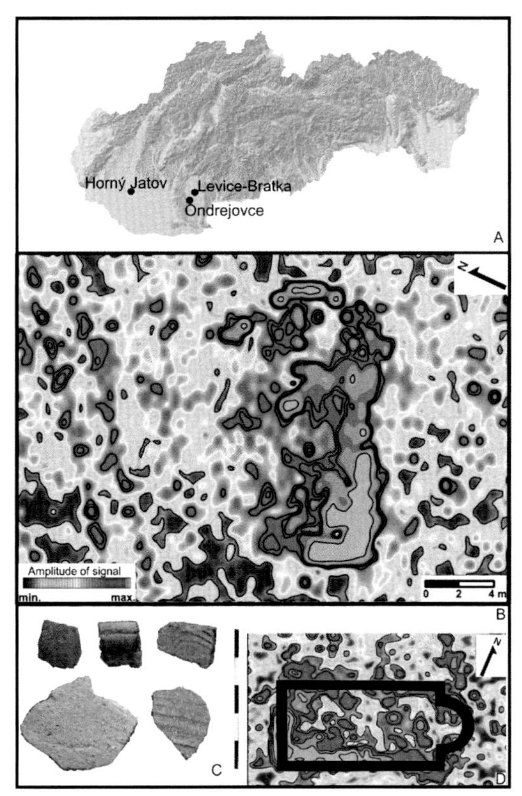

Figure 1: A) Location of former churches in south Slovakia; 1: B) Ondrejovce-results of GPR measurement-horizontal slide; 1: C) Medieval pottery from survey; 1: D) Interpretation of Romanesque church.

Ondrejovce, Locality: And

The aim of non-intrusive geophysical survey (GPR measurements) was to confirm the presence of foundations of the Church of St. Martin, which was mentioned in written sources.

GPR survey measured an area of overall extent 30 x 20m with the density of scans 0,05m (along the profile) x 0,5m (between the individual profiles) (Fig. 1:B). On the figure we see inserted a rudimentary/simple reconstruction of the footprint of circumferential walls of the former one nave church with a semi-circular apse. The former Roman church was built with a divergence of 10 degrees in orientation of the longitudinal axis positioned south-west/north-east. The length of the church was approximately 15 metres and width ca. 6 metres. The thickness of the foundation wall is around 100 centimetres (Fig. 1:D). A field survey was also completed, during which we obtained a smaller amount of the ceramic fragments, dated to 12-13th century (Fig. 1:C). This proves to be evidence for the medieval character of a settlement in the locality. This medieval village was situated in direct proximity of religious building.

Levice, Locality: Bratka

In the years 1958-63 an excavation, under the supervision of the archaeologist A. Habovstiak, was conducted on the deserted medieval village Bratka, during which were unearthed foundations of two small roman churches from 12th and 13th century, and a portion of graves from the cemetery. The older from the two structures was identified as the church of St. Martin and Stephen, which was according to written sources built by local nobleman Eusidinus (Fig. 2:B). GPR survey measured an area of overall extent 30 x 30m with the density of scans 0.05m (along the profile) x 0.5m (between the individual profiles) (Fig.2:B). After processing and interpretation of measured data we can see that on the excavated area right under the surface of topsoil are deemed zonal anomalies, which indicate encroachment/direction of a foundation wall of the tower, naves and church shrines in depths from 40 to 160 centimetres. For this purpose we chose horizontal GPR survey for depths from 75 to 90 centimetres and the footprint/outline of the church's foundation walls, extension and tower, which were excavated by A. Habovstiak (Fig. 2.A). The length of the church was approximately 14 metres and width ca. 6 metres. The thickness of the foundation wall is around 100 centimetres. During the archaeological excavation ceramics from 12-13th century were recovered, which helped with dating of the earliest medieval settlement, and also with the earliest building/construction stage of the church (Fig. 2:C).

Trnovec nad Vahom, Part Horny Jatov, Locality Eastwards from JRD

During the field survey eastwards from JRD remnants of human bones were identified in addition to medieval ceramics. After more detailed survey we identified fractions most likely from the older stone building. At the place of highest concentration of stone artefacts we carried out the GPR measurements. Interpretation of geophysical results was based on results on horizontal and vertical GPR slide (Fig. 3:B). With regards to the preservation of a foundation wall, it is visible that the upper part of foundation masonry was destroyed by tillage, especially the shrine and northern part of church foundations. On Figure 3:A we attempted to outline a simple footprint reconstruction of the circumference walls of a former/destroyed single nave church with a rectangular presbytery. The Medieval church was built with divergence of 30 degrees in orientation of the longitudinal axis directed to South-West/North-East. The length of the church was approximately 13 metres and width ca. 6 metres. Ceramics obtained during the survey were dated to 11th-15th century (Vozak and Kormosi 2013: 148-150). On this site a medieval settlement existed for a long period

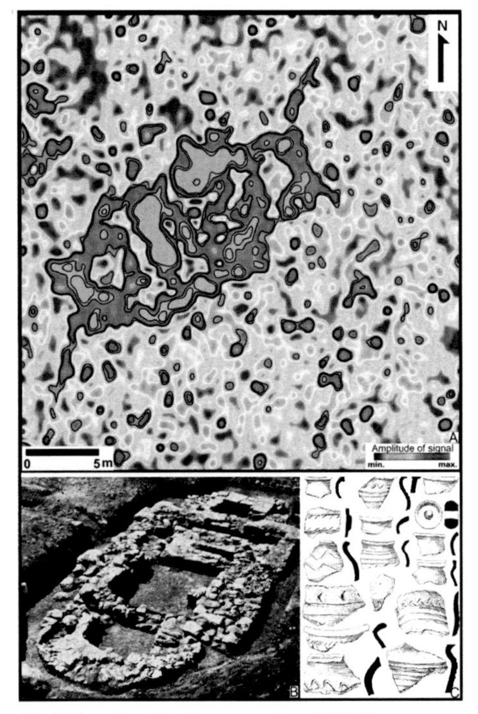

Figure 2: A) Levice-Bratka-results of GPR measurement; 2: B) Pohľad na vykopané základy kostola v roku 1963 (*A. Habovšiak 1963*); 2: C) Medieval pottery from the excavation (*A. Habovšiak 1963*).

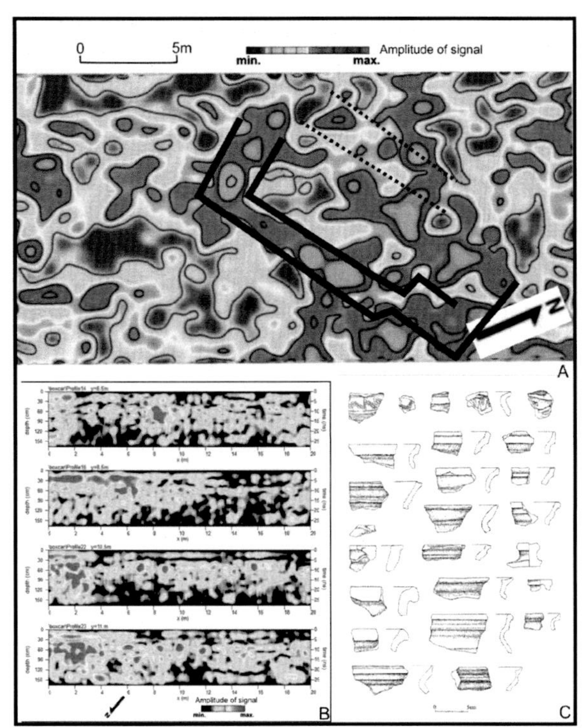

Figure 3: A) Horný Jatov -results of GPR measurement horizontal slide; 3: B) Horný Jatov -results of GPR measurement vertical slide; 3: C) Medieval pottery from the survey (Vozák & Kormoši 2013, tab.148).

of time. The age of the church, which is strongly damaged by agricultural activity, would be possible to specify chronologically only in the event that a full archaeological excavation were carried out.

To sum up, we would like to state that the research of medieval architecture by non-intrusive archaeological methods in Slovakia achieved a considerable upturn in the last few years. The main reason is an intensive cooperation of an archaeologist, geophysicist and archivist. A few dozen geophysical surveys were carried out, which discovered a high number of medieval churches whose preservation and dating will be verified only by an archaeological excavation.

Bibliography

Habovšiak, A. (1963) Zaniknutá stredoveká dedina Bratka pri Leviciach. *Slovenská Archeológia* **XI-2**, (1963): 407-458.

Tirpák, J. (2012) Stredoveká sakrálna architektúra vo svetle archeogeofyzikálneho výskumu na Slovensku. I. Diel. Nitra 2012.

Tirpák, J. (2016) Stredoveká sakrálna architektúra vo svetle archeogeofyzikálneho výskumu na Slovensku. II. Diel. Nitra 2016.

Vozák, Z. and Kormoši, J. (2013) Okres Šaľa. Archeologické nálezy a náleziská. Šaľa 2013.

Acknowledgments

Translated by Mgr. Marek Krnač

From integrated interpretative mapping to virtual reconstruction - a practical approach on the Roman town of Carnuntum

Juan Torrejón Valdelomar[(1)]*, Mario Wallner[(1)]*, Klaus Löcker[(1,2)]*, Christian Gugl[(3)]*, Wolfgang Neubauer[(1,5)], Michael Klein[(4)], Nika Jancsary-Luznik[(1)], Tanja Trausmuth[(1)], Alexandra Vonkilch[(1)], Tomas Tencer[(1,5)], Lisa Aldrian[(1)] and Michael Doneus[(5,1)]

[(1)]Ludwig Boltzmann Institute for Archaeological Prospection and Virtual Archaeology (LBI ArchPro), Vienna, Austria; [(2)]Central Institute for Meteorology and Geodynamics (ZAMG), Vienna, Austria; [(3)] Austrian Academy of Science, Vienna, Austria; [(4)]7Reasons Medien GmbH, Vienna, Austria; [(5)] University of Vienna, Vienna, Austria

Equal contributors

Juan.Torrejon-Valdelomar@archpro.lbg.ac.at

Introduction

The Roman town of Carnuntum, close to Vienna (Austria), is an outstanding archaeological site, having been the capital of the Roman province of Pannonia Superior. As such it has been the subject of a wide range of archaeological investigations (e.g. airborne remote sensing, aerial archaeology, excavation, sampling and historical data). More than a century of excavations could reveal only a patchwork of detailed information, which hardly allowed a comprehension of structure and function of both *canabae legionis* and civil town. Fifty years of aerial archaeology gave an overview of the complete archaeological landscape of this Roman town, but due to its limitations (visibility of buried structures depend on a broad range of factors), some areas were lacking detailed information while others were totally unsuccessful. Therefore, a detailed geophysical prospection approach covering the entire 12 square kilometres (832 ha magnetics, 245 ha ultra-high-resolution GPR, ca. 1200 ha ALS & aerial photography) was the most suitable solution.

The ArchPro Carnuntum Project

This situation leaded to the *ArchPro Carnuntum project* hosted by the LBI ArchPro and the ZAMG. It allowed a comprehensive investigation of the archaeological landscape of the Roman town of Carnuntum and its surrounding area. Over recent years, high-resolution archaeological prospection data using ground-penetrating radar (GPR) and magnetometry surveys were acquired, visualized, combined with datasets from earlier investigations,

Figure 1: 2D integrated interpretation of a selected area, showing individual features as well as distinct areas.

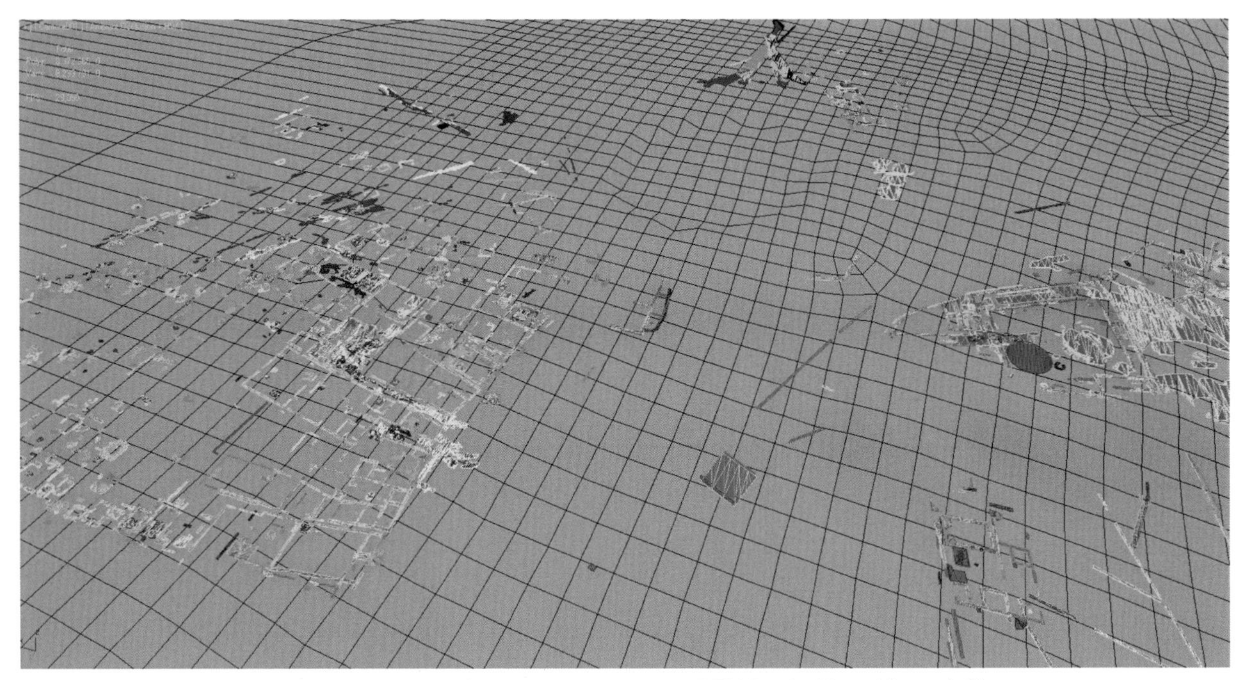

Figure 2: 3D archaeological interpretation of the chosen area and DTM loaded in a 3D modelling program.

and interpreted. The result is an almost complete picture of the nowadays buried remains of this important provincial capital. The acquisition, processing and interpretation of this truly big dataset has been accomplished within only 36 months.

Integrated Interpretation Workflow

It is reasonable to assume that any resulting archaeological interpretation cannot be based on the analysis of one individual method alone. Only by integrating all available data into an integrative archaeological interpretation is it possible to gain a complete and exhaustive understanding of the buried archaeology, while attempting to compensate the individual weaknesses of the different methods. However, some land owners did not allow the investigation of their fields, which resulted in void areas (especially in the *canabae legionis*), which could be filled with information from aerial archaeology. Results from excavation and systematic field surveys allowed the production of a detailed chronological framework for the interpretation of the non-invasive prospection data. Therefore, a combination of all the different data and information sources is imperative. This is, however, a complex task demanding adequate data management and an integrated archaeological interpretation that is accurate, readable and ready to use for the target readers. The main platform for this purpose is a georeferenced GIS-based archaeological information system (AIS).

In Carnuntum, a specific AIS named *ArchaeoAnalyst* was used, which was developed by the LBI ArchPro. As a first step, an integrated interpretative mapping was conducted, which resulted in discrete features. These can be compared with known archaeological and historical analogies. This process leads to a better definition of specific areas in the prospected data, which form the basis for further spatio-temporal analysis and, subsequently, may result in virtual volumetric reconstruction suggestions.

Applying this workflow to the Roman town of Carnuntum, it was possible to define different areas, such as the one that surrounds the amphitheatre of the civil town. The available, highly detailed datasets were combined with archaeological and historical analogies, generating a more enhanced and complete understanding of the buried remains, consequently leading to the definition of former purpose or use of certain building or architectural complexes.

From 2D Interpretation to Virtual Reconstructions

The georeferenced archaeological interpretation resulted in 2D or 2.5D datasets. In our interpretation approach, 3D information from GPR datasets was embedded into the attribute table of each mapped

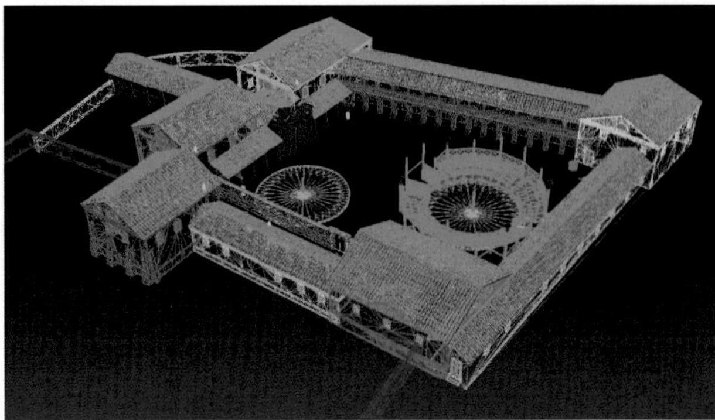

Figure 3: CAD model of the reconstructed school of gladiators of Carnuntum.

feature. This information can be converted into actual 3D objects that, together with a Digital Terrain Model, provide a basis for vertical extrusions based on the archaeological interpretation.

Highly detailed digital terrain models (DTM) derived from airborne and terrestrial laser scanning allows further (infra-)structural and architectural analysis of the different units, as for example entrances, openings, roofing, static issues, open spaces, fresh-water and drainage systems.

The detailed constructive analysis can generate a backflow that may be transferred to the archaeological interpretation and, consequently can lead to a better understanding of any archaeological prospection data. The final product in this stage of research would be a Computer-Aided Design (CAD) model.

The next step focuses on a first texturing of the 3D models, which basically comprises the adding of material properties to the already existing volumes. These textures provide visual information of a certain material that at the same time results in a more realistic visualisation of the reconstruction suggestions, allowing enhanced discussion processes among the specialists.

At this stage, the generated digital 3D models are very suitable for spatio-temporal analysis, as well as for further scientific discussions focusing on specific topics. They will serve as the basis for popular or scientific dissemination and further consequent purposes, such as animations, augmented-reality applications, or the generation of still images.

The virtual archaeology approach permits the cost-efficient generation of different versions of the same archaeological structures – from archaeological artefacts to entire landscapes, which from a scientific point of view is one of its main advantages.

Within this reconstructive process, it is imperative to document every step and stage, in order to create a framework that fulfils both the *London Charter* and the *Principles of Seville* concerning scientific, computer aided reconstructions.

The main argument for the generation and use of such virtual models and visualisations is that they provide a highly flexible tool to communicate ideas within the scientific community as well as to the wide public. Three-dimensional models are also a useful means for further in-depth analysis of the data, and do not only present the final output.

Bibliography

Denard, H. (2009) *The London Charter*, Draft 2.1, King's College London.

Doneus, M., Gugl, C. and Doneus, N. (2013) Die Canabae von Carnuntum – eine Modellstudie der Erforschung römischer Lagervorstädte. Von der Luftbildprospektion zur siedlungsarchäologischen Synthese. *Der Römische Limes in Österreich* **47**.

International Forum of Virtual Archaeology, Principles of Seville, Final Draft.

Torrejón Valdelomar, J., Wallner, M., Trinks, I., Kucera, M., Luznik, N., Löcker, K. and Neubauer, W. (2016) Big Data in Landscape Archaeological Prospection. In *Proceedings of the 8th International Congress on Archaeology, Computer Graphics, Cultural Heritage and Innovation "Arqueológica 2.0" in Valencia (Spain), Sept, 5-7, 2016.*

Neubauer, W., Gugl, C., Scholz, M., Verhoeven, G., Trinks, I., Löcker, K., Doneus, M., Saey, T. and Van Meirvenne, M. (2014) The discovery of the school of gladiators at Carnuntum, Austria. *Antiquity* **88**: 173-190.

Neubauer, W., Doneus, M., Trinks, I., Verhoeven, G., Hinterleitner, A., Seren, S. and Löcker, K. (2012) Long-term Integrated Archaeological Prospection at the Roman Town of Carnuntum/Austria. In P. Johnson and M. Millett (eds.) *Archaeological Survey and the City*. Oxford: Oxbow (Monograph Series, No. 2), 202–221.

Extensive high-resolution ground-penetrating radar surveys

Immo Trinks[(1)], Alois Hinterleitner[(1,2)], Klaus Löcker[(1,2)], Mario Wallner[(1)], Roland Filzwieser[(1)], Hannes Schiel[(1)], Manuel Gabler[(3)], Erich Nau[(3)], Julia Wilding[(1)], Viktor Jansa[(1)], Petra Schneidhofer[(1)], Tanja Trausmuth[(1)] and Wolfgang Neubauer[(1,4)]

[(1)]Ludwig Boltzmann Institute for Archaeological Prospection & Virtual Archaeology, Vienna (LBI ArchPro), Austria; [(2)]Central Institute for Meteorology and Geodynamics, Vienna, Austria; [(3)]Norwegian Institute for Cultural Heritage Research, Oslo, Norway; [(4)]University of Vienna, Vienna, Austria

immo.trinks@archpro.lbg.ac.at

Introduction

Traditionally, ground-penetrating radar (GPR) measurements for near-surface geophysical archaeological prospection are conducted with single-channel systems using a GPR antenna mounted in a manually operated cart similar to a pushchair, or towed like a sledge behind the operator. The spatial data sampling of GPR surveys using such devices for the non-invasive detection and investigation of buried cultural heritage is, with few exceptions, at best 25 cm in cross-line direction of the measurement. With this relatively dense measurement spacing and 2-3 persons participating in the fieldwork, coverage rates between 1/4 hectare and 1/2 hectare per day are common, while considerably smaller survey areas at often coarse measurement spacing are being reported still.

Advancing GPR Prospection

Even though GPR prospection may already be regarded as the most advanced geophysical prospection methods in archaeology, still there exist several ways how this method can be further improved in order to render it more efficient and accurate, and to enhance its imaging capabilities. In analogy to the suggestions made by magnetic archaeological prospection pioneer Helmut Becker, who in July 2006 in Grosseto postulated the three "*S*" (*Speed*, *Sensitivity* and *Spatial Resolution*) regarding the advancement of magnetometry (Becker 2009, 130), the same approach is applicable to near-surface GPR prospection. The overall goal of the advancements is twofold: to increase the efficiency of the data collection procedure, leading to an increase in data quantity, as well as to enhance the quality of the generated data and resulting images.

Survey Speed

An increase in survey speed results in larger spatial coverage, thus increased fieldwork efficiency at constant sample spacing, or increased spatial resolution at constant coverage rate. Greater survey speed can be achieved through three different measures:

- the use of multiple GPR antennae in simultaneous, parallel operation, through arrangement in antennae arrays,
- the use of motorized survey vehicles permitting continuous operation at greater speed, compared to manually operated systems, and
- the use of automatic positioning systems tracking the location of the measurement system with great accuracy, removing the labour and time intensive requirement for exactly spaced survey lines/grids on the ground.

As a consequence of an increase in survey speed it becomes possible to operate more cost efficiently, since less time is needed in the field in order to cover a specific area. The use of automatic positioning systems reduces the number of staff required in the field. Generally, the chance to detect structures of archaeological interest increases when larger areas can be investigated, respectively the detection chances increase when a greater spatial sample resolution can be applied.

Measurement Sensitivity

The characteristics of individual GPR antennae and systems performance can differ greatly in terms of signal-penetration-depth, transmitted and received frequency content, and signal-to-noise ratio. The design of a GPR antennae, its shielding, as well as the high-frequency signal generation and recording appear to be more an art than deterministic science. Comparative GPR measurements performed at test sites under specific circumstances in terms of

Figure 1: Traditional GPR survey with 25 cm profile spacing with a 500 MHz Sensors & Software PulseEkko Pro mounted in a SmartCart.

Figure 2: Motorized high-definition GPR survey using a 6-channel 500 MHz Sensors & Software SPIDAR system with 25 cm profile spacing.

geology and buried structure can result in images of rather different quality (Seren *et al.* 2007). It is recommended that different GPR systems are tested, and that wherever possible, GPR antennae with different frequency characteristics are employed in order to increase the information content of the resulting data and images. Determining for a specific situation the most suitable and sensitive GPR antenna(e) or system has the potential to considerably enhance the quality of the resulting data images. Newly developed GPR systems in terms of ground coupling, signal sampling, frequency coverage, or shielding offer hope for future improvements regarding measurement sensitivity and imaging capabilities.

Spatial Sampling

Considering that most GPR surveys in archaeological prospection are still conducted at rather coarse sample spacing, there exists still considerable potential to improve the resolution of the resulting images by decreasing the spatial sampling. 15 years ago in Vienna, the introduction of a standard GPR profile spacing by *Archeo Prospections®* of no more than 50 cm had been termed '*high-resolution GPR'*. The routine reduction of the GPR crossline spacing to 25 cm in 2005 was hence called '*high-definition GPR*', in analogy to the Wikipedia definition that '*high-definition video is video of higher resolution and quality than standard-definition*'. Grasmueck *et al.* (2005) coined the term '*full-resolution 3D GPR'* for frequency dependent imaging in agreement with the Nyquist theorem for un-aliased sampling. Novo *et al.* (2008) applied this approach to archaeological prospection. Using single-channel systems such grid spacing leads to prohibitively slow survey progress and rather limited coverage. Here multichannel GPR antenna arrays with dense channel spacing offer great potential for improvements, as they do for increased survey speed.

Figure 3: Three motorized 16-channel 400 MHz MALÅ Imaging Radar Arrays from MALÅ Geoscience offering 8×4 cm GPR trace spacing.

State-of-the-art

Over the past years, the advent of novel multichannel GPR antennae array systems has permitted a substantial increase in survey efficiency and spatial sampling density. Using motorized GPR antennae arrays with up to 20 parallel operating channels, in combination with automatic positioning solutions based on real-time kinematic global navigation satellite systems or robotic total-stations, it has become possible to map several hectares per day with as little as 8 cm cross-line and 4 cm in-line GPR trace spacing and 8 mm single-sample slice thickness. The increased spatial sampling for the first time permits the high-resolution imaging of relatively small archaeological structures, such as for example only 20 cm wide post-holes, the brick pillars of Roman floor heating systems, or the variations in reflectivity inside Viking Age coffins, allowing for improved archaeological interpretations of the collected data. The possibility to cover large areas efficiently increases the chances to detect structures of archaeological, geological and geo-archaeological interest, and to image and understand them in their context. Instead of being forced to select individual buildings or structures for investigations using GPR surveys, the technologically and methodologically extensive survey approach permits the comprehensive blanket investigation of entire settlements and extensive archaeological sites as well as their surroundings. Survey areas that measure in square kilometres rather than hectares (over 12.7 square kilometres of area have been mapped by the LBI ArchPro since 2010 in ultra-high resolution) provide detailed information not only on the buried archaeology but also the space in-between.

Large-scale archaeological prospection data examples as well as specifically developed tools for the efficient processing and GIS-based interpretation of the huge high-resolution GPR data sets will be presented.

Bibliography

Becker, H. (2009) Caesium-magnetometry for landscape archaeology. In S. Campana and S. Piro (eds.) *Seeing the unseen – Geophysics and landscape archaeology.* London: Taylor & Francis, 129–165.

Grasmueck. M., Weger, R. and Horstmeyer, H. (2005) Full-resolution 3D GPR imaging. *Geophysics* **70**(1): K12–K19.

Novo, A., Lorenzo, H., Rial, F. I., Pereira, M. and Solla, M. (2008) *Ultra-dense grid strategies for 3D GPR in Archaeology.* In 12th International Conference on Ground Penetrating Radar, June 16-19, 2008, Birmingham, UK.

Seren, S., Eder-Hinterleitner, A., Neubauer, W., Löcker, K. and Melichar, P. (2007) Extended comparison of different GPR systems and antenna configurations at the Roman site of Carnuntum. *Near Surface Geophysics* **5**(6): 389–394.

The challenge of investigating the tumulus of Kastas in Amphipolis (northern Greece)

G N Tsokas[1], P I Tsourlos[1], J-H Kim[2], M-Z Yi[2] and G Vargemezis[1]

[1]Exploration Geophysics Lab., Aristotle University, Thessaloniki, 54124 Greece; [2]Mineral Resources Research Division, Korea Institute of Geoscience and Mineral Resources, 92 Gwahang-no, Yuseong-gu, Daejeon 305-350, South Korea

gtsokas @geo.auth.gr

Introduction

Tumuli are monuments consisting of soil and stone embankments and constitute common burial practice for various civilizations which flourished at different eras and in different places. Clearly, they were erected as burial monuments to cover tombs of important persons.

The findings provide valuable information on the funeral customs, the social structure, the administration, the economic life, etc., of the civilizations who erected the tumuli.

The issue is how to image the interior of the tumuli and detect the concealed one, or more, tombs and the other constructions connected with them like "dromoi", altars, etc. The aim is to lead the subsequent excavation directly to these buried structures, avoiding the destruction of as much of the tumulus as possible. Clearly, archaeological prospection offers the means, and consequently an invaluable service, towards saving these monuments.

Many geophysical techniques have been employed in this respect (Tsokas, 2012). However, the relatively recent advanced geophysical tomographic methods and, in particular Electrical Resistivity Tomography (ERT), provide a unique tool for imaging the interior of tumuli (Tonkov and Loke, 2006; Papadopoulos *et al.* 2010). Whatever geophysical method is used, the investigations on tumuli comprise challenging issues.

The Kastas Hill in Amphipolis (Region of Macedonia, Northern Greece)

The Kastas hill is a tumulus formed after the modification of a pre-existing topographic rise by a huge amount of soil and gravel. These materials constitute the embankment that created the tumulus of about 498 m in circumference and 22 m tall. The tumulus is shown in Figure 1 in a satellite photo of the Hellenic Cadastre before the excavations which commenced in 2013.

A revetment runs around the periphery of Kastas, having marble lining which survives for a certain length. This supporting wall was the first to be revealed and constituted the first target of the excavation.

An important monument has been unearthed so far which is 28 m long, 4.5 wide and 6 m tall. Two sphinxes are guarding the entrance, while two caryatides adorn the entrance of one of the chambers, and an excellent mosaic constitutes the floor of another. Clearly the monument is unique both for its size and its typology.

Since it was suspected that other monuments may also be concealed in the embankment of Kastas, a geophysical survey took place to investigate its interior.

The Challenge of a Geophysical Survey on a Huge Structure

The achievements of modern ERT approaches for investigating the interior of tumuli consist in adjusting inversion algorithms, testing different electrode arrays and proposing measuring modes (Tsourlos *et al.* 2014). However, due to its dimensions, none of these approaches could be employed for the case of Kastas.

At first, we focused on delineating the undulations of the ancient hill which had been covered by the embankment that created the formal spherical shape of a tumulus. Some initial ERT grids were established on the surface of the tumulus whereas some long tomographies were also performed as

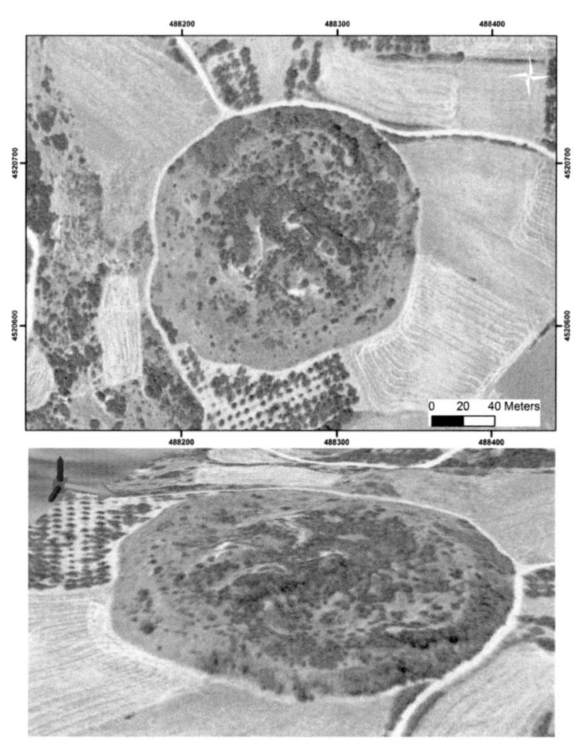

Figure 1: Satellite image of Kastas hill on behalf of the Hellenic Cadastre before the excavations which started at 2013.

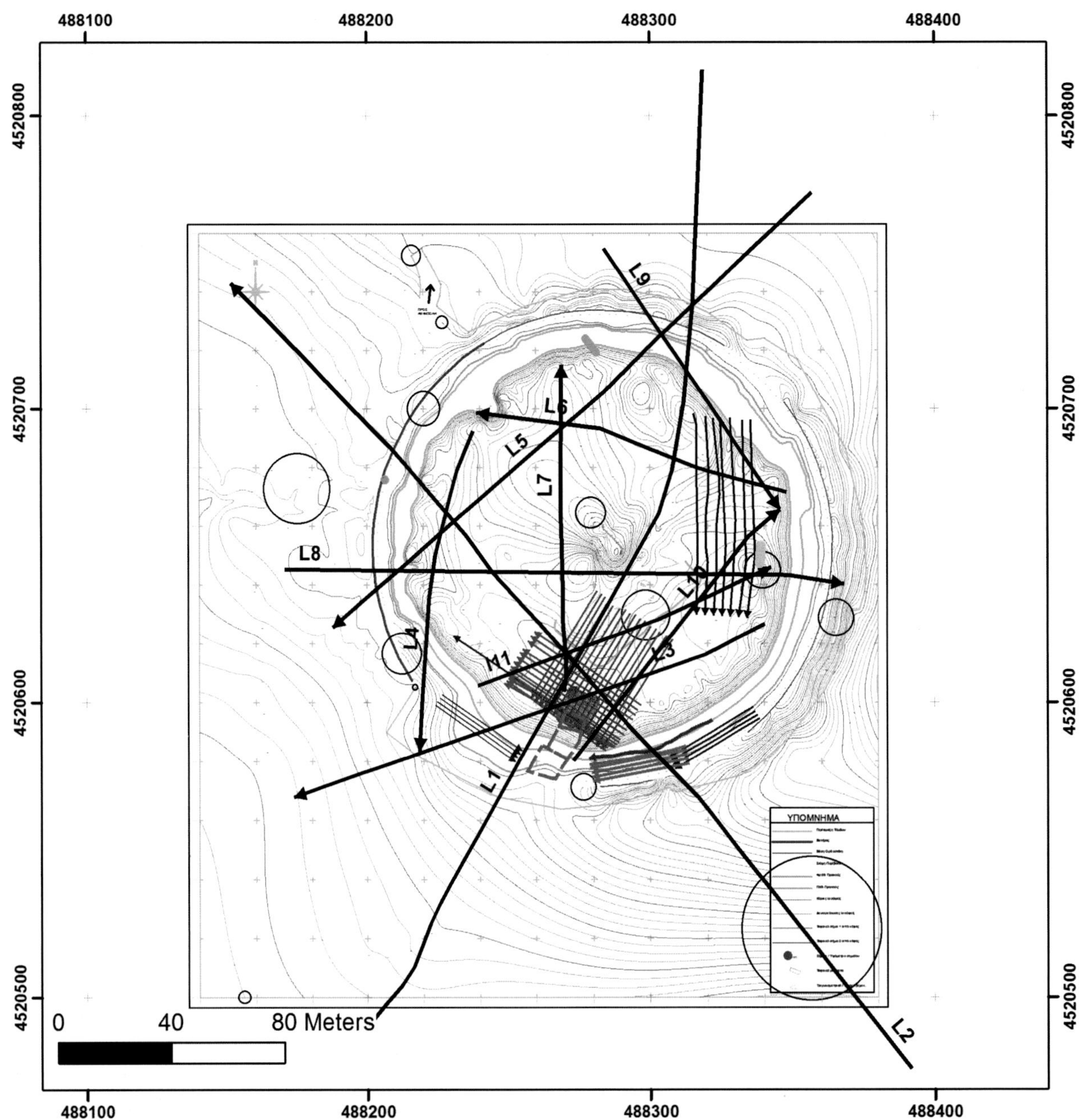

Figure 2: Lay out of the ERTs carried out initially on Kastas in order to delineate the undulations of the hill who was the feature forming the kernel of the tumulus.

shown in Figure 2. Processing and interpretation of these tomographies, either 2 or 3 Dimensional, were performed using conventional software (Tsourlos and Ogilvy 1999, Yi and Kim, 2003).

Next, it was decided to undertake a full 3D survey. The electrodes were laid out on the surface of Kastas as shown in Figure 3. They were deployed in about 10880 m^2 in an area of land having a perimeter of about 400 m. The survey was performed both in the 2D and the 3D context. In other words, some traverses were measured in the 2-dimensional context while a full 3D approach was used for a great portion of readings. At the end, all readings were inverted as a unique data set.

Results

The morphology of the pre-existing hill was assessed by the initial ERTs whose interpretation was assisted by the layering exposed at the flanks of Kastas after the excavation for the revelation of the revetment (*peribolos* in archaeological parlance). Further, the continuing excavation provided also valuable information which was incorporated to our interpretation.

Several features concealed in the embankment were detected and imaged by the ERTs. Some of them are speculated to be ancient structures but ground truthing is needed. Also, conclusions concerning the morphology of the ancient hill are inferred.

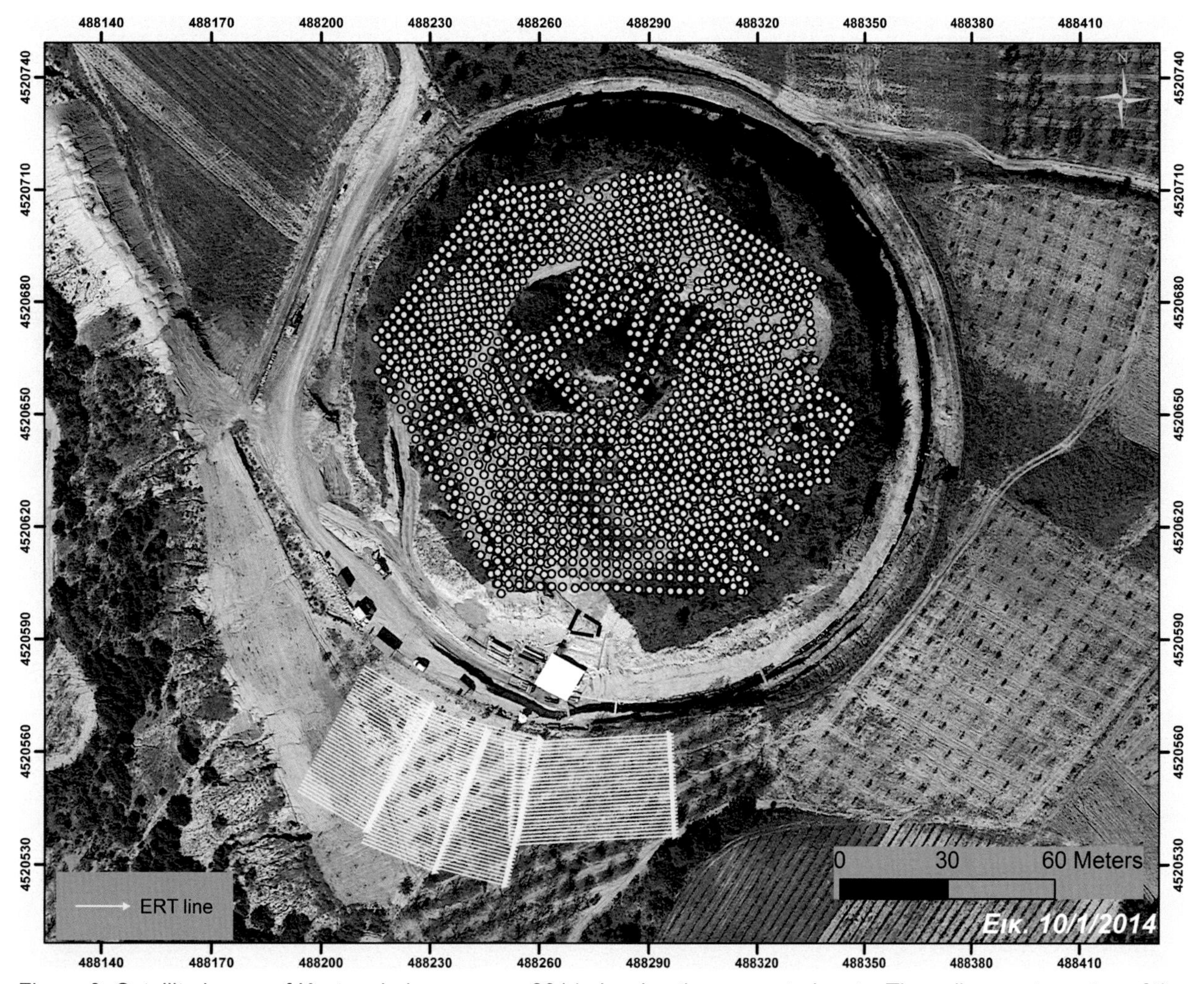

Figure 3: Satellite image of Kastas during summer 2014 showing the excavated parts. The yellow spots on top of the tumulus represent positions of electrodes used for the 3D ERT survey. Further, 3 rectangular grids were established south of Kastas, to investigate for possible concealed structures associated with the monument.

Conclusions

The ERT investigations of Kastas in Amphipolis proved a powerful tool in imaging its interior though no conventional measuring approach was followed. The flexibility and efficiency of the ERT method allowed the specific treatment attempted.

Several detections of possible ancient structures buried in the embankment are reported in this paper.

Bibliography

Papadopoulos N., Yi M-J, Kim J-H, Tsourlos P.& Tsokas G. (2010) Geophysical investigation of tumuli by means of surface 3D Electrical Resistivity Tomography. *Journal of Applied Geophysics* **70**: 192–205.

Tonkov, N. & Loke, M.H. (2006) A Resistivity Survey of a Burial Mound in the "Valley of the Thracian Kings", *Archaeological Prospection* **13**: 129-136.

Tsokas, G.N. (2012) Geophysical investigations in tumuli: Examples from N. Greece. *In Tumulus and Prospection.* National Research Institute of Cultural Heritage of Korea. ISBN 958-89-6325-761-7 93910, 165-205, 2012.

Tsourlos, P.I., and Ogilvy, R. (1999) An algorithm for the 3-D Inversion of Tomographic Resistivity and Induced Polarisation data: Preliminary Results. *Journal of the Balkan Geophysical Society* 2**:** 2, 30–45.

Tsourlos P., Papadopoulos N., Yi M-J, Kim, J-H & Tsokas G. (2014) Comparison of measuring strategies for the 3-D electrical resistivity imaging of tumuli, *Journal of Applied Geophysics* **101**: 77-85

Yi, M.J. and Kim, J.H. (2003) Enhancing the resolving power of least-squares inversion with active constraint balancing. *Geophysics* **68**: 931-941

Deserted fortified Medieval villages in South Moravia

Michal Vágner[1], Tomáš Tencer[1], Petr Dresler[1], Michaela Prišťáková[1], Jakub Šimík[1] and Jan Zeman[1]

[1]Department of Archaeology and Museology, Masaryk University, Brno, Czech Republic

vagnermichal@mail.muni.cz

A key aspect in the research of the rural medieval society concerns deserted medieval villages (DMV). These archaeological time capsules present a unique opportunity to investigate settlement structure development and economic and social relationships in the Middle Ages.

Previous research in the Moravia region focused on the area of south-west (DMV Mstěnice, Fig. 1.1) or central Moravia (DMV Bystřec, Fig. 1.2; DMV Konůvky, Fig. 1.3). These extensive excavations and surveys provided important information about the distribution of individual homesteads and their relationship with the landscape. The knowledge about various types of ground plans, construction techniques, design forms of rural houses, material culture, etc., were acquired. Unfortunately, the area where we focus our research, the valleys of the rivers Dyje, Svratka and Lower Morava in the South Moravian Region has been neglected for a long time. Only minimal rescue excavations or fieldwalking have previously been made. This was mainly due to the fact that in this area the main focus was on researching the early medieval period, especially large-scale and long-term excavations of early medieval centres - hillforts.

We are still missing some basic information from this region, which is necessary for understanding the rural society. We don't know what was the most common type of village ground plan, or how the houses were constructed. Evidence from excavations in other regions of Moravia suggest that rural houses were built of stone and logs or combined techniques. There is still very little information about the region of South Moravia. However, it is assumed that mudbricks were the prevailing building component.

In order to complete the picture, we decided to use non-destructive archaeological methods. Recent developments in the field of remote sensing together with publicly available aerial imagery datasets and historical maps provide the opportunity to fill the knowledge gap in this area. The study of these datasets has so far helped to locate fifteen deserted medieval villages in the south Moravia region. Many of them were enclosed by a single or double ditch. Some of these ditches are depicted on historical maps of the Second and Third Military Survey

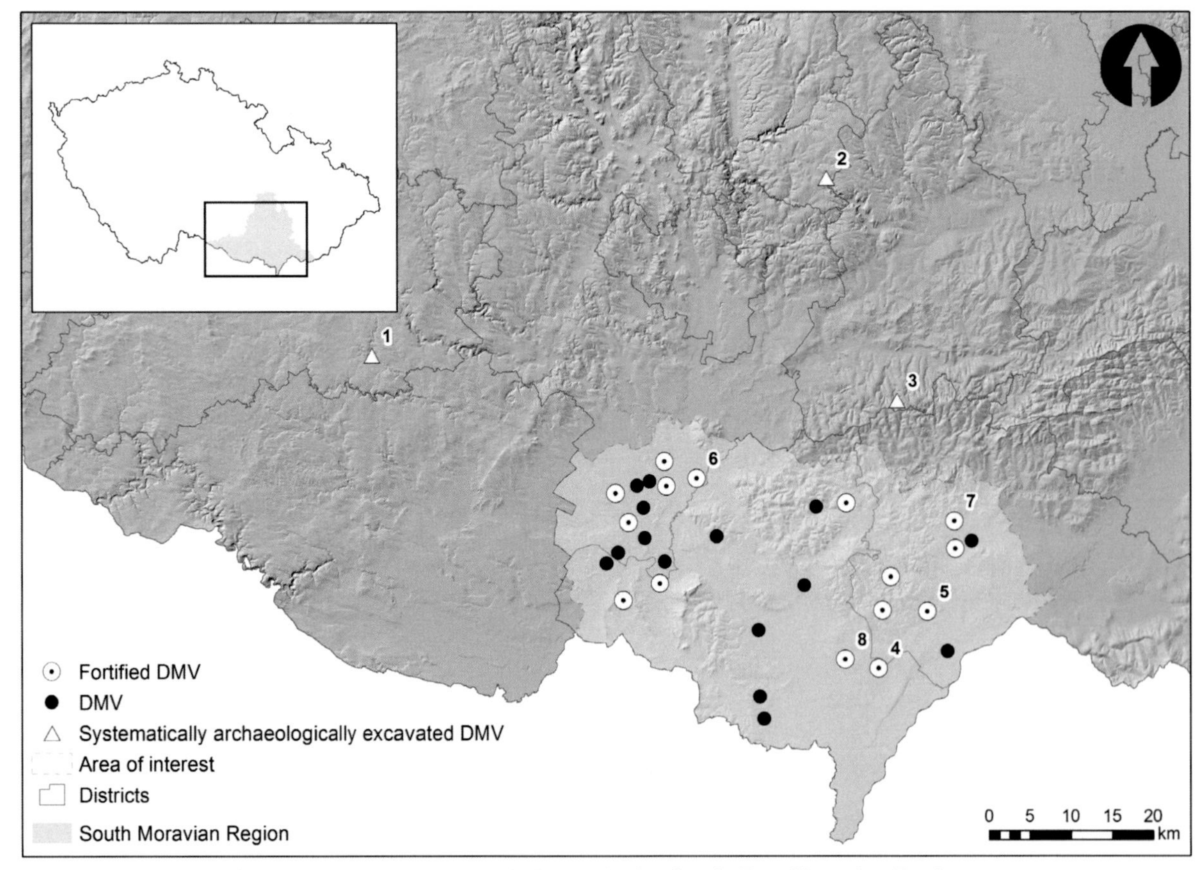

Figure 1: Distribution of located deserted medieval villages in the South-East Moravian Region.

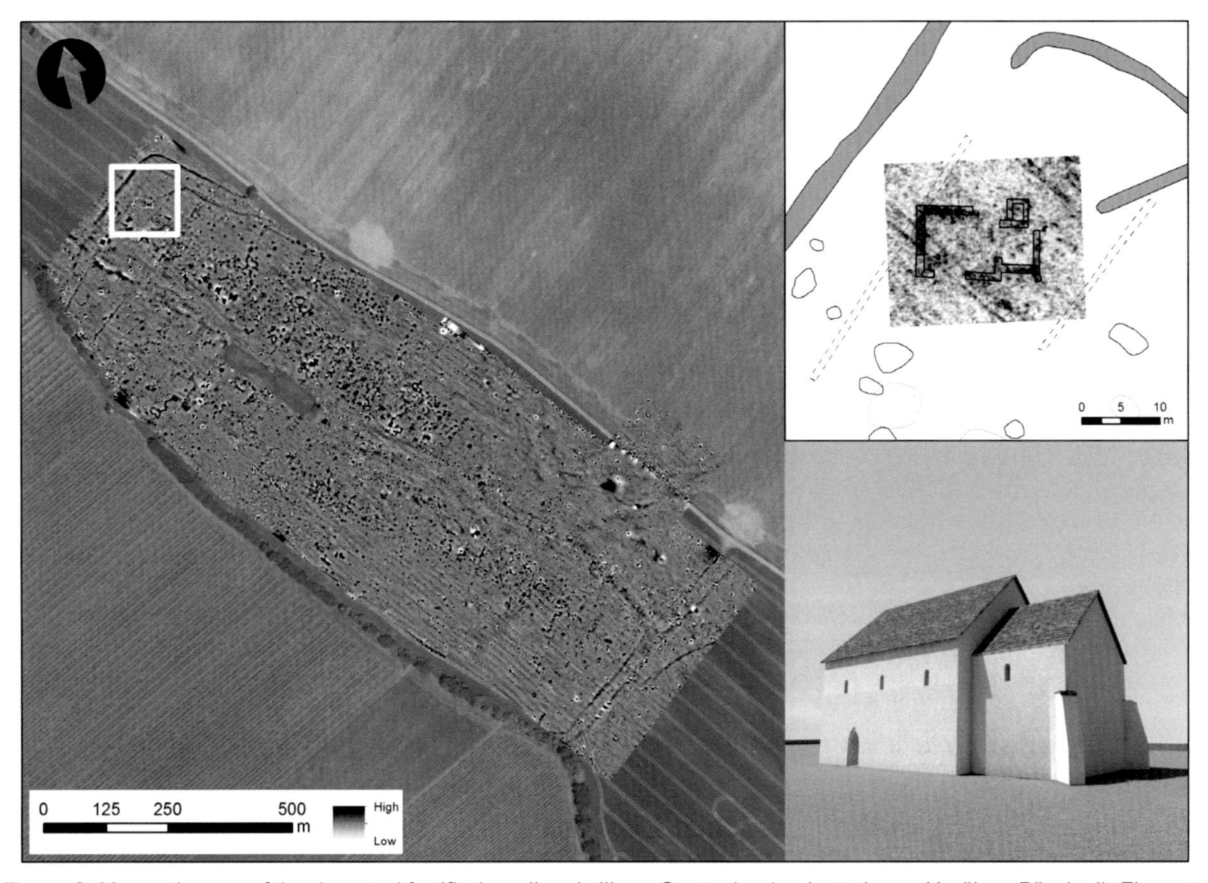

Figure 2: Magnetic map of the deserted fortified medieval village Opatovice (cadastral area Hrušky u Břeclavi). Fluxgate magnetometers Ferex (Förster) and LEA MAX (Eastern Atlas) with FEREX CON 650 probes, sampling interval 0,5 x 0,25 m, (dynamic range [-4,4] nT -> [white , black]) (left). GPR depth slice of stone rectangular building, whose layout, orientation and previous mention in old written sources indicates that the discovered structure is a medieval church (upper right). 3D visualization of the reconstructed church (lower right).

(1841 and 1882) and on the Imperial Imprints of the Stable Cadastre (1826-1843).

The combination of large-scale magnetometry and targeted GPR prospection allows an effective identification and investigation of the large settlement area. So far, we fully surveyed three villages (DMV Opatovice, Fig. 1.4; DMV Vsisko, Fig. 1.5; DMV Želice, Fig. 1.6) and partially another two (DMV Starý Mistřín, Fig. 1.7; DMV Prechov, Fig. 1.8). The overall prospected area is in excess of 30 hectares. So far, we have identified and interpreted more than 2000 features of various kinds such as: remains of individual households, ditches, underground corridors ("lochy" in Czech), etc.

Results of this research indicate that the prevalent construction techniques for individual houses had been based on the use of clay. The unique building identified through magnetometry and confirmed by GPR prospection at DMV Opatovice (Fig. 2) was the only stone building detected. Based on the layout and the orientation we have interpreted it as medieval church. This was later validated by trial trenching and fieldwalking in the neighbouring area, which indicates presence of a burial ground.

Bibliography

Bálek, M. & Unger J. (1996) Ohrazené středověké vesnice na jižní Moravě, *Archeologica Historica* **21**: 429–442.

Dresler, P., Tencer, T. and Vágner, M. (2015) Prospekce zaniklé středověké vesnice Opatovice, k.ú. Hrušky – A Survey of the Deserted Medieval Village of Opatovice, Cadastral District of Hrušky, *Studia Archaeologica Brunensia* **E20**: 113–132.

Dresler, P. and Tencer, T. (2016) Neznámé opevněné sídlo v Dolních Bojanovicích, *Archeologica Historica* **41**: 23-31.

Nekuda, V. (1961) *Zaniklé středověké osady na Moravě v období feudalismu*, Brno.

The Guaquira-Tiwanaku project (Bolivia): a multidisciplinary approach of ancient societies/ environment interactions

M-A Vella[(1)], G Bievre[(2)], R Guerin[(3)], J Thiesson[(3)] and C Camerlynck[(3)]

[(1)]IFEA, UMIFRE 17 CNRS/MAEDI, La Paz, Bolivie; [(2)]ISTerre, Grenoble, France; [(3)]UMR 7619 Milieux environnementaux, transferts et interactions dans les hydrosystèmes et les sols, UPMC, Paris, France

mav.vella@gmail.com

The Tiwanakota culture developed on the shores of Lake Titicaca between the end of the second millennium BC and disappeared between AD 1100 and 1200. The paleoenvironmental data acquired on Lake Titicaca attest to numerous fluctuations during The Tiwanaku cultural period. The work carried out during the 1990s on the evolution of the level of Lake Titicaca proposed the hypothesis of a drought that affected the whole of the Altiplano and which would have led to the disappearance of the Tiwanakota culture. However, few palaeoenvironmental and palaeogeographic studies have been carried out on the Guaquira-Tiwanaku river basin, the ecosystems constituting the geographical center of the Tiwanaku civilization. The contribution of magnetic surveys provided new information on the architectural organization of urban and semi-urban sites; EM surveys and seismic tomography allowed us to reconstruct the organization and nature of the alluvial sedimentary formations; finally, the study of cores on Lake Titicaca and natural sedimentary cuts in the alluvial plain of the Rio Guaquira-Tiwanaku indicates that changes in the lake level are accompanied by significant changes in the landscape during the period of development and decline of the Tiwanaku culture. All the works carried out within the framework of the Franco-Bolivian Mission "Palaeoenvironment and Archeology of the Rio Guaquira-Tiwanaku" attempt to reconstruct the evolution of the landscape of the Bolivian Altiplano in contact with one of the most emblematic cultures of the Andes.

Figure 1: Arerial photography of the Akapana Pyramid of the Tiwanaku archaeological site (photography: Guaquira-Tiwanaku Project).

Semi-automated object detection in GPR data using morphological filtering

Lieven Verdonck[(1)], Alessandro Launaro[(2)], Martin Millett[(2)], Frank Vermeulen[(1)] and Giovanna Bellini[(3)]

[(1)]Department of Archaeology, Ghent University, Belgium; [(2)]Faculty of Classics, University of Cambridge, UK; [(3)]Soprintendenza Archeologia del Lazio e dell'Etruria Meridionale

lieven.verdonck@ugent.be

From 2015 until 2017, a GPR survey of two complete Roman towns in Lazio, Italy (Interamna Lirenas and Falerii Novi) has been conducted within the 'Beneath the surface of Roman Republican cities' project, a collaboration between the University of Cambridge, Ghent University, the British School at Rome and the Soprintendenza Archeologia del Lazio e dell'Etruria Meridionale. The results (Fig. 1) confirm the main street layout, and add details in terms of the internal articulation of the *insulae* (Ballantyne *et al.* 2016; Verdonck *et al.* 2016).

The total survey area was ~60 ha; data were recorded with a sample density of 0.05 m × ~0.1 m. The manual interpretation of such large-area, high-resolution data sets is time-consuming. In a variety of domains (e.g. medical imaging, navigation), increasingly automated procedures for object detection are used. Also in archaeology, there

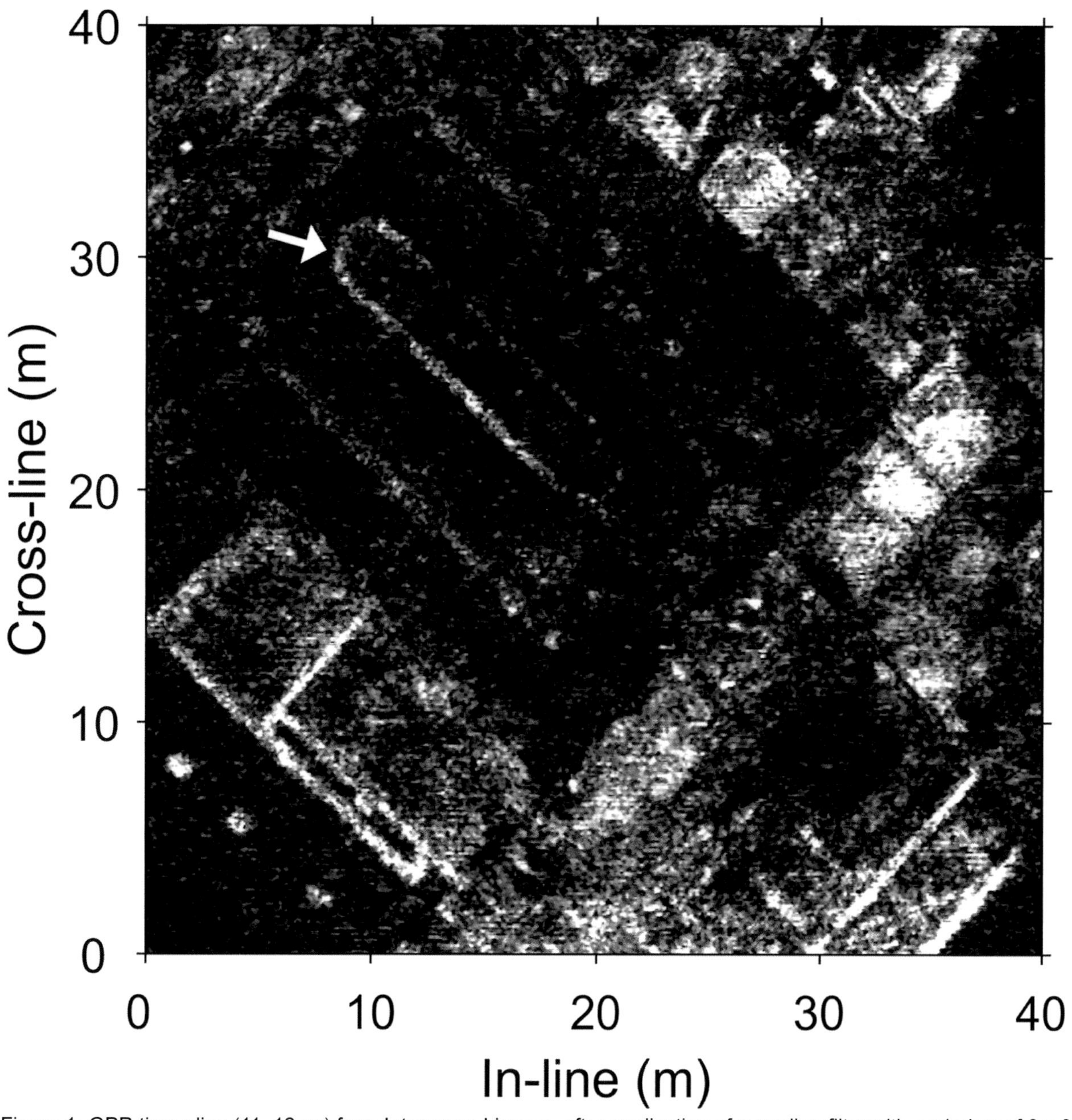

Figure 1: GPR time-slice (11–12 ns) from Interamna Lirenas, after application of a median filter with a window of 3 x 3 samples (0.15 m x 0.15 m). The arrow indicates an apse structure, detected by means of the rank hit-or-miss transform, as described in the text.

seems a growing understanding between those who emphasize the experience and cognitive ability of the archaeologist, and those who recognize the strength of computer vision for pattern extraction (Traviglia *et al.* 2016). Especially in archaeological remote sensing, a number of papers have been published in recent years. In archaeological geophysics, semi-automated detection techniques have so far been debated less intensively, although these approaches have been incorporated in the work of e.g. Pregesbauer *et al.* (2013), Schmidt and Tsetskhladze (2013), and Verdonck (2016).

A few operators belonging to the field of mathematical morphology proved useful when processing the GPR data from Lazio. Mathematical morphology is a set of image processing operations that apply a structuring element (SE) to an image. The size and shape of the SE makes the operation sensitive to particular objects in the image. Two fundamental operations are dilation and erosion. Erosion shows where the SE fits the objects in the image, whereas dilation shows where the SE hits them. These operators were used by Leckebusch *et al.* (2008) to enhance signal-to-noise ratio of GPR data before applying a feature extraction algorithm based on plane fitting.

Erosion removes objects that cannot contain the SE, and shrinks the other objects. By dilating the eroded image with the same SE, the objects are restored, except those that had completely been eliminated by the erosion. Therefore, the morphological opening (MO) operator (erosion followed by dilation) can filter the image and extract objects larger than the SE.

A problem can arise when applying MO to noisy images. For example, using a long line as SE, some of the wall structures in Figure 1 that are interrupted by noise were not extracted as they could not contain the complete SE. Using a shorter SE would result in the extraction of smaller objects that may not belong to walls. Therefore, in this study the rank-max opening (RMO) was used for the detection of walls and floors. The RMO is based on a rank order filter: it is the elementwise minimum of the identity transform (which results in an unaltered image) and the dilation by the reflection of the SE (i.e. the symmetric of the SE with respect to its origin) applied to the rank order filter that uses the SE as its window and $n - k + 1$ as its rank, where k is a number pixels between 1 and the number of elements n in the SE (Soille 2002). A rank order filter replaces each pixel in the image by the rth element of the sorted array corresponding to the SE (where r is the rank).

Using the RMO for the extraction of wall objects from the GPR data (Fig. 2), the values for four parameters needed to be specified: the length of the SE, its orientation (obtained by applying the

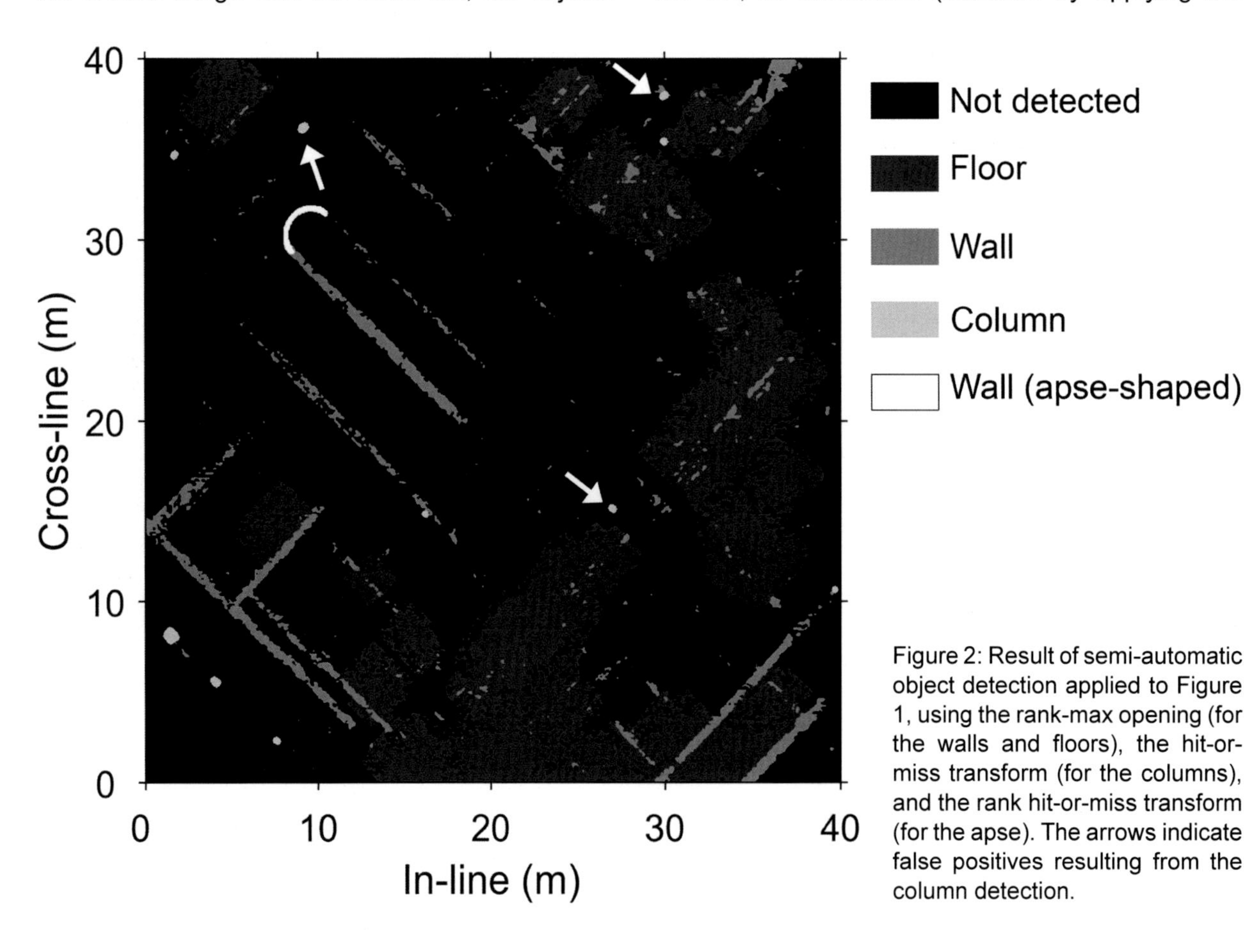

Figure 2: Result of semi-automatic object detection applied to Figure 1, using the rank-max opening (for the walls and floors), the hit-or-miss transform (for the columns), and the rank hit-or-miss transform (for the apse). The arrows indicate false positives resulting from the column detection.

Hough transform), the rank, and a threshold applied in order to obtain a binary image. For the detection of the floors, the SE was a square instead of a line.

For the detection of columns, MO and RMO were not successful because the application of a small circular SE to GPR slices containing surfaces with relatively high reflection strength caused many false positives. Therefore, the hit-or-miss transform (HMT) was used, which is the morphological expression of template matching (Naegel *et al.* 2007). The HMT requires a foreground SE (FSE), which fits the objects to be recognized, and a background SE (BSE), which fits the background (in other words: which must not fit the objects to be recognized). We chose a circular FSE with a radius somewhat smaller than the smallest column in the GPR data. A ring-shaped BSE was selected with an inner radius somewhat larger than the largest column. This resulted in a number of false positives. Because in Figure 1 all columns stand apart from other structures such as walls or floors, most of the false positives were eliminated by not allowing a column on the location of a floor or wall. In Figure 2, some false positives remain (indicated with arrows).

For the detection of complex objects affected by a high level of noise, such as the apse near the arrow in Figure 1, the HMT can be too rigid. One method to improve the tolerance to noise is to replace the erosion and dilation operators in the HMT by rank operators (rank hit-or-miss transform, RHMT). In this way, the objects can be detected also when their shape and size cannot be described exactly (Soille 2002). This implies that an appropriate rank has to be determined for the FSE and BSE. In the case of the apse in the GPR data, we took as FSE a half-ring shape, and for the BSE two half-rings situated inside and outside the FSE. To perform the RHMT, beside the ranks, the inner and outer radius of the FSE and BSE had to be determined.

For the detection of objects that are not numerous or require a large user input (in our example the apse and the columns), manual object extraction remains more efficient. This is an illustration of the fact that computer vision techniques exist beside traditional ones in an iterative relationship (Traviglia *et al.* 2016). Nevertheless, powerful processing tools such as the RMO, which require relatively limited input by the human operator, may constitute a step towards a more automated object detection in archaeological geophysics. The operators described can be extended to three dimensions; their application to 3-D GPR data cubes is currently being investigated.

Bibliography

Ballantyne, R., Bellini, G., Hales, J., Launaro, A., Leone, N., Millett, M., Verdonck, L. and Vermeulen, F. (2016) Interamna Lirenas and its territory. *Papers of the British School at Rome* **84**: 322–325.

Leckebusch, J., Weibel, A. and Bühler, F. (2008) Semi-automatic feature extraction from GPR data for archaeology. *Near Surface Geophysics* **6**: 75–84.

Naegel, B., Passat, N. and Ronse, C. (2007) Grey-level hit-or-miss transforms. Part I: unified theory. *Pattern Recognition* **40**: 635–647.

Pregesbauer, M., Trinks, I. and Neubauer, W. (2013) Automatic classification of near surface magnetic anomalies: an object oriented approach. In W. Neubauer, I. Trinks, R. B. Salisbury and C. Einwögerer (eds) *Archaeological Prospection. Proceedings of the 10th International Conference on Archaeological Prospection, Vienna, Austria, 29 May–2 June 2013*. Vienna: Austrian Academy of Sciences, 350–353.

Schmidt, A. and Tsetskhladze, G. (2013) Raster was yesterday: using vector engines to process geophysical data. *Archaeological Prospection* **20**: 59–65.

Soille, P. (2002) On morphological operators based on rank filters. *Pattern Recognition* **35**: 527–535.

Traviglia, A. Cowley, D. and Lambers, K. (2016) Finding common ground: human and computer vision in archaeological prospection. *AARGnews* **53**: 11–24.

Verdonck, L. (2016) Detection of buried wall remains in ground-penetrating radar data using template matching. *Archaeological Prospection* **23**: 257–272.

Verdonck, L., Bellini, G., Launaro, A., Millett, M. and Vermeulen, F. (2016) Beneath the Surface of Roman Republican Cities: Large-Scale GPR survey of Falerii Novi and Interamna Lirenas (Lazio, Italy). In *Recent Work in Archaeological Geophysics (London, 6 December 2016)*. London: NSGG–Historic England, 59–61.

Acknowledgements

The presented research was conducted in the framework of a Postdoctoral Fellowship of the Research Foundation - Flanders (FWO) (12G2217N – Lieven Verdonck). It is being made possible by the support of the AHRC (Grant Ref. AH/M006522/1 – Beneath the Surface of Roman Republican Cities), the Department of Archaeology (Ghent University), the Soprintendenza Archeologia del Lazio e dell'Etruria Meridionale and the Comune di Pignataro Interamna.

The diverse role of electromagnetic induction survey in development-led alluvial (geo-) archaeology: prehistoric and (post-)Medieval landscape archaeology at Prosperpolder Zuid (north-west Belgium)

Jeroen Verhegge[1,2], Timothy Saey[3,4], Pieter Laloo[5], Machteld Bats[5] and Philippe Crombé[1]

[1]Ghent University-Department of Archaeology-Research Group Prehistory, Gent, Belgium; [2]Geosonda Environment nv, Gent, Belgium; [3]Ghent University- Department of Soil Management-Research Group Soil Spatial Inventory Techniques, Ghent, Belgium; [4]3Dsoil, Beerse, Belgium; [5]GATE bvba, Bredene, Belgium

Jeroen.Verhegge@Ugent.be

Introduction

In 2015-2016 a development-led archaeological evaluation of a 170 ha polder, situated between the village of Doel (Beveren) and the Belgium-Dutch border, was conducted ahead of its conversion into a nature reserve, which could form a threat to the buried archaeological heritage. In the close surroundings, well-preserved prehistoric sites, ranging from the Final Palaeolithic to Early Neolithic period and from the (Post-)Medieval period are known from rescue excavations during past harbour expansions. On the one hand, peaty and alluvial sediments below the embanked floodplain conserve the archaeological sites but on the other hand they impede detection. Therefore, an archaeological evaluation study was done prior to the installation of the nature reserve (Saey *et al.* 2016a).

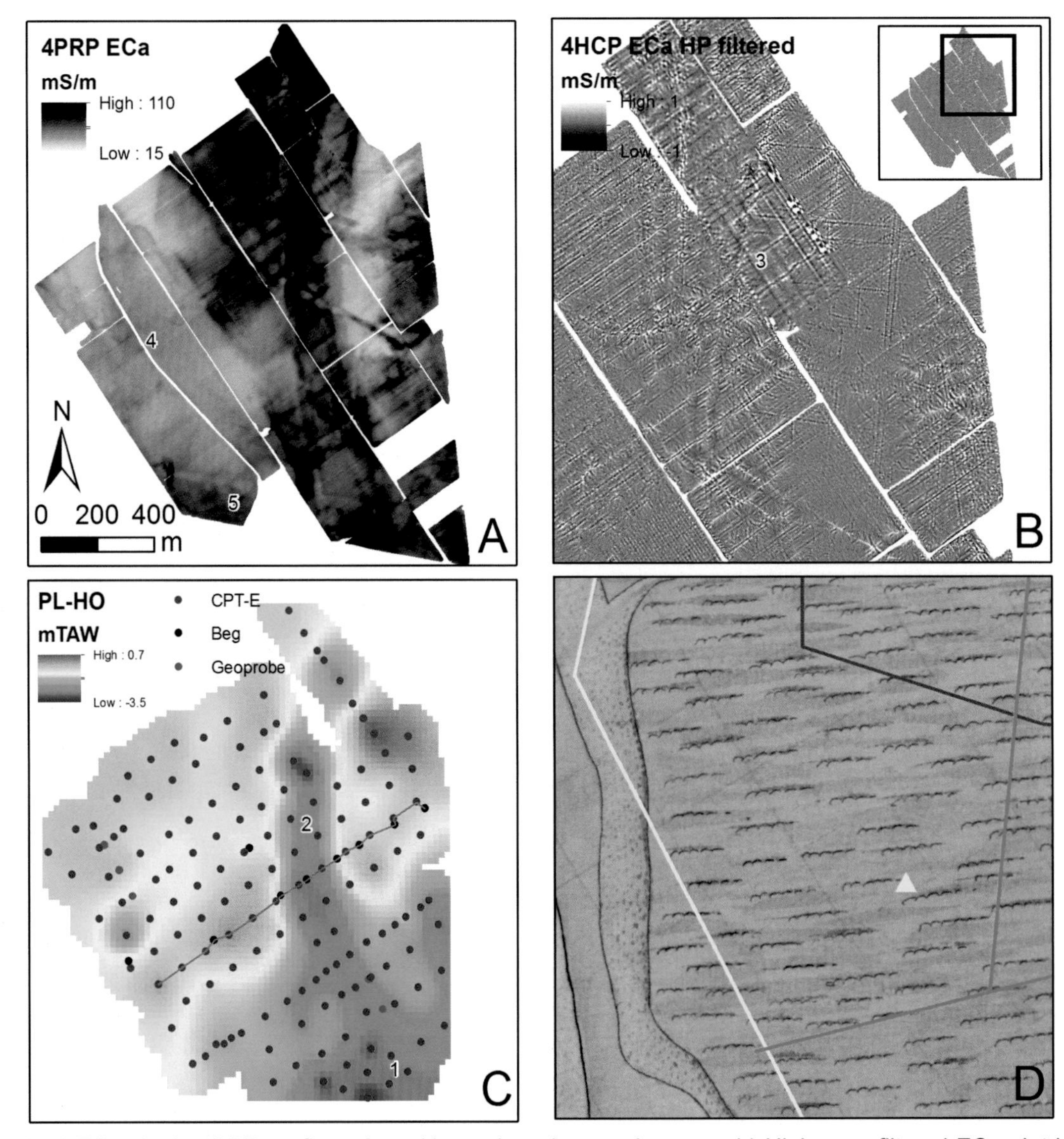

Figure 1: a) ECa plot in 4PRP configuration with numbered anomaly zones; b) High pass filtered ECa plot in 4HCP configuration; c) Location of CPT-E's and corings and derived PL-HO transition; d) Late Medieval embankment features from historic maps on top of the Early Modern Ferraris map (1744 CE) showing the reflooded area.

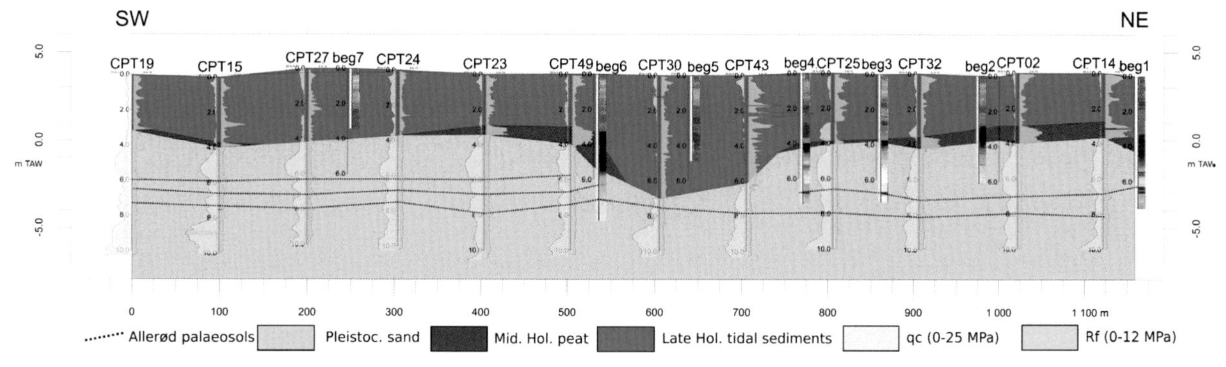

Figure 2: transect of CPT-E's and scanned Begemann corings and derived lithostratigraphic units

Method

From the initiation of the project, the subsurface conditions implied a deviation from the conventional archaeological evaluation approaches in Flanders (systematic trial trenching and hand coring). Therefore, alternative evaluation strategies including electromagnetic induction surveying (EMI) with multiple coil separations, Cone Penetration Testing (CPT-E) and mechanical coring were employed (Verhegge *et al.* 2016). EMI survey was used to model the subsurface in multiple sequence types using a 1D inversion approach (Saey *et al.* 2016b), but this required additional calibration and validation data. Therefore, 146 Electrical Cone Penetration Tests (CPT-E) were performed up to a depth of 10 m, collecting point resistance (qc), sleeve friction (fs) and the friction number (Rf, %), as the ratio between both measures. Mechanical coring provided a pedological and lithostratigraphic interpretative framework for both CPT-Es and EMI. Based on the reconstruction of the buried prehistoric landscape and the planned soil interventions, three zones were selected for intensive archaeological soil sampling (following Crombé and Verhegge 2015) by discontinuous coring of the Holocene-Pleistocene stratigraphic transition, i.e. the basis of the peat and the top of the Pleistocene sand (PL-HO). The samples were sieved and checked for prehistoric archaeological indicators. Trial trenching targeted mostly linear and potential archaeological apparent electrical conductivity (ECa) anomalies from EMI.

Figure 3: a) profile of bank and ditch system in trial trench; b) historical painting of embankment works, detected in the south of the survey area.

Results

The natural sedimentological variation was horizontally mapped using the EMI and vertically through CPT-E and coring profiles. The dataset is characterized by three general profile types (Fig. 2), included in the EMI modelling procedure. While the bottom layer within the entire area consists of Pleistocene sands with a fixed ECa, the upper layers are quite variable due to the sedimentation of different clayey and peaty layers. The first profile type is situated within the EMI zones with moderately low ECa and consists of a peat layer below sand and (clayey) sandy tidal flat deposits. In areas with high ECa, two profile types are present: the first consists of peat below tidal deposits that are predominantly clayey. Alternatively, a tidal gully has eroded the peat on the PL-HO transition and is filled with muddy and clayey sediment. The modelled PL-HO transition depth was used to guide discontinuous archaeological sampling to detect prehistoric artefact scatters.

With additional information from CPT-E's and coring (Fig. 2), the interpretation can be broadened. Within the Pleistocene sands, several thin (<20 cm) layers of peaty and organic rich Allerød soil horizons were detected as small Rf peaks (Fig. 2). In the south the deepest part of the prehistoric landscape coincides with a possible Late-Glacial/Early Holocene gully (Fig. 1.c-1). A long period of non-sedimentation during the Early-Middle Holocene allowed podzolic soil formation on the undulating PL-HO transition. The archaeological core samples retrieved from this paleosol revealed 3 areas which yielded lithic artefacts. These could be preserved in situ by adapting the excavation plans. During the Middle Holocene, peat formed under the influence of relative sea level rise. From the (Early) Middle ages onwards, renewed tidal activity eroded the top of the peat layer in most of the survey area (Fig. 1.a). First, a large (Late-)Medieval gully with a N-S orientation was eroded centrally (Fig. 1.c-2). In the 16th-17th century, a dyke system was erected to reclaim the flooded land. Raised roads, parcelling- and drainage ditches were constructed during the reclamation of the landscape. Buried relicts of these features were detected within the EMI measurements (Fig. 1.b-3). During a next flooding phase, a new tidal channel developed in the East and existing ditches and fields were buried (Fig. 1.a-4). At the southern edge, late 18th century dyke construction features are noticeable (Fig. 1.a-4, Fig. 3.b). The polder developed within its current state during the embankment of 1846. The Late Medieval and Early Modern features were excavated partially with difficulty (Fig. 3.a) but mapped more easily as linear (double peaked) anomalies in 4HCP ECa (Fig. 3.b).

Discussion and conclusion

EMI has shown its suitability to map both natural and cultural landscapes ranging from the Early Holocene to modern times in alluvial plains. First, the large scale ECa variability was modelled to guide archaeological sampling of deeply buried artefact scatters by means of cheaper and faster discontinuous archaeological coring. Secondly, the lithostratigraphic interpretations of the EMI data were strengthened through a combination of CPT-E and coring which revealed important new pedological information. Due to the challenging subsurface conditions, the Late to Post Medieval exploitation landscapes could only be evaluated efficiently by combined EMI and targeted trial trenching. Nevertheless, this approach does not allow the evaluation of more subtle, discrete and deeply buried features. It should be stressed that experience and knowledge of the limitations of all available archaeological prospection methods and their research question oriented application is of primary importance for an optimal archaeological evaluation. This study demonstrates that large scale EMI provides a good base dataset in a development-led archaeological context, if backed up by other methods. Future projects will integrate the archaeological evaluation further within the planning and construction process e.g. through multipurpose use of EMI or CPT-E data.

Bibliography

Crombé, P. and Verhegge, J. (2015) In search of sealed Palaeolithic and Mesolithic sites using core sampling: the impact of grid size, meshes and auger diameter on discovery probability. *Journal of Archaeological Science* **53**: 445-458.

Saey, T., Laloo, P., Bats, M., Cryns, J., Vergauwe, R., Deconinck, J.-F., Verhegge, J. and Cruz, F. (2016a) *OC2719 Prosperpolder Zuid: uitvoeren van een archeologisch onderzoek in opdracht van het gemeentelijk havenbedrijf Antwerpen-Eindrapport.* Gent: Onderzoeksgroep Ruimtelijke Bodeminventarisatie (ORBit) - Universiteit Gent (UGent). Ghent Archaeological Team bvba - GATE.

Saey, T., Verhegge, J., De Smedt, P., Smetryns, M., Note, N., Van De Vijver, E., Laloo, P., Van Meirvenne, M. and Delefortrie, S. (2016b) Integrating cone penetration testing into the 1D inversion of multi-receiver EMI data to reconstruct a complex stratigraphic landscape. *Catena* **147**: 356-371.

Verhegge, J., Missiaen, T. and Crombé, P. (2016) Exploring Integrated Geophysics and Geotechnics as a Paleolandscape Reconstruction Tool: Archaeological Prospection of (Prehistoric) Sites Buried Deeply below the Scheldt Polders (NW Belgium). *Archaeological Prospection* **23**(2): 125-145.

Multi-channel GPR surveys for the detection of buried Iron-Age settlement remains: a case study from Bårby ring fort, Öland, Sweden

Andreas Viberg[1]

[1]Archaeological Research Laboratory, Department of Archaeology and Classical Studies, Stockholm University, Wallenberglaboratoriet, 106 91, Stockholm, Sweden

andreas.viberg@arklab.su.se

Introduction

The island of Öland, situated east of the Swedish mainland, is home to several large ring-forts dated to about AD 200-700 (Fig. 1). 18 ringforts are currently known from historical maps and sources but only 15 are still visible in the landscape today. The best documented ring fort is undoubtedly Eketorp, which was excavated between 1964 and 1974. During the excavations 53 stone foundations to houses were discovered, and it was concluded that Eketorp was used during three different phases (AD 300-1240) (Borg *et al.* 1976). Additional archaeological inventories and excavations have confirmed the presence of similar stone house foundations in at least 10 other forts on the island (Fallgren 2008). Eketorp, as the only completely excavated ring fort on the island, has been seen as a model for all of the other Ölandic forts, but as the available archaeological information regarding the other forts are scarce, this view has been challenged (e.g. Viberg *et al.* 2014: 413). Many forts are also too large for traditional archaeological excavations and, as a consequence, few archaeological investigations are initiated.

During 2010 one of the smaller forts, Sandby borg, was targeted for a feasibility study using ground-

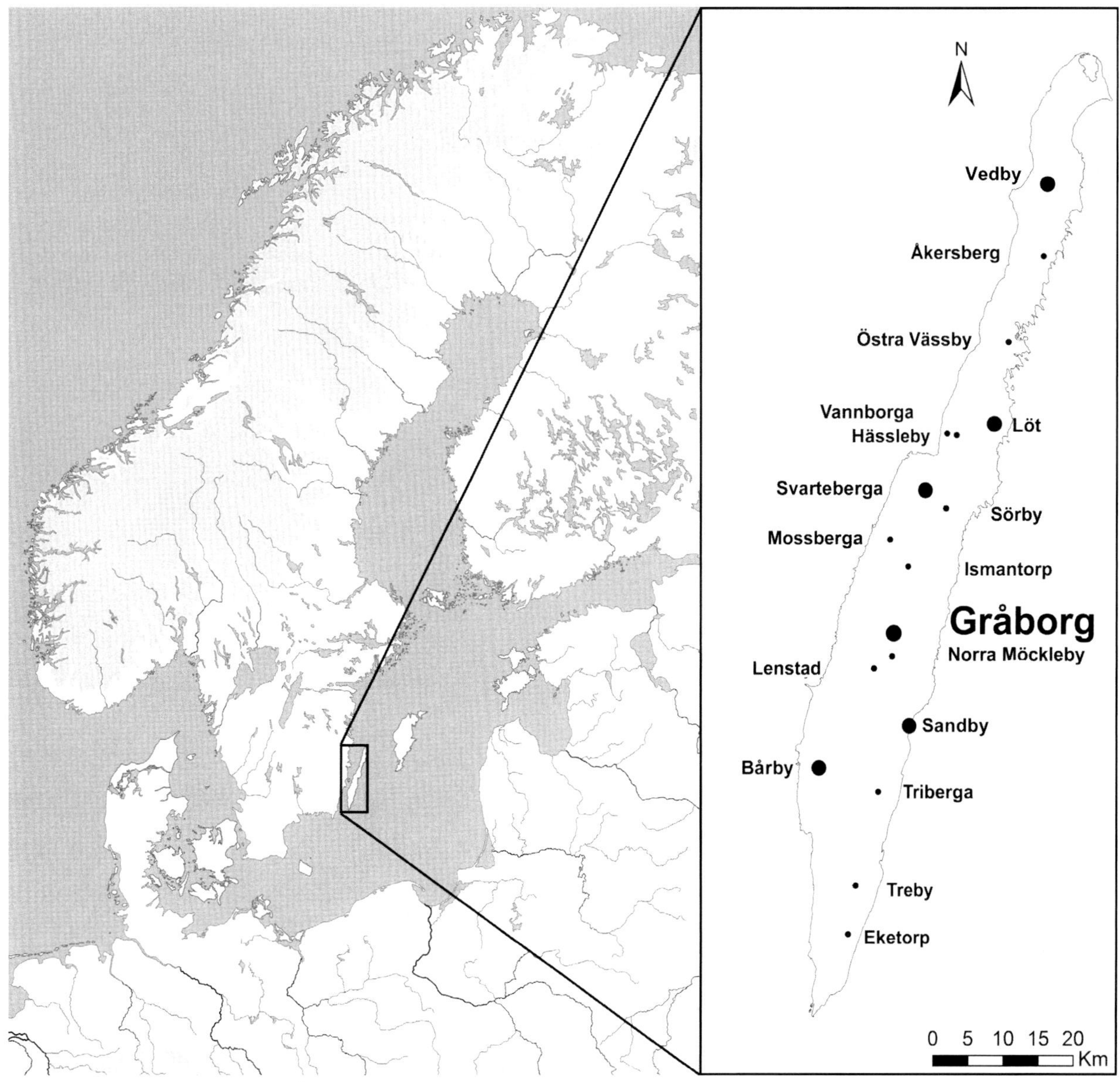

Figure 1: Map of the island of Öland and the location of its ring-forts.

Figure 2: (left) Aerial photo of Bårby borg (Photo by: J, Norrman 1992 ©RAÄ), (right) Map showing the spatial layout of Bårby borg and the location of the excavation trench from 1930.

penetrating radar (GPR) (Viberg *et al.* 2014). The results confirmed the presence of 52 stone house foundations and yearly excavations have since highlighted the benefits of geophysical surveys as the starting point and foundation for archaeological research (e.g. Victor 2015).

The Big Five

Strengthened by the successful survey at Sandby borg the project "The Big Five" was initiated in 2014 (see Viberg 2015). The project aims to investigate several large Ölandic ring forts with the purpose of providing an empirical foundation for future archaeological research. An additional purpose has also been to evaluate the impact of the extensive agricultural activities within the forts. Surveys have subsequently been carried out in four large forts: Bårby, Löt, Gråborg and Vedby.

The forts have been surveyed using the Malå Imaging Radar Array (MIRA) (400MHz, 16 channels), with an inline and crossline data point distance of 8cm (see Trinks *et al.* 2010 for an extended discussion of the system).

Bårby Ring Fort

Bårby ring fort, the islands only semi-circular fort, is situated on the west coast of Öland at the edge of a lime stone cliff (Fig. 2). It is the smallest fort included in the project with an inner area of approximately 7500m^2. The inner area of the fort is enclosed by a partly demolished stone wall, approximately 11-13m wide and 2-3.5m high (Swedish registry of ancient monuments).

One smaller excavation has been carried out in 1930 by Swedish archaeologist Mårten Stenberger. The excavations and subsequent field inventories

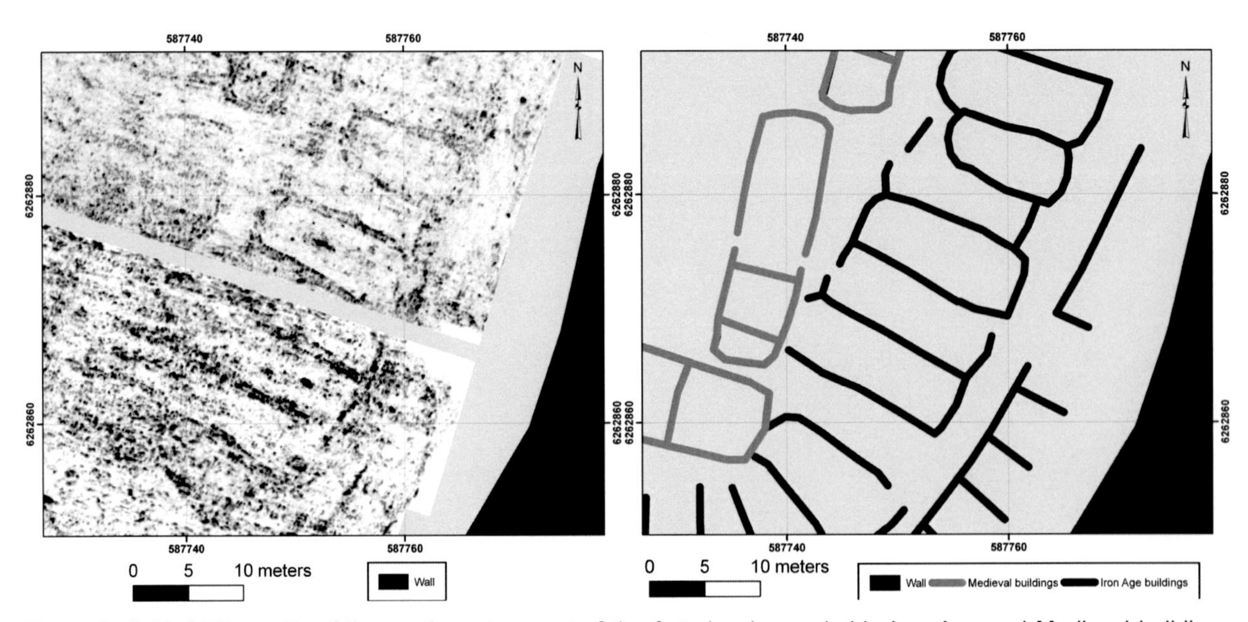

Figure 3: (left) GPR results of the south-eastern part of the fort showing probable Iron Age and Medieval buildings (depth slice at 0.2-0.3m below ground). Image is showing the envelope of the GPR data (strong reflections depicted in black). (right) Schematic interpretation of the different houses in the GPR data.

produced finds belonging to two different chronological phases, one Migration period/early Vendel period phase (ca. AD 375-600) and one Medieval phase (Stenberger 1933).

The land owner of the Bårby fort in 1930 regularly ploughed the inner area and noted that the south-eastern area of the fort seemed to contain large quantities of animal bones, fire cracked rock and larger stones, and Stenberger suggested that this might be the remains of Iron Age houses (Unpublished report no 1082, ATA, 1930). He furthermore suggested that additional investigations should be carried out within the fort as soon as possible (Unpublished report no 4793, ATA, 1930).

Nothing further would, however, be done in the fort until 2012, when a metal detector survey was carried out (Erlandsson 2015: 68). The metal detector survey further strengthened the chronology suggested by Stenberger and clearly shows that we are dealing with a settlement used over a long period of time and that we should expect to find both Iron Age and medieval houses within the fort. The surveys at Bårby were carried out in May 2014 and took roughly 3 hours to carry out.

Results and Interpretation

The result of the GPR survey clearly shows several well preserved houses (Fig. 3). The south-eastern part of the fort, previously identified as a possible location for well-preserved Iron Age houses, contains a distinct cluster of buildings radially expanding from the perimeter wall towards the forts central parts. These buildings seem to be followed by a small road and then a similar radial cluster of houses. The pattern of clustering along with the size of the detected houses indicates that these buildings, most likely, have an Iron Age origin. They, for example, share the same spatial layout as houses identified at Eketorp and Sandby borg.

There are also longer buildings identified in the data and their spatial layout is similar to the medieval buildings of the Eketorp III phase. They seem to be placed in a U-shape enclosing an open square in the central westernmost part of the fort.

Conclusion

The results clearly show both well-preserved Iron Age and Medieval buildings within the fort. The result clearly exemplifies the benefits of using large scale geophysical surveys for the detection of Iron Age remains on Öland which can be used as the foundation for future investigations.

Bibliography

Borg, K., Näsman, U. and Wegraeus, E. (eds.) (1976) *Eketorp. Fortification and Settlement on Öland/Sweden. The Monument.* Stockholm: Royal Academy of Letters History and Antiquities.

Erlandsson, K.-O. (2015) Nya inblickar i några Öländska fornborgar. In K.-H. Arnell, and L. Papmehl-Dufay (eds.) *Grävda minnen. Från Skedemosse till Sandby borg.* Kalmar: Kalmar läns museum, 61-71.

Fallgren, J. H. (2008) Fornborgar, bebyggelse och odlingslandskap. In G. Tegnér (ed.) *Gråborg på Öland. Om en borg, ett kapell och en by.* Stockholm: Kungliga Vitterhets historie och antikvitets akademien, 223.

Stenberger, M. (1933) *Öland under äldre järnåldern. En bebyggelsehistorisk undersökning.* Stockholm: Kungl. Vitterhets, Historie och Antikvitets Akademien.

Trinks, I., Johansson, B., Gustafsson, J., Emilsson, J., Friborg, J., Gustafsson, C., Nissen, J. and Hinterleitner, A. (2010) Efficient, Large-scale Archaeological

Prospection using a True Three-dimensional Ground-penetrating Radar Array System. *Archaeological prospection* **17**: 175-186.

Viberg, A., Victor, H., Fischer, S., Lidén, K. and Andrén, A. (2014) The ringfort by the sea : Archaeological geophysical prospection and excavations at Sandby borg (Öland). *Archäologisches Korrespondenzblatt* **44**(3): 413-428.

Viberg, A. (2015) The Big Five. Mapping the Subsurface of Iron Age forts on the Island of Öland Sweden. *Archaeologia Polona* **53:** 521-525.

Victor, H. (2015) Sandby borg - ett fruset ögonblick under folkvandringstid. In K.-H. Arnell and L. Papmehl-Dufay (eds.) *Grävda minnen. Från Skedemosse till Sandby borg.* Kalmar: Kalmar läns museum: 97-115.

Acknowledgements

I would like to thank the Swedish Research Council, the Swedish Royal Academy of Letters, History and Antiquities and the Berit Wallenberg foundation for financial support. Thanks also to Karl-Oskar Erlandsson, Robert Danielsson, Börje Karlsson, Christer Gustafsson and Vasily Kharitonov (MALÅ Geoscience) for fieldwork support. and the Swedish National Heritage Board for survey permissions.

Unique details on the structural elements of a Neolithic site in Velm, Lower Austria - the necessity of integrated prospection and visualization in archaeological prospection

Mario Wallner[1], Juan Torrejón Valdelomar[1], Immo Trinks[1], Michael Doneus[3,1], Wolfgang Neubauer[1,3], Hannes Schiel[1], Tanja Trausmuth[1], Alexandra Vonkilch[1] and Alois Hinterleitner[1,2]

[1]Ludwig 'Boltzmann Institute for Archaeological Prospection and Virtual Archaeology, Vienna, Austria; [2]Zentralanstalt für Meteorologie und Geodynamik, Vienna, Austria; [3]University of Vienna, Vienna, Austria

mario.wallner@archpro.lbg.ac.at

Introduction

Modern archaeology is nowadays widely understood as a multi-methodological discipline, which, as an interdisciplinary approach, often produces a large amount of mostly non-uniform information. This melange of data generates new challenges concerning its joint interpretation; the generated information has to be understood, interpreted and interlinked within its given archaeological setting. In particular, non-invasive high-resolution archaeological prospection projects, offer a great potential for the detailed investigation of archaeological sites. The combination of remote sensing and near-surface geophysical survey methods is a specifically useful approach as the methods complement each other without damaging the archaeological heritage.

One of the LBI ArchPro case studies is to investigate the phenomenon of Neolithic circular enclosures (Kreisgrabenanlagen) in eastern Austria. These monuments are usually built on loess, a soil type that is favourable for magnetometry but provides no or only very limited penetration for ground-penetrating radar (GPR) pulses. Therefore, these prehistoric monuments were overwhelmingly discovered and investigated using aerial photography and magnetic prospection. Both methods are well suited to detect and map these sites. However, three-dimensional depth information could only be gathered through cost intensive and destructive excavations.

The presented site of the circular ring ditch system of Velm (parish of Himberg in Lower Austria) is situated on gravel, deposited as river sediments. Discovered by aerial photography in 2000 and re-photographed in 2001 (Fig. 1), it offered itself as an ideal candidate for a multi-methodological prospection approach. Due to the unique ground conditions, several non-invasive archaeological prospection methods, including GPR, could be successfully applied and generated spectacular results.

Methodology and Results

The Neolithic monument of Velm was investigated using a combination of non-invasive prospection techniques. In addition to remote sensing investigations (in this case aerial photography and airborne laser scanning), several geophysical prospection methods were used to supplement the aerial archaeology results. Applying multichannel motorized magnetometer and ground-penetrating radar systems together with exact RTK-GNSS positioning systems it has become possible to map large areas quickly and with very high spatial sampling resolution (down to 4×8 cm).

Especially in the years 2000, 2001, 2009 and 2015 vegetation marks documented on aerial photographs revealed extraordinary details on the monument, showing three massive ditches and a triple palisade with two entrance corridors. In the near vicinity of the ring ditch system, the traces of rectangular buildings, at least one with two rooms and central posts were clearly visible as cropmarks. These details have also been identified and interpreted on data visualisations of caesium magnetometry conducted in 2003 (Melichar and Neubauer 2010, 360). In addition to this manually collected caesium magnetometry data, a motorised Foerster/fluxgate survey was carried out in late autumn 2015, in order to monitor the state of preservation and to gain a reference survey for comparison between these two magnetic survey methods.

GPR surveys provide detailed three-dimensional information on the approximate depth, shape and location of archaeological structures with a high spatial resolution. Until recently, GPR surveys have

Figure 1: Oblique aerial photograph from June 2001 showing the location of a Neolithic circular enclosure with three ditches, each accompanied by a palisade. In the back of the monument, traces of rectangular houses can be seen (© Luftbildarchiv, IUHA, University of Vienna).

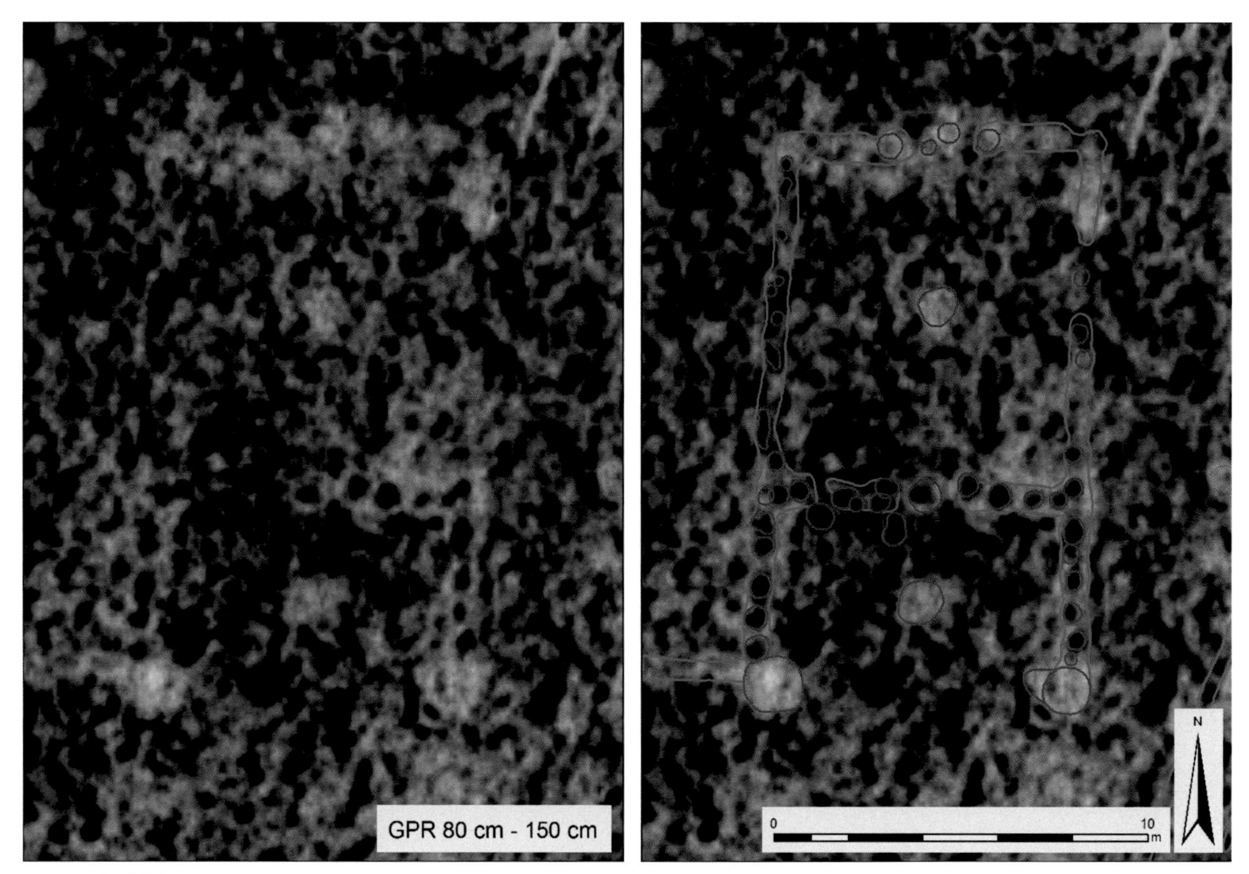

Figure 2: GPR depth slice and, superimposed, the integrated interpretation of a Neolithic house structure at Velm (blue = wall gully, red = post holes).

been limited to relatively small-area applications. New motorised multichannel GPR arrays, such as the 16-channel 400 MHz MALÅ Imaging Radar Array (MIRA), permit a considerably increased spatial coverage with simultaneously greatly improved sample spacing, resulting in images of the subsurface of unprecedented resolution and structural clarity.

Unique Structural Details of Palisade and Buildings

For an archaeological interpretation, structures with a higher reflective surface, such as stones, stonewalls or gravelly road paving's, are typically easier to recognise than absorbing features, like pits, postholes or ditches. From this perspective, it seems reasonable that remains from cultures that used stones as preferred building material – like the Roman – are better suited for GPR surveys than sites dating to the Neolithic or Bronze Age. Nevertheless, having suitable soil conditions, it is possible to map also absorbing structures to the level of individual postholes (Fig. 2).

Although the archaeological information of the site based on the earlier acquired data looked almost complete, the more recently conducted survey generated unexpected new information on the structural elements of the triple palisade and several houses. Only due to the accurate positioning and the high resolution of the GPR measurements was it possible to depict the precise outline of the wall gullies and even individual postholes constructing the remains of walls of a middle Neolithic house of the Lengyel culture (Fig. 2). This is the first time that individual posts can be identified along the exterior and interior walls of this type of building. Additionally, the settings of the wooden posts that once formed the triple palisade can be individually identified (Fig. 3).

From Prospection Results to an Integrated Archaeological Interpretation

The respective archaeological prospection methods provide specific information on physical and chemical properties of structures buried in the subsurface. Only by the combination and integration of all applied individual prospection methods an almost complete representation of the underlying archaeological structures emerges. In this way, it is possible to compensate the individual disadvantages of the different methods applied, and to ensure that the most complete archaeological information is gained.

The proposed multi-method approach produces complex datasets of variable archaeological information content. These demand adequate data

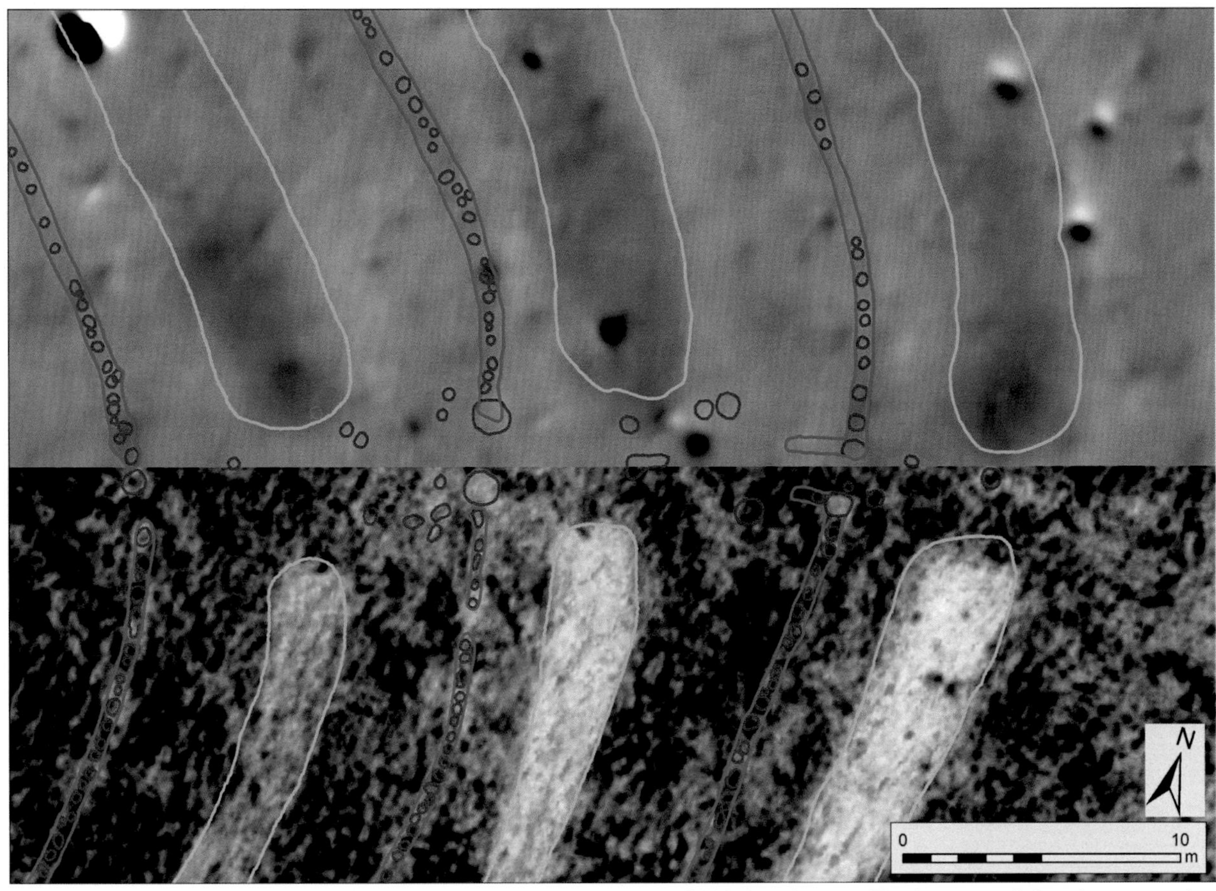

Figure 3: Upper: Greyscale image of magnetometry data. Lower: GPR data image. The integrated data interpretation has been superimposed (yellow = ditch, blue = palisade, red = posthole).

management and an integrated archaeological interpretation. The main challenge hereby is the transformation of the acquired and processed prospection data into interpretative archaeological information that is accurate, readable and ready to use for the scientific community. The main platform for this purpose is a georeferenced GIS based archaeological information system (AIS), extended by appropriate tools for dynamic data visualization, spatial analysis for integrated archaeological interpretation, data archiving, data retrieval, and long term maintenance. The LBI ArchPro has developed for this purpose a specific AIS named ArchaeoAnalyst, which is tested and developed further within large-scale archaeological case studies such as the one presented here (Wallner *et al.* 2015).

Once all the different georeferenced prospection data have been interpreted within the AIS, a general overview on the archaeological structures emerges. Through the integration of all the different sources of archaeological information it is possible to investigate the prospected area in greater detail and to understand the temporal development of the archaeological site and its surrounding landscape in a regional and historical context.

References

Melicharm, P. and Neubauer, W. (2010) *Mittelneolithische Kreisgrabenanlagen in Niederösterreich*. Vienna: ÖAW.

Wallner, M., Kucera, M., Neubauer, W., Torrejón Valdelomar, J., Brandtner, J. and Sandici, V. (2015) *Application of Georeferenced Archaeological Information Systems for Archaeological Digital Heritage - The Auxiliary Fortress of Carnuntum (Lower Austria)*, Digital Heritage Conference: Granada.

Castra Terra Culmensis - results of non-invasive surveys of the Teutonic Order's strongholds in the Culmerland (Poland)

Marcin Wiewióra[(1)], Krzysztof Misiewicz[(2)], Wiesław Małkowski[(2)] and Miron Bogacki[(2)]

[(1)]Uniwersytet Mikołaja Kopernika, Instytut Archeologii ,Toruń, Poland; [(2)] Uniwersytet Warszawski, Instytut Archeologii, Warsaw, Poland

kmisiewicz@uw.edu.pl, wmalkowski@uw.edu.pl

In June 2016, we began a research project carried out under a grant from the National Programme for the Development of Humanities titled 'Castra Terrae Culmensis' – on the outer reaches of the Christian world. The project involves conducting an interdisciplinary research of five Teutonic castles, so far not investigated, which may answer some key research questions regarding the history of this region of Poland in the Middle Ages.

Castles built by the Teutonic Knights in Culmerland between the second half of the 13th and the second half of the 14th centuries became one of the most characteristic features of the landscape. They played different functions and roles in accomplishing political, military, propagandistic and economic goals. They have become a visible testimony of a well-organized state.

For many years, attempts have been made towards a holistic view on the problems of border organization of the Order's state, but the status of our knowledge is uneven. A number of important issues still have not been documented. This particularly concerns such matters as the chronology of the subsequent stages of construction, layout and spatial arrangement of fortifications, the function of individual buildings and the presence of settlements prior to construction work. Major problems are also issues concerning the relationship between the effect of construction activity and the natural environment at the region. However, many of those question could be answered by collecting data with the use of non-invasive methods, such as:

- analysis of aerial photographs and satellite images
- verification of cartographic materials (historical and contemporary)
- use of data from LiDAR and ground laser scanning
- documentation of the sites from the air
- geophysical prospection by magnetic, electromagnetic and electrical methods.

The first stage of the study included the evaluation of the most important sites:

Castle in Lipienko

No archaeological works were carried out on this site. However, the literary sources indicate the presence of a small Teutonic fort. The specific location of the fortification remains – on a natural, defended peninsula – may indicate that the facility was built at the site of an older defensive system.

Castle in Starogrod

A large complex consisting of a High Castle and perhaps three lower fortifications, the first of which was made of brick, and the other two of wood (Torbus 1998, 70-72). According to 19th century German scholars, Starogrod could be the greatest defense system in Culmerland. Its irregular High Castle was (with a stronghold in Thorn) one of the oldest brick castles in all Teutonic Order's state (Steinbrecht 1888, 19, Abb. 23). Description of the Castle in the literature says that it was built in the place of an older (Slavic?) fortification. Also, the first city of Culm was located in its vicinity.

Castle in Unisław

Built in the third quarter of the 14th century is one of the castles about which we know practically nothing; it has never been investigated. It is even unknown whether the stronghold was built from bricks or wood. The alleged castle hill encompassed a prairie and farmland, and the place is suitable to carry out more extensive geophysical surveys.

Bierzgłowski Castle

Despite the good state of preservation of the monument it has never been the subject of any archaeological research. We have information that in 1236, the Teutonic Knights captured a Prussian stronghold existing here, called Pipinsburg and replaced it with their own castle complete with heavily fortified tower. However, one cannot be sure whether a stone castle was built exactly in the place of older wooden strongholds (Torbus 1998, 77-79, 361-368).

The Project goal is to answer questions conserving the above mentioned monuments:

- When was construction activity started?
- Were castles erected according to the previously adopted model of wooden structures, through irregular brick buildings, to a form and regular (quadrilateral) shape?
- Can the chronology of the strongholds classified today as the oldest examples of military architecture (Starogrod), which in the literature are dated relatively early – in fact dated back to the middle of the 13th century, be verified and what is, therefore, the origin and form of the oldest Teutonic castles?

Figure 1: Field works carried out in the project. Aerial photography, magnetic, electric and ERT measurements.

- Did the construction of regular brick castles occur at previously fortified settlements (Prussian, Slavic) or were strongholds were built in previously unsettled areas?
- What was the original form of some of these buildings (Unisław)?

The first non-invasive surveys began at castles in Starogrod and Unisław. Magnetic and electric surveys were used in conjunction with drone based aerial photography for mapping the surface of sites (Fig. 1).

During the magnetic survey, values of full vector of the magnetic field intensity were recorded, using a cesium magnetometer Geometrics MapMag 858-G with two sensors controlled by the Topcon Hiper-pro GPS RTK. As a result, we obtained data not only about the changes of recorded intensity of magnetic field and calculated (on the base of the difference between the sensors) components of the total vector field strength, but also direct elevation measurements that allows for the preparation of the 3-D model of the surface within the study areas. Electric measurements have been performed with the use of the Elmes ADA 07 AC device (Herbich et al. 1998) with dipole-dipole array allowing for the registration of changes in the value of apparent resistivity with a range of a current penetration of over 5 m below the present ground level. In the case of multi-layer objects (such as Bierzgłowski Castle) ERT (Electrical Resistivity Tomography) measurements with the ABEM Terrameter LS device with 1 m electrode spacing and gradient type measurement protocol (Dahlin, Zhou, 2006) were planned.

The hitherto obtained results open an interdisciplinary discussion about state of preservation of the buried remains. Maps of anomalies in distribution of the apparent resistivity values recorded in Starogrod (Fig. 2) showed the possibility of the presence of brick structures in the places where the High Castle was located.

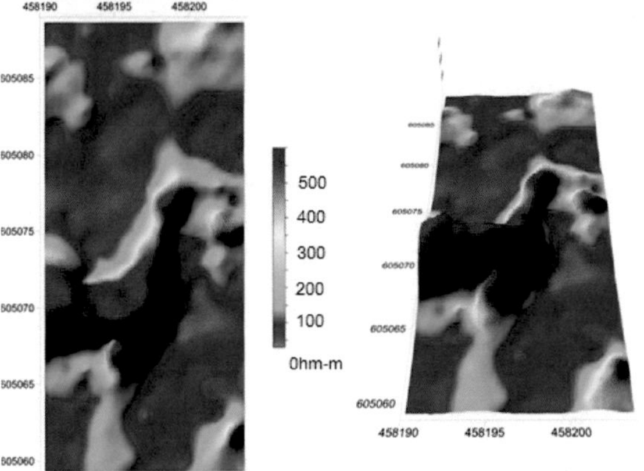

Figure 2: Starogrod, Map of the distribution of values of apparent resistivity (ohm/m).

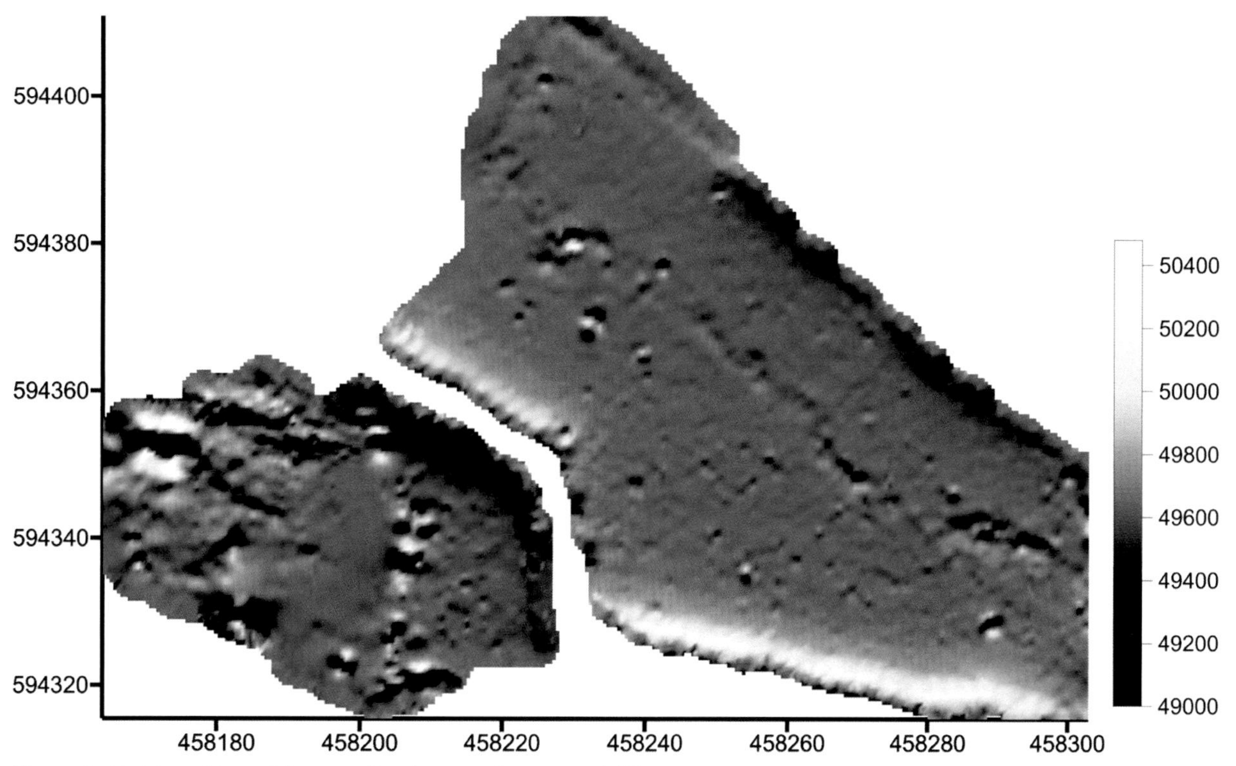

Figure 3: Unisław. Map of the results of magnetic survey (nT).

Also in Unisław where, according to written sources, remains of the castle was demolished in the late 19th century, the results of magnetic survey (Fig. 3) suggest that the relics of buildings completely invisible at the surface are still preserved under the ground.

Non-invasive surveys will continue, and its results will be verified by trial pits and large scale excavations.

Bibliography

CLASEN K. H. (1927) *Die mittelalteriche Kunst im Gebiete des Deutschordensstaates Preussen, Bd. 1 Die Burgbauten*, Königsberg.

DAHLIN T., ZHOU B. (2006) Multiple-gradient array measurements for multichannel 2D resistivity imaging, *Near Surface Geophysics* **4** (2): 113-123.

HEISE J. (1887) *Die Bau- und Knstdenkmäler der Provinz Westpreussen – Der Kreis Kulm, H. 5,* Danzig.

HERBICH T., MISIEWICZ K., MUCHA L. (1998), The ARA Resistivity Meter and its Application, *Materialhefte zur Archäologie* **41**: 127-131.

Imaging a Medieval shipwreck with 3D marine reflection seismics

Dennis Wilken[(1)], Hannes Hollmann[(1)], Tina Wunderlich[(1)], Clemens Mohr[(1)], Detlef Schulte-Kortnack[(1)] and Wolfgang Rabbel[(1)]

[(1)]Institute for Geosciences Christian-Albrechts-University, Kiel, Germany

dwilken@geophysik.uni-kiel.de

Introduction

The geophysical investigation of ancient harbour areas means working at the edge of waterbodies. These areas show water depths of only a few 10cm to meters and often a high salinity. DC electric and electromagnetic depth resolving prospection methods like GPR/ERT/EMI fail in these waters because of high attenuation. Offshore magnetic gradiometry suffers from the decrease of resolution caused by the enlarged distance between sensors floating near the water surface and targets located below the seafloor. However, hydroacoustic or reflection seismic methods offer the possibility to prospect large areas with a depth resolution of a decimeter scale. The requirements to these systems are:

- multichannel/array acquisition to be able to cover large areas,
- low draught to be able to access shallow water areas,
- low weight for easy handling,
- easy and stable steering behavior, and
- operating source frequencies of a few kHz.

In the framework of the DFG Priority Program Harbours from the Roman Period to the Middle Ages (von Carnap-Bornheim and Kalmring 2011) a marine seismic acquisition system was developed that complies with these requirements. We explain the properties of this system and its imaging capabilities using a case study of a medieval shipwreck. The study area is located at the innermost part of the Baltic fjord Schlei, Germany (see Fig. 1.a). In September 2014, divers confirmed the existence of a shipwreck in this area, completely covered by mud. Findings indicate that the wreck is similar to the Karschau-wreck, a Scandinavian transport ship which dates to the middle of the 12th century (Englert *et al.* 2002). The ship is probably related to Hedeby, which is located 2 km south-west of the study area.

The *PingPong* System

In the last 10 to 15 years, many high resolution acoustic sub-bottom profiler systems have been developed and used in archaeological prospection. Most systems use single channel acoustics. For example Grøn *et al.* (2015) image sub-bottom cross sections through shipwrecks using a single channel system. The horizontal resolution of unmigrated seismic data is defined by the Fresnel volume, and thus depends on the depth of the reflector and the wavelength. Two dimensional acquisition which suffers from the smearing of the images inside the Fresnel volume, which causes a severe loss of horizontal resolution. Migrating 2D profile data can only improve the in-line horizontal resolution but not the crossline resolution. Therefore, efforts have been made to construct and use 3D acquisition and imaging/migration for archaeological prospection.

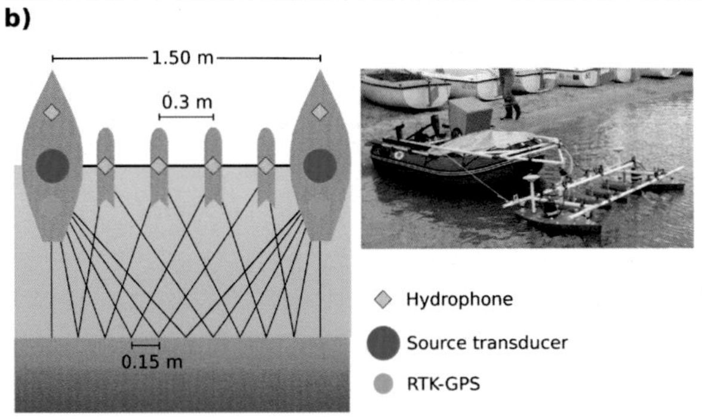

Figure 1: a) Map showing the case study investigation area. b) right: picture of the seismic reflection system PingPong. Left: Sketch of the system's acquisition geometry and setup.

Bull *et al.* (2005) presented the first results of a 3D sub-bottom profiler called Chirp3D. Later, Plets *et al.* (2009) demonstrated Chirp3D's potential by 3D imaging the Grace Dieu, a wooden English ship that sunk in 1439. In the same year, the SEAMAP-3D system was developed (Mueller et al. 2009; Mueller *et al.* 2013). Both systems showed a high potential of imaging quality in decimeter resolution, but quite high weight, and thus probably difficulties, when being used in very shallow waters with difficult access.

Besides 3D seismic approaches, the INNOMAR SES-2000 system became widely used in subbottom profiling. The system avoids the 2D resolution problem by creating a very narrow beam of approximately 3° width, but covering large areas in an adequate density is undoubtedly difficult.

Therefore, we developed a new shallow water 3D seismic system called "Pingpong" specifically following the above listed requirements. The system consists of two piezoelectric sources, six hydrophones and two RTK-GPS, installed on a buoyant frame (Fig. 1.b). The array is attached to the front of the boat via a hinge instead of being towed, as most seismic systems are. This reduces the noise coming from the ship's backwash. PingPong can be operated with a common rubber dinghy. The array has a size of 1 m x 2 m, a footprint of 1.5m, and 0.15m horizontal cross-line resolution (Fig. 1.b). It is operable in areas as shallow as 0.3m, weights roughly 30kg, and takes 2 persons to operate. Sources alternately send a pulse wavelet with a bandwidth from 2Hz to 6kHz. Repetition rate for each transducer is 4Hz.

The system was found to work well in areas with good sea conditions, being able to cover approximately 1 hectare per day. Processing included geometry calculation, bandpass filtering, spike-deconvolution, Normal-Moveout-correction, trace normalization, binning to a 15cm horizontal resolution datacube, 3D semblance based smoothing, and 3D Kirchhoff Migration using water velocity.

Results

Figure 2 shows two example profiles crossing the wreck in North-South and East-West directions (Fig. 2.a) and three example timeslices showing the major features of the data cube.

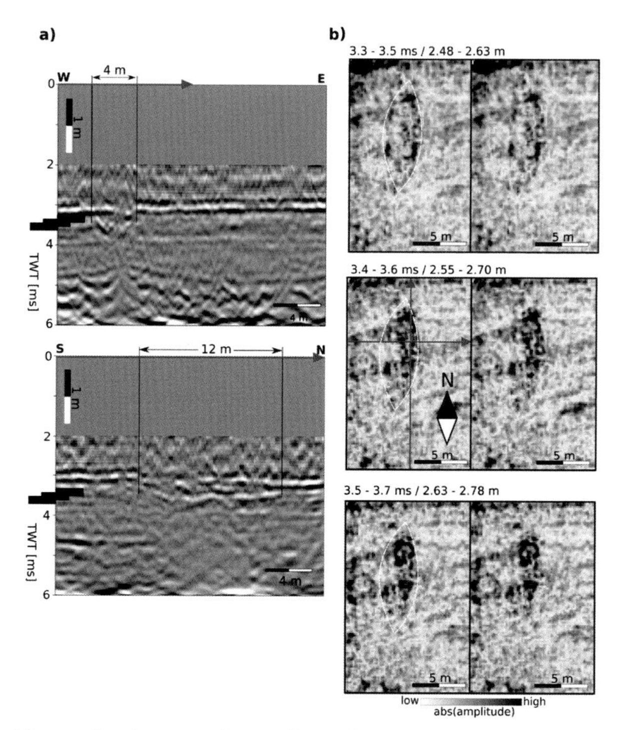

Figure 2: a) example profiles of the 3D data cube. Black rectangles to the left indicate the depth of the three example timeslices in b). b) example timeslices showing that half of the datacube that includes the wreck. Red arrows show the position of the profiles; Dashed white line estimates the dimensions of the wreck.

The results show that the remains of the ship's hull concentrate in the first meter below seafloor. Reflections in the profiles as well as high reflection areas in the timeslices show a ship shaped structure. The ship's overall dimensions can be estimated at 4m by 12m (long-dashed white line). Both zero offset sections and timeslices show a high noise level (e.g. some shiptrack artifacts/ residual static effects) which is due to the weather conditions during the measurements. Nevertheless, the PingPong system was able to image the main parts of the wooden medieval shipwreck at the seafloor and below. The timeslices comprise several reflection events that can clearly be assigned to the shipwreck. The one question left is, whether the complete remains are imaged. This depends on the impedance contrast of wooden constructions, which is definitely a matter of preservation.

Bibliography

von Carnap-Bornheim, C. and Kalmring, S. (2011) *DFG-Schwerpunktprogramm 1630 "Häfen von der Römischen Kaiserzeit bis zum Mittelalter. Zur Archäologie und Geschichte regionaler und überregionaler Verkehrssysteme"*. Schleswig: Annual Report Centre for Baltic and Scandinavian Archaeology, 28-31.

Bull, J. M., Gutowski, M., Dix, J. K., Henstock, T. J., Hogarth, P., Leighton, T. G., and White, P. R. (2005) Design of a 3D Chirp sub-bottom imaging system. *Marine Geophysical Researches,* **26**(2): 157-169.

Englert, A., Kühn, H.-J. and Nakoinz, O. (2002) Das Wrack von Karschau – ein nordisches Lastschiff aus dem 12. Jahrhundert. *Enogtyvende tværfaglige Vikingesymposium Kiel 2002, Forlaget Hikuin og Afdelig for Middelalderarkæologi, Aarhus University*

Grøn, O., Boldreel, L.O., Cvikel, D., Kahanov, Y., Galili, E., Hermand, J.P., Naevestad, D., and Reitan, M. (2015) Detection and mapping of shipwrecks embedded in seafloor sediments. *Journal of Archaeological Science,* **4**: 242-251.

Mueller, C., Woels, S., Ersoy, Y., Boyce, J., Jokisch, T., Wendt, G., and Rabbel, W. (2009) Ultra-high-resolutionmarine 2D–3D seismic investigation of the Liman Tepe/Karantina Island archaeological site (Urla/ Turkey). *Journal of Applied Geophysics,* **68**: 124-134.

Mueller, C., Woelz, S., and Kalmring, S. (2013) High-Resolution 3D Marine Seismic Investigation of Hedeby Harbour, Germany. *Nautical Archaeology,* **42**(2): 326-336.

Plets, R. M. K., Dix, J. K., Adams, J. R., Bull, J. M., Henstock, T. J., Gutowski, M., and Best, A. I. (2009) The use of a high-resolution 3D Chirp sub-bottom profiler for the reconstruction of the shallow water archaeological site of the Grace Dieu (1439), River Hamble, UK. In: *Journal of Archaeological Science,* **36**: 408-418.

Seeing is believing? Non-destructive research of the western Lesser Poland upland, 2010-2017

Piotr Wroniecki[(1)]

[(1)]Institute of Archaeology, University of Wrocław, Poland

piotr.wroniecki@gmail.com

Introduction

The Western Lesser Poland Upland is located in southern Poland (near Kraków) and contains a large course of the Vistula river. It has been a topic of professional archaeological interest for over a century, including large-scale field walking surveys, small-scale research digs and, more recently, numerous motorway rescue excavations. The area's fertile loess soils and undulating natural landscape have attracted a broad spectrum of steady settlement activities since the Neolithic.

The abundance of archaeological resources and advanced state of research was deemed an excellent testing ground for a macro-regional landscape study and non-destructive classification of the area (Fig. 1). Based on the sole application of (aerial and terrestrial) remote sensing techniques (especially light aircraft aerial prospection), it was envisioned as a proof-of-concept in 2009, and first surveys were conducted in 2010 (see Brejcha and Wroniecki 2010; Brejcha 2010; Dulęba et al. 2015). Since 2015 funding has been acquired via the "Hidden cultural landscapes of the Western Lesser Poland Upland" project (Wroniecki 2016).

A no digging approach was treading new ground as it stood in direct opposition to established (cultural-historical) modes of conduct. As a secondary objective, it served as a backdrop to challenge various widespread fables about non-invasive methods which were based not on factual data but opinions and hearsay. For instance, aerial archaeology was (and still is) deemed as too expensive and ineffective in Poland. It has not found a place in the cold, uncaring hearts of national policy makers (despite attempts to change this state of things, Nowakowski et al. 2005).

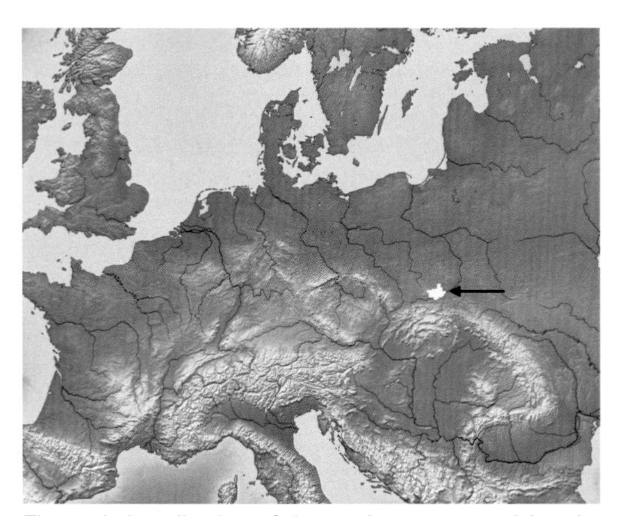

Figure 1: Localisation of the study area comprising the Lesser Poland Upland (over 2500 km2)

Methodology

The research methodology has been based on the acquisition of data through non-destructive surveys coupled with a Geographical Information System (GIS) approach. This includes remote sensing (Airborne Laser Scanning, satellite imagery), geophysics (mostly magnetic gradiometry), analytical surface surveys and archival data query. Inspiration was drawn primarily from Czech archaeological circles (Gojda 2004). These experiences served as a backbone due to common ground (historical background, geographical vicinity). Of the research objectives, the most important can be surmised as:

- Understanding the function and archaeological chronology of landscape components through their morphology
- Valuation of soil units for the study of archaeological resources
- Assessing and comparing of non-destructive methods of data collection to traditional (mostly field-walking) methods and datasets

Research is not aimed at specific chronological periods or cultural units. Its mainstay is a total, non-discriminatory approach to the landscape and archaeological resources hidden within it. The use of the proposed methods is not treated as purely pre-excavation activity. Focus is put into the analysis and interpretation of aspects of prehistoric communities in a regional dimension beyond the "site" perspective. Finally, the assessment of these procedures and methodology (i.e. possibilities, limitations, application sequence) in order to develop or adapt existing approaches to integrated landscape research is a prospective end result.

Overview of Results

From 2010 field work has included 46 hours of aerial prospection, over 35 hectares of geophysical surveys, and 9 analytical field walk surveys. The bulk of discoveries was made through aerial survey, numbering close to 2000 crop/soil mark sites. Due to being either partially revealed or difficult to attribute, most of them do not decisively contribute to the archaeological record. Of the 2000 sites, around 20% are of higher cognitive potential (Fig. 2).

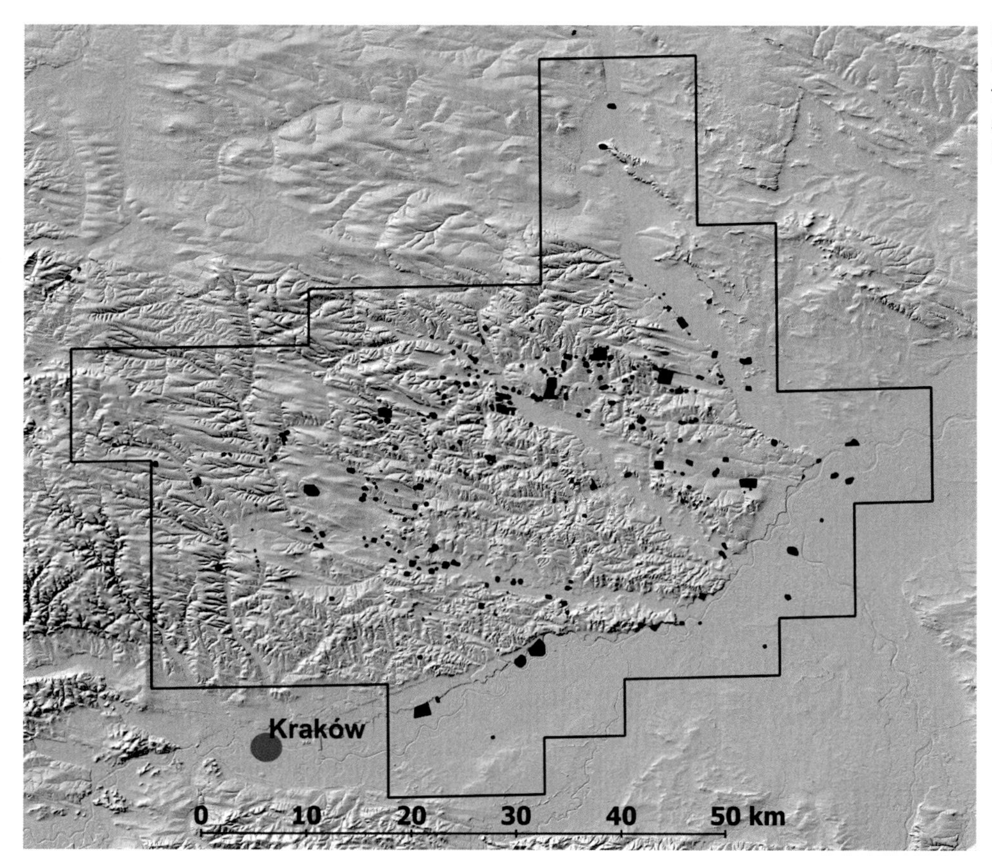

Figure 2: Archaeological resources registered through remote sensing imposed on a hillshaded DTM.

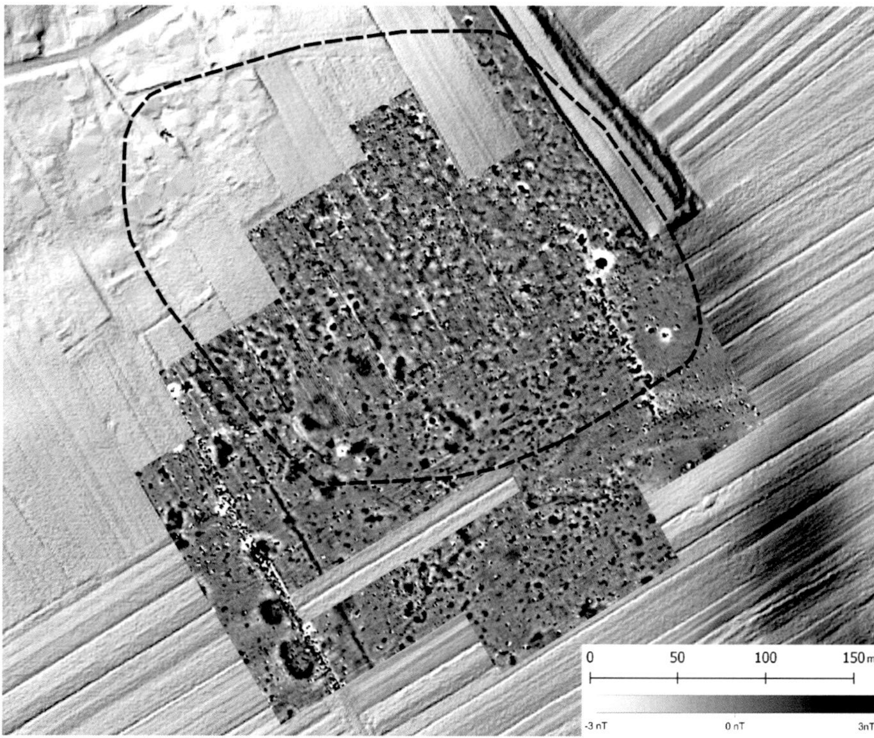

Figure 3: Cognitive dissonance: black dashed line is a registered archaeological "site" (based on surface finds) whilst the magnetic gradiometry reveals a larger palimpsest of natural & man-made structures hidden in the Nieprowice landscape.

Aerial survey is ideal for facilitating a non-discriminatory approach, as the discovery process is random and imposed through chance on the aerial observer. The abundance of monument types that are not topics of mainstream research focus such as post-medieval farmsteads (traces of Poland's past economic power) and relicts of WW1/WW2 (trenches, field fortifications) cannot be ignored. Enclosed prehistoric settlements, thought to be extremely rare in Poland, have sprouted in numerous places. Eight have been registered in the study area, changing the recorded amount by 800%. Other categories revealed are open settlements, oval or rectangular shaped pit features, monumental graves, barrows, former field boundaries, medieval strongholds, roads, abandoned villages, modern infrastructure and natural features.

Only selected areas may be continued with terrestrial surveys due to limited funding. Geophysical work often verifies the extent of archaeological resources as aerial archaeology is too dependent on weather conditions (especially on loess soils). Analytical field walking wraps up the dataset regarding site erosion and broad archaeological chronological and taxonomical spread.

Conclusions

Data collected through the use of multi-method surveys in southern Poland prove the sensibility of the applied methodology. It is cost-effective, time-efficient and has in only a few years significantly changed our understanding of the archaeological record. The benefits of this approach outweigh drawbacks, even in the unfavourable terrain and soil conditions of the loess upland.

With the acquired knowledge new problems have arisen encompassing a broad spectrum of topics. The fiercest is the lack of implementation of non-invasive solutions as top-down government backed projects (such as obligatory pre-rescue excavation surveys). Similarly, a need arises to review various theories on past human settlement activity in the study area.

The end result of these surveys unveils the fallacy of stewardship based on pottery dispersal - representing rather a state of mind than state of things (Fig. 3). The understanding of non-invasive data and proper interpretation is based on almost a decade of intensive work which, if continued, will allow us to study and efficiently decipher the marvelous lost landscapes of the Lesser Poland Loess Upland. Last, but not least, as seeing is believing, the collected case studies will hopefully be a stepping stone to furthering the understanding and appreciation of such research projects.

Bibliography

Brejcha, R. (2010) Využití volně dostupných dat dálkového průzkumu Země k identifikaci archeologických komponent: čtyři příklady z polského území, In M. Gojda (ed.), *Studie k dálkovému průzkumu v archeologii*. Plzeň: Západočeská univerzita v Plzni, 60–68.

Brejcha, R, and Wroniecki, P. (2010) Usefulness of non destructive methods in determining and documenting the state of preservation of an archaeological site liable to destruction. In: L. Gardeła and L. Ciesielski (eds.), *Na marginesie. W kręgu tematów pomijanych – In the margin. Among the omitted topics*. Poznań, 369 – 385.

Dulęba, P., Wroniecki, P. and Brejcha, R. (2015) Non-destructive survey of a prehistoric fortified hill settlement in Marchocice, Little Poland. *Sprawozdania Archeologiczne*, **67**: 245-258

Gojda, M. (2004) *Ancient landscape, settlement dynamics and non-destructive archaeology: Czech research project 1997-2002*. Prague.

Nowakowski, J., Prinke, A. and Rączkowski, W. (eds.) (2005) *Biskupin... i co dalej? Zdjęcia lotnicze w polskiej archeologii*. Poznań.

Wroniecki, P. (2016) Hidden cultural landscapes of the Western Lesser Poland Upland. Project overview and preliminary results. In P. Kołodziejczyk and B. Kwiatkowska-Kopke (eds.) *Cracow Landscape Monographs*, vol. 2. Kraków, 21-32.

Acknowledgements

Hidden Cultural Landscapes of the Western Lesser Poland Upland. Non-destructive methods applied to settlement studies project financed by the National Science Centre Preludium (2014/15/N/HS3/01719); Mirosław Furmanek, Roman Brejcha, Przemysław Dulęba, Krzysztof Wieczorek, Jan Bulas, Marcin M. Przybyła, Marcin Jaworski, Maksym Mackiewicz, Bartosz Myślecki, Gábor Mesterházy, Martin Krajňák, David Kušnirák & Kevin Barton.